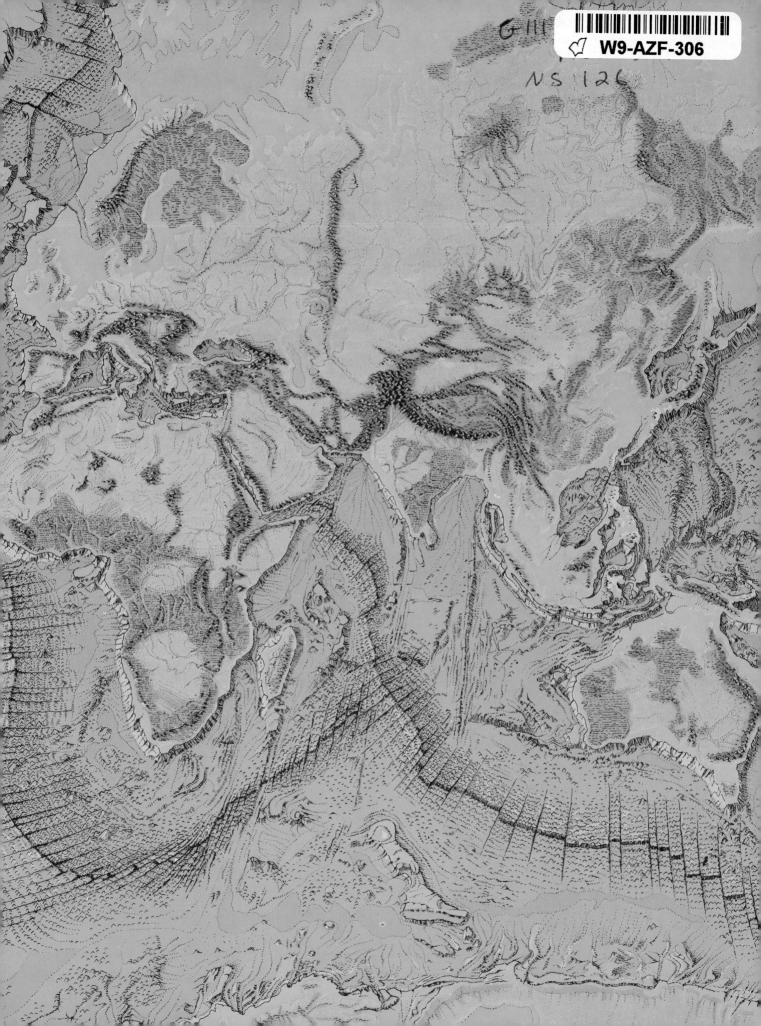

THE EARTH'S
DYNAMIC SYSTEMS

THE EARTH'S DYNAMIC SYSTEMS

SYSTEMS **Third Edition**

A Textbook in Physical Geology

by W. KENNETH HAMBLIN Brigham Young University

BURGESS PUBLISHING COMPANY • MINNEAPOLIS, MINNESOTA

Editor   Wayne Schotanus
Production Manager   Morris Lundin
Assistant Production Manager   Pat Barnes
Copy Editor   J. D. Wicklatz

Design and layout   Dennis Tasa, Tasa Graphic Arts

Cover photo   J. A. McAlonen/PhotoCraft Inc.
Frontispiece   Dennis Tasa

Burgess Publishing Company
7108 Ohms Lane
Minneapolis, Minnesota 55435

J   I   H   G   F   E   D   C

CONTENTS

v

PREFACE

Geology is in a "golden age"—a time of new enlightenment, with major discoveries and revolutionary theories giving us profound new insights into the origin of our planet and how it works. For the first time in the history of the discipline, a unifying theory of the earth's dynamics has been developed, which permits an all-encompassing synthesis of much of the geological knowledge that has been gained since the birth of the science.

This new theory concerns the dynamics of our planet and relates many major features of the earth, such as continents, ocean basins, mountain belts, earthquakes, and volcanoes, to a system of slow-moving material below the crust. It was developed largely as a result of our newly acquired ability to study the characteristics of the ocean floor, map its landforms, and measure its magnetic properties. These new data clearly show that the earth's crust is not rigid and fixed but is actually in motion, with individual fragments, or plates, moving thousands of kilometers. As the plates move, the continents split and move apart, collide, and are sutured together in various patterns. In contrast, the sea floors are temporary features, opening and closing, continually being created and destroyed as the crust moves. The theory of plate tectonics answers many questions about the earth, but it also poses new problems, which are being tackled by geologists in a framework of thinking that is new and exciting.

The revolutionary theory of plate tectonics, however, is not the only new development in geology. Discoveries made through the space program during the last few years have given us a new perspective of the earth and have brought most of the planetary bodies in the solar system well within the study of earth scientists. Detailed photographs of the moon, Mercury, Mars, and the four large moons of Jupiter are analyzed much as we study aerial photography of the earth's surface. Radar observations have finally parted the veil of clouds that surrounds Venus and have given us our first comprehensive glimpse of its surface. Geologists are thus able to make topographic and geologic maps of these planets, using ordinary techniques of photogeology. Indeed, we have learned more about the surface of Mars in the past few years than was known about the earth only a few hundred years ago. From this newly acquired knowledge, we are able for the first time to compare planets in terms of their surface features and geologic processes, and we can formulate new ideas about how a planet originates and evolves.

With this new era of enlightenment comes a new awareness of how fragile our planet is and how impossible it is for exponential growth to occur indefinitely. Geologists will find more oil, gas, and coal with new or improved methods, but we know that nature required more than a million years to concentrate the oil we are now consuming in 12 months.

In addition to stimulating an awareness of the problems of energy and resources, an understanding of geology invites serious reflection on our space and environment. For example, the amount of sediment transported by rivers is smaller than the amount of material dumped into them as waste. What effect does this have on our environment and on the future of life itself?

Many new tasks with new dimensions and new discussions await us, all of which require a clear understanding of the nature of the earth and how it works.

The changes in the third edition of this text were made to incorporate these new developments and to make the book more timely and more useful to students. Every chapter has been revised and updated. Several have been completely rewritten. Our objective in this edition is to use the unifying theory of plate tectonics in the study of the earth as a dynamic system that is constantly changing toward a state of equilibrium.

We have refined the "Statement" and "Discussion" sections of each chapter with a definite purpose—to state summaries of the main ideas and to support them with organized discussion, repetition, and review to enhance and facilitate learning.

We have added questions and problems at the end of each chapter in an effort to help students review the material and to test retention. Many of the questions simply test knowledge of facts, but some are more thought provoking and may help students grasp the broader meaning of concepts discussed in the text.

The most obvious improvements in the third edition are the illustrations. We have revised many of the line drawings and have utilized colors in most of the major diagrams to emphasize basic concepts. We have included many new line drawings and photographs, because these visual aids are of considerable value in helping students understand new, and sometimes complex, material. In addition, explanatory sketches have been added to many photographs to help focus students' attention on the major features being illustrated. Sixteen full-color space photographs were selected to provide a new perspective of the earth, from which students can develop an accurate concept of the earth's surface and tectonic features. Students should use these plates as a basic visual reference as they study the facts and concepts in each chapter.

ACKNOWLEDGMENTS

Sincere thanks go to many colleagues and others who have helped with this book by discussing ideas and problems, critically reading the manuscript, and supplying unpublished information. For the first edition, these colleagues included James L. Baer, Willis H. Brimhall, Myron G. Best, H. M. Davis, Thomas Hendrix, Douglas Hill, Robert Lakemacher, Wade E. Miller, Joseph R. Murphy, John Reid, James A. Rhodes, and Morris S. Petersen. Many of their suggestions were incorporated in the later editions. For the second edition, thanks are also due to Kenneth Van Dellen, of Macomb County Community College, who read the entire manuscript and offered many helpful suggestions. Others who reviewed the entire manuscript were E. Julius Dash, Oregon State University; Douglas Brookins, University of New Mexico; Roger Hoggen, Ricks College; and Chester Johnson, Gustavus Adolphus College. Early reviewers included H. Wayne Leimer, Tennessee Technical University; Arthur L. Reesman, Vanderbilt University; and Donald C. Haney, Eastern Kentucky University. Other colleagues reviewed portions of the manuscript at an early stage.

Special thanks are extended to those who reviewed the manuscript and offered many helpful suggestions for the third edition. Technical reviewers were Cornelius Klein, Indiana University at Bloomington; George Davis, University of Arizona; David Nash, University of Cincinnati; Paul Lowman, Goddard Space Flight Center; and Ted V. Jennings, Purdue University. General reviewers were Burton A. Amundson, Sacramento City College; Erwin J. Mantei, Southwest Missouri State University; M. James Aldrich, North Carolina State University at Raleigh; Calvin James, University of Notre Dame; C. Patrick Ervin, Northern Illinois University; Perry B. Wigley, Eastern Kentucky University; Peter S. Dahl, Kent State University; Warren D. Huff, University of Cincinnati; H. C. Clark, Rice University; and William H. Forbes, University of Maine.

The revision of the artwork was done by Tasa Graphic Arts of Minneapolis, Minnesota under the direction of Dennis Tasa. New illustrations were done by Robert Pack, whose talent and efforts are greatly appreciated.

The assistance of Paula Landon and Jill Rhodes, who typed the manuscript and aided in the revision, is also sincerely appreciated.

TO THE STUDENT

OUR APPROACH

In studying the processes in the earth's system, we are continually confronted with the problem of scale. How can we begin to understand the nature and interrelation of features as large as rivers, mountains, continents, and ocean basins? One approach is to use models, a common method in all branches of science and engineering for studying and analyzing things too large to see or handle conveniently. A model is a formulation that simulates a real-world phenomenon and has value in analyzing and predicting behavior of the real thing. Models can be physical or abstract representations of the structure and function of real systems. Throughout this text, we have made extensive use of graphic models (diagrams, photographs, and maps) in an effort to provide you with an accurate image of the features and processes described. These models, especially the block diagrams, are as important as the text in providing information, and as much time should be devoted to studying them as to reading the text.

ORGANIZATION OF THE CHAPTERS

One of the most difficult problems a student faces in beginning a course in a new subject is to identify fundamental facts and concepts and separate them from supportive material. This problem is often expressed by the question, What do I need to learn? We have attempted to overcome this problem by presenting the material in each chapter in a manner that will help you recognize immediately the essential concepts. In each chapter, the major subdivisions are introduced under the subheading "Statement." The "Statement" presents major concepts, principles, or theories. It is not in-tended to be a summary or abstract of the section, but it expresses the facts and concepts of the subject in one all-embracing view. These may be difficult for you to comprehend fully the first time you read the "Statement," but you will gain further insight from the paragraphs under the subheading "Discussion." Here, a more complete discussion of the major concepts and principles is presented and illustrated. Terms that are defined in the glossary are introduced in boldface type. Evidence supporting the statement of principles is given, and if it is pertinent to an understanding of a concept, a brief history of how it developed is included. This material is designed to help you grasp the ideas presented in the "Statement."

The idea is to learn the material in the "Statement" and clearly understand it. In this way you can use your study time effectively, and in reviewing for a test you need only to make sure you understand the "Statement."

PHYSIOGRAPHIC MAP OF THE EARTH

The physiographic map of the surface of the earth printed on the inside cover was compiled from the best sources available and represents, in perspective, the surface features of our planet. These features are the net result of processes operating in the earth's systems, and you should study them carefully as you read the text.

A larger, more detailed physiographic map of the earth, together with a physiographic map of the moon, Mercury, and Mars, is an optional accession to the book and is available from Burgess Publishing Company. For the most effective results, hang this model on the walls of your study and identify and analyze each feature as it is discussed. Much can be gained if you annotate the map as you study.

1

THE PLANET EARTH

One of the most significant advances made in the twentieth century concerns our view of the earth and the dynamics of how it changes. Until recently, we obtained most of our knowledge of the earth by looking horizontally from viewpoints at or near its surface. This perspective was nearsighted and limited, and in an attempt to overcome it, local observations were plotted on maps and charts, which served as models of the real world. Rivers, mountains, shorelines, and weather patterns were surveyed and studied from hundreds of observation points, but observers were never able to see the earth in a regional, panoramic view or to observe how it functions as a planet in space. Now, as the poet Archibald MacLeish declared in 1968, "for the first time in all of time men have seen the earth from the depths of space—seen it whole and round and beautiful and small." The horizontal view no longer dominates; the new vertical dimension of space photography has extended the limits of vision so that we can look down on the earth in a way that seems almost godlike.

Space photography permits us to see the earth as it truly is, to view the large-scale structures of the atmosphere, the oceans, and the continents, and to observe the regional relationships of river systems, mountain belts, and coastal features. We are now able to see the actual physical features of the planet and can view the earth functioning as a dynamic system on a scale that may at first be difficult to grasp.

To help you understand and appreciate the earth, this chapter includes a section of space photographs that illustrate many of its landforms and structural features (*plates 1–16*). These photographs are, in essence, a pictorial essay on the earth's dynamic systems. They are intended to be a visual reference for material discussed in this and all of the following chapters.

MAJOR CONCEPTS

1. The surface fluids of the earth (air and water) are in constant motion.
2. Continents and ocean basins are the principal surface features of the earth.
3. The continents consist of three major components: (*a*) shields, (*b*) stable platforms, and (*c*) belts of folded mountains. All of these components show the mobility of the crust.
4. The ocean floor contains several major topographic and structural divisions: (*a*) the oceanic ridge, (*b*) the abyssal floor, (*c*) seamounts, (*d*) trenches, and (*e*) continental margins.
5. The earth is a differentiated planet, with the following major structural units: (*a*) a central core, (*b*) a thick mantle, (*c*) a soft upper mantle (asthenosphere), (*d*) a rigid lithosphere, and (*e*) surface fluids (hydrosphere and atmosphere).

INTERPRETING THE SURFACE FEATURES OF A PLANET

Statement

The surface features of a planet usually preserve some imprint of the processes by which they were formed. In some cases, the record is clear and easily understood. Sand dunes of the great desert regions obviously are formed by wind action; the beaches of the Atlantic coast are the result of wave action; the islands of Hawaii are clearly a record of volcanic eruptions. The origin of other landforms, however, can be more obscure and only partly understood. The fascinating thing about the surface of a planet is that it can be explained by natural, physical, chemical, and geologic processes. As Einstein once said, "The eternal mystery of the world is its comprehensibility."

Discussion

The Moon

Consider first the moon and what its cratered surface tells about its history and internal dynamics (*plate 1*). It is obvious from telescopic views, as well as from space photographs, that the moon's surface is practically saturated with craters, and thus the major geologic process on the moon has been the impact of meteorites. The moon does not have water or an atmosphere, so that there are no stream valleys or windblown sand. The surface of the moon has been modified only by the impact of meteorites coming from space and by the internal energy that once deformed its crust and produced volcanic features. From rock samples brought back by the Apollo missions and from close-up satellite photographs, we now know that the smooth, dark areas called maria (Latin, "seas") are covered by vast floods of lava. As is shown in *plate 1*, lava flows filled the large, circular basins and adjacent lowlands and spread out over parts of the rugged, densely cratered terrain. Thus, clearly, the densely cratered terrain was formed before the lava flows occurred—an obvious, but fundamental, temporal relationship between two major events in lunar history.

Two additional facts are apparent from *plate 1*. First, the maria, or lava plains, have relatively few craters. The rate of meteorite impact must have decreased greatly since the lava was extruded. Second, the outlines of all craters, young and old, remain circular and undeformed. Therefore, the crust of the moon has not undergone enough compression to produce major buckling and folding.

From these observations of the moon's surface features, we can recognize the following sequence of events in lunar history:

1. A period of intense meteorite bombardment, when the densely cratered terrain was formed
2. A major thermal event, resulting in widespread volcanic activity and great floods of lava
3. A period of light bombardment, with relatively few meteorite impacts

Except for the impact of a few meteorites, the moon has not changed significantly since the lava plains were formed. Its crust has not been deformed by internal forces, and there is no wind or water to modify its surface. Nothing moves on the moon except on rare occasions when a meteorite strikes the surface. No motion was observed by the Apollo astronauts except for what they themselves created. The footprints they left behind will remain fresh and unaltered, modified only by particles from space, for literally millions of years.

Mars

The surface of Mars is similar to that of the moon, and yet the two planets are different. Much of the Martian surface is pockmarked with craters, and there are large, circular basins and vast floods of lava. Mars, however, has an atmosphere, and the Martian winds have significantly eroded the crater rims and have piled sand into large dune fields (see *figure 24.32*). There is also water on Mars, but it is now frozen in the pore spaces of the soil and rocks or is locked up in the polar ice caps. At times in the past, however, conditions must have been different. There is evidence that water once flowed on the surface and eroded numerous stream channels (see *figure 24.29*). Also, the crust of Mars has been fractured by tensional forces. Huge canyons have been formed, and giant volcanoes may still be active.

From the cratered surface and lava plains, it is clear that the early history of Mars is similar to that of the moon. But the winds of Mars continually shift sand over its surface. Running water has eroded networks of stream channels. Relatively recent volcanic activity has built up huge volcanic mountains, and the planet's crust has been deformed by upwarps and fracturing.

Mars, then, is a more dynamic planet than the moon. But neither of them has continents, ocean basins, folded mountain belts, and great river systems like those on the earth. These features are the result of much greater dynamic activity.

THE EARTH FROM A PERSPECTIVE IN SPACE

Statement

A view of the earth from space, like the one in *plate 2*, shows many features that provide an insight into our planet's history and how it changes. The atmosphere is the thin, gaseous envelope that surrounds the earth, and its bright, swirling clouds are the most conspicuous feature seen from space. The lithosphere, the outer, solid part of the earth, is visible as continents and islands. The hydrosphere, the planet's discontinuous water layer, is seen in the vast surface of the oceans. Even parts of the biosphere, the organic realm, which includes all of the earth's living things, can be seen from space, in the dark green tropical forests of equatorial Africa.

Each of these major realms of the earth's surface is in constant motion and changes continually. The atmosphere and the hydrosphere move in ways that are dramatic and obvious. Movement, growth, and change in the biosphere are readily appreciated and easily understood. But the solid, rigid lithosphere—the earth's crust—is also in motion and has been throughout most of the earth's history.

Discussion

The Atmosphere and the Hydrosphere

Everyone knows that air and water are the fundamental elements of the earth's environment, but few people give much thought to the fact that these surface fluids are as much parts of the planet as the solid rocks are. Yet the continually moving **atmosphere** and **hydrosphere,** which blanket the globe, are responsible for many of the **landforms** and surface features of the continents. Without surface waters, there would be no rivers, valleys, glaciers, or even continents as we know them. Without a moving atmosphere, there would be no desert dunes and no mechanism to transport water from the oceans to the land.

The circulation patterns of the atmosphere are expressed in *plate 2* by the shape and orientation of the clouds. At first glance, the patterns seem confusing, but upon close examination, they are found to be well organized. If the details of local weather systems are smoothed out, the global atmospheric circulation becomes apparent. Solar heat, which is greatest in the equatorial regions, causes the water in the oceans to evaporate, and the moist air rises. This warm, humid air forms an equatorial cloud belt, bordered on the north and south by relatively cloud-free zones. At higher latitudes, low-pressure systems develop where the warm air from the low latitudes meets the polar air masses. The pattern of circulation around the resulting low-pressure cells produces winds moving counterclockwise in the Northern Hemisphere and clockwise in the Southern Hemisphere. Such patterns are apparent in the swirls of clouds shown in *plate 2*.

The Continents and the Ocean Basins

If the earth had neither an atmosphere nor a hydrosphere, two principal regions—**continents** and **ocean basins**—would stand out as its dominant features. These major structural and topographic divisions of the earth differ markedly, not only in elevation, but also in rock types, density, chemical composition, age, and history. The ocean basins, which occupy about two-thirds of the earth's surface, are characterized by spectacular **topography**, most of which originated from extensive volcanic activity and earth movements that continue today. The continents rise above the ocean basins as large platforms, but the waters of the oceans more than fill their basins and flood about 18% of the continental surface. The present shoreline, so important geographically and so carefully mapped, has no great structural significance. Indeed, it has fluctuated greatly throughout the earth's history.

From a geological point of view, the elevation of the continents, which rise high above the ocean basins, is much more significant than the position of the shore. The difference in elevation of continents and ocean basins represents a fundamental difference in rock **density.** Continental rocks are less dense than the rocks of the ocean basins. That is, a given volume of continental rock weighs less than the same volume of oceanic rock. This difference causes the **continental crust** to float above the denser **oceanic crust.**

The elevation and area of the continents and ocean basins have been mapped with precision, and the data can be summarized in various forms. The data presented graphically in *figure 1.1* show that the average elevation of the continents is 880 m above sea level, and the average depth of the ocean floor is about 3700 m below sea level. Only a relatively small percentage of the earth's surface rises above the average elevation of the continents or drops below the average depth of the ocean floor. This difference in elevation is significant because, if the continents did not rise quite so high above the ocean floor, the entire surface of the earth would be covered with water.

Why does the earth have ocean basins and continents instead of a cratered surface like that of the moon and Mars? By their very existence, the continents pose one of the most fundamental questions about the earth. Theoretically, a large, rotating planet like the earth with a strong gravitational field would mold itself into a smooth spheroid covered with a layer of water approximately 2.4 km deep. Continents, as they are known today, have not been found on other planets in this solar system. The origin and evolution of continents and ocean basins will be studied in chapters 21 and 22, but it should be emphasized at this point that the continents and ocean basins constitute the two fundamental topographic and structural features of the earth, features not found on the moon, Mercury, or Mars.

Figure 1.1 A graph of the elevation of the continents and ocean basins shows that the average height of the continents is 800 m above sea level and the average depth of the ocean floor is 3700 m below sea level. Only a small percentage of the earth's surface rises above the average elevation of the continents or drops below the average elevation of the ocean floor.

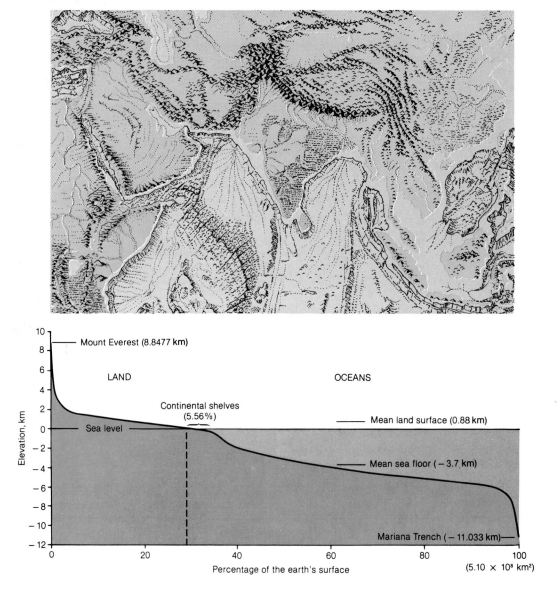

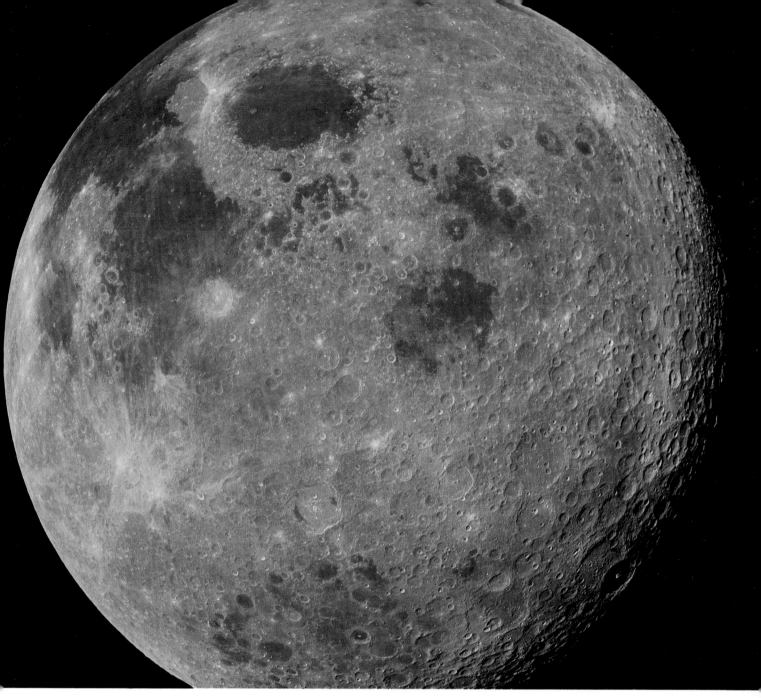

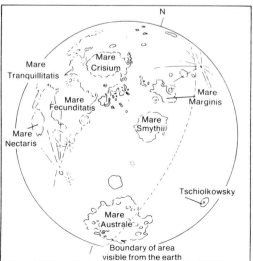

Plate 1 A view of the moon as seen by *Apollo 8* astronauts shows two contrasting types of surface features: densely cratered highlands, called terrae, and dark, smooth areas, called maria. We know from rock samples brought back from the Apollo missions that the maria resulted from great floods of lava, which filled many large craters and spread out over the surrounding area. Thus, the volcanic activity occurred after the formation of the densely cratered terrain. These relationships between surface features imply that the moon's history involves three major events: (1) a period of intense bombardment by meteorites, (2) a period of volcanic activity, and (3) a subsequent period of relatively light meteorite bombardment (resulting in young, bright-rayed craters). The lunar surface has a very low level of erosion and has not been modified by wind, water, or glaciers.

Plate 2 A view of the earth from space is dominated by the blue ocean and the swirling patterns of clouds, which underline the importance of water in the earth's system. In this view, a large part of Africa and Antarctica are visible, and climatic zones are clearly delineated. Much of the vast tropical forest of central Africa is seen beneath the discontinuous cloud cover. Several complete cyclonic storms, spiraling over hundreds of square kilometers, can be seen pumping huge quantities of water from the ocean to the atmosphere. Much of this water is precipitated on the continents and erodes the land as it flows back to the sea. A large part of the South Polar ice cap, which covers the continent of Antarctica with a glacier more than 3000 m thick, can also be seen. Of particular interest in this view is the rift system of the Red Sea, a large fracture in the African continent separating the Arabian Peninsula from the rest of Africa (see plate 14 for details).

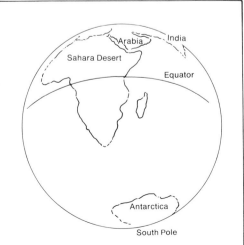

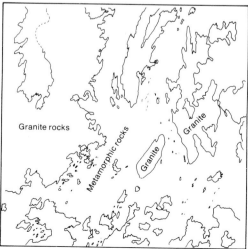

Granite rocks

Metamorphic rocks

Granite

Granite

Plate 3 The Canadian Shield is a fundamental structural component of North America. It is composed of complexly deformed crystalline rock bodies, eroded down to a nearly flat surface near sea level. Throughout much of the Canadian Shield, the topsoil has been removed by glaciers, and different rock bodies are etched out in relief by erosion. The resulting depressions commonly are filled with water and form lakes and bogs, which emphasize the structure of the rock bodies. Dark tones indicate areas of metamorphic rock. Light pink tones indicate areas of granitic rock.

Plate 4 The stable platform in the vicinity of Nashville, Tennessee, is dominated by a broad domal upwarp, although vegetation covers many details. Erosion has removed the younger rocks from the central part of the Nashville dome, so that the eroded edges of major rock formations are exposed in large, elliptical outcrop patterns, which can be identified in this photograph by dark tones and a delicate erosional texture. These outcrops of resistant rock form a ridge, known as the Highland Rim, rising from 30 to 60 m above the adjacent area. The older rocks exposed in the center of the dome are nonresistant limestone, and the surface there has been lowered by erosion so that the region, although structurally a dome, is expressed topographically as a basin, or lowland. To the southeast, younger sedimentary rocks mark the western margins of the Cumberland Plateau, a broad flatland dissected by stream erosion, which rises 150 m above the surrounding area.

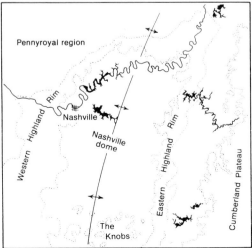

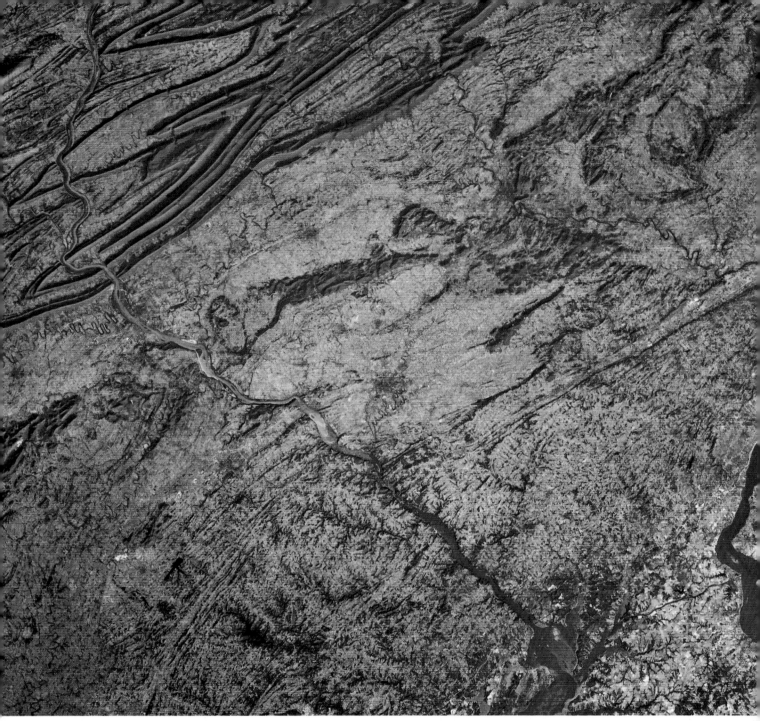

Plate 5 The Appalachian Mountains in central Pennsylvania show the typical style of deformation in a folded mountain belt. The folded strata are expressed by long, narrow ridges of resistant sandstone, which rise about 300 m above the surrounding area. Erosion has removed the upper part of many of the folds, so that their resistant limbs (which form the ridges) are exposed in elliptical or zigzag outcrop patterns. This area is part of the Ridge and Valley province, one of the physiographic provinces of North America, which extends from central Alabama to New England. Similar styles of deformation are found in the Ouachita Mountains of Oklahoma, the Marathon Mountains of Texas, and most other mountain belts. Such deformation is an important structural feature of the earth's crust and clearly shows that the crust is mobile and has moved throughout most of geologic time.

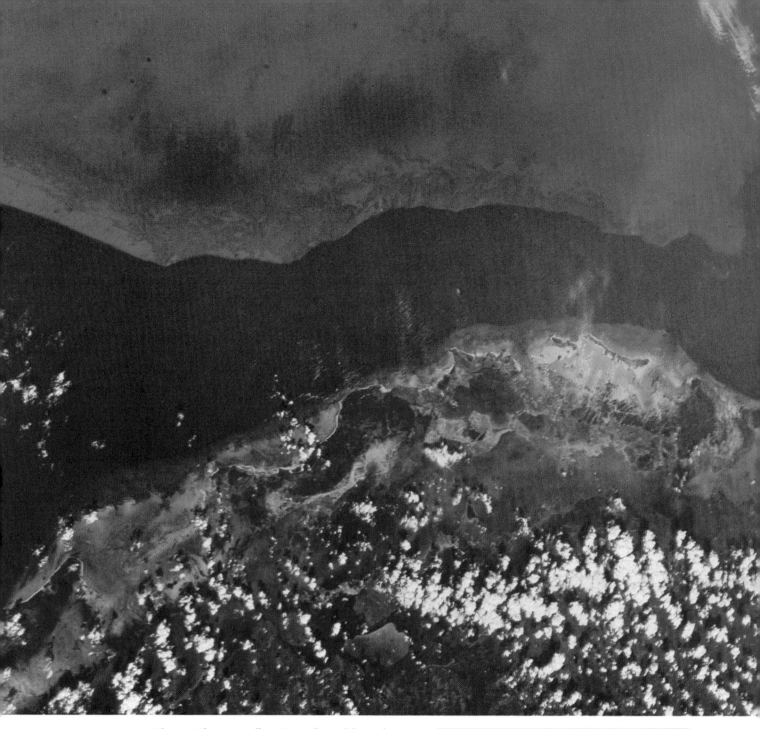

Plate 6 The ocean floor is not flat and featureless but contains many fascinating and important landforms. These have been mapped from oceanographic ships by special instruments that plot a profile of the sea floor. Some submarine landforms can be seen in satellite photographs; this view of the Bahamas Platform, which rises abruptly from depths below 3700 m, is an example. The light tones of blue and purple are the shallow Bahama Banks, covered by water that is only from 20 to 100 m deep. The dark tones indicate deep water. The shallow platform is the site of extensive deposition of carbonate sediment, much of which is produced by marine organisms. Shallow seas such as this covered much of the continental platforms in past geologic ages and resulted in the deposition of extensive limestone formations.

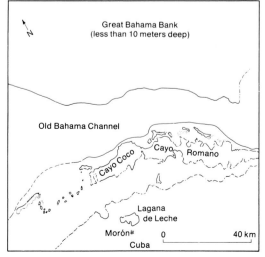

Great Bahama Bank
(less than 10 meters deep)

N

Old Bahama Channel

Cayo Coco

Cayo Romano

Lagana
de Leche

Morón#

Cuba

0   40 km

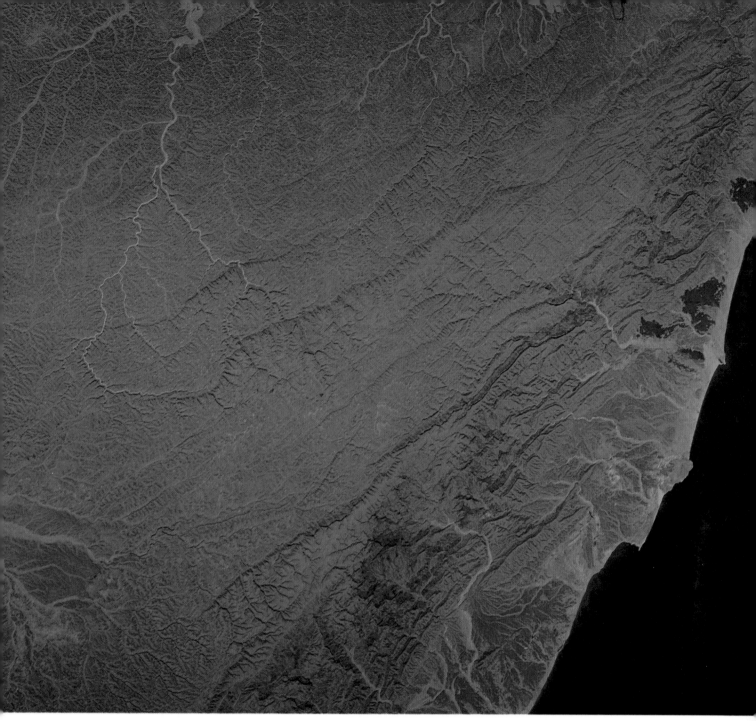

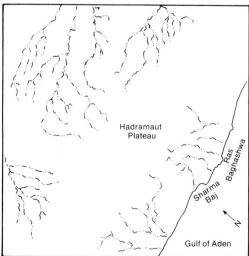

Hadramaut
Plateau

Ras
Baghashwa

Sharma
Baj

N

Gulf of Aden

Plate 7 Drainage systems are a clear record of how surface runoff has sculptured the land and testify to the magnitude of the earth's hydrologic system. Few areas of the land are untouched by stream erosion. In this photograph of a desert region, details of the delicate network of tributaries are clearly shown. On the moon, Mercury, and Mars, craters dominate the landscape, but on the earth, stream valleys are the most abundant landform.

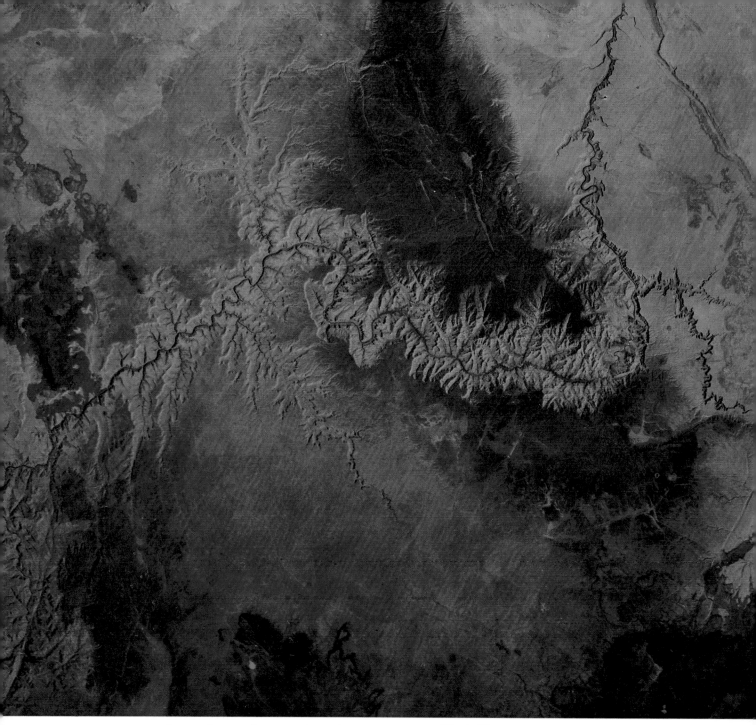

Plate 8 The Grand Canyon of the Colorado River, in Arizona, vividly illustrates the capacity of a stream to erode the earth's surface. The Colorado River drains a large part of the Rocky Mountains and flows across the Colorado Plateau to the Gulf of California. The plateau has been uplifted several thousand meters, so that the river has been able to cut a system of deep canyons across the region. Where uplift is greatest, the canyons are deepest. The Grand Canyon, incised into essentially horizontal sedimentary rock, is roughly 1600 m deep and 19 km wide. The innermost gorge is cut into the ancient cystalline rocks of the shield. Two volcanic fields are seen in this view: the San Francisco field to the southeast and the Uinkaret field to the west, where frozen lava falls cascade 1000 m into the inner gorge of the canyon.

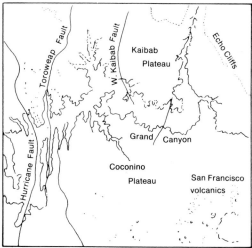

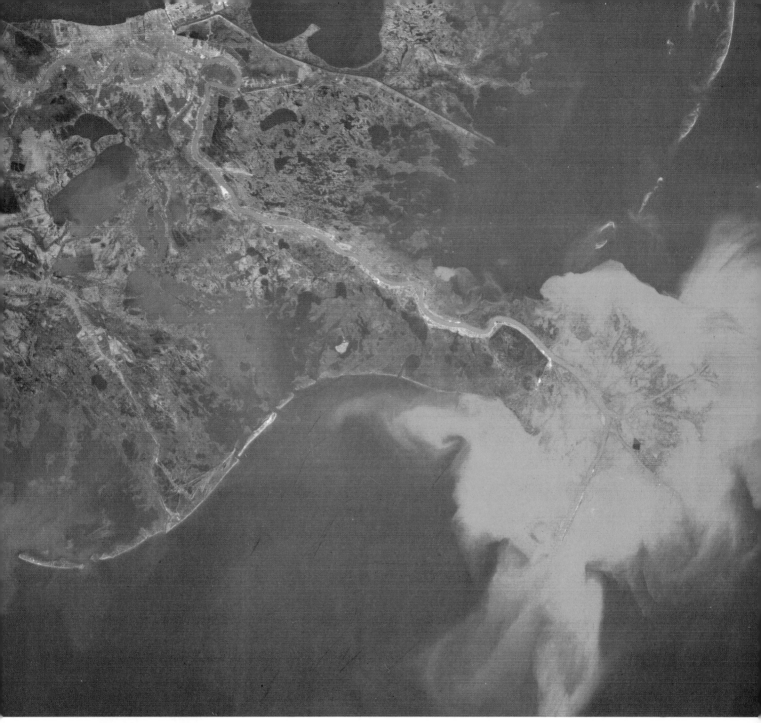

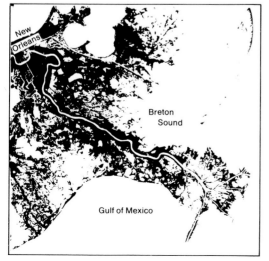

Plate 9 The Mississippi Delta, like deltas of other major rivers, is a record of erosion due to the hydrologic system. Sediment eroded from the land is transported by a river system and deposited in the sea. The dynamics of delta building are displayed vividly in this photograph. The cloud of mud and silt delivered to the ocean colors the water a light tone around the mouth of the river. This material, derived from the erosion of the continent, is building banks of mud, sand, and clay over the continental shelf as the delta grows seaward, at a rate of nearly 20 km per 100 years. Measurements indicate that the Mississippi River pours more than a million metric tons of sediment into the Gulf of Mexico each day. In the process of deltaic growth, the river builds up a projection of new land out into the ocean, where the channel breaks up into numerous distributaries. Eventually, the river finds a shorter route to the ocean and abandons its channel for the shorter course. The active subdelta ceases to grow and is eroded back by wave action. Abandoned river channels and inactive subdeltas can be seen clearly on each side of the present river.

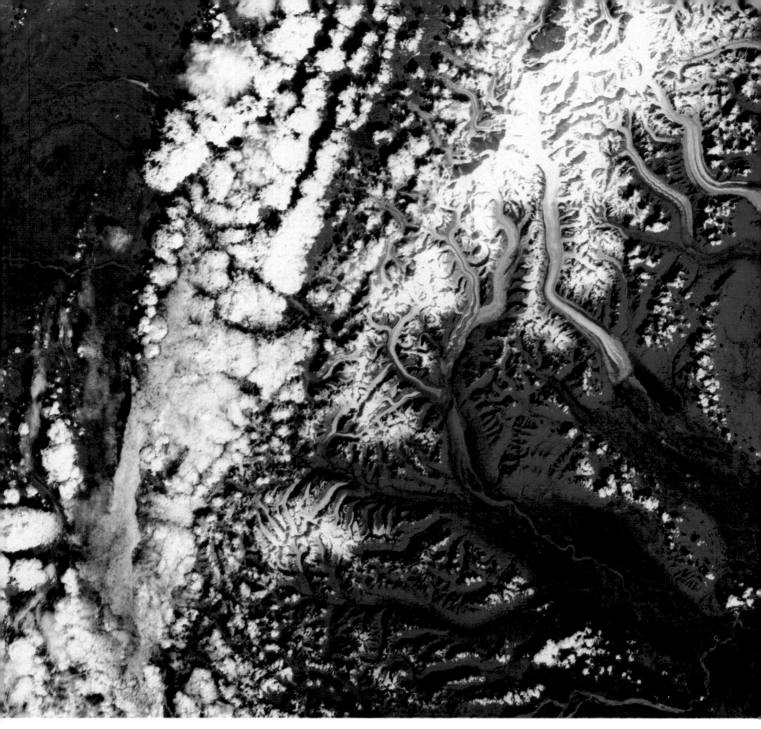

Plate 10 Valley glaciers, such as these in the vicinity of Mount McKinley, Alaska, occur where more snow accumulates each year than is melted during the summer months. Valley glaciers originate in the snowfields of high mountain ranges and flow as large tongues of ice down preexisting stream valleys. The moving ice is an effective agent of erosion and modifies the valleys in which it flows. Thus, glaciers cause local modifications of the normal hydrologic system. The dark lines on the glaciers are rock debris derived from the valley walls. Their wavy appearance indicates surging (rapid movement), with rates as high as 2 km per hour. Nonsurging glaciers have straight and uniform lines of debris and move only a few centimeters per day.

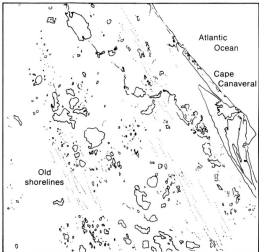

Plate 11 Ground water is a largely invisible part of the hydrologic system, since it occupies pore spaces in the soil and rocks beneath the surface. It can, however, dissolve soluble rocks, such as limestone, to form complex networks of caves and subterranean passageways. As the caverns enlarge, their roofs can collapse, so that circular depressions called sinkholes are formed. Sinkholes create a pockmarked surface called karst topography. The hundreds of lakes shown in this photograph of the area around Cape Canaveral, Florida, occupy sinkholes and testify to the effectiveness of ground water as a geologic agent. The linear features of the shoreline are offshore bars and islands, formed by waves and currents along the South Atlantic coast.

Plate 12 The Atlantic coastline of the southern states is dominated by offshore bars and barrier islands. A considerable amount of sediment eroded from the continent is brought seaward by rivers and reworked by wave action to form long, narrow, sandy barrier beaches, which fringe the mainland. This photograph shows part of the shoreline of North Carolina. The lagoon of Pamlico Sound and the Great Dismal Swamp, protected by the barrier island from the vigorous wave action of the Atlantic, will eventually become filled with sediment, so that the mainland will extend seaward to the present position of the barrier beach. The edge of the continental shelf is delineated by puffy clouds, which form where the cold surface water to the east meets the warmer surface water of the shallow nearshore zone. The edge of the shelf is only about 120 m deep.

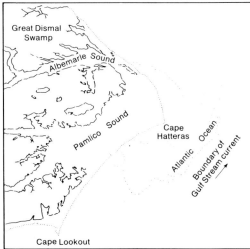

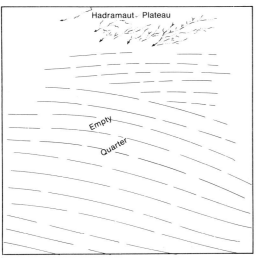

Plate 13 "Sand seas" of the great deserts of the earth form in arid low-latitude regions where there is not enough precipitation for the hydrologic system to operate in its normal manner. This view of sand dunes in the Arabian Peninsula looks to the southwest, toward the Hadramaut Plateau. The long parallel stripes of dunes are part of one of the largest sand seas in the world. The dunes are typical of much of the Empty Quarter of Arabia and are remarkable in that some extend more than 200 km without a break. The vast areas of migrating sand dunes in the world's deserts illustrate the effectiveness of the circulating atmosphere alone as a geologic agent, continually transporting enormous quantities of sediment over the surface of the earth.

Plate 14 The Red Sea rift, which separates Africa from Asia, is a major fracture system in the earth's crust. This feature is an important part of the earth's rift system in that it represents the incipient stages of crustal fragmentation and the movement of continental plates. The rift extends up the Red Sea and splits at its northern end, with one branch forming the Gulf of Suez and the other extending up the Gulf of ʿAqaba, into the Dead Sea, and up the Jordan valley. Movement of the Arabian plate in a northeasterly direction, away from Africa, has created a new ocean basin, which is in the initial stage of its formation. As the rift widens, the edges of the continental block break off and form a series of steps leading down toward the depression. These step blocks can be seen along the Gulf of Suez as distinct parallel lines in the bedrock and as linear trends in the offshore islands.

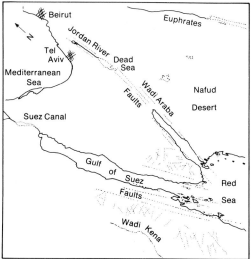

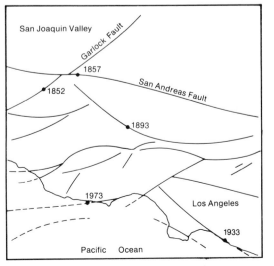

Plate 15 The San Andreas Fault is one of the most significant geologic structures in California. This view shows the intersection of the San Andreas Fault and the Garlock Fault, northwest of Los Angeles. These two major structural features intersect to form the boundary of the Mojave Desert, the light-colored triangular area in the upper right-hand quarter of the photograph. At least a dozen major fault systems can be seen as linear trends in the landscape in this view. Most have been active during historical times (in the last 30,000 or 40,000 years). Movement along the San Andreas Fault is horizontal; that is, one block of the earth's crust slides laterally past the other. This type of movement is associated with areas where the crust is being pulled apart. The annotated map shows the system of faults and the location of historical earthquakes whose magnitudes on the Richter scale were 6.0 or greater.

Plate 16 Volcanoes of the Indonesian island arc are formed by volcanic activity produced where two oceanic plates converge and one is thrust under the other and is assimilated into the mantle. As the descending plate melts, the molten rock rises to form a chain of volcanoes. If the upper plate contains oceanic crust, the volcanic activity produces an arc of volcanic islands. This area, just east of Bali, is dotted with both active and extinct volcanic cones built up from the sea floor. Recent eruptions have occurred at Tambora, whose central caldera rises 2755 m. Young flows are visible where vegetation has not yet been established. If plates converge and the upper plate contains continental crust, volcanic material is extruded on a folded mountain belt. An excellent example of this type of volcanic activity occurs in the Andes Mountains of South America.

MAJOR FEATURES OF THE CONTINENTS

Statement

The broad, flat continental platforms, which rise above the ocean basins, present a great diversity of surface features, with an almost endless variety of hills and valleys, plains and plateaus, and mountains. From a regional perspective, however, the continents are remarkably flat. Most of their surface lies within a few hundred meters of sea level. Extensive geologic studies during the past hundred years have revealed several striking facts about the continents.

1. The continents are roughly triangular in shape.
2. They are concentrated in the Northern Hemisphere.
3. Although each may seem unique, all continents have three basic components: (a) a shield, (b) a stable platform, and (c) folded mountain belts. Geologic differences between continents are mostly in size, shape, and proportions of these components.
4. Continents consist of rock that is less dense than the rock in the ocean basins.
5. The continental rocks are old, some as old as 3.8 billion years.
6. The climatic zone occupied by a continent usually determines the style and variety of landforms developed on it.

Discussion

Shields

Without some firsthand knowledge of a shield, it is difficult to visualize the nature and significance of this important part of the continental crust. *Plate 3*, showing part of the Canadian Shield of the North American continent, will help you comprehend the extent, the complexity, and some of the typical features of shields. You should also study *figures 7.1, 22.1*, and *22.3*.

Characteristically, a **shield** is a regional surface of low **relief,** whose elevation is generally within a few hundred meters of sea level. The only features of relief are the resistant rocks that rise 50 or 100 m above their surroundings.

A second characteristic of shields is their complex structure and rock types. Many rock bodies were once liquid; others have been compressed and extensively deformed. All of the rocks are crystalline and were formed several kilometers below the surface. They are now exposed only because the shields have been subjected to extensive erosion.

Stable Platforms

Large parts of the continental shields are covered with a veneer of sedimentary rocks. These areas have been relatively stable throughout the last 600 million or 700 million years. That is, they have never been uplifted a great distance above sea level or submerged far below it, and hence the term *stable platform.* In North America, the stable platform lies between the Appalachian Mountains and the Rocky Mountains and extends northward to the Lake Superior region and into western Canada. Throughout most of this area, the sedimentary rocks that cover the shield are nearly horizontal, but locally they have been warped into broad domes and basins (*plate 4*).

Folded Mountains

One of the most signficant features of the continents is the young, **folded mountain belts,** which typically occur along their margins. Most people think of a **mountain** as simply a high, more or less rugged landform, in contrast to flat plains and lowlands. Mountains, however, are much more than high country. To a geologist, the term *mountain belt* refers to a long, linear zone in the earth's crust where the rocks have been intensely deformed by horizontal stresses and generally intruded by molten rock material. The topography can be high and rugged, or it can be worn down to a surface of low relief. It is not the topography of mountain belts that is important but the extent and style of deformation. The great folds and fractures in mountain belts, therefore, provide evidence that the earth's crust is, and has been, in motion.

Plate 5 illustrates some of the characteristics of folded mountains and the extent to which the margins of continents have been deformed. The layers of rock shown in this photograph have been deformed by compression and are folded like wrinkles in a rug. Erosion has removed the upper part of the fold, so that the resistant layers form zigzag patterns similar to those that would be produced if the crests of wrinkles in a rug were cut off.

We now know that the moon, Mars, and Mercury lack this type of deformation; all of their impact craters (regardless of their age) are circular, and thus they have not been deformed by compressive forces. The crusts of these planets, unlike that of the earth, appear to have been fixed and immovable throughout their histories.

MAJOR FEATURES OF THE OCEAN FLOOR

Statement

The ocean floor, not the continents, is the typical surface of the solid earth, and it is the ocean floor that holds the key to the evolution of the earth's crust. Yet, not until the 1960s did we obtain enough data about the ocean floor to get a clear picture of its regional characteristics. This new knowledge caused a revolution in geologists' ideas about the nature and evolution of the crust.

Before 1947, most geologists believed that the ocean floor is simply a submerged version of the continents, with huge areas of flat abyssal plains covered with sediment derived from the landmass. Although echo sounding was used to determine depths as early as 1922, the primitive devices were limited to shallow water. An observer with headphones timed the interval between the transmission of a sound pulse and the return of its echo. A major breakthrough occurred in 1953 with the development of a precision depth recorder that could automatically plot a continuous profile of the ocean floor in any depth of water. Since then, millions of kilometers of profiles have been made, and the information has been used to map the surface features of the ocean basins.

Profiles of the ocean floor show that submarine topography is as varied as that on the continents and, in some respects, is more spectacular. Among the most significant facts that we have learned about the oceanic crust are the following:

1. The oceanic crust is mostly basalt, a dense volcanic rock, and its major topographic features are somehow related to volcanic activity. The oceanic crust, therefore, is entirely different from the continental crust.
2. The rocks of the ocean floor are young in a geologic time frame. Most are less than 150 million years old, whereas the ancient rocks of the shields are more than 700 million years old.
3. The rocks of the ocean floor have not been deformed by compression. Their undeformed structure stands out in marked contrast to the complex deformation of rocks in the folded mountains and shields of the continents.
4. The major provinces of the ocean floor are
   a. The oceanic ridge
   b. The abyssal floor
   c. Seamounts
   d. Trenches
   e. Continental margins

Discussion

Although most of the topography of the ocean floor can be seen only indirectly, through profile records, some features are visible in satellite photographs. *Plate 6*, for instance, shows a deep channel on the ocean floor north of Cuba, which cuts across the shallow Bahamas Platform. But **physiographic maps**, such as the one on the endpapers of this book, provide the best visual reference for regional features of the ocean basins. You should refer to this map continually as you study the following material.

The Oceanic Ridge

The **oceanic ridge** is perhaps the most striking and important feature on the ocean floor. It extends continuously from the Arctic basin down the center of the Atlantic, into the Indian, and across the South Pacific. The oceanic ridge is essentially a broad, fractured swell, generally more than 1400 km wide. Its higher peaks rise as much as 3000 m above the ocean floor. A huge, cracklike valley, called the **rift valley**, runs along the axis of the ridge throughout most of its length. In addition, great fracture systems, some as long as 4000 km, trend across the ridge.

The Abyssal Floor

The oceanic ridge divides the Atlantic and Indian oceans roughly in half and traverses the southern and eastern parts of the Pacific. On both sides of the ridge are vast areas of broad, relatively smooth, deep-ocean basins known as the **abyssal floor**. This surface extends from the flanks of the oceanic ridge to the continental margins and generally lies at depths of about 3000 m.

The abyssal floor can be subdivided into two sections, the abyssal hills and the abyssal plains. The **abyssal hills** are relatively small hills, rising as much as 900 m above the surrounding ocean floor. They cover from 80% to 85% of the Pacific sea floor, and thus they are the most widespread landforms on the earth. Near the continental margins, land-derived sediment completely covers the abyssal hills, forming flat, smooth **abyssal plains**.

Seamounts

Isolated peaks of submarine volcanoes are called **seamounts**. Some seamounts rise above sea level and form islands, but most are completely submerged and are known only from oceanographic soundings. Although many seem to occur at random, others form chains along well-defined lines, such as the Hawaiian chain. Islands and seamounts testify to the extensive and continuous volcanic activity that has occurred throughout the ocean basins. They also provide an important insight into the dynamics of the inner earth.

Trenches

The deep-sea **trenches** are the lowest areas on the earth's surface. The Mariana Trench, in the Pacific Ocean, is the deepest part of the world's oceans—11,000 m below sea level—and many other trenches are more than 8000 m deep. Trenches have attracted the attention of geologists for years, not only because of their depth, but because they represent fundamental structural features of the earth's crust. As is illustrated in *figure 1.2*, the trenches are invariably adjacent to island arcs or coastal mountain ranges of the continents. We will see in subsequent chapters how the trenches are involved in the most intense volcanic and seismic (earthquake) activity on the planet.

Continental Margins

The zone of transition between a continental mass and an ocean basin is called a **continental margin.** The submerged part of a continent is referred to as a **continental shelf.** Geologically, it is part of the continent, not part of the ocean basin. The continental shelves presently constitute 18% of the earth's total surface, but at times in the geologic past, the ocean covered much more of the continental platform.

The sea floor descends in a long, continuous slope from the outer edge of the continental shelf to the deep-ocean basin. This **continental slope** marks the edge of the continental rock mass. Continental slopes are found around the margins of every continent and also around smaller fragments of continental crust, such as Madagascar and New Zealand. Look at the physiographic map on the endpapers of this book and study the continental slopes—especially those surrounding North America, South America, and Africa—and you can see that they form one of the earth's major topographic features. On a regional scale, they are by far the steepest, longest, and highest slopes on the earth. Within this zone, from 20 to 40 km wide, the average relief above the sea floor is 4000 m. Along the marginal trenches, relief is as great as 10,000 m. In contrast to the shorelines of the continents, the continental slopes are remarkably straight over distances of thousands of kilometers.

In many areas, the continental slopes are cut by deep **submarine canyons,** which are remarkably similar to canyons cut by rivers into continental mountains and plateaus. As is shown on the physiographic map, submarine canyons cut across the edge of the continental shelf and terminate on the deep abyssal floor, some 5000 or 6000 m below sea level.

Figure 1.2 The major features of the ocean floor are related to plate boundaries. The oceanic ridge coincides with divergent plate margins. Trenches form where plates converge. The abyssal floor is the deep part of the sea floor, off the flanks of the oceanic ridge.

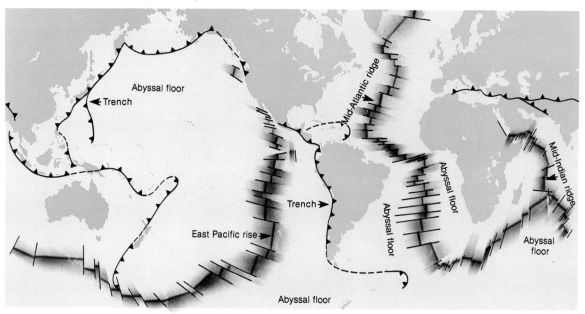

THE STRUCTURE OF THE EARTH

Statement

Determining the internal structure of the earth presents some of the most difficult problems faced by geologists and geophysicists. The deepest boreholes penetrate no more than 8 km, and erosion exposes rocks that were created no more than 20 or 25 km below the surface. Volcanic eruptions provide samples of material from greater depths, possibly as much as 200 km. These limited data provide our only direct knowledge about the nature of the earth's interior. How, then, are we able to determine the structure and composition of the earth's interior? The evidence comes largely from studies of the physical characteristics of the earth—its density, the nature of its magnetic field, and the way it transmits seismic waves. These characteristics will be considered in later chapters. Here we will only summarize the present-day under-standing of the structure, composition, and physical properties of the earth's interior.

The earth is a differentiated planet. That is, its constituent materials are separated and segregated according to density. Heavy material is concentrated near the center, and light material near the surface. The major structural units are as follows:

1. A solid inner core, composed predominantly iron and nickel
2. A liquid outer core, also composed of iron and nickel
3. A thick, surrounding mantle, composed of silicate minerals rich in iron and magnesium
4. A soft, plastic asthenosphere, which constitutes the upper part of the mantle
5. A rigid lithosphere, composed of relatively light silicate minerals, which includes the crusts of the continents and the ocean basins
6. A hydrosphere, or system of circulating water, and an atmosphere, composed of various gases (including water vapor) (*figure 1.3*)

Figure 1.3 The internal structure of the earth is known from studies of its density, its magnetic field, and the way in which it transmits seismic waves. The major components are a dense, solid inner core; a liquid outer core; a thick, solid mantle; and a rigid lithosphere. The upper mantle contains a significant layer, the asthenosphere, which is soft and plastic. Surrounding the entire planet is a blanket of surface fluids—water and air.

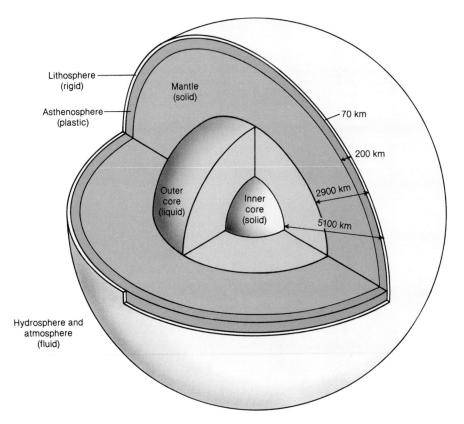

Discussion

The Core

The **core** of the earth is a central mass about 7000 km in diameter. Its density increases with depth but averages about 10.78 g/cm³. It is nearly twice as dense as the mantle, and though it constitutes only 16.2% of the earth's volume, it accounts for 31.5% of the earth's mass. Most scientists believe that the core is mostly iron and nickel and that it consists of two distinct parts—a solid inner core and a liquid outer core. The rotation of the earth probably causes the liquid core to circulate, and its circulation generates the earth's magnetic field.

The Mantle

The next major structural unit of the earth, the **mantle,** surrounds or covers the core. This zone constitutes the great bulk of the earth (82.3% of its volume and 67.8% of its mass). The upper part of the mantle is called the **asthenosphere.** It is approximately 200 km thick, and its upper boundary is about 70 km below the earth's surface. The asthenosphere is distinctive because its temperature and pressure are in delicate balance, so that most of its material is near the melting point. It is therefore believed to be partly molten and structurally weak and thus capable of flow. Movement within this layer is responsible for the volcanic activity and crustal deformation observed at the surface.

The Lithosphere

The **lithosphere** is the outermost shell of the earth. It is a rigid, solid layer about 70 km thick, which rests upon the weak, partly molten asthenosphere. The continental crust can be thought of as raftlike slabs of granitic rock embedded in the lithosphere (*figure 1.4*).

The Surface Fluids

The gases and liquids of the atmosphere and hydrosphere—the earth's surface fluids—are as much a part of the earth's systems as the rocks of the crust. Not only are they part of the planet, but they are of prime importance in the origin and development of the landscape. The earth's atmosphere and water are concentrated above its solid part because of their lower density.

The atmosphere is a mixture of highly mobile gases, held to the earth by gravitational attraction. The gases are dense close to sea level, but they thin rapidly at higher altitudes, so that 97% of the mass of the atmosphere lies within 29 km of the surface. There is no sharp boundary marking the outer limit of the atmosphere. The density of gas molecules decreases almost imperceptibly into interplanetary space. Many scientists consider the outer boundary to be about 9500 km above the earth's surface, a distance nearly as great as the diameter of the earth. Only the lower part of the atmosphere, where water vapor forms clouds, is visible.

Water on the earth is present as oceans, lakes, rivers, glaciers, ground water, and vapor in the atmosphere. It is in continual motion, and as it moves over the earth's surface, it erodes, transports, and deposits weathered rock material and constantly modifies the landscape. It is water that is responsible for many of the earth's unique features and permits the development of life upon its surface.

Figure 1.4 The outermost layers of the earth are the asthenosphere and the lithosphere. The asthenosphere is near the melting point and is capable of flow. The lithosphere above it is rigid. It includes two types of crust: a thin, dense oceanic crust and a thick, lighter continental crust.

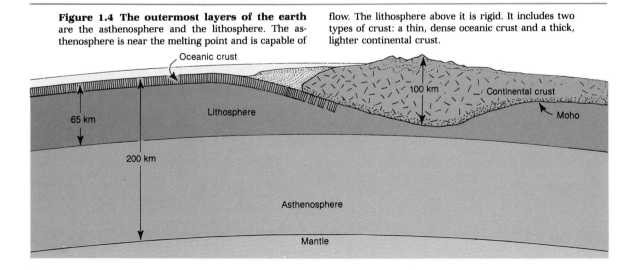

Figure 1.5 The major surface features of the earth reflect the structure of the crust. The continental crust rises above the ocean basins and forms continents, whose major topographic features are shields, stable platforms, and folded mountain belts. The oceanic crust, composed primarily of basalt, forms the ocean floor, whose major features include the oceanic ridge, the abyssal floor, seamounts, and trenches.

SUMMARY

Perhaps the most overwhelming impression gained from seeing the earth from space and observing the major features of its surface is that the earth is a dynamic planet and has been changing throughout all geologic time. Not only are the surface fluids extremely mobile, but the crust is in motion, having been deformed by mountain-building forces, volcanic activity, and earthquakes. No such activity has occurred on the moon or Mercury, and Mars has been much less dynamic than the earth.

The following facts about the earth have been discussed and illustrated by means of space photographs:

1. The atmosphere and the hydrosphere constitute a system of moving surface fluids in which tremendous volumes of water are in constant motion.
2. The two major structural and topographic features of the earth are (a) the continents and (b) the ocean basins.
3. The continents consist of three major components: (a) the shields, which are composed of complexly deformed rock eroded down to near sea level; (b) stable platforms, or areas where the shield is covered with a veneer of horizontal sedimentary rocks; and (c) mountain belts, in which sedimentary rocks have been compressed into folds.
4. The major features on the ocean floor are (a) the oceanic ridge, (b) the abyssal floor, (c) seamounts, (d) trenches, and (e) continental margins.
5. The major structural units of the earth are (a) a solid inner core, (b) a liquid outer core, (c) a thick mantle, (d) a soft asthenosphere, (e) a rigid lithosphere, and (f) a fluid hydrosphere and atmsophere.

Key Words

landform	oceanic crust
continent	relief
ocean basin	topography
density	physiography
continental crust	physiographic map

shield	continental margin
stable platform	continental shelf
folded mountain belt	continental slope
mountain	submarine canyon
rift valley	core
oceanic ridge	mantle
abyssal floor	asthenosphere
abyssal hills	lithosphere
abyssal plains	hydrosphere
seamount	atmosphere
trench	biosphere

Review Questions

1. Draw a diagram of the internal structure of the earth and briefly describe the core, mantle, asthenosphere, and lithosphere.
2. Why are the atmosphere and the oceans considered as much a part of the earth as solid rock is?
3. What are the major differences between the continents and the ocean basins?
4. Briefly describe the distinguishing features of continental shields, stable platforms, and folded mountain belts.
5. Briefly describe the distinguishing features of the oceanic ridge, the abyssal floor, seamounts, trenches, and continental margins.
6. Study the physiographic map of the earth on the inside cover. Make a sketch map of the major structural features of the earth by tracing or sketching the continental crust, shields, stable platforms, and folded mountain belts, together with the oceanic ridge, the abyssal floor, and deep-sea trenches.
7. Study the view of the earth in *plate 2* and make a sketch map by tracing or sketching the following features: (a) major patterns of atmospheric circulation, (b) low-latitude deserts, (c) the tropical belt, (d) the Red Sea rift, and (e) the Antarctic ice cap.

Additional Readings

Bates, D. R., ed. 1964. *The planet Earth*. Elmsford, N.Y.: Pergamon Press.
Bolt, B. A. 1973. The fine structure of the earth's interior. *Scientific American* 228(3):24–33.
Bott, M. H. P. 1971. *The interior of the earth*. New York: St. Martin's Press.

THE EARTH'S DYNAMIC SYSTEMS

We saw in the previous chapter that the earth is a differentiated planet; its materials are segregated and separated by density into a series of concentric layers. We speak of the earth as a dynamic planet because the materials of the various layers are in motion. The most obvious motion is that of the surface fluids (air and water), which circulate in response to solar energy.

The complex cycle by which water moves from the oceans to the atmosphere, to the land, and back to the oceans again is called the hydrologic system. Water precipitated on the earth furnishes a fluid medium for many processes that shape the earth's surface features. Rivers systematically erode the land as they flow to the sea. Ground water percolating through the pore spaces in rocks dissolves and carries away soluble minerals. Where temperatures are low enough, glaciers form and spread over large parts of a continent, modifying its surface by erosion and deposition.

The earth's own internal energy also plays a major role in moving material in the earth from place to place. Even though the earth's crust appears to be fixed and stable, there is convincing evidence that it too is in constant motion, and as you continue in your study of geology, the tremendous significance of the earth's internal energy will become increasingly clear. Thermal energy is believed to be transferred from the asthenosphere to the earth's surface by convection, to produce what is known as the tectonic system. As material in the asthenosphere is heated, it rises and then moves laterally, causing the rigid lithosphere to arch upward and split apart. Thus the lithosphere is broken into a series of plates, which are moved about by the slow convection flow in the asthenosphere. This motion is expressed by volcanic activity, earthquakes, mountain building, continental drift, and sea-floor spreading.

This chapter summarizes the basic elements of the hydrologic and tectonic systems, which constantly modify the earth. The remaining chapters largely concern details of the physical processes of these systems.

MAJOR CONCEPTS

1. The earth's system of moving water—the hydrologic system—involves movement of water in rivers, as ground water, in glaciers, and in oceans. The volume of water in motion is almost incomprehensibly large. As it moves, it erodes, transports, and deposits sediment. Heat from the sun is the source of energy for the earth's hydrologic system.
2. The tectonic system involves the movement of material in the earth's interior, which results in sea-floor spreading, the creation of new crust, continental drift, volcanism, earthquakes, and mountain building. Heat in the upper mantle is the source of energy driving the tectonic system.
3. The earth's lithosphere floats on the denser, plastic mantle beneath and rises and sinks in attempts to establish and maintain isostatic equilibrium.

THE HYDROLOGIC SYSTEM

Statement

In the broadest sense, the hydrologic system includes all possible paths of motion of the earth's surface fluids, both air and water. The basic elements of the system, diagrammed in *figure 2.1*, are simple. The system operates as heat from the sun evaporates water from the oceans, the principal reservoir for the earth's water. Most of the water returns directly to the oceans as rain. Atmospheric circulation carries the rest over the continents, where it is precipitated as rain or snow. Water that falls on the land can take a variety of paths back to the oceans. The greatest quantity returns to the atmosphere by evaporation, but the most obvious return is by surface runoff in river systems, which funnel water back to the oceans. Some water also seeps into the ground and moves slowly through the pore spaces of the rocks. Plants use part of the ground water and then expel it into the atmosphere by transpiration. Much of the ground water slowly seeps into streams and lakes or migrates through the subsurface back to the oceans. Water can be temporarily trapped upon a continent as glacial ice, but glaciers move from cold centers of accumulation into warmer areas, where they ultimately melt, returning their water to the system as surface runoff.

Water in the hydrologic system—moving as surface runoff, ground water, glaciers, and waves and currents—erodes and transports surface rock material and ultimately deposits it as deltas, beaches, and other types of sedimentary formations. In this way, the surface material is in motion—motion that results in a continually changing landscape.

Discussion

The idea of a complete cycle in the movement of water was recorded in biblical times (Eccles. 1:7), but such a cycle was not demonstrated as a fact until the mid-seventeenth century. The discovery was made by two French scientists, Pierre Perrault (1608–1680) and Edme Mariotte (1620?–1684), who independently measured precipitation in the drainage basin of the Seine River and then measured the discharge into the ocean during a given interval of time. Their measurements proved that precipitation alone produces not only enough water for the river to flow but also enough for springs and seeps. Precipitation then was recognized as the basic source of all surface water.

Figure 2.1 The circulation of water in the hydrologic system operates by solar energy. Water evaporates from the oceans, circulates around the globe with the atmosphere, and is eventually precipitated on the surface as rain or snow. The water that falls on the land returns to the ocean by surface runoff and ground-water seepage. Variations in the major flow patterns of the system include the temporary storage of water in lakes and glaciers. Within this major system are many smaller cycles, shortcuts, and two-way paths. Some occur in just minutes, as when rain evaporates before it falls to the surface of the earth. Others endure for millions of years, as in the case of water locked up in minerals or deposits of sediments.

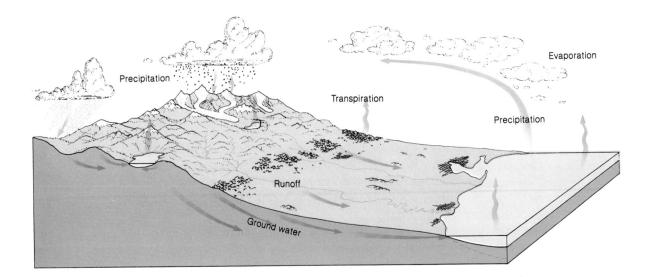

Precipitation
Evaporation
Transpiration
Precipitation
Runoff
Ground water

The **hydrologic system** is perhaps the most fundamental and significant geologic system operating on the earth's surface. Its influences on the development of surface features are sometimes subtle, sometimes dramatic. As a student new to geology, your problem will be not so much understanding the process as conceiving the system's worldwide scope.

The Hydrologic System as Seen from Space

The best way to gain an accurate conception of the magnitude of the hydrologic system is to study the space photograph in *plate 2*, which permits a view of the system in operation on a global scale. A traveler arriving from space would observe that the earth's surface is predominantly water (an obvious conclusion from the view in *plate 2*). The movement of water from the oceans to the atmosphere is expressed in the flow patterns of the clouds. This motion is one of the most distinctive features of the earth as viewed from space, and it stands out in marked contrast to conditions on the moon, Mars, and Mercury.

The swirling, moisture-laden clouds carry an enormous amount of energy. For instance, the kinetic energy produced by a hurricane amounts to roughly 100 billion kilowatt-hours per day. That is considerably more than the energy used by man in one day.

Another way to grasp the magnitude of the hydrologic system is to consider the volume of water involved. From measurements of rainfall and stream discharge, together with measurements of heat and energy transfer into bodies of water, scientists have calculated that 400,000 km³ of water evaporates each year (or 1100 km³ per day). Of this, about 336,000 km³ is taken from the oceans, and 63,000 km³ evaporates from bodies of water on the land. Of the total 400,000 km³ of water evaporated annually, about 101,000 km³ falls as rain or snow on the continents. Most of the precipitation (from 60% to 80%) returns directly to the atmosphere by evaporation. About 38,000 km³ of water flows back to the oceans over the surface or beneath the ground.

At these rates, if the hydrologic system were interrupted and water did not return to the oceans (by precipitation into the oceans and by surface **runoff** from the continents), sea level would drop 1 m per year. All of the ocean basins would be completely dry within 4000 years. The recent glacial epoch demonstrates this point clearly: the hydrologic system was partly interrupted, as much of the water that fell upon the Northern Hemi-

sphere froze and formed huge continental glaciers, which prevented the water from flowing immediately back to the oceans as surface runoff. Consequently, during the ice age, sea level dropped more than 100 m.

As a final observation, consider the volume of water running off the earth's surface. The circulation of water in the hydrologic system provides a continuous supply of surface runoff over the land. At any given instant, about 1260 km³ of water is flowing in the world's rivers. Gauging stations on the major rivers of the world indicate that, if the ocean basins were empty, runoff from the continents would fill them completely in 40,000 years. (This estimate assumes no precipitation falling directly into the oceans.) Moreover, if all the evaporated water returned by runoff, with no return to the atmosphere by immediate evaporation and the **transpiration** of plants, then the ocean basins would be filled in slightly more than 3000 years.

Gravity plays an important role in the hydrologic system. It is the force causing rain to fall on the land and move downslope. It causes rivers to flow and transport sediment from higher to lower levels. It pulls loose rock debris down slopes in the form of landslides, rockfalls, and soil creep. Indeed, the movement of water in each part of the hydrologic cycle is a gravity flow system.

River Systems

As would be expected, the vast volumes of water that move over the earth's surface produce a landscape dominated by features formed by running water. From viewpoints on the ground, we cannot appreciate the prevalence of channels on the earth's surface. From space, however, we readily see that stream valleys are the most abundant landform on the planet. In arid regions, where vegetation and soil cover do not obscure details, the intricate network of stream valleys is most impressive (*plates 7* and *8*). Most of the surface of every continent is somehow related to a slope of a stream valley, which collects and funnels surface runoff toward the ocean.

The important point is that each continent has one or more major **river systems,** and the valley slopes formed by the network of tributary streams are the dominant landforms. We can see from space photographs that the surface drainage of a continent is an important segment of the hydrologic system and is a vivid record of how moving water has continually modified and shaped the surface of the land.

Another important aspect of the hydrologic

system is that it provides the fluid medium that transports huge amounts of sand, silt, and mud to the oceans. This material forms the great deltas of the world and records the amount of material washed off the continents by rivers. The Mississippi Delta (*plate 9*) is a classic example. The river is confined to a single channel far downstream from New Orleans. It then breaks into a series of distributaries, which build extensions of new land into the ocean. A large cloud of suspended sediment forms in the ocean where the river empties into the Gulf of Mexico and is actively building new land seaward. Ultimately, the main channel shifts its course to seek a more direct route to the ocean, and the extension of land (subdelta) is eroded back by waves and currents. Previous courses of the Mississippi can be seen on both sides of the present river.

Glacial Systems

In cold climates, precipitation falls in the form of snow, which remains frozen and does not return immediately to the ocean as surface runoff. If more snow falls each year than melts during the summer months, huge bodies of ice build up to form **glaciers.** *Plate 10* shows the glacial field in the vicinity of Mount McKinley, in Alaska. Large valley glaciers originate from snowfall in the high country and slowly flow down valleys as rivers of ice. They melt at their lower end and return their water to the hydrologic system as surface runoff.

At the present time, the continent of Antarctica is covered with a continental glacier, a sheet of ice from 2.0 to 2.5 km thick. It covers an area of 13,000,000 km²—an area larger than the United States and Mexico. A large part of the Antarctic glacier can be seen in *plate 2*. The formation of a continental glacier completely modifies the normal hydrologic system, because the water does not return immediately to the ocean as surface runoff but moves slowly as flowing ice. An ice sheet similar to that now on Antarctica recently covered a large part of North America. As the ice moved, it modified the landscape by creating numerous lakes and other landforms.

Ground-Water Systems

Another segment of the hydrologic system is the water that seeps into the ground and moves slowly through the **pore spaces** in the soil and rocks. As it moves, **ground water** dissolves soluble rocks and creates caverns and caves, which can enlarge and collapse to form depressions called sinkholes. This type of solution-generated landform is common in

Kentucky and Florida and is easily recognized in space photographs (*plate 11*). The sinkholes are filled with water and create a pockmarked surface, somewhat resembling the cratered surface of the moon.

Shoreline Systems

The hydrologic system also operates along shores, where water and sediment are moved by waves and currents to form beaches, bars, lagoons, and other coastal features (*plate 12*). The sediment, derived from erosion of the continents, is reworked and deposited in the shallow offshore areas.

Eolian Systems

The hydrologic system also operates in the desert regions of the world. In many deserts, river valleys are still the dominant landform, because there is no completely dry place on the earth. Even in the most arid regions, some rain falls, and climatic patterns change over the years. River valleys can be obliterated, however, by the great seas of windblown sand that cover the desert landscape (*plate 13*). Air circulating in the atmosphere constitutes **eolian** (wind) systems, which can transport enormous quantities of loose sand and dust, leaving a distinctive record of the operation of the wind. But in the broadest sense, the wind itself is part of the hydrologic system, a moving fluid on the planet's surface.

The energy source for the hydrologic system is heat from the sun, which evaporates water from the oceans and causes the atmosphere to circulate. Water vapor moves with the circulating atmosphere and eventually condenses, under proper conditions of temperature and pressure, and falls as rain or snow. Under the force of gravity, it then flows back to the oceans in several subsystems (rivers, ground water, and glaciers), all of which involve gravity flow from higher to lower levels. Without solar heat, there would be no hydrologic system, because there would be no energy to transport water from the oceans to the continents.

The motion involved in the hydrologic system is one of the distinguishing features of the earth. It causes surface materials to be in constant motion and to change the landscape continually, on a scale so great that it is difficult to comprehend. Although the surface of the earth may appear to be stable and unchanging from year to year, or even throughout a human lifetime, it is continually being altered by the hydrologic system. Indeed, change is a fundamental law of the universe.

THE TECTONIC SYSTEM

Statement

Motion in the hydrologic system can be easily observed, especially from viewpoints in space, but there is also convincing evidence that the earth's crust is in motion and that it has moved throughout all of geologic time. The evidence comes from many phenomena, including earthquakes, volcanoes, huge fractures in the crust, folded mountain belts, and characteristics of the ocean floor, but it was not until the late 1960s that a unifying theory of the earth's dynamics was developed. This theory, known as plate tectonics, provides for the first time a master plan into which everything geologists know about the earth seems to fit. The term *tectonics*, like *architecture*, comes from the Greek *tektonikos* and refers to building or construction. In geology, tectonics is the study of the formation and deformation of the earth's crust that result in large-scale structural features. Plate tectonics is a theory that explains the structure of the crust as a result of the division of the lithosphere into a series of semirigid plates, which move about by convection currents in the asthenosphere.

The basic elements of the plate tectonic theory are quite simple and can be understood by studying the diagram in *figure 2.2*. The lithosphere, which includes the earth's crust, is rigid, but the underlying asthenosphere yields to plastic flow. The principle behind plate tectonics is that the rigid lithosphere moves in response to flow in the asthenosphere below it. Although there are many unanswered questions, it is believed that the basic energy source for tectonic movement is radiogenic heat (heat generated by radioactivity) in the upper mantle. This heat causes material in the asthenosphere to move slowly by convection: the hot material rises to the base of the lithosphere, where it then moves laterally, cools, and descends to become reheated, so that the cycle begins again. (A

Figure 2.2 The tectonic system operates by the earth's internal energy. The material in the asthenosphere is thought to be moving in a series of convection cells resulting from radiogenic heat. Where the convecting mantle rises and moves laterally, it pulls the overlying, rigid lithosphere apart and creates new crust in the rift zone. Each segment of lithosphere is carried by a convection cell and moves as a single mechanical unit. Where converging plates meet, one descends into the mantle and forms a deep-sea trench. Some plates contain blocks of continental crust, and plate motion can cause the continents to split and drift apart. Being less dense than the mantle, continental crust cannot sink into the asthenosphere. As a result, where a plate carrying continental crust collides with another plate, the continental margins are deformed into mountain ranges. The plate margins are the most active areas of the earth—the sites of the most intense volcanism, seismic activity, and crustal deformation.

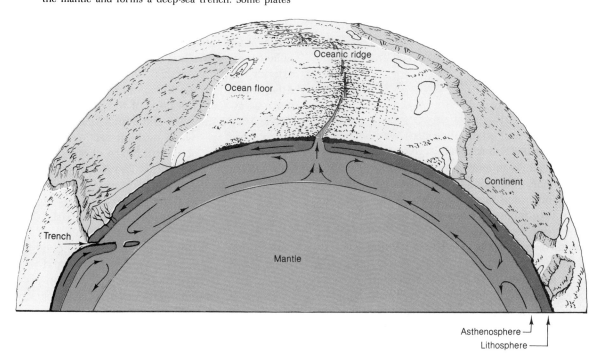

familiar example of convection can be seen in the heating of a pan of soup. Heat applied to the base of the pan warms the soup, which therefore expands and becomes less dense [*figure 2.3B*]. The warm fluid rises to the top and is forced to move laterally. It then cools, becomes more dense, and sinks.) The regular flow circuit of rising warm fluid and sinking cold fluid is called a convection current.

Where the convecting mantle rises, it causes the lithosphere to arch upward and form the oceanic ridge, as illustrated in *figure 2.3A*. As the asthenosphere flows laterally, it pulls the rigid lithosphere apart. Thus, the lithosphere is broken into a series of fragments, or plates, which are several thousand kilometers across (*figure 2.4*). As the plates move apart, molten rock from the hot asthenosphere rises into the rift zone and cools to form new oceanic crust. The continental blocks, composed of relatively light rock, float passively on the denser, lower part of the lithosphere, sometimes splitting (if a convection current rises beneath a continent) and sometimes colliding. Since the earth is a sphere, the shifting plates collide with each other. Plates containing dense oceanic crust move down into the asthenosphere at the deep-sea trenches and are consumed; this process is called subduction. In contrast, plates containing light continental crust cannot sink back into the mantle. Instead, continental margins that are adjacent to descending plates are deformed into linear, folded mountain belts. It should be emphasized that most plate boundaries are *not* associated with boundaries between oceans and continents. Rather, plate boundaries coincide with oceanic ridges, trenches, and young mountain belts and are characterized by zones of earthquakes and volcanic activity.

Discussion

The theory of **plate tectonics** was developed during the 1960s. It resulted largely from newly acquired abilities to study the characteristics of the ocean floor, map its surface features, and measure its magnetic and seismic properties. Evidence for the revolutionary theory of crustal movement comes from many sources and includes data on the structure, topography, and magnetic patterns of the ocean floor, the locations of earthquakes, the patterns of **heat flow** in the crust, the locations of volcanic activity, the structure and geographic fit of the continents, and the nature and history of mountain belts. (See chapters 17 through 22 for details.)

Consider the present structural features of the

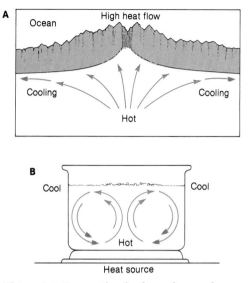

Figure 2.3 Convection in the asthenosphere can be compared to convection in a pan of soup. Heat from below causes the material to expand and become less dense. The warm material rises by convection and spreads laterally. It then cools, becomes more dense, and sinks. It is reheated as it descends, and the cycle is repeated.

earth and how the plate tectonic theory accounts for them. The boundaries of the **plates** are delineated with dramatic clarity by belts of active earthquakes and volcanoes (*figures 2.4* and *2.5*). Seven major lithospheric plates are recognized—the North American, South American, Pacific, Australian, African, Eurasian, and Antarctic plates—together with several smaller ones. The spreading center, where the lithosphere is being pulled apart, is marked by the oceanic ridge, which extends from the Arctic down through the central Atlantic and into the Indian and the Pacific. Plates move away from the crest of the ridge.

The North American and South American plates are moving westward and interacting with the eastern Pacific, Cocos, and Nazca plates along the west coast of the Americas. The Pacific plate, consisting only of oceanic crust, is moving northwestward from the oceanic ridge, toward the system of deep trenches in the western Pacific basin. The Australian plate includes Australia, India, and the northeastern Indian Ocean. It is moving northward, causing India to collide with Asia and producing the Himalaya Mountains. The African plate includes the continent of Africa, plus the southeastern Atlantic and western Indian oceans. It is moving eastward and northward. The Eurasian plate is the largest, and it moves eastward.

Many **global tectonic** features of the earth are

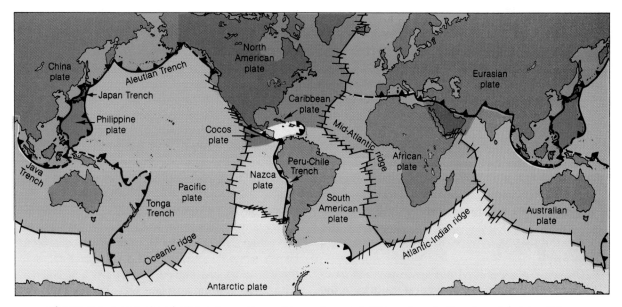

Figure 2.4 A mosaic of plates forms the earth's lithosphere, or outer shell. The plates are rigid, and each moves as a single unit. There are three types of plate boundaries: (1) the axis of the oceanic ridge, where the plates are diverging and new oceanic crust is generated; (2) transform faults, where the plates slide past each other; and (3) subduction zones, where the plates are converging and one descends into the asthenosphere.

Figure 2.5 Plate margins are outlined with remarkable fidelity by zones of earthquake activity and recent volcanism. Shallow earthquakes, submarine volcanic eruptions, and tensional fractures occur along the oceanic ridge, where plates are being pulled apart. Deep earthquakes, volcanic eruptions, and folded mountain belts occur along margins where plates collide.

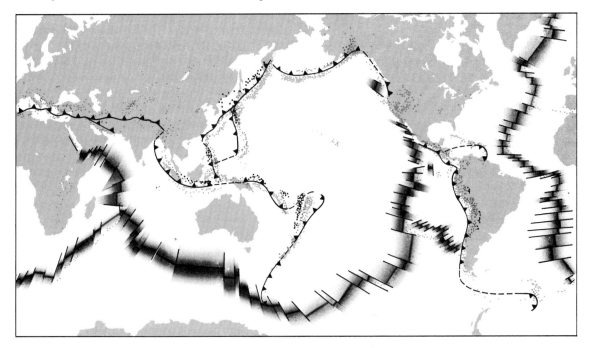

explained nicely by this theory. The mountain chains of the Rockies and the Andes result from the encounter of the American and eastern Pacific plates. The earthquakes that frequently rock Chile, Peru, and Central America result from this same encounter. The mountain systems of the Alps and the Himalayas result from the collisions of the African plate and the Australian plate with the Eurasian plate. This collision also produces the volcanism and earthquakes that torment the Mediterranean and the Near East. The great system of deep-sea trenches in the Pacific marks the zone where the Pacific plate descends into the mantle. Earthquakes and volcanic activity also mark this plate boundary, where lithosphere is being destroyed.

Our new perspective from space permits us to see many of the regional structural features of the earth in one synoptic view and to see for the first time on a global scale the results of a moving crust. For example, the great rift of the Red Sea, shown in *plate 14*, is an extension of the spreading center of the Indian Ocean, which splits the Sinai and Arabian peninsulas from Africa. The structure of the area shown in *plate 14* is dominated by two long, linear trenches. One is occupied by the Gulf of 'Aqaba, the Dead Sea, and the Jordan River, and the other by the Gulf of Suez. These rifts dramatically express the tensional or shear stresses in the crust and show how they affect the shape of the earth's surface. The long, straight lines extending through the Jordan River valley and the Gulf of 'Aqaba are major fractures produced by the rift. Volcanic rocks extruded along the fracture zone are visible as a dark tone, especially evident near the Sea of Galilee, in the upper part of the photograph. Oceanographic studies show that the Red Sea is an extension of the oceanic ridge, which has split the upper part of Africa and caused the Arabian Peninsula to drift to the north. It represents the initial stages of sea-floor spreading and the development of a new ocean basin. As the Arabian Peninsula moved northward, a large fracture developed along its left flank to form the long, narrow trench of the Jordan valley and the Dead Sea.

Another type of fracture in the crust results if plates slide past each other horizontally. The great San Andreas Fault in California, parts of which are clearly shown in *plate 15*, results from this type of movement.

Where the moving plates collide, the crust is compressed and folded. The space photograph in *plate 5* shows the type of deformation that most vividly expresses motion in the crust. Sequences of sedimentary rocks, which were originally deposited in horizontal layers, now are tightly folded as a result of plate collision. The folded belt of the Appalachian Mountains extends from Newfoundland to central Alabama, a distance of more than 4000 km. It is buried by sediment in much of the Gulf Coast region, but the fold belt continues through Arkansas and Oklahoma (the Ouachita Mountains), Texas (the Marathon Mountains), and into Mexico. Many of the other folded mountain belts of the world are similar.

Where two plates containing continental crust collide, one can override the other to produce a double layer of continental crust. This type of movement results in an exceptionally high mountain range. The Himalaya Mountains are believed to be an example. According to the plate tectonic theory, India has moved northward and collided with Asia. The Indian plate was thrust under the Asian plate to produce the great Himalaya, the highest mountain range in the world.

The mountain belts just discussed are all relatively young in a geologic time frame. Much older and more highly eroded mountain belts are found in the shield areas of the continents (*plate 3*). These areas are essentially the roots of ancient mountains. Their compressed and folded rock layers vividly illustrate the mobility of the crust.

Volcanism is another product of the tectonic system; it, too, expresses the motion of material in the earth. During volcanic activity, molten rock rises to the surface to form new crust. Evidence of recent floods of lava is widespread in southern Idaho and in Iceland, areas where the plates are splitting and moving apart. The greatest volcanic activity on the planet, however, is generally not seen. It occurs underwater, along the oceanic ridge.

Where plates collide, the descending plate is partly melted, and the lighter liquids rise to the surface to produce a distinctive type of volcanism. This type of volcanic activity occurs throughout the Andes Mountains and in the island arcs of the Pacific (*plate 16*). Most of the great volcanic mountains of the world are the result of colliding plates.

In this summary of the tectonic system, one point stands out. Space photographs show striking evidence of constant changes in the crustal structure of the earth—changes on a scale so great that it is difficult to comprehend from limited viewpoints on the ground. The great central theme that these and other examples illustrate is that the earth is a dynamic planet. Although its surface may appear stable and unchanging, it is continually modified by two great dynamic systems, the hydrologic and tectonic systems.

GRAVITY AND ISOSTASY

Statement

In addition to changes brought about by the hydrologic and tectonic systems, the earth's crust continually responds to the force of gravity, as it tries to maintain a gravitational balance, or isostasy (Greek *isos*, "equal," *stasis*, "standing"). Isostasy occurs because the lithospheric plates are buoyed up on the softer, plastic asthenosphere beneath them, and each portion of the crust displaces the mantle according to its bulk and density (*figure 2.6*). Denser crustal material sinks deeper into the mantle than less dense crustal material.

Isostatic adjustment in the earth's crust can be compared to adjustments in a sheet of ice floating on a lake as you skate on it. The layer of ice bends down beneath you, displacing a volume of water whose weight is equal to your weight. As you move ahead, the ice rebounds behind you, and the displaced water flows back.

As a result of isostatic adjustments, high mountains and plateaus having a great vertical thickness sink deeper into the mantle than areas of low elevation do. Any change in an area of the crust—such as removal of material by erosion or addition of material by sedimentation, volcanic extrusion, or accumulations of large continental glaciers—causes an isostatic adjustment.

The concept of isostasy, therefore, is fundamental to studies of the major features of the crust, such as the continents, ocean basins, and mountain ranges, and also to understanding the response of the crust to erosion, sedimentation, and glaciation.

Discussion

Artificial Reservoirs and Isostasy

The construction of the Hoover Dam, on the Colorado River, provides an excellent, well-documented illustration of isostatic adjustment, because the added weight of water and sediment in the reservoir was sufficient to cause measurable **subsidence**. From the time of its construction in 1935, 24 billion metric tons of water, plus an unknown amount of sediment, accumulated in Lake Mead, behind the dam. In a matter of years, this added weight caused the crust to subside in a roughly circular area around the lake. Maximum subsidence was 1.7 m.

Ice Sheets and Isostasy

Continental glaciers provide another clear exam-

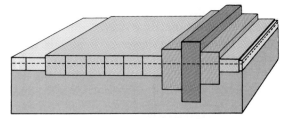

(A) *In 1859, J. H. Pratt proposed that mountains are high because they are composed of lighter materials than the surrounding lowland.*

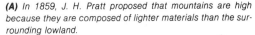
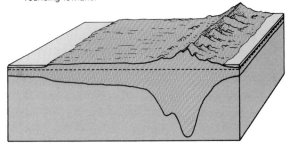

(B) *In 1855, G. B. Airy concluded that mountains are of the same density as the adjacent lowland but are high because they are thicker.*

Figure 2.6 Isostasy is the universal tendency of segments of the earth's crust to establish a condition of gravitational balance with one another. Differences in both density and thickness can cause isostatic adjustments in the earth's crust.

ple of isostatic adjustment of the crust. The weight of an ice sheet several thousand meters thick disrupts the crustal balance and causes the crust beneath to be depressed. In both Antarctica and Greenland, the weight of the ice has depressed the central part of the landmasses below sea level. A similar isostatic adjustment occurred during the last ice age, when continental glaciers existed in Europe and North America. Parts of both continents are still below sea level. Now that the ice is gone, however, the crust is rebounding at a rate of 5 or 10 m per 1000 years. Geologists can measure the extent and rates of rebound by mapping the tilted shorelines of ancient lakes (see *figure 14.21*).

Ancient Lakes and Isostasy

The tilted shorelines of ancient lakes provide a means of documenting isostatic rebound (*figure 2.7*). An example is Lake Bonneville, a large lake that existed in Utah and Nevada during the ice age but has since dried up to small remnants, such as Utah Lake and Great Salt Lake. Lake Bonneville was 305 m deep and covered a much smaller area than a continental glacier, but even this relatively small weight was sufficient to depress the crust. The shorelines of the lake were level when they were formed, but they have been tilted in response to unloading as the water has been removed. Shorelines near the deepest part of the lake have rebounded nearly 60 m.

These examples illustrate several important facts about gravity and isostatic adjustments.

1. Gravity is the driving force for all isostatic adjustments. Therefore, all types of loading and unloading cause vertical movements. **Isostasy** is involved in all of the processes that shift material on the earth's surface. Some of the more obvious isostatic adjustments to be expected are as follows:

   a. In mountains and highlands, as erosion removes material, the crust should rebound.

   b. In deltaic areas, where sediment is deposited, the added weight should cause the crust to subside.

   c. In areas of volcanic activity, the added weight should cause the crust to subside.

   d. In regions of glaciation, the growth of the ice sheet should cause the crust to subside; as the ice is removed, the crust should rebound.

2. Very small loads, such as water a few hundred meters deep, are sufficient to cause isostatic adjustments.

3. Isostatic adjustments can occur very rapidly (60 m in less than 20,000 years, in the case of Lake Bonneville).

4. Gravity plays a fundamental role in the earth's dynamics, since it is intimately involved with

   a. Isostatic adjustments

   b. The differentiation of the planet's interior

   c. The continental crust's floating on the denser lithospheric plates

   d. Gravity flow systems such as rivers, glaciers, ground water, and atmospheric circulation

Gravity is thus an underlying force in all geologic systems.

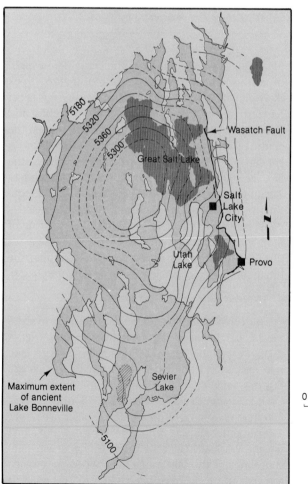

Figure 2.7 Isostatic rebound of the crust occurred after the load of the water in ancient Lake Bonneville was removed. The map shows the maximum size of Lake Bonneville and the present remnants, Great Salt Lake and Utah Lake. The contour lines indicate the present elevations of the shoreline features of ancient Lake Bonneville. Note that the shorelines around islands in the central part of the lake are now nearly 60 m higher than those along the margins. This difference in elevation is interpreted as the result of the rebound of the crust, which occurred as the lake dried.

SUMMARY

This chapter has outlined the mechanics of the two major dynamic systems of the earth.

The hydrologic system includes all possible paths by which water moves through the atmosphere and oceans, over the land, and below the ground. It is a gravity flow system. Solar heat is its source of energy.

The tectonic system is an internal energy system. The upper mantle moves as a result of radiogenic heat inherited from the time of the earth's formation. Heat within the upper mantle produces convection. As a convection current moves, it splits the rigid lithosphere and carries it laterally. New crust is created where molten mantle material moves into the fracture zone and cools. As the convection current descends, it carries with it the lithosphere, which is ultimately consumed in the hotter mantle.

The tectonic system is an irreversible system in which energy inherited from the beginning of the earth is continually dissipated into space. Ultimately, this energy will run out, just as the sun's energy must someday be consumed, and the earth will cease to change by its own internal heat. Until then, the tectonic system will continue to create ocean basins, cause continental drift, build mountains, and produce new crust by volcanic activity.

As the hydrologic and tectonic systems operate, isostatic adjustment in the earth's crust continually attempts to establish gravitational equilibrium.

In order for the earth's dynamic systems to operate, matter changes back and forth from solid to liquid or gaseous states. Rocks are created and destroyed, in processes that basically involve the growth and destruction of minerals. A knowledge of the nature and origin of minerals and rocks is therefore fundamental, not only to understanding the composition of the earth, but also to understanding the processes that operate upon it and within it. These topics are the subject of the next five chapters.

Key Words

hydrologic system	plate
runoff	radiogenic heat
river system	heat flow
glacier	convection
ground water	convection current
pore space	global tectonics
shore	isostasy
eolian system	subsidence
plate tectonics	

Review Questions

1. Diagram the major paths by which water circulates in the hydrologic system.
2. Draw a diagram of the asthenosphere with arrows showing the major paths of circulation of material.
3. Show on a map the major plate boundaries. How are these boundaries related to volcanism, earthquakes, and young, folded mountain ranges?
4. Explain the concept of isostasy in the earth's system and give two examples of isostatic adjustment of the earth's crust in the recent geologic past.

Additional Readings

Bates, D. R., ed. 1964. *The planet Earth*. Elmsford, N.Y.: Pergamon Press.
Cailleux, A. 1968. *Anatomy of the earth*. New York: McGraw-Hill.
Gaskell, T. F. 1970. *Physics of the earth*. Scranton, Pa.: Funk and Wagnalls.
Phillips, O. M. 1968. *The heart of the earth*. San Francisco: Freeman, Cooper.
Sumner, J. S. 1969. *Geophysics, geologic structures and tectonics*. Dubuque, Iowa: William C. Brown.
Takeuchi, H., S. Uyeda, and H. Kanamori. 1970. *Debate about the earth*. Rev. ed. San Francisco: Freeman, Cooper.

3

ENERGY AND MATTER

All of the complex geologic processes operating on or within the earth and causing it to change involve transformation of energy and matter. Without energy there can be no change, no motion. Without energy flow systems, the earth would not be a dynamic planet, and there would be no wind, rain, rivers, volcanoes, or earthquakes. All motion would cease and the planet would be still, lifeless, and unchanging. Therefore, to understand the earth, you must understand something about the various forms of energy and how energy transformations occur. You must also understand some of the fundamentals of the structure of matter and how it changes from one state to another.

The study of energy and matter is immense in scope and involves every branch of physical and biological science. An elementary survey of some of the fundamental concepts, however, is a necessary first step in a study of geology.

MAJOR CONCEPTS

Energy

1. The forms of energy most important in geologic processes are (a) kinetic energy, (b) gravitational potential energy, (c) thermal energy, (d) chemical energy, and (e) nuclear energy.
2. Energy can change from one form to another but can neither be created nor destroyed.
3. Energy is necessary for all geologic processes.

Matter

4. An atom is the smallest particle of an element.
5. An atom consists of a nucleus of protons and neutrons and a surrounding cloud of electrons.
6. An atom of a given element is distinguished by the number of protons in its nucleus.
7. Isotopes are varieties of an element, distinguished by the number of neutrons in the nucleus.
8. Ions are electrically charged atoms, produced by the gain or loss of electrons.
9. In many minerals, atoms combine by ionic bonding.
10. Matter exists in three states: (a) solid, (b) liquid, and (c) gas. The differences among the three states are related to the degree of ordering of the atoms.

ENERGY

Statement

All of the processes in the hydrologic and tectonic systems result from the flow of energy on or within the earth. Indeed, without energy, the materials of the earth would be totally inert and motionless. There would be no rain, rivers, earthquakes, or volcanoes. There would be no changes in the states of matter. There would be no change at all. But what is energy? In simplest terms, it is a measure of motion or the capacity for producing motion. The motion can be visible (like that of a speeding train) or invisible (like the motion in an atom), but all motion is associated with energy.

The forms of energy most important in geologic processes are

1. Kinetic energy—the energy of motion
2. Gravitational potential energy—the energy resulting from the position of a body in a gravitational field
3. Thermal energy—the energy resulting from the random movement of molecules and atoms
4. Chemical energy—the energy resulting from the transfer of outer electrons in atoms and molecules
5. Nuclear energy—the energy resulting from changes in the arrangement of protons and neutrons in nuclei

One of the fundamental laws of natural science is the law of conservation of energy, also known as the first law of thermodynamics, which states that *energy can be neither created nor destroyed; it can be changed from one form to another, but the total amount of energy remains constant.* This law has been tested and observed over a wide range of conditions and has always held true.

Discussion

Although the study of energy is fairly abstract, an introduction to the basic relationships between energy and geologic processes can be better understood through intuitive reasoning and by considering examples of familiar events on the earth's surface (see *figure 3.1*).

Kinetic Energy

Kinetic energy, or the energy of motion, is probably the most familar form. The word comes from the Greek *kinetikos* ("to move"). Every moving object has kinetic energy. The earth possesses kinetic energy as it moves around the sun and spins on its axis. All of the processes in the hydrologic system, such as running water, moving glaciers, rushing ocean waves, and even the tiny, swirling rock particles in a dust storm, involve obvious kinetic energy. All involve matter in motion, and all have the capacity to change and modify the earth.

The kinetic energy of an object is proportional to its quantity of mass multiplied by the square of its velocity. For example, a large boulder rolling down a hill has more energy than a small pebble moving at the same velocity. A small meteorite moving at a high velocity, however, can have much more energy than a large boulder moving only a few meters per second.

Gravitational Potential Energy

Gravitational potential energy is energy that is stored in an object and associated with the object's position in a gravitational field. It equals the kinetic energy the object would attain if it were allowed to fall under the influence of gravity. Moisture in the atmosphere has potential energy because of its position. As rain falls, it loses gravitational potential energy and gains kinetic energy. When it finally hits the earth, part of the kinetic energy is converted to heat and some is transferred into the soil particles with which it collides. If the rain falls on highland or mountain areas, it retains some potential energy, which is transformed into kinetic energy as it flows downslope in the drainage system.

Thermal Energy

Thermal energy is a special type of kinetic energy; it is internal rather than the external form seen in a moving mass. Thermal energy is the energy of the motion of atoms. All atoms are constantly moving, and the faster their motion, the greater their thermal energy. Atoms in a crystalline solid are locked into a fixed geometric position, and their motion is one of vibration, without one atom moving past another. If the atoms move faster, they eventually break out of their fixed position to move freely, and the crystal melts. Additional heat, and thus faster motion, causes the liquid to vaporize.

Heat is transferred in three principal ways: (1) conduction, (2) convection, and (3) radiation. *Conduction* is the process by which the vibration of an atom is transmitted to adjacent atoms. It is the method by which heat is transmitted through solids. *Convection* is the process by which heated material expands and moves upward. It occurs in liquids and gases and transfers thermal energy much faster than conduction does. As was seen in chapter 2, convection is the principal way in which thermal energy is transferred in both the hydrologic system and the tectonic system. *Radiation* involves the emission of electromagnetic waves from the surface of the hot body to its cooler surroundings. The waves travel in a straight line through space at 300,000 km per second (186,000 miles per second) and occur in a wide range of wavelengths, which together constitute the electromagnetic spectrum. The transfer of energy from the sun to the earth by radiation is obviously the most important energy transformation in the earth's system.

Chemical Energy

Chemical energy is the energy involved in chemical reactions. It binds atoms together into molecules and becomes available if atoms lose or gain electrons. Processes of rock formation in which material changes state from liquid to solid involve the transformation of large quantities of chemical energy.

Another example is the chemical energy stored in plants. Energy from the sun is absorbed by molecules within plant cells and is used to combine carbon dioxide and water to form carbohydrates. This energy, stored in carbohydrate compounds of plant tissue, can later be released through burning (chemical oxidation) and radiated as heat to the surroundings.

Nuclear Energy

Nuclear energy results from the arrangement of protons and neutrons in atomic nuclei. It was acquired by the elements at the time of their formation. The radioactive elements (such as uranium, thorium, and potassium) release this type of energy spontaneously by ordinary chemical reactions.

Transformation and Conservation of Energy

In all natural processes, energy is changed from one form to another, but it is never created or destroyed. Several examples of natural energy systems illustrate this transformation and conservation.

Example 1. Solar radiation is absorbed by land and water at the earth's surface. Air in contact with the surface warms by conduction, becomes less dense, and rises. This process results in winds, which may blow sand into migrating dunes. In this case, radiant thermal energy from the sun passes into thermal energy of the land and water and then into kinetic energy of the moving air. The wind transports sand and changes the earth's surface.

Example 2. Nuclear energy in the asthenosphere is transformed into thermal energy by radioactive disintegration of uranium or potassium. The heat melts some of the minerals in the rocks of the mantle, and the liquid rises in a convection system. When molten material reaches the surface, it is ejected in a volcanic eruption. The explosion accelerates some of the lava fragments to a high speed; that is, thermal energy is transformed into kinetic energy. As a fragment rises, it gains gravitational potential energy at the expense of kinetic energy. The fragment reaches a maximum height when all of its kinetic energy is transformed into potential energy, and it stops rising. As it falls back to the earth, its potential energy is converted back to kinetic energy. When the fragment strikes the ground, some of the kinetic energy is converted to thermal and the rest sets up vibrations that pass through the ground and the air. As the lava cools, some heat is lost to the surroundings through convection, conduction, and radiation, and some is transformed into chemical energy as the minerals crystallize. In each transfer, some energy is removed from the volcanic system and is dispersed into the surroundings.

Example 3. Radiation from the sun strikes the ocean surface and heats the water. The thermal energy causes greater motion in the water molecules, and evaporation occurs. The heat from the water also warms the surrounding air, which rises by convection and carries the evaporated water with it. The water vapor is precipitated as rain or snow, and its gravitational potential energy changes into kinetic energy as it falls. When it reaches the surface, the impact transforms some of this kinetic energy into thermal energy. The water still retains some potential energy, and as it flows to lower levels, it washes down sand and mud with it. Gravitational potential energy of both the water and the solid particles is transformed into kinetic energy. Friction and resistance to flow transform some of the kinetic energy into thermal energy, which is dispersed by conduction or radiation.

Example 4. Phobos and Deimos, the two moons of Mars, are small, potato-shaped satellites pockmarked with craters. The only significant energy acting on these two asteroid-size planetary bodies is thermal energy from the sun, which they absorb at their surfaces, but the amount of energy that they receive in this way is too small to generate motion in the surface material. Early in the history of the two moons, kinetic energy from the impact of meteorites modified their surfaces by forming impact craters, but most meteorites in the solar system have since been swept up by the orbiting planets and satellites. Now, without sufficient energy to cause further alterations in their surfaces, Phobos and Deimos are dead and will remain unchanged for eons.

Figure 3.1 Forms of energy operating on the earth include kinetic energy from the impact of meteorites, thermal energy from the sun, and thermal energy from the earth's interior. Energy from these sources drives the earth's dynamic systems.

(A) Meteorite impacts introduce the kinetic energy of moving bodies coming from space.

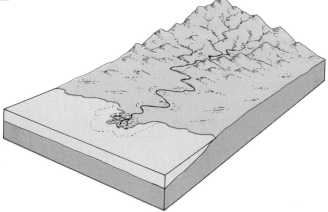

(B) Rivers are part of the hydrologic system, in which thermal energy from the sun is transformed into the kinetic energy of flowing water, through a cycle of evaporation, precipitation, and runoff.

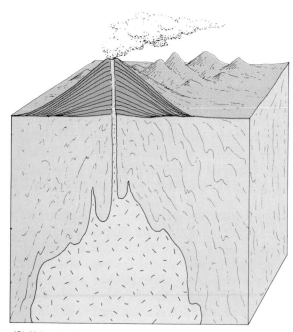

(C) Volcanism is a result of thermal energy from the earth's interior.

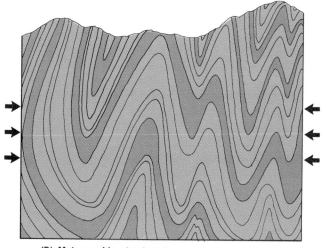

(D) Metamorphism involves thermal energy from the earth's interior and mechanical energy associated with the deformation of rocks.

MATTER

Statement

To understand the dynamics of the earth and how rocks and minerals are formed and changed through time, you must have some knowledge of the fundamental structure of matter and how it behaves under various conditions. The solid materials that make up the earth's outer shells are called rocks. Most rock bodies are mixtures, or aggregates, of minerals. A mineral is a naturally occurring inorganic compound with a definite chemical formula and a specific internal structure. Minerals, in turn, are composed of atoms, so that it is necessary to understand something about atoms and the ways in which they combine.

The atomic theory that is the basis for the present-day understanding of matter was developed by John Dalton in 1805. He proposed that all matter is composed of tiny, individual particles called atoms. Since then, many momentous discoveries have greatly increased our understanding of matter. A simplified description of modern atomic theory includes the following fundamental concepts:

1. An atom is the smallest fraction of an element that can exist and still show the characteristics of that element.
2. The most fundamental subatomic particles are protons, neutrons, and electrons, although many others have been identified in recent years.
3. A typical atom consists of a nucleus of protons and neutrons and a cloud of electrons surrounding the nucleus.
4. The distinguishing feature of an atom of an element is the number of protons in the nucleus. The number of electrons and neutrons in an atom of a given element can vary, but the number of protons is constant.
5. Normal atoms are electrically neutral, because they have one negatively charged electron for every positively charged proton.
6. Electrically charged atoms, called ions, are produced by the gain or loss of electrons.
7. Varieties of a given atom (element), called isotopes, are produced by variations in the number of neutrons in the nucleus.
8. Atoms combine, mostly through ionic bonding, to form minerals.

Discussion

Atoms

Individual **atoms** are far too small to be seen with any microscope. They are best described by abstract models constructed from mathematical formulae involving probabilities. In its simplest approximation, an atom is characterized by a relatively small nucleus of tightly packed protons and neutrons, with a surrounding cloud of electrons. Each proton carries a positive electrical charge, and the mass of a proton is taken as the unit of mass. The neutron, as its name indicates, is electrically neutral and has approximately the same mass as the proton. The electron is a much smaller particle, with a mass approximately $\frac{1}{1850}$ times that of the proton. It carries a negative electrical charge equal in intensity to the positive charge of the proton. Since the electron is so small, for practical purposes, the entire mass of the atom can be considered to be concentrated in the protons and neutrons of the nucleus.

Hydrogen is the simplest of all elements; it consists of one proton in the nucleus and one orbiting electron (*figure 3.2*). The next heavier atom is helium, with two protons, two neutrons, and two electrons. Each subsequently heavier element contains more protons, neutrons, and electrons. The distinguishing feature of an element is the number of protons in the nucleus of each of its atoms. The number of electrons and neutrons in an atom of a given element can vary, but the number of protons is constant.

Atoms normally have the same number of electrons as protons and do not carry an electrical charge. As the number of protons increases in the succeedingly heavier atoms, the number of electrons also increases. The electrons fill a series of

Figure 3.2 The atomic structures of hydrogen and helium illustrate the major particles of an atom. Hydrogen has one proton and one orbiting electron. Helium has two protons, two neutrons, and two orbiting electrons.

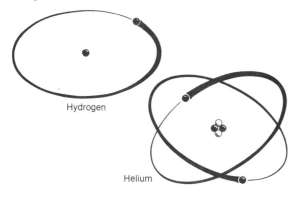

Hydrogen

Helium

energy-level shells, each of which has a maximum capacity.

Isotopes

Although the number of protons in the atoms of a given element is constant, the number of neutrons can vary. This means that atoms of a given element are not necessarily all alike. Iron atoms, for example, have 26 protons but can have 28, 30, 31, or 32 neutrons. These varieties of iron are examples of **isotopes.** They all have the properties of iron but vary from one another only in mass number. Most common elements exist in nature as a mixture of isotopes.

Ions

Atoms that have as many electrons as protons are electrically neutral, but atoms of many elements can gain or lose electrons in their outermost shells. If this occurs, an atom loses its electrical neutrality and becomes charged. These electrically charged atoms are called **ions.** The loss of an electron results in a positively charged ion, since the number of protons exceeds the number of negatively charged electrons. If an electron is gained, the ion has a negative charge. The electrical charges of atoms are important, because the attraction between positive ions and negative ions is the bonding force that holds matter together.

Bonding

Atoms are most stable if their outermost shells are filled to capacity with electrons. Except for the inner shell, which can always hold 2, each shell can hold 8 electrons. Neon, for example, has 10 protons in the nucleus and 10 electrons, of which 2 are in the first shell and 8 are in the second shell. A neon atom does not have an electrical charge, and its two electron shells are complete. As a result, neon is stable: it does not interact chemically with other atoms. Helium, argon, and the other noble gases also have 8 electrons in their outermost shells, and they normally do not combine with other elements. Most elements have incomplete outer shells. Their atoms readily lose or gain electrons to achieve a structure like that of neon and the other inert gases, with 8 electrons in their outermost shells.

For example, an atom of sodium has only one electron in its outer shell but eight in the shell beneath (*figure 3.3*). If it could lose the lone outer electron, it would have a stable configuration like that of the inert gas neon. Chlorine, in contrast, has seven electrons in its outer shell, and if it could gain an electron, it too would attain a stable configuration. Thus, whenever possible, sodium gives up an

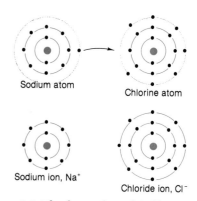

Figure 3.3 The formation of sodium and chloride ions by transfer of an electron from the outermost shell of a sodium atom to the outermost shell of a chlorine atom results in a stable outer shell for each ion.

electron and chlorine gains one; the sodium atom becomes a positively charged sodium ion, and the chorine atom becomes a negatively charged chloride ion. With opposite electrical charges, the sodium ions and chloride ions attract each other and bond together to form the compound sodium chloride (common salt, also known as the mineral halite). This type of bond, between ions of opposite electrical charge, is called an **ionic bond.**

Atoms also can attain the electronic arrangement of a noble gas (and thus stability) by sharing electrons. No electrons are lost or gained, and no ions are formed. An electron simply orbits both nuclei in a figure-eight pattern (*figure 3.4*). This type of bond is called a **covalent bond.**

A third type of bond is called a **metallic bond.** In a metal, each atom contributes one or more electrons to a sea of electrons that moves relatively freely throughout the entire aggregate of atoms. A given electron is not attached to a specific pair of atoms but moves about, orbiting one nucleus and then another. This sea of negatively charged electrons holds the positive metallic ions together in a crystalline structure and is responsible for the special characteristics of metals.

Figure 3.4 In a covalent bond, electrons orbit two nuclei, instead of only one nucleus.

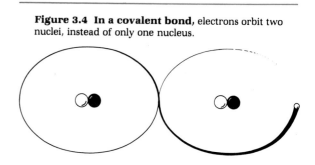

STATES OF MATTER

Statement

A group of atoms can exist in one of three states of aggregation, as a gas, a liquid, or a solid.

Gases. The principal properties of gases are as follows:

1. Gases have no definite shape or volume; that is, they assume the shape and volume of their containers.
2. In gases, particles are in constant, random motion and are widely spaced.
3. Gases can be compressed or expanded.
4. Gases exert pressure, which increases as temperature increases.
5. Gases can be converted into liquids by subjecting them to increased pressure, decreased temperature, or both.

Liquids. The principal properties of liquids are as follows:

1. Liquids have a definite volume but no definite shape. A liquid therefore tends to flow.
2. In liquids, the atoms or molecules are packed more closely together than they are in gases, but they are not set in a rigid framework.
3. Liquids are more difficult to compress than gases.
4. The density of liquids is greater than that of gases.
5. Liquids can be changed into gases by an increase in temperature, a decrease in pressure, or both.

Solids. The principal properties of solids are as follows:

1. Solids have a definite volume and a definite shape.
2. In crystalline solids, the atoms are bound tightly together into a fixed, rigid framework with a definite arrangement of atoms.
3. Most solids become liquids if they are heated sufficiently, although solids can pass directly into a gaseous state. If the temperature of a solid is close to the melting point, a decrease in pressure causes melting.
4. Liquids can be converted to solids by a decrease in temperature or an increase in pressure. The process is called freezing, and

Figure 3.5 Degrees of atomic ordering in the three states of matter are illustrated by ice, water, and steam. In a crystalline solid, the atoms are arranged in a rigid, geometric framework. In a liquid, the atoms are packed closely together but are free to move and glide past one another. In a gas, the atoms are in rapid motion and move apart.

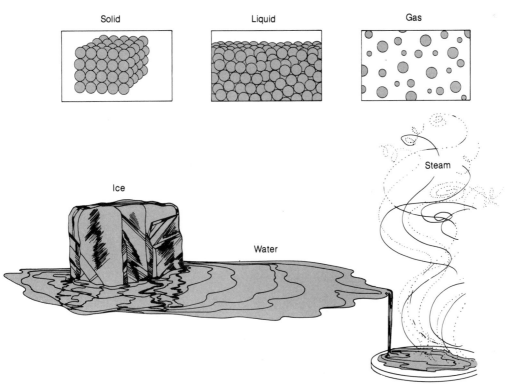

the temperature at which solidification occurs is called the freezing point.

A given sample of matter can change from one state to another, so that it has completely different physical properties but retains the same chemical composition. To change matter from one state to another, there must be a gain or a loss of energy. That is, heat or pressure (or both) must be increased or reduced.

Discussion

The principal differences between solids, liquids, and gases involve the degree of ordering of the constituent atoms.

In a typical **solid** (*figure 3.5*), atoms are arranged in a rigid framework. The arrangement in crystalline solids is a regular, repeating, three-dimensional pattern called a **crystal structure,** but in **amorphous solids** the arrangement is random. Each particle occupies a more or less fixed position but has a vibrating motion. As the temperature rises, the vibration increases, and eventually the particles become free and are able to glide by one another. Melting ensues, and the matter passes into the liquid state.

In a **liquid,** the basic particles are in random motion, but they are packed closely together. They slip and glide past one another or collide and rebound, but they are held together by forces of attraction greater than those in gases. This explains why density generally increases and compressibility decreases as matter changes from gas to liquid to solid. If a liquid is heated, the motion of the particles increases, and individual atoms or molecules become separated as they move about at high speeds.

In a **gas,** the individual atoms or molecules are separated by empty spaces and are comparatively far apart. They are in rapid motion and travel in straight lines until their direction is changed by collision. This explains why a gas exerts pressure and can be markedly compressed. Gases have the ability to expand indefinitely, and the continuous rapid motion of the particles results in rapid diffusion.

Water is undoubtedly the most familiar example of matter changing through the three basic states. At pressures prevailing on the earth's surface, water changes from a solid to a liquid to a gas in a temperature range of only 100 °C. Other forms of matter in the solid earth are capable of similar changes, but usually their transitions from solid to liquid to gas occur at high temperatures. At normal room temperature and pressure, 93 of the 106 elements are solids, 2 are liquids, and 11 are gases.

Most people are familiar with the effects of temperature changes on the state of matter because of their experience with water as it freezes, melts, and boils. Fewer people are familiar with the effects of pressure. At high pressures, water remains liquid at temperatures as high as 371 °C. The combined effects of temperature and pressure on water are shown in the phase diagram in *figure 3.6.* Similar diagrams, constructed from laboratory work on other minerals, can provide important insight into the processes operating at the high temperatures and pressures below the earth's surface.

Figure 3.6 Temperature and pressure are the major factors that determine the state in which matter exists. In this diagram, the ranges of temperature and pressure for the various phases of water are shown. The triple junction is a single point at which all three phases are in equilibrium. The critical temperature (T_c) and the critical pressure (P_c) cross at the critical point (the end of the gas-liquid phase line) beyond which the liquid and gas phases cannot be distinguished. Similar diagrams can be constructed for other minerals.

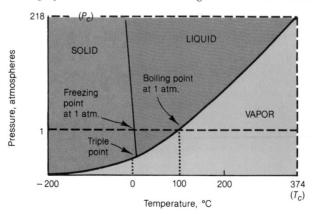

SUMMARY

All change requires the interaction of energy and matter. The forms of energy most significant in geologic and planetary systems are

1. Kinetic energy, or the energy of motion
2. Gravitational potential energy, or the energy resulting from position in a gravitational field
3. Thermal energy, or the energy resulting from random molecular or atomic motion
4. Chemical energy, or the energy resulting from the transfer of outer electrons
5. Nuclear energy, or the energy resulting from changes in the arrangement of protons and neutrons in nuclei

In all natural processes, energy is transformed from one type to another, but it is never created or destroyed.

The atomic theory is the basis for our present-day understanding of matter. A simplified description of the fundamental ideas in this theory follows.

1. An atom is the smallest fraction of an element that can exist and still show the characteristics of that element.
2. Although many subatomic particles have been identified, the most fundamental are (a) protons, (b) neutrons, and (c) electrons.
3. An atom consists of a nucleus of protons and neutrons and a surrounding cloud of electrons.
4. The atoms of each element have a unique number of protons in the nucleus. The number of electrons and neutrons can vary from atom to atom of the same element, but the number of protons is constant.
5. Isotopes are varieties of an atom (element) produced by variations in the number of neutrons in the nucleus.
6. Normal atoms are electrically neutral, because they have one negatively charged electron for each positively charged proton.
7. Electrically charged atoms are called ions. They are produced by the gain or loss of electrons.
8. Atoms combine to form minerals by ionic bonding, covalent bonding, and metallic bonding.

There are three states of matter: gas, liquid, and solid. The three states are distinguished by completely different physical properties, but matter retains the same chemical composition in each state.

Processes in the earth's dynamics mostly involve matter changing from one state to another.

Key Words

kinetic energy	ionic bond
gravitational potential energy	covalent bond.
thermal energy	metallic bond
chemical energy	gas
nuclear energy	liquid
atom	solid
ion	crystal structure
isotope	amorphous solid

Review Questions

1. Define kinetic energy and cite several examples of this form of energy in the geologic system.
2. How is heat transferred?
3. What heat sources are important in the earth's dynamics?
4. Cite several examples of the transformation and conservation of energy in the natural systems of the earth.
5. Explain why all changes that occur in the earth's system involve the transformation of energy.
6. Describe the structure of an atomic nucleus.
7. Which is the most distinguishing feature of an atom of an element—the number of protons in the nucleus, the number of neutrons, or the number of electrons?
8. Briefly describe the distinguishing characteristics of gases, liquids, and solids.
9. What causes matter to change from one state to another?
10. What is the relationship between an atom and an ion of a given mineral?
11. What is an isotope?

Additional Readings

Asimov, I. 1961. *Building blocks of the universe.* New York: Abelard-Schuman.

Azaroff, L. V. 1960. *Introduction to solids.* New York: McGraw-Hill.

Fyfe, W. S. 1964. *Geochemistry of solids: An introduction.* New York: McGraw-Hill.

Krauskopf, K. B., and A. Beiser. 1971. *Fundamentals of physical science,* 6th ed. New York: McGraw-Hill.

Lapp, R. E. 1969. *Matter.* Alexandria, Va.: Time-Life Books.

Wilson, M. 1967. *Energy.* Alexandria, Va.: Time-Life Books.

4

MINERALS

All of the earth's dynamic processes involve matter changing from one state to another. Minerals grow, melt, and dissolve or are broken and modified by physical forces. As the earth's surface weathers and erodes, some minerals are destroyed and others grow in their place. As sediments accumulate in the oceans, new minerals grow from solution, cementing mineral fragments into a coherent mass to form solid rock. Other minerals grow from the hot, liquid lava extruded from volcanoes. Deep below the earth's surface, high pressure and temperature remove atoms from the crystal structure of minerals and recombine them into new structures, to form minerals that are more stable in that environment. As tectonic plates move and continents drift, minerals are created and destroyed by a variety of chemical and physical processes. Some knowledge of the earth's major minerals, therefore, is essential to an understanding of the earth's dynamics.

In this chapter, we will survey the general characteristics of minerals and the physical properties by which they can be identified. Then we will explore the silicate mineral group (the major rock-forming minerals) in preparation for a study of the major rock types in chapters 5, 6, and 7.

MAJOR CONCEPTS

1. A mineral is a natural, inorganic solid possessing a specific internal structure and a definite chemical composition that varies only within certain limits.
2. Minerals grow when atoms are added to the crystal structure as matter changes from the gaseous or the liquid state to the solid state. Minerals dissolve or melt when atoms are removed from the crystal structure as solids change to liquids or gases.
3. All specimens of a given mineral type have a limited range of physical and chemical properties (such as crystal structure, cleavage or fracture, hardness, and specific gravity).
4. Silicate minerals form more than 95% of the earth's crust.
5. The most important rock-forming minerals are feldspars, micas, olivines, pyroxenes, amphiboles, quartz, clay minerals, and calcite.

THE NATURE OF MINERALS

Statement

People commonly think of minerals only as exotic crystals in museums or as valuable gems and metals. But grains of sand, snowflakes, and salt particles are also minerals, and they have much in common with gold and diamonds.

Minerals are the major solid constituents of the earth. A precise definition is complicated, but for a substance to be considered a mineral, the following conditions must be met:

1. It occurs naturally as an inorganic solid.
2. It has a specific internal structure. That is, its constituent atoms are precisely arranged into a crystalline solid.
3. It has a chemical composition that varies within definite limits and can be expressed by a chemical formula.
4. It has definite physical properties that result from its composition and crystalline structure.

The differences between minerals arise from the kinds of atoms they contain and the way the atoms are arranged.

Discussion

Inorganic Solids

By definition, only naturally occurring inorganic solids are **minerals,** that is, natural elements or inorganic compounds in a solid state. Thus, synthetic products, such as artificial diamonds, are not minerals in the strict sense. Nor are organic compounds, such as coal and petroleum, because they are organic materials and are not crystalline solids.

Minerals can consist of a single element, such as gold, silver, copper, and sulfur, but most are compounds of two or more elements, because the abundant elements in the earth have a strong tendency to combine.

The Structure of Minerals

Perhaps the key words in the definition of *minerals* are *internal structure.* The component atoms of a mineral have a unique and orderly internal arrangement, and every specimen of a given mineral has the same internal structure, regardless of when, where, and how it was formed. This property of minerals was suspected long ago by mineralogists who observed the many expressions of order in **crystals.** Nicolaus Steno (1638–1687), a

Danish monk, was among the first to note this property. He found from numerous measurements that each mineral has a characteristic **crystal form.** Although the size and shape of a mineral crystal form may vary, similar pairs of crystal faces always meet at the same angle. This is known as the *law of constancy of interfacial angles.*

Later, René Haüy (1743–1822), a French mineralogist, accidently dropped a large crystal of calcite and observed that it broke only along three sets of planes, so that all the fragments had a similar shape (see *figure 4.1*). He then proceeded to break other calcite crystals in his own collection, plus many in the collections of his friends, and found that all of the specimens broke in exactly the same manner. All of the fragments, however small, had the shape of a rhombohedron. To explain his observations, he assumed that calcite is built of innumerable, infinitely small rhombohedra packed together in an orderly manner, and he concluded that the cleavage of calcite relates to the ease of parting of such units from adjacent layers. This was a remarkable advance in the understanding of crystals.

Today, we know that cleavage planes are planes of weakness in the crystal structure, which are not necessarily parallel to the crystal faces. They do, however, constitute a striking expression of the orderly internal structure.

With modern methods of **X-ray diffraction,** we can precisely determine a mineral's internal structure and reveal much about its atomic arrange-

Figure 4.1 Cleavage of calcite occurs in three planes that do not intersect at right angles. Each cleavage fragment is bounded by similar sets of cleavage planes.

ment. The technique is illustrated in *figure 4.2*. If a thin beam of X rays is passed through a mineral, it is diffracted (or dispersed) by the framework of atoms. The dispersed rays produce an array of dots on photographic film placed behind the crystal. From measurements of the relationships among the dots, the systematic orientation of planes of atoms within the crystal can be deduced and expressed in mathematical formulae. Thus, detailed models of crystal structures can be constructed and analyzed. The X-ray diffraction instrument is now the most basic device for determining the internal structure of minerals, and geologists use it extensively for precise mineral identification and analysis.

To understand the importance of structure in a mineral, consider the characteristics of diamond and graphite. These two minerals are identical in chemical composition; both consist of a single element, carbon (C). Their crystal structures and physical properties, however, are very different. In diamond, the carbon atoms are packed closely together and the bonds between the atoms are very strong. This explains why diamonds are extremely hard—the hardest substance known—and form only under high pressure. In graphite, the carbon atoms are loosely bound in a layered structure. The layers separate easily, which accounts for graphite's slippery, flaky property. Because of its softness and slipperiness, graphite is used as a lubricant and is also the main constituent in common "lead" pencils. The important point to note is that different structural arrangements of the same elements produce different minerals with different properties. This ability of a specific chemical substance to crystallize with more than one type of structure is known as *polymorphism*.

The Composition of Minerals

A mineral has a definite chemical composition, in which specific elements occur in definite proportions. The chemical composition of some minerals,

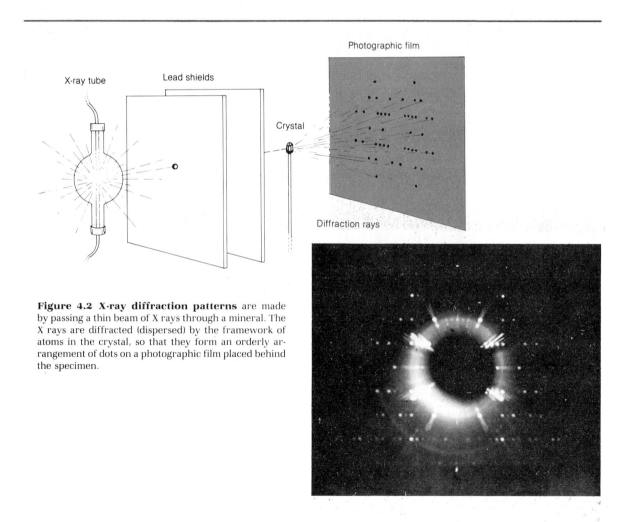

Figure 4.2 X-ray diffraction patterns are made by passing a thin beam of X rays through a mineral. The X rays are diffracted (dispersed) by the framework of atoms in the crystal, so that they form an orderly arrangement of dots on a photographic film placed behind the specimen.

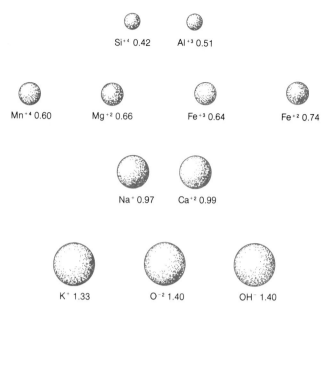

Si^{+4} 0.42   Al^{+3} 0.51

Mn^{+4} 0.60   Mg^{+2} 0.66   Fe^{+3} 0.64   Fe^{+2} 0.74

Na^{+} 0.97   Ca^{+2} 0.99

K^{+} 1.33   O^{-2} 1.40   OH^{-} 1.40

Figure 4.3 The relative size and electrical charge of ions are important factors governing the suitability of one ion to substitute for another in a crystal structure. Iron could be replaced by magnesium; sodium by calcium; and silicon by aluminum. If the substituting ions do not have exactly the same electrical charge, the difference in charge must be balanced or compensated for by other substitutions in the structure.

however, can vary within specific limits. In these minerals, two or more kinds of ions can substitute for each other in the mineral structure, a process called **ionic substitution.** Ionic substitution results in a chemical change in the mineral without a change in the crystal structure, but there are definite limits within which substitution can occur. Thus, the composition of a mineral can be expressed by a chemical formula that specifies ionic substitution and how the composition can change.

The suitability of one ion to substitute for another is determined by several factors, the most important being the size and the electrical charge of the ions in question (*figure 4.3*). Ions can readily substitute for one another if their ionic radii differ by less than 15%. If a substituting ion differs in charge from the ion for which it is substituted, the charge difference must be compensated for by other substitutions in the same structure.

Ionic substitution is somewhat analogous to substituting a plastic brick for an equal-sized clay brick. Because the substitute brick is the same size as the original, the structure of the wall is not affected, but there is a change in composition.

Ionic substitution occurs widely in the common rock-forming minerals and is responsible for mineral groups, the members of which have the same structure but varying composition. For example, in the olivine group, with the formula $(Mg,Fe)_2SiO_4$, ions of iron (Fe) and magnesium (Mg) can substitute freely for one another. The total number of Fe and Mg atoms is constant in relation to the number of silicon (Si) and oxygen (O) atoms in the olivine, but the ratio of Fe to Mg can vary in different samples. The common minerals feldspar, pyroxene, amphibole, and mica each constitute a group of related minerals in which atomic substitution produces a range of chemical composition.

Physical Properties of Minerals

Because a mineral has a definite chemical composition and internal crystalline structure, all specimens of a given mineral, regardless of when or where they were created, have the same physical and chemical properties, which vary only within certain limits. This means that one piece of quartz, for example, is as hard as any other piece, that it has the same specific gravity, and that it breaks in the same manner, regardless of when, where, or how it was formed.

The more significant and readily observable physical properties of minerals are crystal form, cleavage, hardness, specific gravity, color, and streak.

Crystal Form. If a crystal is allowed to grow in an unrestricted environment, it develops natural

crystal faces and assumes a specific geometric form. The shape of a crystal is a reflection of the internal structure and is an identifying characteristic for many mineral specimens. Quartz, for example, typically forms elongate, hexagonal crystals (*figure 4.4*). If the space for growth is restricted, however, smooth crystal faces cannot develop.

Cleavage. Cleavage is the tendency of a crystalline substance to split or break along smooth planes parallel to zones of weak bonding in the **crystal lattice** (*figure 4.5*). If the bonds are especially weak in a given direction (as in mica and halite), perfect cleavage occurs with ease, and it is difficult to break the mineral in any direction other than along a cleavage plane. In other minerals, the difference in bond strength is not great, so that cleavage is poor or imperfect. Cleavage can occur in more than one direction, but the number and direction of cleavage planes in a mineral species are always the same.

Some minerals have no weak planes in their crystalline structures, so that they lack cleavage and break along various types of fracture surfaces. Quartz, for example, characteristically breaks by **conchoidal fracture,** that is, along curved surfaces, like the curved surfaces of chipped glass.

Hardness. Hardness is a measure of a mineral's resistance to abrasion. This property is easily determined, and it is used widely for field identification of minerals. More than a century ago, Friedrich Mohs (1773–1839), a German mineralogist, assigned arbitrary relative numbers to 10 common minerals in order of their hardness. He assigned the number 10 to diamond, the hardest mineral known. Softer minerals were ranked in descending order, with talc, the softest mineral, assigned the number 1. The Mohs hardness scale (*table 4.1*) provides a standard for testing minerals for preliminary identification.

Specific Gravity. Specific gravity is the ratio be-

Figure 4.4 Quartz crystals form prisms in which the crystal faces meet at 120°, regardless of the gross shape and size of the crystals or when and where they were formed. The constancy of interfacial angles is an expression of the atomic structure of the mineral. Every mineral has a characteristic crystal form as a result of its crystalline structure.

Figure 4.5 Perfect cleavage in one direction is illustrated by the mineral mica.

tween the weight of a given volume of a substance and the weight of an equal volume of water. For example, a liter of solid lead weights a little over 11 times more than a liter of water, and thus the specific gravity of lead is 11.

Specific gravity is one of the more precisely defined properties of a mineral. It depends on the kinds of atoms making up the mineral and how closely they are packed into the crystal structure. Clearly, the more numerous and compact the atoms, the higher the specific gravity. Most com-

mon rock-forming minerals have specific gravities from 2.65 (for quartz) to about 3.37 (for olivine).

With a little experience, most people can estimate the specific gravity of a mineral merely by lifting a specimen. Most metallic minerals feel heavy, whereas most common nonmetallic minerals seem relatively light.

Color. Color is one of the more obvious properties of a mineral. Unfortunately, it is *not* diagnostic because most minerals are found in various hues, depending on such factors as subtle variations in composition and the presence of inclusions and impurities. Quartz, for example, ranges through the spectrum from colorless, clear crystals to purple, red, white, and jet black.

Streak. The color of a mineral in powder form, referred to as streak, is usually more diagnostic than the color of a large specimen. For example, the mineral pyrite (fool's gold) has a gold color but a black streak, whereas real gold has a gold streak—the same color as that of larger grains. To test for streak, a mineral is vigorously rubbed against the surface of an unglazed piece of white porcelain. Minerals softer than the porcelain leave a streak, or line of fine powder. For minerals harder than porcelain, a fine powder can be made by crushing a fragment, and the powder is then examined against a white background.

Table 4.1 The Mohs Hardness Scale

Hardness	Mineral	Test
1	Talc	Fingernail
2	Gypsum	
3	Calcite	Copper coin
4	Fluorite	Knife blade or
5	Apatite	glass plate
6	K-feldspar	
7	Quartz	Steel file
8	Topaz	
9	Corundum	
10	Diamond	

THE GROWTH AND DESTRUCTION OF MINERALS

Statement

Minerals are characteristically susceptible to chemical change. They grow as matter changes from a gaseous or liquid state to a solid state and are destroyed as the solid changes back to a liquid or gas. All minerals came into being because of specific physical and chemical conditions, and all are subject to change as the physical and chemical conditions change. Minerals, therefore, provide an important means of interpreting the changes that have occurred in the earth throughout its history.

Discussion

Crystal Growth

A single crystal, containing all of the physical and chemical properties of the mineral, may be so small that it cannot be identified with a high-power microscope (*figure 4.6*). **Crystallization** occurs by the addition of atoms to the crystal face. This is possible because the outer layers of atoms on a crystal are never completed and can be extended indefinitely.

An environment suitable for crystal growth includes (1) proper concentration of the kinds of atoms or ions required for a particular mineral and (2) proper temperature and pressure.

The time-lapse photographs in *figure 4.7* show how crystals grow in an unrestricted environment. Although the size of the crystal increases, its form and internal structure remain the same. New atoms are added to the outer edge of the crystal, parallel to the plane of atoms in the basic structure, and thus produce a perfectly symmetrical crystal form.

Some crystal faces, however, grow faster than others, so that in an environment where spaces are restricted, a crystal face may not grow symmetrically. Where a crystal face encounters a barrier, it ceases to grow. A crystal growing in a restricted

Figure 4.6 Submicroscopic crystals growing in the pore spaces between sand grains can be seen through an electron microscope. Each crystal contains all of the physical and chemical properties of the mineral.

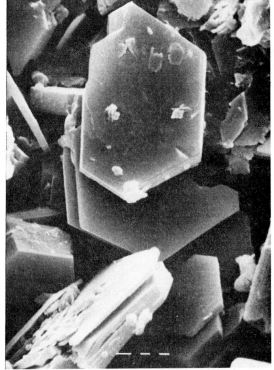

Figure 4.7 Crystal growth can be recorded by time-lapse photography. Each crystal grows as atoms lock on to the outer face of the crystal structure.

space assumes the shape of the confining area, and well-developed crystal faces do not form. This process is illustrated in *figure 4.8.* Thus, the external form of the crystal can be practically any shape, but the internal structure of the crystal is in no way modified. The mineral's internal structure remains the same, its composition is unaffected, and there is no change in its physical and chemical properties. The only modification is a change in the relative size of the crystal faces.

The process of crystal growth in restricted space is especially important in rock-forming minerals. In a **melt** or a solution, many crystals grow at the same time and must compete for space. As a result, in the latter stages of growth, crystals in rocks commonly lack unique, well-defined crystal faces and typically interlock with adjacent crystals to form a strong, coherent mass (*figure 4.9*).

Most crystals are rather small, measuring from a few tenths of a millimeter to several centimeters in diameter. Some are so small that they can only be seen enlarged thousands of times with an electron microscope. Minerals composed of such minute crystals are said to have a **cryptocrystalline texture.** In an unrestricted environment, however, crystals can grow to enormous sizes (*figure 4.10*).

Destruction of Crystals

Minerals dissolve or melt by removal of outer atoms from the crystal structure, so that the matter

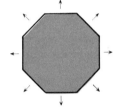

(A) *Where growth is unrestricted, all crystal faces grow with equal facility, and the perfect crystal is enlarged by growth on each crystal face.*

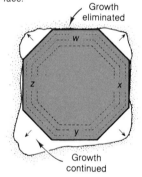

(B) *In a restricted environment, growth on certain crystal faces, such as w, x, y, and z, is terminated as available space is filled.*

(C) *The final shape of the crystal is determined by the geometry of the available space in which it can grow. In the example, a normally octagonal crystal develops an irregular form. The internal (atomic) structure of the crystal, however, remains the same, regardless of the space in which the crystal grows.*

Figure 4.8 Crystals growing in a restricted environment do not develop perfect crystal faces.

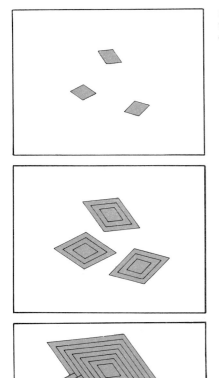

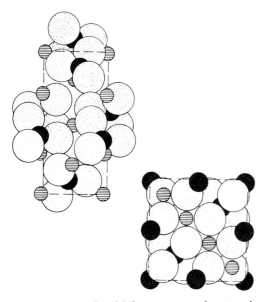

Figure 4.9 Interlocking texture develops if crystals grow in a restricted environment and are forced to compete for space.

Figure 4.10 Large crystals can form where there is ample space for growth, such as in caves. These crystals of gypsum are more than 1 m long.

returns to a fluid state (liquid or gas). The heat that causes a crystal to melt increases atomic vibrations enough to break the bonds holding an atom to the crystal structure. Similarly, atoms can be "pried" loose and carried away by a solvent, usually (in geologic processes) water. The breakdown or **dissolution** of a crystal begins at the surface and moves inward.

The growth and destruction of crystals are of paramount importance in geologic processes, because rocks originate and change as minerals are formed and destroyed. Crystals grow when lava cools and solidifies. They grow in the ocean, where they form from solution as limestone, salts, and other mineral deposits. They grow at the surface, where minerals in common rocks react chemically with elements in the atmosphere. They also form and are destroyed deep within the earth's crust. There, heat and pressure cause some crystal structures to break down and new minerals to form in their place, with a more dense, compact atomic structure (*figure 4.11*).

Figure 4.11 Under high pressure, the atomic structure of a mineral can collapse into a more dense form, in which the atoms are more closely packed.

THE SILICATE MINERALS

Statement

Although more than 2000 minerals have been identified, 95% of the volume of the earth's crust is composed of a group of minerals called the silicates. This should not be surprising, because silicon and oxygen constitute nearly three-fourths of the mass of the earth's crust (*table 4.2*) and therefore must predominate in most rock-forming minerals. Silicate minerals are complex in both chemistry and crystal structure, but all contain a basic building block called the silicon-oxygen tetrahedron. This is a complex ion [$(SiO_4)^{-4}$] in which four large oxide ions (O^{-2}) are arranged to form a four-sided pyramid with a smaller silicon ion (Si^{+4}) fitted into the cavity between them (*figure 4.12*). The major groups of silicate minerals differ mainly in the arrangement of silica tetrahedra in their crystal structure.

Discussion

Perhaps the best way to understand the unifying characteristics of the **silicates,** as well as the reasons for the differences, is to study the models shown in *figure 4.13*. These were constructed on the basis of X-ray studies of silicate crystals. **Silicon-oxygen tetrahedra** combine to form minerals in two ways. In the simplest combination, the oxygen ions of the tetrahedra form bonds with other elements such as iron or magnesium. Olivine is an example. Most silicate minerals, however, are formed by the sharing of an oxygen ion between

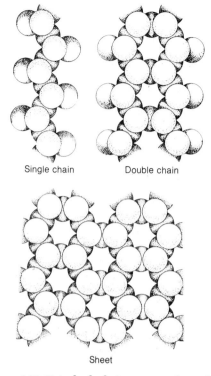

Single chain   Double chain

Sheet

Figure 4.13 Tetrahedral groups can form single chains, double chains, and sheets, by the sharing of oxide ions among silicon ions.

two adjacent tetrahedra. In this way, the tetrahedra form a larger ionic unit, just as beads are joined to form a necklace. The sharing of oxygen ions by the silicon ions results in several fundamental configurations of tetrahedral groupings: single chains, double chains, hexagonal sheets, and three-dimensional frameworks. These structures define the major silicate mineral groups.

1. Single chains—pyroxenes
2. Double chains—amphiboles
3. Hexagonal sheets—micas, chlorites, and clay minerals
4. Three-dimensional frameworks—feldspars and quartz

The unsatisfied electrons are balanced by various metallic ions, such as ions of calcium, sodium, potassium, magnesium, and iron. Thus, the silicate minerals contain silica tetrahedra linked into various patterns by metallic ions.

Considerable ionic substitution can occur in the basic crystal structure; for example, sodium can substitute for calcium or iron can substitute for magnesium. Thus, minerals of a major silicate group can differ chemically among themselves but have a common silicate structure.

Figure 4.12 The silicon-oxygen tetrahedron is the basic building block of the silicate minerals. Four large oxygen atoms are arranged in the form of a pyramid (tetrahedron) with a small silicon atom fitted into the central space between them. This is the most important building block in geology, because it is the basic unit for 95% of the minerals in the earth's crust.

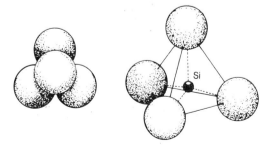

ROCK-FORMING MINERALS

Statement

Fewer than 20 kinds of minerals account for the great bulk of the earth's crust and most of the upper mantle. The most important minerals in each of the major rock types are as follows:

1. In igneous rocks—feldspars, micas, amphiboles, pyroxenes, olivines, and quartz
2. In sedimentary rocks—quartz, calcite, dolomite, clays, halite, gypsum, and feldspars
3. In metamorphic rocks—quartz, feldspars, amphiboles, pyroxenes, micas, garnet, and chlorites

The identification of minerals in rocks presents some special problems, because the mineral grains are usually small and rarely have well-developed crystal faces. You will find, however, that the distinguishing properties of the rock-forming minerals are easy to learn if you examine a specimen while you study the written description.

Discussion

A careful examination of the minerals that make up granite is a good beginning. The polished surface of a granite in *figure 4.14A* shows that the rock is composed of myriads of mineral **grains** of different sizes, shapes, and colors. Although the minerals interlock to form a tight, coherent mass, each one has distinguishing properties.

Table 4.2 Concentrations of the Most Abundant Elements in the Earth's Crust (by Weight)

Element	Percentage
O	46.60
Si	27.72
Al	8.13
Fe	5.00
Ca	3.63
Na	2.83
K	2.59
Mg	2.09
Ti	0.44
H	0.14
P	0.12
Mn	0.10
S	0.05
C	0.03

After Mason, 1958, p. 48.

Feldspars

A large part of granite consists of a pink, porcelainlike mineral that has a rectangular shape, and a milky white, porcelainlike mineral that is somewhat smaller but similarly shaped. These are **feldspars** (German, "field crystals"), the most abundant mineral group in the earth's crust. The feldspars have good cleavage in two directions, a porcelainlike luster, and a hardness of about 6 on the Mohs hardness scale.

The crystal structure of the feldspars permits considerable ionic substitution, giving rise to two major types—potassium feldspar (K-feldspar) and plagioclase feldspar.

Potassium feldspar ($KAlSi_3O_8$), most commonly pink in rocks, is shaded brown in *figure 4.14.* **Plagioclase** (shown in white in the sketch) permits complete substitution of sodium (Na) for calcium (Ca) in the crystal structure, which gives rise to a compositional range from $NaAlSi_3O_8$ to $CaAl_2Si_2O_8$. White plagioclase in granite is rich in sodium.

Feldspars are common in most igneous rocks, in many metamorphic rocks, and in some sandstones.

Micas

The tiny, black, shiny grains in *figure 4.14* are **mica.** This group of minerals is readily recognized by its perfectly one-directional cleavage, which permits breakage into thin, elastic flakes. Mica is a complex silicate, whose sheet structure is responsible for its perfect cleavage. Two common varieties occur in rocks: **biotite,** which is black mica, and muscovite, which is white or colorless. Mica is abundant in granites and in many metamorphic rocks and is also a significant constituent in many sandstones.

Quartz

The glassy, irregularly shaped grains in *figure 4.14* are **quartz.** Because quartz is usually the last mineral to form in a granite, it lacks well-developed crystal faces. The mineral simply fills the spaces between the early-formed feldspars and micas. Quartz is abundant in all three major rock types. It has the simple composition SiO_2 and is distinguished by its hardness (7), its conchoidal fracture, and its glassy luster. Pure quartz crystals are colorless, but slight impurities produce a variety of colors. Where crystals are able to grow freely, they form elongate, six-sided crystals that terminate at a point (*figure 4.4*), but well-formed crystals are rarely found in rocks. In sandstone, quartz is abraded into rounded sand grains. Quartz is stable both

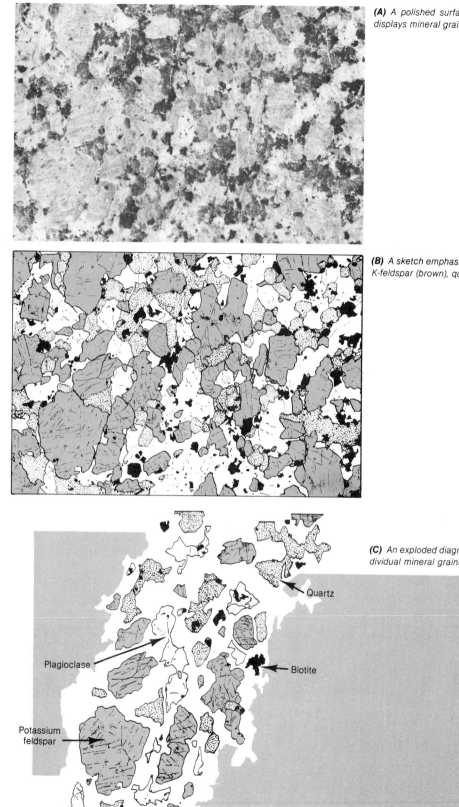

(A) A polished surface of a granite, shown at actual size, displays mineral grains of different sizes, shapes, and colors.

(B) A sketch emphasizing grains of individual minerals shows K-feldspar (brown), quartz (stippled), and plagioclase (white).

(C) An exploded diagram of **B** shows the size and shape of individual mineral grains.

Quartz

Plagioclase

Biotite

Potassium feldspar

Figure 4.14 Mineral grains in a granite form a tight, interlocking texture because each mineral is forced to compete for space as it grows.

mechanically (it is very hard and lacks cleavage) and chemically (it does not react readily with elements at or near the earth's surface). Therefore, it is a difficult mineral to alter or destroy once it has formed.

Ferromagnesian Minerals

In addition to the minerals found in a granite, there are a number of silicate minerals referred to as **ferromagnesian minerals,** because they contain appreciable amounts of iron and magnesium. These minerals generally range from dark green to black in color and have a high specific gravity. Biotite is classified in this general group, together with the olivines, pyroxenes, and amphiboles. (Biotite is common in granite, but the other ferromagnesian minerals are rare or absent.) The ferromagnesian minerals are common, however, in basalt (*figure 4.15*).

Olivines. The only mineral clearly visible in the hand specimen in *figure 4.15* is the green, glassy mineral called olivine. The olivine family is a group of silicates in which iron and magnesium substitute freely in the crystal structure. The composition is expressed as $(Mg,Fe)_2SiO_4$. This hard mineral is characterized by an olive green color (if magnesium is abundant) and a glassy luster. In rocks, it rarely forms crystals larger than a millimeter in diameter. Olivine is a high-temperature mineral and is common in basalt (volcanic rock). It is probably a major constituent of the material beneath the earth's crust.

Pyroxenes. Pyroxenes are high-temperature minerals commonly found in many igneous and

Figure 4.15 Mineral grains in basalt are microscopic in size.

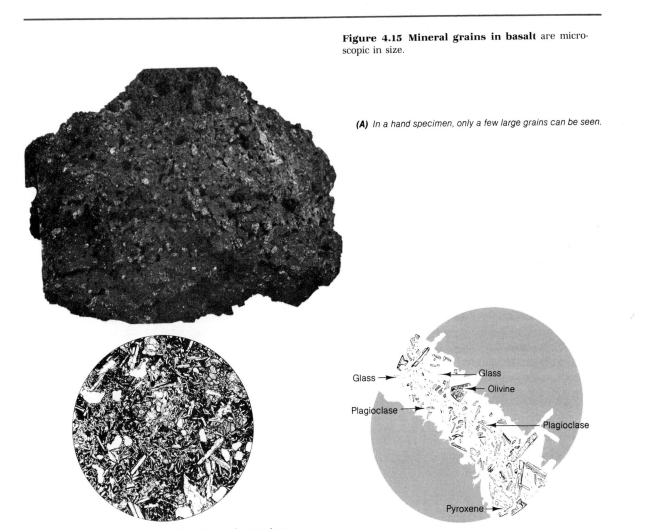

(A) In a hand specimen, only a few large grains can be seen.

(B) Viewed under a microscope, the grains can be seen forming an interlocking texture. Plagioclase crystals typically form lathlike grains.

(C) An exploded diagram of **B** shows the size and shape of individual mineral grains.

Figure 4.16 Amphibole crystals are among the first to crystallize in "granitic" rocks and therefore have well-developed crystal faces.

metamorphic rocks. In *figure 4.15*, pyroxene occurs as microscopic crystals, but some samples contain larger grains of this mineral, whose color typically runs from dark green to black.

Amphiboles. Figure 4.16 shows a rock that is similar to a granite but has less quartz and contains appreciable amounts of the mineral amphibole. Amphiboles have much in common with the pyroxenes. Their chemical compositions are similar, except that amphibole contains hydroxyl ions (OH^-) and pyroxene does not. They differ in structure, however; the amphiboles produce elongate crystals that cleave perfectly in two directions, which are not at right angles. The color of amphibole ranges from green to black. This mineral is common in rocks closely related to granite and may be more abundant in that rock than biotite. It is especially common in the metamorphic rock known as amphibolite. **Hornblende** is the most common variety of amphibole.

Clay Minerals

The **clay minerals** constitute a major part of the soil and are thus encountered more frequently than other minerals in everyday experience. Clay minerals form at the earth's surface where air and water interact with the various silicate minerals. Like the micas, the clay minerals are sheet silicates, but their crystals are microscopic and usually can be detected only with an electron microscope (*figure 4.6*). More than a dozen clay minerals can be

distinguished on the basis of their crystal structures and variations in composition.

Calcite

Calcite is calcium cabonate ($CaCO_3$), the principal mineral in limestone. It can be precipitated directly from seawater, removed from seawater by organisms and used to make their shells, or dissolved by ground water and reprecipitated as new crystals in caves and fractures in rock. It is usually transparent or white in color, but the aggregates of calcite crystals that form limestone contain various impurities, which give them gray or brown hues. Calcite is common at the earth's surface and is easy to identify. It is soft enough to scratch with a knife (hardness of 3), and it effervesces in dilute hydrochloric acid. It has perfect cleavage in three directions, which are not at right angles, so that cleaved fragments form rhombohedra (*figure 4.1*). Besides being the major constituent of limestone, calcite is the major mineral in the metamorphic rock marble.

Dolomite

Dolomite is a carbonate of calcium and magnesium. Large crystals form rhombohedra, but most dolomite occurs as granular masses of small crystals. Dolomite is widespread in sedimentary rocks, where it forms if calcite reacts with solutions of magnesium carbonate in seawater or ground water. Dolomite can be distinguished from calcite because it effervesces in dilute hydrochloric acid only if it is in powdered form.

Halite and Gypsum

Halite and gypsum are the two most common minerals formed by evaporation of seawater or saline lake water. **Halite,** common salt (NaCl), is easily identified by its taste. It also has one of the simplest of all crystal structures—the sodium and chloride ions are united in a cubical array. Most physical properties of halite are related to this structure. Halite crystals cleave in three directions, at right angles, to form cubic or rectangular fragments. Salt, of course, is very soluble and readily dissolves in water.

Gypsum is composed of calcium sulfate and water ($CaSO_4 \cdot 2H_2O$). It forms crystals that are generally clear white, with a glassy or silky luster. It is a very soft mineral and can be scratched easily with a fingernail. It cleaves perfectly in one direction to form thin, nonelastic plates. Gypsum occurs as single crystals, as aggregates of crystals in compact masses (alabaster), and as a fibrous form (satin spar).

SUMMARY

A mineral is a natural inorganic solid with a specific internal structure and a chemical composition that varies only within specific limits. All specimens of a given mineral, regardless of where, when, or how they were formed, share certain physical properties. Some of the more readily observable physical properties are cleavage, crystal form, hardness, specific gravity, color, and streak, all of which aid in mineral identification.

Minerals are the building blocks of rocks. They grow or are destroyed by chemical reactions as matter changes to and from the solid state. Minerals, therefore, provide an important means of interpreting the changes that have occurred in the earth throughout its history.

More than 95% of the earth's crust is composed of the silicate minerals, a group of minerals containing silicon and oxygen linked together in tetrahedral units, with four oxygen atoms to one silicon atom. Several fundamental configurations of tetrahedral groupings—single chains, double chains, hexagonal sheets, and three-dimensional frameworks—result from the sharing of oxygen ions among silicon ions.

Fewer than 20 minerals form the great bulk of the earth's crust. The most important minerals in igneous rocks are feldspars, micas, amphiboles, pyroxenes, olivines, and quartz. The most important among the sedimentary rocks are quartz, calcite, dolomite, clay minerals, and feldspars. The most important among the metamorphic rocks are quartz, feldspars, amphiboles, pyroxenes, micas, chlorites, and garnet.

Key Words

mineral	conchoidal fracture
X-ray diffraction	hardness
ionic substitution	specific gravity
crystal	streak
crystal form	crystallization
crystal face	dissolution
crystal lattice	melt
cryptocrystalline	grain
cleavage	silicates

silicon-oxygen tetrahedron	pyroxenes
feldspars	amphiboles
plagioclase	hornblende
micas	clay minerals
biotite	calcite
quartz	dolomite
ferromagnesian minerals	halite
olivines	gypsum

Review Questions

1. Give a brief but adequate definition of a mineral.
2. Explain the meaning of "the internal structure of a mineral."
3. Why does a mineral have a definite chemical composition?
4. Briefly explain how minerals grow and are destroyed.
5. Explain the origin of cleavage in minerals.
6. Describe the silicon-oxygen tetrahedron. Why is it important in the study of minerals?
7. Why does mica have excellent cleavage whereas quartz and olivine lack cleavage?
8. What are the silicate minerals? List the silicate minerals that are important in the study of rocks.
9. Give three examples of how silicon-oxygen tetrahedra are arranged in silicate minerals.
10. Study *figure 4.14* and explain why most of the mineral grains in a granite have an irregular shape.

Additional Readings

Ahrens, L. H. 1965. *Distribution of the elements in our planet.* New York: McGraw-Hill.

Deer, W. A., R. A. Howie, and J. Zussman. 1966. *An introduction to the rock forming minerals.* New York: John Wiley and Sons.

Ernst, W. G. 1969. *Earth materials.* Englewood Cliffs, N.J.: Prentice-Hall.

Mason, B. 1958. *Principles of geochemistry.* New York: John Wiley and Sons.

Simpson, B. 1966. *Rocks and minerals.* Elmsford, N.Y.: Pergamon Press.

U.S. Geological Survey. 1971. *Atlas of volcanic phenomena.* Reston, Va.: U.S. Geological Survey.

Zim H. S., and P. R. Shaffer. 1957. *Rocks and minerals.* New York: Golden Press.

5

THE IMPORTANCE OF ROCK TEXTURES

Statement

The texture of a rock refers to the size, shape, and arrangement of its constituent mineral grains. This characteristic of the rock is separate and distinct from its composition. Texture is important because the mineral grains bear a record of the energy involved in the rock-forming process and the conditions existing at the time the rock originated. Igneous rocks have distinctive textures, characterized mostly by the interlocking grains that grow from cooling magma.

The major textures of igneous rocks are related to the size of crystals:

1. Glassy—no crystals
2. Aphanitic—crystals too small to be seen without a microscope
3. Phaneritic—crystals large enough to be seen without a microscope
4. Porphyritic—two different crystal sizes
5. Pyroclastic—volcanic fragments

Discussion

The genetic imprint left upon the **texture** of a rock is commonly clear and easy to read. For example, the texture of lunar rocks that were formed by the impact of meteorites is characterized by various sizes of angular rock fragments interspersed with bits and pieces of glass fused by the impact. A sandstone formed on a beach consists of well-rounded smooth grains, all approximately the same size, which result from the abrasion and washing action of waves. In contrast, rocks formed from a cooling liquid have a texture characterized by interlocking grains.

To illustrate the importance of texture, we will consider five examples of rocks that have essentially the same chemical and mineralogical composition but different textures. In each rock, a chemical analysis would disclose about 48% O, 30% Si, 7% Al, and between 3% and 4% each of Na, K, Ca, and Fe. In glassy rocks, crystallization has not taken place, so that the constituent atoms are randomly dispersed rather than arranged in a crystalline structure. In the other specimens, the elements occur in crystals of feldspar, quartz, mica, and amphibole. On the basis of chemical composition alone, these rocks would be considered the same—the difference is *only* in texture. It is the texture that provides the most information about how the specimen was formed.

Glassy Texture

The nature of volcanic **glass** is illustrated in *figure 5.1*. The hand specimen displays a conchoidal fracture with the sharp edges typical of broken glass. No distinct crystals are visible, but, viewed under a microscope, distinct flow layers are apparent. These result from the uneven concentration of innumerable, minute, embryonic crystals.

In the laboratory, melted rock or synthetic lava hardens to glass if it is quenched (or quickly cooled) at temperatures above those at which crystals form. We can conclude that a **glassy texture** is produced by very rapid cooling. The randomness of the ions in a high-temperature melt is "frozen in," because the ions do not have time to migrate and organize themselves into an orderly, crystalline structure. Field observations of glassy rocks in volcanic regions support the hypothesis that rapid cooling produces glass. Small pieces of lava blown from a volcanic vent into the much cooler atmosphere harden to form glassy ash. A glassy crust forms on the surface of many lava flows, and glassy fragments form if a flow enters a body of water.

Aphanitic Texture

If crystal growth from a melt requires time for the ions to collect together and organize themselves, then a crystalline rock indicates a slower rate of cooling than that of a glassy rock. The texture illustrated in *figure 5.2* is crystalline but extremely fine grained—a texture referred to as **aphanitic** (Greek *a*, "not," *phaneros*, "visible"). In hand specimens, few, if any, crystals can be detected in aphanitic textures. Viewed under a microscope, however, many crystals of feldspar and quartz can be recognized.

An aphanitic texture indicates relatively rapid cooling, but not nearly as rapid as that which produces glass. Aphanitic textures are typical of the interior of lava flows, in contrast to the glassy texture that forms on the surface or crust.

A feature in many aphanitic and glassy rocks is the presence of numerous small, spherical or ellipsoidal cavities called **vesicles.** These are produced by gas bubbles trapped in the solidifying rock. As hot magma rises toward the earth's surface, the confining pressure diminishes and dissolved gas (mainly steam) separates and collects into bubbles. The process is similar to the effervescence of champagne and soda pop when their bottles are opened. Vesicular textures typically develop in the upper

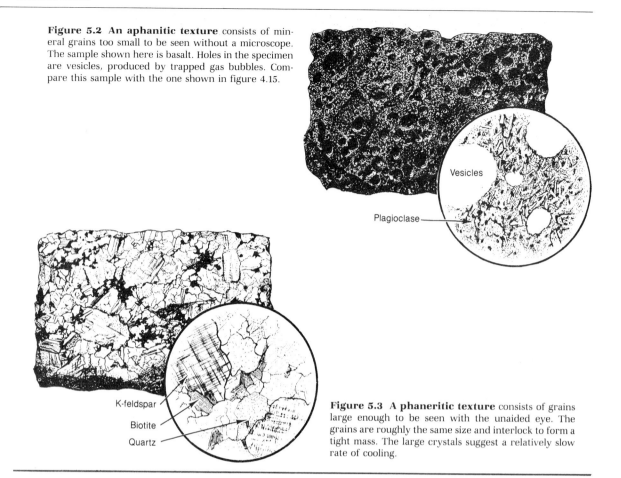

Figure 5.2 An aphanitic texture consists of mineral grains too small to be seen without a microscope. The sample shown here is basalt. Holes in the specimen are vesicles, produced by trapped gas bubbles. Compare this sample with the one shown in figure 4.15.

Vesicles

Plagioclase

K-feldspar

Biotite

Quartz

Figure 5.3 A phaneritic texture consists of grains large enough to be seen with the unaided eye. The grains are roughly the same size and interlock to form a tight mass. The large crystals suggest a relatively slow rate of cooling.

part of a lava flow, just below the solid crust, where the upward-migrating gas bubbles are trapped. Though vesicles change the outward appearance of the rock and indicate the presence of gas in a rapidly cooling lava, they do not change the basic aphanitic texture.

Phaneritic Texture

The specimen shown in *figure 5.3* is composed of grains large enough to be recognized without a microscope, a texture known as **phaneritic** (Greek *phaneros,* "visible"). The grains are approximately equal in size and form an interlocking mosaic. The equigranular texture suggests a uniform rate of cooling, and the large size of the crystals shows that cooling was very slow.

In order for cooling to take place at such a slow rate, the magma must have cooled far below the surface. Field evidence supports this conclusion, since volcanic eruptions produce only aphanitic and glassy textures. Rocks with phaneritic textures are exposed only after deep erosion has removed the covering rock.

Porphyritic Texture

In some igneous rocks, two distinct sizes of crystals are apparent. The larger, well-formed crystals are referred to as **phenocrysts,** and the smaller crystals constitute the **matrix,** or the **groundmass.** This texture is referred to as **porphyritic.** It can occur in either aphanitic or phaneritic rocks.

A porphyritic texture usually indicates two stages of cooling. An initial stage of slow cooling, in which the large crystals developed, is followed by a period of more rapid cooling, in which the smaller grains formed. The aphanitic matrix shown in *figure 5.4* indicates that the cooling melt had sufficient time for all its material to crystallize. The initial stage of relatively slow cooling produced the larger crystals, and the later stage of rapid cooling, when the magma was extruded, produced the smaller grains. Similarly, a phaneritic matrix indicates two stages of cooling. An initial stage of very slow cooling was followed by a second stage, when cooling was more rapid but not rapid enough to form an aphanitic matrix.

6. Describe some of the common surface features of basaltic flows.

7. Why does a magma tend to rise upward toward the earth's surface?

8. Draw a series of diagrams showing the form and internal structure of (*a*) a cinder cone, (*b*) a composite volcano, and (*c*) a shield volcano.

9. Describe the events that are typically involved in the origin of a caldera.

10. Describe the extrusion of an ash flow.

11. Describe and illustrate the major types of igneous intrusions. What is the textural difference between intrusive rocks and extrusive rocks?

12. Draw a diagram showing the following types of contacts between igneous intrusions and the surrounding rock: (*a*) discordant, (*b*) concordant, and (*c*) sedimentary.

13. What is magmatic differentiation?

14. Draw a simple diagram (label *figure 5.32*) and explain how basaltic magma may originate from partial melting of the asthenosphere.

15. Draw a simple diagram (label *figure 5.33*) and explain how granitic and andesitic magma originate from partial melting of the lithosphere as it descends into a subduction zone.

Additional Readings

Ernst, W. G. 1969. *Earth materials.* Englewood Cliffs, N.J.: Prentice-Hall.

Hyndman, D. W. 1972. *Petrology of igneous and metamorphic rocks.* New York: McGraw-Hill.

Schumann, W. 1974. *Stones and minerals.* London: Lutterworth Press.

Simpson, B. 1966. *Rocks and minerals.* Elmsford, N.Y.: Pergamon Press.

Tindall, J. R., and R. Thornhill. 1975. *Rock and mineral guide.* London: Blandford Press.

Williams, H., F. J. Turner, and C. M. Gilbert. 1954. *Petrography.* San Francisco: W. H. Freeman.

6

SEDIMENTARY STRUCTURES

Statement

Most sediment is transported by current systems in streams, along shorelines, and in shallow seas before it finally is deposited. For this reason, sedimentary rocks commonly show layering and other structural features that formed as the material was moved, sorted, and deposited by currents. These features, called primary sedimentary structures, provide key information about the conditions under which the sediment accumulated. The most important sedimentary structures are

1. Stratification
2. Cross-bedding
3. Graded bedding
4. Ripple marks and mud cracks

Discussion

Stratification

Stratification occurs on many scales and reflects the changes that occur during the formation of a sedimentary rock. Large-scale stratification is expressed by major changes in rock types, which can be seen in large exposures, such as the Grand Canyon (*figure 6.1*), where cliffs of limestone or sandstone alternate with slopes of weaker shale. Within each of these major rock units, bedding occurs on several smaller scales, expressed by differences in the texture, color, and composition of the rock.

An important aspect of stratification is that the rock layers do not occur in a random fashion but overlie one another in definite sequences and patterns. One of the more simple and common patterns in a vertical sequence is the cycle of sandstone, shale, and limestone. This pattern can be produced by the advance and retreat of shallow seas over the continental platform (*figure 6.9*). As is shown in diagram *A* of the figure, the sea begins to expand over a lowland drained by a river system. Sand accumulates along the shore, mud is transported in suspension offshore, and limestone is precipitated farther offshore, beyond the mud zone. All three types of **sediment** are deposited simultaneously, each in a different environment. Stream deposits (not shown in the figure) accumulate on the flood plain of the river system.

As the sea expands over the lowland, each environment shifts landward, following the shoreline (*figures 6.9A, B,* and *C*). Consequently, beach sands are deposited over the stream sediments, offshore mud is deposited over the previous beaches, and lime is deposited over the mud. As the sea continues to expand, the layers of sand, mud, and limestone are deposited further and further inland.

As the sea withdraws (*figure 6.9D*), the mud is deposited over the limestone and the nearshore sand over the mud. The net result is a wedge of limestone encased in a wedge of shale, which in turn is encased in a wedge of sandstone. Below and above the marine deposits are **fluvial** sediments deposited by the river system. Subsequent uplift and erosion of the area produces a definite pattern of rock (*figure 6.9E*). Beginning at the base, beach sand is overlain by shale and limestone, and these in turn are overlain by mud and beach sand. (See the lower rock layers in *figure 6.2.*)

Cross-bedding

The formation of **cross-bedding** is shown in *figure 6.10*. As clastic particles are moved by currents (either wind or water), they form ripples or dunes called **sand waves.** These range in scale from small ripples less than a centimeter high to giant sand dunes several hundred meters high. Typically, they are asymmetrical, with the steep slope facing in the direction of the moving current. As the particles migrate up and over the sand wave, they accumulate on the steep downcurrent face and form inclined layers. The direction of flow of the ancient currents that formed a given set of cross-strata can be determined by measuring the direction in which they are inclined. It is possible to determine the patterns of ancient current systems by mapping the direction of cross-bedding in sedimentary rocks.

Figure 6.10 Cross-bedding is formed by the migration of ripple marks, sand waves, and dunes. Particles of sediment, carried by currents, travel up and over the sand wave and are deposited on the steep downcurrent face to form inclined layers.

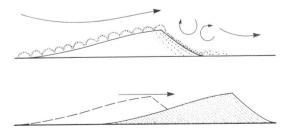

Graded Bedding

A distinctive type of stratification, called **graded bedding,** is characterized by a progressive decrease in grain size upward through the bed (*figure 6.11*). This type of stratification commonly is produced on the deep-ocean floor by **turbidity currents,** which transport sediment from the continental slope to the adjacent deep ocean. A turbidity current is generated by turbid (muddy) water, which, being denser than the surrounding clear water, sinks beneath it and moves rapidly down the submarine slope (*figure 6.12*).

Turbidity current flow can easily be demonstrated in the laboratory by pouring muddy water down the side of a tank filled with clear water. The mass of muddy water moves down the slope of the tank and across the bottom at a relatively high speed, without mixing with the clear water. Turbidity currents can also be observed where streams discharge muddy water into a clear lake or reservoir. The denser, muddy water moves out along the bottom of the basin and can flow for a considerable distance.

Turbidity currents can also be generated by an earthquake or a submarine landslide, during which mud, sand, and even gravel can be thrown into suspension. A turbidity current triggered by an earthquake near the Grand Banks, off Newfoundland, broke a series of transatlantic cables as it swept across the continental slope. The speed of the current, determined from the intervals between the times when the cables broke, was from 80 to 95 km

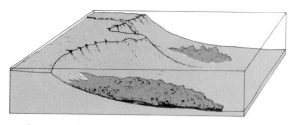

Figure 6.12 The movement of turbidity currents down the slope of the continental shelf can be initiated by a landslide or an earthquake. Sediment is moved largely in suspension. As the current slows down, the coarse grains are deposited first, followed by successively finer-grained sediment. A layer of graded bedding is thus produced from a single turbidity current.

per hour. This mass of muddy water formed a graded layer of sediment over a large area on the Atlantic floor.

As a turbidity current moves across the flat floor of a basin, its velocity at any given point gradually decreases. The coarser sediment is deposited first, followed by successively smaller particles. After the turbid water ceases to move, the sediment remaining in suspension gradually settles out. One pulse of sedimentation, therefore, deposits a single layer of sediment, which exhibits continous gradation from coarse material at the base to fine material at the top. There is no sudden change in grain size and no internal lamination. Subsequent turbidity flows can deposit more layers of graded sediment, with sharp contacts between layers. The result would be a succession of widespread horizontal layers, each being a graded unit deposited by a single pulse of sedimentation from a turbidity flow.

Ripple Marks, Mud Cracks, and Other Surface Impressions

Ripple marks are commonly seen in modern stream beds, in tidal flats, and along the shores of lakes and the sea. Preserved in rocks, they provide information concerning the environment of deposition, such as depth of water, ancient current directions, and trends of ancient shorelines. **Mud cracks** are also commonly preserved in sedimentary rock and show that the sedimentary environment occasionally was exposed to the air during the process of deposition. Mud cracks in rocks suggest deposition in shallow lakes, tidal flats, and exposed stream banks.

Tracks, trails, and borings of animals are typically associated with ripple marks and mud cracks and can provide additional important clues about the environment in which the sediment accumulated.

Figure 6.11 Graded bedding, produced by turbidity currents, occurs in widespread layers, generally less than a meter thick. Deep-marine environments commonly produce a great thickness of graded units, which can easily be distinguished from sediment deposited in most other environments.

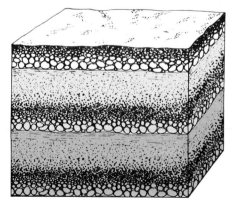

Figure 6.18 The delta environment (see also plate 9). **(A)** A small delta formed in a lake in Switzerland. **(B)** Ancient deltaic deposits in Tertiary rocks of the Colorado Plateau. One of the most significant environments of sedimentation occurs where the major rivers of the world enter the oceans and deposit most of their sediment in marine deltas. A delta environment can be very large, covering areas of more than 36,000 km². Commonly, deltas are very complex and involve various distinct subenvironments, such as beaches, bars, lagoons, swamps, stream channels, and lakes. Because deltas are large features and include a number of both marine and nonmarine subenvironments, a great variety of sediment types accumulate in them. Sand, silt, and mud dominate. Therefore, a deltaic deposit can be recognized only after considerable study of the sizes and shapes of the various rock bodies and their relationships to each other. Both marine and nonmarine fossils can be preserved in a delta.

Figure 6.19 The beach environment (see also plate 12). **(A)** A modern gravel beach along the southern coast of France. **(B)** Ancient beach deposits of Cambrian age in northern Michigan. Much sediment accumulates in the zone where the land meets the ocean. Within this zone, a variety of subenvironments occur, including beaches, bars, spits, lagoons, and tidal flats. Each has its own characteristic sediment. Where wave action is strong, mud is winnowed out and only sand or gravel accumulates as beaches or bars. Beach gravels accumulate along shorelines, where high wave energy is expended. The gravels are well sorted and well rounded and commonly are stratified in low, dipping cross-strata. Ancient gravel beaches are relatively thin and widespread. They commonly are associated with clean, well-sorted sand deposited offshore.

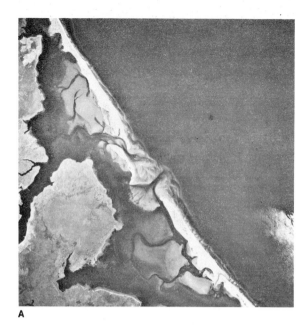

A

A

B

B

Figure 6.20 The lagoon environment (see also plate 12). **(A)** A lagoon along the central Atlantic coast of the United States. **(B)** Lagoonal deposits of Cretaceous age in central Utah. Offshore bars and reefs commonly seal off part of the coast, forming lagoons. A lagoon is protected from the high energy of waves, so that the water is typically calm and quiet. Fine-grained sediment, rich in organic matter, accumulates as black mud. Eventually, the lagoon becomes filled with sediment and evolves into a swamp. Where the vegetation provides enough organic matter, a coal deposit can form. The rise and fall of sea level shifts the position of the barrier bar. Thus, the organic-rich mud or coal formed in the lagoon or swamp is interbedded with sand deposited in the barrier island.

Figure 6.21 The tidal-flat environment. (A) A modern tidal flat in the Gulf of California. **(B)** Tidal-flat deposits of Triassic age in southern Utah. The tidal-flat environment is unique in being alternately covered with a sheet of shallow water and exposed to the air. Tidal currents are not strong. They generally transport only fine silt and sand and typically develop ripple marks over a broad area. Mud cracks commonly form during low tide and are subsequently covered and preserved. Ancient tidal-flat deposits are thus characterized by accumulations of silt and mud in horizontal layers with an abundance of ripple marks and mud cracks.

A

A

B

B

Figure 6.22 The organic-reef environment.
(A) A modern organic reef in the Bahama Islands.
(B) Ancient reefs of Devonian age in Australia. An organic reef is a solid structure of calcium carbonate constructed of shells and secretions of marine organisms. The framework of most reefs, consisting of a mass of colonial corals, forms a wall that slopes steeply seaward. Wave action continually breaks up part of the seaward face, and blocks and fragments of the reef accumulate as debris on the seaward slope. A lagoon forms behind the reef, toward the shore or toward the interior of the atoll, and lime mud and evaporite salts are deposited there. Gradual subsidence of the sea floor permits continuous upward growth of reef material to a thickness of as much as 1000 m. Because of their limited ecological tolerance (they require warm, shallow water), fossil reefs are excellent indicators of ancient environments. Fossil reefs commonly are found in shallow-marine limestones.

Figure 6.23 The shallow-marine environment
(see also plates 6 and 12). **(A)** A modern shallow-marine environment in the Bahamas. **(B)** Ancient shallow-marine sediments of Pennsylvanian age in eastern Kansas. Shallow seas border most of the land areas of the world and can extend to the interior of a continent, as do Hudson Bay, the Baltic Sea, and the Gulf of Carpentaria (north of Australia). The characteristics of the sediment deposited in a shallow-sea environment depend on the supply of sediment from the land and the local conditions of climate, wave energy, circulation of water, and temperature. If there is a large supply of land-derived sediment, sand and mud accumulate and ultimately form sandstone and shale, respectively. If sediment from the land is not abundant, limestone generally is precipitated or deposited by biological means. Ancient shallow-marine deposits are characterized by thin, widespread, interbedded layers of sandstone, shale, and limestone.

A

B

Figure 6.24 The deep-marine environment.
(A) A sketch of modern deep-sea fans off the coast of India. **(B)** Ancient deep-marine deposits near the southern coast of France. The part of the deep-ocean floor that lies near a continent is, for the most part, extremely smooth because of the deposition of sediment brought down the continental slope by turbidity currents. The deposits from turbidity currents are characterized by a sequence of graded beds with each layer extending over large areas. Thus, they are easily distinguishable from sediment deposited in most other environments. Sediment also accumulates on the floor of the open ocean, far from the continents. This material consists of fine particles of mud, which are carried in suspension and gradually sink. The most abundant sediment is a fine-grained brown or red clay. Some sediment in the open ocean apparently crystallized directly from seawater. In large parts of the South Pacific, these deposits form the bulk of the sediment.

water seeping through the pore spaces of the grains, is precipitated. This postdepositional crystallization of **cement** (the common minerals are calcite, quartz, and limonite), which holds the grains of sediment together, is a fundamental process in transforming sediment into solid rock.

Significance of Sedimentary Rocks

From an economic point of view, sedimentary rocks are extremely important, because many of our natural resources are formed by sedimentary processes. The most familiar, of course, are petroleum, natural gas, and coal. But many other resources, such as sand and gravel for construction, building stone, clay for bricks and ceramics, salt, uranium, iron, gold, and aluminum, are concentrated by sedimentary processes.

From a scientific point of view, sedimentary rocks are important for the geologic history they record. Since they are produced by interactions of the hydrologic system and the earth's crust, sedimentary rocks record the history of physical and biological events on the earth. From the record of sedimentary rocks, geologists are able to interpret such things as ancient mountain building, the erosion of continents, climatic changes, the evolution of life, and the constantly changing patterns of land and sea.

SUMMARY

Sedimentary rocks originate from fragments of other rocks, minerals precipitated chemically, and organic matter. Typically, they occur in layers, or strata, which cover large parts of the continents. Sedimentary rocks preserve a record of the erosional history of the earth and also a history of life on the planet over a period of at least 2 billion or 3 billion years.

Two main types of sedimentary rocks are recognized: (1) clastic rocks, consisting of rock and mineral fragments (clay, silt, sand, and gravel) and (2) chemical and organic rocks.

Important structures in sedimentary rocks include stratification, cross-bedding, graded bedding, ripple marks, and mud cracks. These features are primary sedimentary structures, which originate when the sediment is deposited.

Four major processes are involved in the genesis of sedimentary rocks: (1) weathering, (2) transportation, (3) deposition, and (4) compaction and cementation. During weathering, transportation, and deposition, the sedimentary material is differentiated (sorted and concentrated according to grain size and composition).

A sedimentary environment is a place where sediment is deposited and the physical, chemical, and biological conditions that exist there. The major sedimentary environments include (1) fluvial, (2) alluvial-fan, (3) eolian, (4) glacial, (5) delta, (6) shoreline, (7) organic-reef, (8) shallow-marine, and (9) deep-marine environments. Each of these is characterized by certain physical, chemical, and biological conditions and therefore develops distinctive rock types and fossil assemblages.

Key Words

sedimentary rock	sandstone
sediment	shale
formation	limestone
geologic cross section	chalk
stratum	oolite
bedding plane	dripstone
bed	dolostone
lamina	evaporite
fossil	stratification
cement	cross-bedding
clastic	sand waves
conglomerate	graded bedding

primary sedimentary structure
turbidity current
ripple marks
mud cracks
weathering
sorting
sedimentary differentiation
sedimentary environment
fluvial environment
alluvial fan
eolian environment
glacial environment
delta
beach
lagoon
tidal flat
reef
shallow-marine environment
deep-marine environment

Review Questions

1. Study the photograph and draw a simple sketch, like the one in *figure 6.2*, showing a cross section of the major rock layers.
2. List the characteristics that distinguish sedimentary rocks from igneous and metamorphic rocks.
3. What is the principal mineral in sandstone? Why does this mineral dominate?
4. How does limestone differ from clastic rocks?
5. What is the mineral composition of limestone?
6. Show by a series of sketches the characteristics of stratification, cross-bedding, and graded bedding.
7. What rock types form in the following sedimentary environments: (a) delta, (b) lagoon, (c) alluvial fan, (d) eolian environment, (e) organic reef, (f) deep-marine environment?

Additional Readings

Dunbar, C. O., and J. Rogers. 1957. *Principles of stratigraphy.* New York: John Wiley and Sons.
Kuenen, P. H. 1960. Sand. *Scientific American* 202(4):21–34. Reprint no. 803. San Francisco: W. H. Freeman.
Laporte, L. F. 1968. *Ancient environments.* Englewood Cliffs, N.J.: Prentice-Hall.
Pettijohn, F. J., and P. E. Potter. 1964. *Atlas and glossary of primary sedimentary structures.* New York: Springer-Verlag.
Selley, R. C. 1976. *An introduction to sedimentology.* New York: Academic Press.

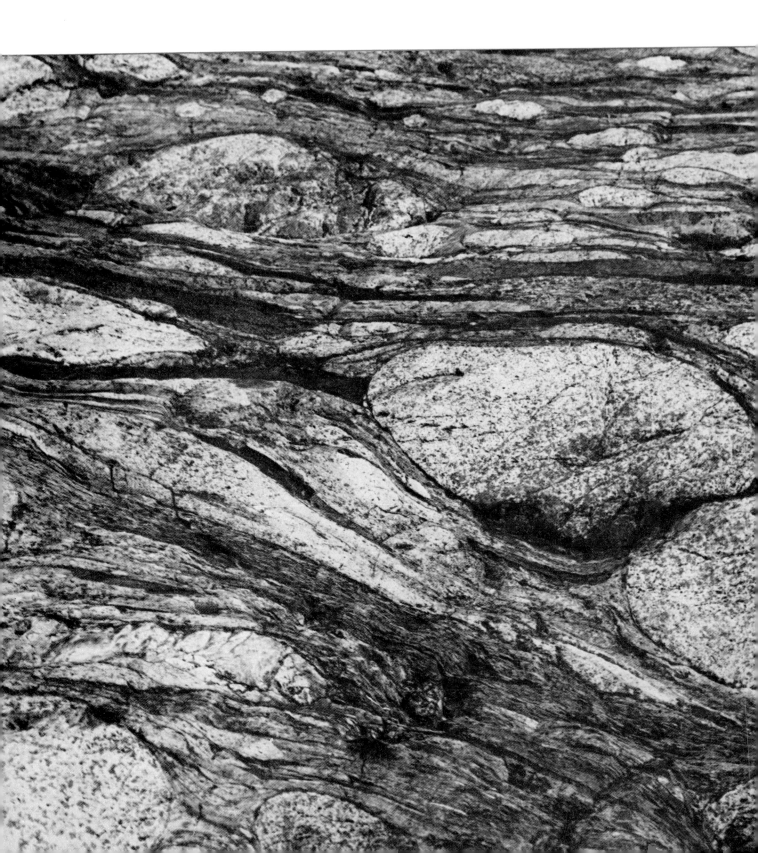

7

METAMORPHIC ROCKS

Most of the rocks exposed in the continental shields and in the cores of mountain belts show evidence that their original texture and composition have been changed. Many have been plastically deformed, as is indicated by contorted parallel bands of minerals, resembling swirling layers in a marble cake. Others have recrystallized and developed larger mineral grains, and the constituent minerals of many rocks have a strong fabric with a specific orientation. The mineral assemblages in these rocks are also distinctive. They are characterized by mineral types that form only under high temperature and pressure.

Geologists interpret these and other lines of evidence to mean that the original rock has been recrystallized in a solid state. The resulting new rock type has a distinctive texture and fabric and, in some cases, a new mineral composition. These rocks are called metamorphic rocks. They are important because of their wide distribution in the continental crust, and also because they provide clear evidence about the dynamics of the earth.

In this chapter, we will consider the nature and distribution of metamorphic rocks and the processes by which they originate.

MAJOR CONCEPTS

1. Metamorphic rocks result from changes in temperature and pressure and from the chemistry of their pore fluids. These changes produce new minerals, new textures, and new structures within the rock body.
2. During metamorphism, new minerals grow in the direction of least stress, producing a planar rock structure called foliation. The three main types of foliation are (a) slaty cleavage, (b) schistosity, and (c) gneissic layering.
3. Rocks with only one mineral (such as limestone) do not develop a strong foliation but instead develop a granular texture with large crystals.
4. The major types of metamorphic rocks are slate, schist, gneiss, quartzite, marble, amphibolite, metaconglomerate, and hornfels.
5. Regional metamorphism develops in the roots of mountain belts along convergent plate boundaries. Contact metamorphism is a local phenomenon associated with thermal and chemical changes near the contacts of igneous intrusions.

Figure 7.1 The characteristics of metamorphic rocks are shown in this photograph of the Canadian Shield, taken from an altitude of 15 km. These rocks have been compressed and deformed to such an extent that many original features have been obliterated. Metamorphism occurred at great depths, and the area was then eroded to expose the complex rock sequence. In addition to the tight folding, the rocks have been broken by fractures and intruded by granitic rocks (light tones).

SUMMARY

Metamorphic rocks result mainly from changes in temperature and pressure and changes in the chemistry of pore fluids within a rock body. These changes develop new minerals, new textures, and new structures. The diagnostic features of the original sedimentary and igneous rocks are greatly modified or completely obliterated. Metamorphic rocks are especially significant because they constitute a large part of the continental crust (the shield) below the veneer of sedimentary rocks (the stable platform) and indicate that the continents have been mobile and dynamic throughout most of geologic time. Therefore, they constitute the best record of ancient plate movements.

During metamorphism, new minerals grow in the direction of least stress, so that rocks are produced with a planar element, called foliation. Three major types of foliation are recognized: (1) slaty cleavage, (2) schistosity, and (3) gneissic layering. Rocks with only one mineral (such as limestone and sandstone) do not develop strong foliation. Instead, they develop a granular texture with large mineral grains.

Metamorphic rocks are classified on the basis of texture and composition. Two major groups are recognized: (1) foliated and (2) nonfoliated. The major types of foliated rocks are slate, schist, and gneiss. The major types of nonfoliated rocks are quartzite, marble, amphibolite, metaconglomerate, and hornfels.

Heat is one of the most important factors in metamorphism. As temperatures increase, the original minerals begin to change from solid to liquid. The pore fluid then recrystallizes to form new minerals. Directed pressures reduce pore spaces and can produce new minerals with closer atomic packing. Changes in chemistry can result from the partial melting of low-temperature minerals and the migration of the fluid, which subsequently recrystallizes.

An increase in heat and pressure can result from deep burial or from igneous intrusions. Most metamorphic rocks of regional extent, however, develop in the deep roots of folded mountain belts as a result of plate collision. The metamorphic mountain roots are welded to the continent and become part of the continental shield.

Key Words

metamorphic rock	schistosity
metamorphism	gneissic layering
stress	slate
plastic deformation	schist
pore fluid	gneiss
recrystallization	quartzite
contact metamorphism	marble
low-grade metamorphism	amphibolite
high-grade metamorphism	metaconglomerate
foliation	hornfels
slaty cleavage	

Review Questions

1. Compare and contrast the characteristics of metamorphic rocks with those of igneous and sedimentary rocks.
2. Make a series of sketches showing the changes in texture that occur with a metamorphism of (a) slate, (b) sandstone, and (c) conglomerate.
3. Define *foliation* and explain the characteristics of (a) slaty cleavage, (b) schistosity, and (c) gneissic layering.
4. Describe the major types of metamorphic rocks.
5. Make a generalized flow chart showing the origin of the common metamorphic rocks.
6. What are the agents of metamorphism?
7. Draw an idealized diagram of converging plates to illustrate the origin of metamorphic rocks.

Additional Readings

Ernst, W. G. 1969. *Earth materials.* Englewood Cliffs, N.J.: Prentice-Hall.

Hyndman, D. W. 1972. *Petrology of igneous and metamorphic rocks.* New York: McGraw-Hill.

Winkler, H. G. F. 1974. *Petrogenesis of metamorphic rocks.* 3d ed. New York: Springer-Verlag.

8

GEOLOGIC TIME

Some sciences deal with incredibly large numbers, others with great distances, still others with infinitesimally small particles. In every field of science, students must expand their conceptions of reality, a sometimes difficult, but very rewarding, adjustment to make.

Geology students must expand their conceptions of the duration of time. Because life is short, we tend to think 20 or 50 years is a long time. A hundred years in most frames of reference is a very long time. Yet, in studying the earth and the processes that operate on it, you must attempt to comprehend time spans of 1 million years, 100 million years, and even several billion years.

How do scientists measure such long periods of time? Nature contains many types of time-measuring devices. The earth itself acts like a clock, rotating on its axis once every 24 hours. Rocks are records of time, and from their interrelationships, the events of the earth's history can be arranged in proper chronologic order. Rocks also contain radioactive clocks. These permit us to measure with remarkable accuracy the number of years that have passed since the minerals forming the rocks crystallized. Fossils embedded in rock constitute a separate organic clock, by which we can identify synchronous events in the earth's history.

Geologists use all of these methods to date geologic events. They also have devised a geologic calendar, which makes it possible to organize and correlate long periods of geologic time into workable units, thus permitting a systematic study of the earth's history.

MAJOR CONCEPTS

1. Geologists' interpretation of events in the earth's history is based on the principle that the laws of nature do not change with time.
2. Relative dating (determining the chronologic order of a sequence of events without knowledge of their actual dates) is achieved by applying the principles of (a) superposition, (b) faunal succession, (c) crosscutting relations, and (d) inclusion.
3. The standard geologic column was established from studies of the rock sequence in Europe. It is now used worldwide. Rocks are correlated from different parts of the world on the basis of the fossils they contain.
4. Absolute time designates a specific duration in units of hours, days, or years. In geology, long periods of time can be measured by radiometric dating.

THE CONSTANCY OF NATURAL LAWS

Statement

The interpretation of rocks as products and records of events in the earth's history is based upon one of the fundamental assumptions of scientific inquiry: the principle of uniformitarianism, which states that *the laws of nature do not change with time.* We assume that the chemical and physical laws operating today have operated throughout all time. The physical attraction between two bodies (gravity) acted in the past as it does today. Oxygen and hydrogen, which today combine (under certain conditions) to form water, did so in the past. Although scientific explanations have improved and changed over the centuries, the natural laws and processes are constant and do not change. All chemical and physical actions and reactions occurring presently are produced by the same causes that produced similar events 100 years or 5 million years ago.

Discussion

The principle of uniformitarianism can perhaps best be understood in its historical context. In the late eighteenth century, before modern geology had developed, the Western world's prevailing view of the earth's origin and history was derived from the biblical account of creation. The earth was believed to have been created in 6 days and to be approximately 6000 years old. Creation in so short a time was thought to have involved forces of tremendous violence, surpassing anything experienced in modern times. This type of creation theory is called **catastrophism.** Foremost among its proponents in the eighteenth century was Baron Georges Cuvier (1769–1832), a noted French naturalist. Cuvier, an able student of fossils, concluded that each fossil species is unique to a given sequence of rocks. He cited this discovery in support of the theory that each fossil species resulted from a special creation and was subsequently destroyed by a catastrophic event.

This theory was generally supported by scholars until 1785, when it was challenged by James Hutton (1726–1797). He saw evidence that the earth evolved by uniform, gradual processes over an immense span of time, and he developed a concept that became known as the principle of **uniformitarianism.** According to this principle, past geologic events can be explained by natural processes, such as erosion by running water, vol-

canism, and gradual uplift of the earth's crust. Hutton assumed that these processes occurred in the distant past just as they occur now. On the basis of his observations of the rocks of Great Britain, he visualized "no vestige of a beginning—no prospect of an end." Sir Charles Lyell (1797–1875) based his *Principles of geology* (1830–1833) on Hutton's uniformitarianism. Lyell's book established uniformitarianism as the accepted method for interpreting the geologic and natural history of the earth. Charles Darwin (1809–1882) accepted Lyell's principles in formulating his theory of the origin of species and the descent of man. Lyell, however, also believed that the *rates* at which processes operate do not change with time. More recent studies indicate that such changes may have occurred.

Modern Views of Uniformitarianism (Naturalism)

With the help of modern scientific instruments, geologists have studied much more of the geologic record than Hutton and Lyell did, and they have observed many subtle details that earlier scientists could not see. Modern science is making significant advances in understanding the earth, its long history, and how it was formed. By applying principles of thermodynamics, electromagnetism, chemistry, and related scientific disciplines, geologists are discovering more clues about the earth's genesis and evolution.

Hutton and Lyell saw no vestige of the earth's beginning, no prospect of its end. They envisioned cycles of erosion and sedimentation repeated throughout an indefinitely remote past and continuing indefinitely into the future. Today, however, we do see a beginning, and we can predict with some confidence an end to the earth's dynamic systems. We envision an evolving planet that was formed about 4.5 billion years ago, whose surface has continually been modified by systems of energy operating within it and upon it. From the beginning, the earth has undergone a continuous dissipation of energy, and the rates at which processes operate may have fluctuated and changed. Moreover, some processes (such as meteorite impact, glaciation, and the action of ocean tides) may have been more active in the past than they are now.

Hence the modern concept of uniformitarianism, embraced by geologists today, simply states that *natural laws do not change with time.* Our ability to express them may change, but the laws themselves do not. This principle is not unique to geology. It is a basic assumption of all scientific enterprise.

RELATIVE DATING

Statement

An expanded conception of time is perhaps the main contribution of geology to the history of thought. Geologists, however, are concerned not so much with the philosophical problem of defining time, or analyzing the concept of time, as with measuring it in relation to events in the history of the earth. Two different concepts of time, and hence two different but complementary methods of dating, are used in geology. One method is relative dating, by which the chronologic order of a sequence of events is established even though the actual dates of the events are unknown. Certain events can be found to have occurred before or after certain other events, without reference to how much time elapsed between them or how long ago they occurred. The other general method is absolute dating, by which the dates of certain events or the duration of certain processes are established through the use of various natural clocks that measure long intervals of time.

Relative dating is important in geology because many physical events, such as volcanism, canyon cutting, deposition of sediment, and upwarping of the crust, can easily be arranged in their proper chronologic order. We can deduce the order of events by applying several principles of remarkable simplicity and universality. The most significant are

1. The principle of superposition
2. The principle of faunal succession
3. The principle of crosscutting relations
4. The principle of inclusion

Discussion

Geologic and recorded history are both organized into units of relative time. We speak of human events as occurring in eras B.C. or A.D., which constitute a broad division of time, and historians place events in relation to ancient dynasties, reigns of kings, or major events such as wars. Geologists do the same thing with geologic time. The earth's history is divided into eras, periods, and ages by markers such as the existence of certain forms of life or major physical events (for example, disturbances in the earth's crust), even though the actual dates of the events may be unknown. **Relative dating** implies no quantitative or absolute length of time; an event can only be inferred to have occurred earlier or later than another.

The Principle of Superposition

The principle of **superposition** is the most basic guide in the relative dating of rock bodies. It states that, in a sequence of undeformed sedimentary rock, the oldest beds are on the bottom and the higher layers are successively younger. Thus, the **relative ages** of rocks in a sequence of sedimentary beds can be determined from the order in which they were deposited.

In applying the principle of superposition, we make two assumptions: (1) layers were essentially horizontal when they were deposited, and (2) the rocks have not been so severely deformed that the beds are overturned. (Rock sequences that have been overturned are generally easy to recognize by their sedimentary structures, such as crossbedding, ripple marks, and mud cracks.)

The Principle of Faunal Succession

In addition to superposition, the sequence of sedimentary rocks in the earth's crust contains another independent element that can be used to establish the chronologic order of events: the upward succession of fossil assemblages contained in the rocks. Fossils are the actual remains of ancient organisms, such as bones and shells, or evidence of their presence, such as trails and tracks. Their abundance and diversity are truly amazing. Some rocks (such as coal, chalk, and certain limestones) are composed almost entirely of fossils, and others contain literally millions of specimens. Invertebrate marine forms are most common, but even large vertebrate fossils of mammals and reptiles are plentiful in many formations. For example, it is estimated that over 50,000 fossil mammoths have been discovered in Siberia, and many more remain buried.

The principle of **faunal succession** was recognized some 150 years ago by William Smith (1769–1839), a British surveyor, even before Darwin developed the theory of organic evolution. Smith worked throughout much of southern England and carefully studied the fresh exposures of rocks in quarries, road cuts, and excavations. In a succession of interbedded sandstone and shale formations, the several shales were very much alike, but the fossils they contained were not. Each shale had its own particular groups of fossils. By correlating types of fossils with rock sequences, Smith developed a practical tool that enabled him to predict the location and properties of rocks beneath the surface.

Soon after Smith announced that the fossil assemblages of England change systematically from the older beds to the younger, other investigators discovered the same thing to be true throughout the world—even in countries separated by oceans.

Fossils provide a means of establishing relative dates in much the same way that artifacts do. Both show evolution and change with time. For example, in a city dump where refuse is buried in succession, we could recognize a period of time prior to the automobile by the remains of wagon wheels, saddles, and similar equipment. A layer containing abundant scraps of Model T Fords would be recognized as being older than one containing remains of the Model A, and layers containing new models would be recognized as being younger, even though they might not rest upon layers containing any of the older materials.

Today, the principle of faunal succession has been confirmed beyond doubt. It has been used extensively to locate valuable natural resources, such as petroleum and mineral deposits, and is the foundation for the standard geologic column, which divides geologic time into eras, periods, and ages (see *figure 8.2*).

The Principle of Crosscutting Relations

The relative age of certain events is also shown by **crosscutting relations.** Faults (surfaces along which a rock body has been fractured and displaced) and igneous intrusions are obviously younger than the rocks they cut. Crosscutting rela-

tions can be complex, however, and careful observation may be required to establish the correct sequence of events in complex areas. The scale of crosscutting features is highly variable, ranging from large faults with displacements of hundreds of kilometers to small fractures less than a millimeter long. The sequence of events in *figure 8.1* can be worked out by applying the principles of superposition and crosscutting relations.

The Principle of Inclusion

The relative age of intrusive igneous rocks (with respect to the surrounding country rock) is commonly apparent from inclusions, or fragments of older rocks in the younger. As a magma moves upward through the crust, it dislodges and engulfs large fragments of the surrounding material, which remain as unmelted foreign inclusions.

The principle of inclusion can also be applied to conglomerates in which relatively large pebbles and boulders eroded from preexisting rocks have been transported and deposited in a new formation. The conglomerate is obviously younger than the formations from which the pebbles and cobbles were derived. In areas where superposition or other methods do not indicate relative ages, a limit to the age of a conglomerate can be determined from the rock formation represented in its pebbles and cobbles.

Figure 8.1 The sequence of geologic events can be determined by means of the principles of superposition and crosscutting relations. Each rock body and major geologic feature in the block diagram is marked with conventional symbols and is labeled with a letter. Determine the sequence of geologic events shown in the diagram.

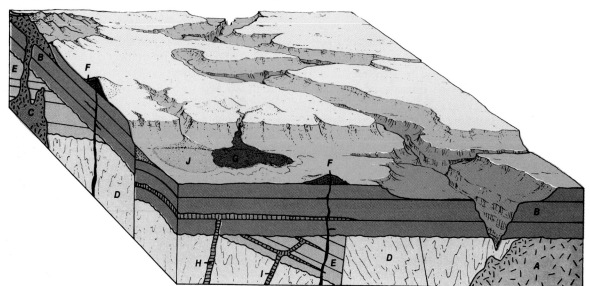

RADIOMETRIC MEASUREMENTS OF ABSOLUTE TIME

Statement

Unlike relative time, which specifies only the chronologic relationships among events, absolute time, or finite time, designates specific durations measured in units of hours, days, or years. Time can be measured by any regularly recurring event, such as the swing of a pendulum or the rotation of the earth. One very useful natural clock measures time by the process of radioactive decay. In this process, atoms of certain elements lose particles from their nuclei and thereby become atoms of other elements. Because the rate at which they decay is unaffected by conditions such as pressure, temperature, and chemical binding forces, it can be used as a very precise and accurate measure of geologic time. The time that has elapsed since a radioactive element was locked into the crystal of a mineral can be determined if the rate of decay is known and the proportions of the original element and the decay product can be measured. The most important radioactive clocks for geologic studies are uranium, thorium, rubidium, and potassium.

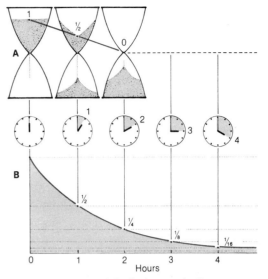

Figure 8.3 Rates of depletion can be linear or exponential. **(A)** Most processes are characterized by uniform, straight-line depletion, like sand moving through an hourglass. If half of the sand is gone in 1 hour, all of it will be gone in 2 hours. **(B)** Radioactive decay, in contrast, is exponential. If half is depleted in 1 hour, half of the remainder, or one-fourth, will be depleted in 2 hours. Rates of radioactive decay are expressed in half-lives, the time required for half of the remaining amount to be depleted. In this case, the half-life is 1 hour.

Discussion

When Henri Becquerel (1852–1908), a French physicist, discovered natural **radioactivity** in 1896, he opened whole new vistas in every field of science. Among the first to experiment with radioactive substances was the distinguished British physicist Lord Rutherford (1871–1937). After defining the structure of the atom, Rutherford made the first clear suggestion that radioactive decay could be used to date geologic events in **absolute time.**

Radioactive atoms are unstable: their nuclei can spontaneously disintegrate, transforming them into completely different atoms. In the process, radiation is given off and heat is liberated. Initially, scientists assumed that each radioactive substance disintegrates at its own rate, and that for many of them the rate is extremely slow. This assumption has been proved by experiment.

The rate of radioactive decay is defined in terms of **half-life,** the time it takes for half of the nuclei in the sample to decay. In one half-life, half of the original atoms decay. In a second half-life, half of the remainder (or a quarter of the original atoms) decay. In a third half-life, half of the remaining

quarter decay, and so on (see *figure 8.3*). The time elapsed since the formation of a crystal containing a radioactive element can be calculated from the rate at which that particular element decays. The amount of the radioactive element remaining in the crystal (parent isotope) is simply compared to the amount of the disintegration product (**daughter isotope**).

There are numerous radioactive isotopes, but most decay rapidly (that is, they have short half-lives) and lose their radioactivity within a few days or years. Some decay very slowly, however, with half-lives of hundreds of millions of years. These can be used as atomic clocks for measuring very long periods of time. The parent isotopes that are most useful for geologic dating and their daughter products are listed in *table 8.1*.

The theory of **radiometric dating** is simple enough, but the laboratory procedures are complex. The principal difficulty lies in the precise measurement of minute amounts of isotopes. The accuracy of the method also depends on the accuracy with which the half-life is known. (Measurements of the decay of uranium 235 to lead 207 are considered accurate within 2%.)

At present, the potassium-argon method is of

Table 8.1 Radioactive Isotopes Used in Determining Geologic Time

Parent Isotope	Daughter Isotope	Half-Life
Uranium 238	Lead 206	4.5 billion years
Uranium 235	Lead 207	713.0 million years
Thorium 232	Lead 208	13.9 billion years
Rubidium 87	Strontium 87	50.0 billion years
Potassium 40	Argon 40	1.5 billion years

great importance. It can be used with the micas and amphiboles, which are widely distributed in igneous rocks. It also can be used on rocks as young as a few thousand years and on older rocks.

Another important radioactive clock uses the decay of **carbon 14** (C^{14}), or **radiocarbon**, which has a half-life of 5730 years. C^{14} is produced continually in the earth's atmosphere as a result of the bombardment of nitrogen 14 (N^{14}) by cosmic rays. This newly formed radioactive carbon becomes mixed with ordinary carbon atoms in carbon dioxide gas (CO_2). The formation of radiocarbon balances its decay, and the proportion of C^{14} in atmospheric CO_2 is essentially stable.

Plants absorb carbon dioxide and animals eat plants. Thus, both maintain a fixed proportion of C^{14} while they are alive. After death, however, no additional C^{14} can replenish what is lost by radioactive decay. The C^{14} steadily reverts to N^{14}. The time elapsed since an organism died can therefore be determined by measuring the remaining proportion of C^{14}. The longer the time elapsed since death, the less C^{14} remains. Since its half-life is 5730 years, the amount of C^{14} remaining in organic matter older than 50,000 years is too small to be measured accurately. Therefore, the radiocarbon method is useful for dating very young geologic events plus most archaeologic material.

To safeguard against errors, radiometric dating is subjected to constant checks. A number of radioactive isotopes are suitable for absolute dating, so that an obvious test is to date a mineral or rock by more than one method. If the results agree, the probability is high that the date is reliable. If the results differ significantly, additional methods must be used to determine which, if any, are correct. Another independent check can be made by comparing the absolute date, determined radiometrically, with the relative age of the rocks, determined from such evidence as superposition and fossils. Through this system of tests and cross-checks, many very reliable radiometric dates have been determined.

THE RADIOMETRIC TIME SCALE

Statement

Absolute dates of numerous geologic events have been determined. These, in combination with the standard geologic column, provide a radiometric time scale from which the absolute age of a geologic event can be estimated. Unfortunately, radiometric dates cannot be determined for every layer of sedimentary rock, because sediments are composed of eroded debris of preexisting rocks from various sources. We can determine the radiometric dates of minerals in sedimentary rocks, but the dates indicate when the minerals were formed, not when the beds of sediment were deposited. Radioactive isotopes can be used to determine when igneous rocks crystallized and when heat and pressure developed new minerals in metamorphic rocks, because in these cases, the mineral and the rock formed together. *The problem in developing a reliable radiometric time scale is accurately placing the radiometric dates of igneous and metamorphic rocks in their proper positions in th relative time scale established by sedimentary ro* Layers of volcanic rocks and bracketed intrus are the most suitable time markers in the sta geologic column. These benchmarks, acc ly placed in the geologic column, constitute ' asis for the radiometric time scale.

Discussion

Layered Volcanics

The best reference points fc radiometric time scale are probably volcan alls and lava flows. They are deposited inst ously, as far as geologic time is concerned. B e they commonly are interbedded with f erous sediments, their exact position in the g gic column can be determined. They provic e main basis for establishing the absolute e scale within the geologic column.

Bracketed Intrusion

As is shown in *figure* , a molten rock can cool within the earth's crust thout ever breaking out to the surface. Subsequ t erosion may expose this rock. Later, younger s iments may be deposited on top. In some cases, e entire sequence of events takes only a few milli years. In others, it may require a much longer t e. The relative age of the igneous rock falls in t *bracket* between the older

SUMMARY

The basic assumption endorsed by essentially all geologists today in studying and interpreting the history of the earth is the principle of uniformitarianism, which states that natural laws do not change with time. This principle is not unique to geology. It is a fundamental law in all fields of science.

Relative dating determines the chronologic order of a sequence of events. The most important methods of relative dating are (1) superposition, (2) faunal succession, (3) crosscutting relations, and (4) inclusions.

The standard geologic column was established during the early 1800s by means of the principles of relative dating.

Radiometric dating provides a method for measuring geologic time directly in terms of a specific number of years (absolute dating). It has been used extensively during the last 50 years to provide an absolute time scale for the events in earth history.

In the next section, we will discuss the details of the hydrologic system and the processes by which the surface features of the earth are formed, modified, and changed with time.

Key Words

uniformitarianism	Cenozoic
catastrophism	absolute time
relative age	radiometric dating
relative dating	radioactivity
superposition	half-life
faunal succession	daughter isotope
crosscutting relations	carbon 14
geologic column	radiocarbon
Precambrian	geologic time scale
Paleozoic	bracketed intrusion
Mesozoic	

Review Questions

1. Explain the modern concept of uniformitarianism.
2. Explain the concept of relative dating.
3. Explain how the following principles are used in determining the relative age of rock bodies: (a) superposition, (b) faunal succession, (c) crosscutting relations, (d) inclusions.
4. Determine the sequence of events illustrated in *figure 8.1.*
5. What is the standard geologic column? How did it originate?
6. Explain the meaning of half-life in radioactive decay.
7. How is the absolute age of a rock determined?
8. How is the half-life of a radioactive isotope used to determine the radiometric age of a rock?
9. Why are most rocks dated with respect to their position in the standard geologic column rather than assigned a definite numerical age, even though accurate methods of radiometric dating are well established?

Additional Readings

Berry, W. B. N. 1968. *Growth of a prehistoric time scale.* San Francisco: W. H. Freeman.

Deevey, E. S., Jr. 1952. Radiocarbon dating. *Scientific American* 186(2):24–28.

Eicher, D. L. 1968. *Geologic time.* Englewood Cliffs, N.J.: Prentice-Hall.

Faul, H. 1966. *Ages of rocks, planets and stars.* New York: McGraw-Hill.

Harbaugh, J. W. 1968. *Stratigraphy and geologic time.* Dubuque, Iowa: William C. Brown.

Hurley, P. M. 1959. *How old is the earth?* Garden City, N.Y.: Doubleday, Anchor Books.

Toulmin, S., and J. Goodfield. 1965. *The discovery of time.* New York: Harper and Row.

9

WEATHERING

A new building gradually deteriorates. The paint chips and peels and is gone in a matter of years. Wood dries and splits, and even bricks, building stone, and cement eventually decay and crumble. Left by themselves, most buildings decompose into a pile of rubble within a few hundred years. This process of natural decay is called weathering. Solid bedrock is also subject to weathering and eventually decomposes into piles of rubble. In fact, some rocks are even less resistant to weathering processes than the paint on a house is.

Weathering is a general term describing all of the changes that result from the exposure of rock materials to the atmosphere. Its effects can be seen wherever rocks are exposed (*figure 9.1*). Weathering occurs because most rocks are in equilibrium at the high temperatures and pressures deep within the earth. If they are exposed to the much lower temperatures and pressures at the surface, to the gases in the atmosphere, and to the elements in water, they become unstable and undergo various chemical changes and mechanical stresses. As a result, the solid bedrock breaks down into loose, decomposed products.

From a geological point of view, the importance of weathering is that it transforms the solid bedrock into small, decomposed fragments and prepares it for removal by the agents of erosion. It is this weathered rock material that is washed away by river systems in the process of eroding the surface of the land. Without weathering, there would be no erosion as it is known today, and the landscape would be strikingly different. In addition, the products of weathering form a blanket of soil over the solid bedrock, and soil is the basis for most terrestrial life. Therefore, weathering should be considered a part of the geologic system with tremendous ecological significance.

MAJOR CONCEPTS

1. The major types of weathering are mechanical disintegration and chemical decomposition.
2. Ice wedging is the most important form of mechanical weathering.
3. The major types of chemical weathering are oxidation, dissolution, and hydrolysis.
4. Joints facilitate weathering, because they permit water and the atmosphere to attack a rock body at considerable depth. They also greatly increase the surface area on which chemical reactions can occur.
5. The major products of weathering are a blanket of soil (regolith) and spheroidal rock forms.
6. Climate greatly influences the type and rate of weathering. The major controlling factors are precipitation and temperature, with their seasonal variations.

(A) Enlargement of fractures

(B) Development of columns and pillars

(C) Shattering

(D) Exfoliation

(E) Piles of rock debris

(F) Granular disintegration

(G) Block separation

Figure 9.1 The effects of weathering are seen wherever rocks are exposed. These photographs show typical examples.

MECHANICAL WEATHERING

Statement

Mechanical weathering is the breakdown of rock into smaller fragments by various physical stresses. It is strictly a physical process, involving no change in chemical composition. No chemical elements are added to or subtracted from the rock. Mechanical weathering simply breaks down the rock into small fragments. The most important types of mechanical weathering are

1. Ice wedging, in which freezing water expands in cracks or bedding planes and wedges the rock apart
2. Sheeting, or unloading, in which a series of fractures is produced by expansion in the rock body itself, as a result of the removal of overlying material by erosion

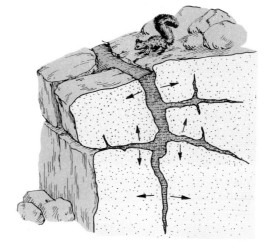

Figure 9.2 Ice wedging occurs where water seeps into fractures and expands as it freezes. The expansion wedges the rock apart and produces loose, angular fragments.

Discussion

Ice Wedging

Figure 9.2 is a simple diagram showing how ice wedging breaks a rock mass into small fragments. Water from rain or melting snow easily penetrates cracks, bedding planes, and other openings in the rock. As it freezes, it expands about 9%, exerting great pressure (similar to that produced by driving a wedge into a crack) upon the rock walls. Eventually, the fractured blocks and bedding layers are pried free from the parent material. The stress generated by each freeze is approximately 110 kg/cm², roughly equivalent to that produced by dropping a shot-putter's shot from a height of 3 m. Stress is exerted with each freeze, so that over a period of time, the rock is literally hammered apart.

Necessary conditions for effective ice wedging include (1) an adequate supply of moisture; (2) fractures, cracks, or other voids within the rock, into which water can enter; and (3) temperatures that rise and fall across the freezing point. Temperature

Figure 9.3 Talus cones are piles of rock debris that accumulate at the base of a cliff as the result of rockfalls. Most rock fragments in talus cones are produced by frost action.

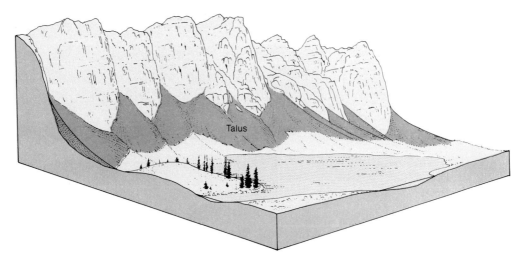

Talus

is especially important, because pressure is applied with each freeze. In areas where freezing and thawing occur many times a year, frost action is far more effective than it is in exceptionally cold areas, where water is permanently frozen. Thus, ice wedging occurs most frequently above the timberline. It is especially active on the steep slopes above valley glaciers, where meltwater produced during the warm summer days seeps into cracks and joints and freezes during the night.

The products of ice wedging commonly accumulate at the bases of cliffs in piles of angular rock fragments, called **talus.** Since most cliffs are notched by steep valleys and narrow ravines, the fragments dislodged from the high valley walls are funneled through the ravines to the base of the cliff, where they accumulate in cone-shaped deposits (*figure 9.3*).

Sheeting

Rocks formed deep within the earth's crust are under high confining pressure from the weight of thousands of meters of overlying rocks. As this overlying cover is removed by erosion, the confining pressure is released, and the buried rock body tends to expand. The internal stresses set up by expansion can cause large fractures, or expansion joints, parallel to the earth's surface (*figure 9.4*). This process is called **sheeting.** It can be directly observed in quarries, where the removal of large blocks is sometimes followed by the rapid, almost explosive expansion of the quarry floor. A sheet of rock several centimeters thick bursts up, and at the same time, numerous new parallel fractures appear deeper in the rock body. The same process occasionally causes rock bursts in mines and tunnels, when the confining pressure is released during the tunneling operation. It can also be seen in many valley walls and in road cuts, where rock slumping due to sheeting can cause serious highway problems.

Other Types of Mechanical Weathering

Animals and plants play a variety of relatively minor roles in mechanical weathering. Burrowing animals, such as rodents, mechanically mix the soil and loose rock particles, a process that facilitates further breakdown by chemical means. Pressure from growing roots widens cracks and contributes to the rock breakdown. Lichens can live on the surface of bare rock and extract nutrients from its minerals by ion exchange. This results in both mechanical and chemical alteration of the minerals. These processes may seem trivial, but the work of innumerable plants and animals for a long period of time can add significantly to the **disintegration** of the rock.

Thermal expansion and contraction caused by daily or seasonal temperature changes were once thought to be an effective process of mechanical weathering. The idea is plausible, but experiments show that stresses developed by alternate heating and cooling over long periods of time are insignificant in comparison with the elastic strength of rock. Even on the moon, where daily temperature changes are much greater than those on the earth, the effects of thermal expansion on rock fragmentation are uncertain.

Figure 9.4 Sheeting in granite of the Sierra Nevada occurs as erosion removes the overlying rock cover and releases the confining pressure. The bedrock expands and large fractures develop parallel to the surface. The fractures can subsequently be enlarged by frost action.

CHEMICAL WEATHERING

Statement

Chemical weathering, or chemical decomposition, involves several important reactions between the elements in the atmosphere and those in the minerals of the earth's crust. In these processes, the internal structures of the original minerals are destroyed, and new minerals are created, with new crystal structures, which are stable under atmospheric conditions. Such reactions cause a significant change in the chemical composition and physical appearance of the rocks.

Water is of prime importance in chemical weathering. It takes part directly in the chemical reactions. It carries elements of the atmosphere into contact with the minerals of the rocks. It removes the products of weathering, to expose fresh rock. The rate of chemical weathering therefore is greatly influenced by the amount of precipitation.

No area of the earth's surface is continually dry. Even in the most arid deserts, some rain falls and causes decomposition. Therefore, chemical weathering is essentially a worldwide process, which is least effective in deserts and in climates where water is frozen the entire year.

The chemical reactions involved in the decomposition of rock are complex, but three main groups are recognized:
1. Hydrolysis
2. Dissolution
3. Oxidation

Discussion

Hydrolysis

The chemical union of water and a mineral is known as **hydrolysis.** The process involves not merely absorption of water, as in a sponge, but a specific chemical change in which a new mineral is produced from the original material. Ions derived from the mineral react with the H^+ or OH^- ions of the water. Hydrolysis is thus an important process in **chemical weathering.**

A good example of hydrolysis is the chemical weathering of K-feldspar. Two substances are essential: carbon dioxide and water. The atmosphere and soil contain carbon dioxide, which unite with rainwater to form carbonic acid. If K-feldspar comes in contact with carbonic acid, a chemical reaction occurs, which can be described in the equation at the bottom of this page. The hydrogen ion of the H_2CO_3 displaces the potassium ion of the feldspar and thus disrupts the crystal structure. It then combines with the aluminum silicate of the feldspar to form a clay mineral. The potassium combines with the carbonate ion to form potassium carbonate, a soluble salt. Silica is also released but may remain in solution. The new clay mineral does not contain potassium, which was present in the orignal feldspar. It also has a new crystal structure, consisting of sheets of silica tetrahedra that form submicroscopic crystals. Hydrolysis is an extremely important weathering process because it acts on the feldspars and ferromagnesian minerals, the dominant minerals in most rocks.

Dissolution

Water is one of the most effective and universal solvents known, and practically all minerals are soluble in water to some extent. This is because of the polarity of the water molecule. The structure of the water molecule requires the two hydrogen atoms to be positioned on the same side of the larger oxygen atom. Thus, the molecule has a concentration of positive charge on the side with the two hydrogen atoms, balanced by a negative charge on the opposite side. As a result, the water molecule is polar and behaves like a tiny magnet. It acts to loosen the bonds of the ions at the surface of minerals with which it comes in contact. **Dissolution** is thus another important process in chemical weathering.

Some rock types can be completely dissolved and **leached** away by water. Rock salt is perhaps the best-known example. It is extremely soluble, surviving at the earth's surface only in the most arid regions. Gypsum is less soluble but is easily dissolved by surface water. There are no large out-

$$2KAlSi_3O_8 \quad + \quad H_2CO_3 \quad + \quad H_2O \quad \longrightarrow \quad K_2CO_3 \quad + \quad Al_2Si_2O_5(OH)_4 \quad + \quad 4SiO_2$$

| (K-feldspar) | (carbonic acid) | (water) | (potassium carbonate— readily soluble) | (a clay mineral) | (soluble hydrated silica or finely divided quartz) |

crops of these rocks in humid regions. Limestone is also soluble in water, especially if the water contains carbon dioxide. Where water is abundant, limestone commonly weathers into valleys, but it forms cliffs in arid regions.

Chemical analysis of river water illustrates the effectiveness of dissolution in the weathering of rocks. Fresh rainwater has relatively little dissolved mineral matter, but running water soon dissolves the more soluble minerals in the rock and transports them in solution. Each year the rivers of the world carry about 3.9 million metric tons of dissolved minerals to the ocean, so that it is not surprising that seawater contains 3.5% (by weight) dissolved salts, all of which were dissolved from the continents by pure rainwater.

Oxidation

Oxidation is the combination of atmospheric oxygen with a mineral to produce an oxide. It is especially important in minerals that have a high iron content, such as olivine, pyroxene, and amphibole. The iron in silicate minerals unites with oxygen to form hematite (Fe_2O_3) or limonite [$FeO(OH)$]. Hematite is deep red, and if it is dispersed in sandstone or shale, it imparts a red color to the entire rock.

The various processes of chemical weathering almost invariably work in combination. They are not isolated, separate processes, but all attack the rock at the same time.

THE IMPORTANCE OF JOINTS IN WEATHERING

Statement

Almost all rocks are broken by systems of fractures called joints. These result from the strain established by earth movements, the release of confining pressure, and the contraction produced by the cooling of lava. Joints greatly influence the weathering of rock bodies in two ways:
1. They effectively cut large blocks of rock into smaller ones and thereby greatly increase the surface area, where chemical reactions take place.
2. They act as channelways through which water can penetrate to break down the rock by frost action.

Discussion

The importance of **joints** in weathering processes can be appreciated by considering the amount of new surface area produced by jointing. For example, consider a cube of rock that measures 10 m on each side (as shown in *figure 9.5*). If only the upper surface of the cube were exposed and the rock were not jointed, weathering could attack only the exposed surface of 100 m². If the block were bounded by intersecting joints 10 m apart, however, the surface area exposed to weathering

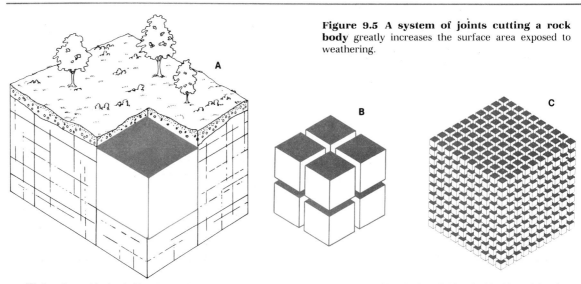

Figure 9.5 A system of joints cutting a rock body greatly increases the surface area exposed to weathering.

(A) A surface of bedrock, 10 m long and 10 m wide, with no joints, exposes a total area of 100 m² to weathering processes. If a set of joints divides the rock into a 10-m cube, the surface area exposed to weathering is increased to 600 m².

(B) Three additional joints, dividing the block into eight cubes, would increase the surface area to 1200 m².
(C) If joints 1 m apart cut the rock, the surface area exposed to weathering would be increased to 6000 m².

Figure 9.6 Joint systems cutting a sandstone formation near Arches National Park, Utah, have been enlarged by weathering into deep, narrow crevasses. Continued weathering will further enlarge the joints. Ultimately, the rock between joint planes will form a long, narrow ridge.

processes would be 600 m². If three additional joints cut the cube into eight smaller cubes, the surface area exposed to weathering would be 1200 m². If joints 1 m apart cut the rock, 6000 m² of rock surface would be exposed to weathering. Obviously, a highly jointed rock body weathers much more rapidly than a solid one.

Besides providing a much larger surface area for chemical decomposition, joints act as a system of channels through which water can penetrate the surface. They thus permit mechanical and chemical weathering processes to attack the rock from several sides, even hundreds of meters below the surface.

Striking examples of the influence of jointing on weathering are found in the Colorado Plateau. The thick sandstone formations there have developed prominent joint systems because the rock is so brittle. Weathering proceeds along each joint surface and cuts the rock into large slabs (*figure 9.6*). In other formations, intersecting joints divide the rock into large columns and play an important role in the sculpture of columns and pillars (*figure 9.7*). Similarly, jointing plays a dominant role in the weathering of large masses of granite, as it provides the initial shape for further rock breakup.

Figure 9.7 Intersecting joints can divide massive rock into rough columns.

THE PRODUCTS OF WEATHERING

Statement

The results of weathering can be seen over the entire surface of the earth, from the driest deserts and the frozen wastelands to the warm, humid tropics. The major products include

1. A blanket of loose, decayed rock debris, known as regolith, which forms a discontinuous cover over the solid bedrock
2. Rock bodies modified into spherical shapes by spheroidal weathering
3. Soluble products carried away by streams and ground water

Discussion

Regolith

The **regolith** is important to human ecology because its upper part—the soil—is essential for plant life, the critical link in the food chain for humans. The thickness of the regolith ranges from a few centimeters to many meters, depending on the climate, the type of rock, and the length of time that uninterrupted weathering has proceeded. The transition from **bedrock** to regolith can be seen in road cuts and steep stream valleys. As is shown in *figure 9.8*, the solid rock is usually divided by numerous joints, which separate it into angular blocks. Above the jointed bedrock is a zone with smaller, highly decomposed blocks. These grade upward into sand-size fragments of rocks and clay and mineral grains. The upper layer commonly is fine, decayed mineral matter with a rich admixture of decaying vegetation.

A more regional view of regolith and its relationship to bedrock is shown in *figure 9.9*. It is clear from the photograph that exposures of bedrock are limited to certain areas of resistant limestone and sandstone strata, which form discontinuous cliffs along the upper part of the mountain front. On the steep canyon walls, little soil is retained, and bedrock is exposed from the base to the top of the canyon. The sketch in diagram *B* was made from the photograph and outlines the surface covered with regolith. In diagram *C*, the outcropping bedrock is not shown, so that the regolith appears as a thin, discontinuous blanket with holes where bedrock is exposed. Sediment fills the valley in the foreground, but the regolith there is not shown in the diagram. If you carefully study the areas of exposed bedrock, you can see that the strata are

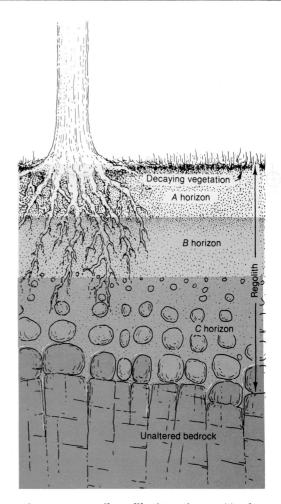

Figure 9.8 A soil profile shows the transition from bedrock to regolith through a sequence of layers, or horizons, consisting of successively smaller fragments of rock.

warped into broad folds, shown in diagram *D*, which form the structure of the mountains.

Soil

Soil is so widely distributed and so economically important that it has acquired a variety of definitions. From a geological point of view, it is the upper part of the regolith. It is composed chiefly of small particles of rocks and minerals, plus varying amounts of decomposed organic matter. Between the surface and the unbroken bedrock, mature **soil profiles** show a rather consistent sequence of zones, called **horizons,** which are distinguished by composition, color, and texture. These are shown in *figure 9.8*. The *A* horizon is the topsoil layer, which often is visibly divided into three layers: A_0 is a thin surface layer of leaf mold, especially obvious

(A) The Wasatch Range, in central Utah, displays contrasting areas of bedrock and regolith.

(B) Outcrops of bedrock appear in cliffs and canyons; slopes are covered with regolith.

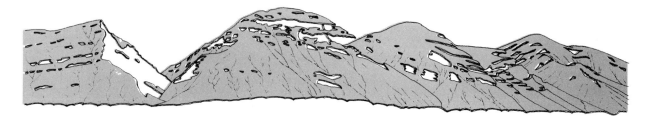

(C) The discontinuous blanket of regolith almost completely covers some formations, while others are exposed as discontinuous cliffs. Outcrops of bedrock form holes in the regolith cover.

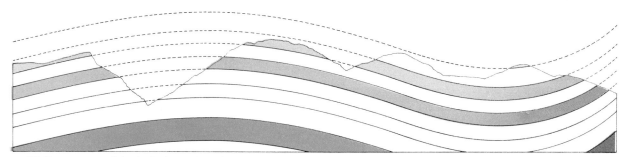

(D) The structure of the bedrock consists of rock layers warped into broad folds, some of which are cut by canyons. Compare with **A**.

Figure 9.9 The relationship between bedrock and regolith is depicted in the photograph and diagrams.

on forest floors; A_1 is a humus-rich, dark layer; and A_2 is a light, bleached layer. The **B horizon** is the subsoil, containing fine clays and colloids washed down from the topsoil. It is largely a zone of accumulation and commonly is reddish in color. The **C horizon** is a zone of partly disintegrated and decomposed bedrock. The individual rock fragments are weathered spheroidal boulders, which may be completely decomposed. The C horizon grades downward into fractured bedrock.

The type of soil depends on a number of factors, the most important of which are climate, parent rock material, and topography. Climate is of major importance, because rainfall, temperature, and seasonal changes all directly affect the development of soil. For example, in deserts, arctic regions, and high mountainous regions, mechanical weathering dominates and organic matter is minimal. The resulting soil is thin and consists largely of broken fragments of bedrock (*figure 9.10*). In equatorial regions, where rainfall is heavy and temperatures are high, chemical processes dominate and thick soils develop rapidly. As a consequence, soil profiles 60 m thick are common in

Figure 9.10 The type of soil found in an area depends on a number of factors, such as climate, parent rock, topography, and time.

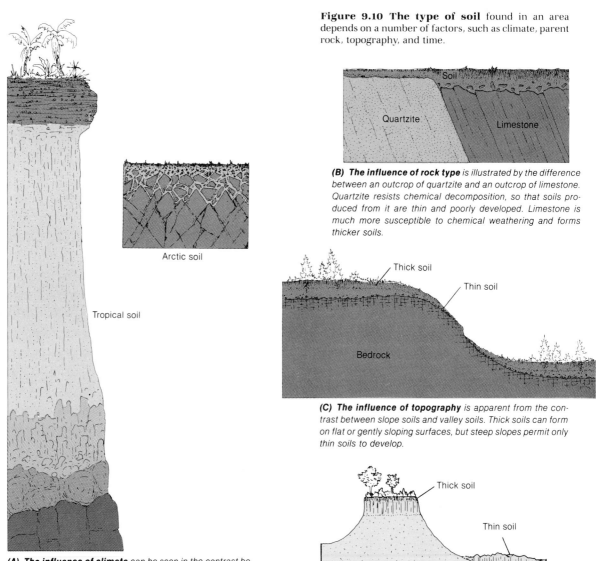

(A) The influence of climate can be seen in the contrast between soils in arctic and desert regions and soils in the tropics. Thin soils, characterized by partly decomposed rock fragments, develop in arctic and desert regions. Thick soils form in tropical regions as a result of extensive chemical decomposition.

(B) The influence of rock type is illustrated by the difference between an outcrop of quartzite and an outcrop of limestone. Quartzite resists chemical decomposition, so that soils produced from it are thin and poorly developed. Limestone is much more susceptible to chemical weathering and forms thicker soils.

(C) The influence of topography is apparent from the contrast between slope soils and valley soils. Thick soils can form on flat or gently sloping surfaces, but steep slopes permit only thin soils to develop.

(D) The influence of time can be seen in areas of volcanism. Thick soils have developed on old lava flows, in contrast to thin soils on younger flows.

the tropics and subtropics. In some areas (such as central Brazil), the zone of decayed rock is more than 150 m thick.

The mineral composition of bedrock exerts a strong influence on the type of soil, because it provides the chemical elements and mineral grains from which the soil develops. Pure quartzite, for example, containing 99% SiO_2, produces a thin, infertile soil (*figure 9.10*).

Topography affects soil development, because it influences the amount and rate of erosion and the nature of drainage. Flat, poorly drained lowlands develop a bog-type soil, whereas steep slopes permit rapid removal of regolith and inhibit the accumulation of weathered materials. Well-drained uplands are conducive to thick, well-developed soils.

Spheroidal Weathering

In the weathering process, there is an almost univeral tendency for rounded (or spherical) surfaces to form on decaying rock. A rounded shape is produced because weathering attacks an exposed rock from all sides at once, and therefore the depth of decomposition is much greater along the corners and edges of the rock (*figure 9.11*). As the decomposed material falls off, the corners become rounded, and the block eventually is reduced to an ellipsoid or a sphere. The sphere is the geometric form that has the least amount of surface area per unit of volume; once the block attains this shape, it simply becomes smaller. This process is known as **spheroidal weathering.**

Although examples of spheroidal weathering can be seen in almost any exposure of rock, it is perhaps best appreciated in the rounded blocks of ancient buildings and monuments, such as the Parthenon, in Greece (*figure 9.12*). The original blocks had sharp corners and were fitted together with precision, as can be seen in the unweathered blocks in the restored section. The edges of the original blocks, however, have completely decomposed and have assumed ellipsoidal or spherical shapes. In nature, spheroidal weathering is produced both at the surface and at some depth.

Exfoliation is a special type of spheroidal weathering in which the rock breaks apart by separation along a series of concentric shells or layers that look like cabbage leaves (*figure 9.13*). The layers, essentially parallel to each other and to the surface, develop by both chemical and mechanical means. Sheeting can play an important part, because rocks like granite, if they are deeply buried, have a tendency to expand upward and outward as the overlying rock is removed. In cold climates, frost wedging along the sheeting joints helps to remove successive layers gradually. The increase in volume associated with the decomposition of feldspar might also promote exfoliation. Exfoliation causes massive rocks like granite to develop a spherical form characterized by a series of concentric layers in a wide range of scales.

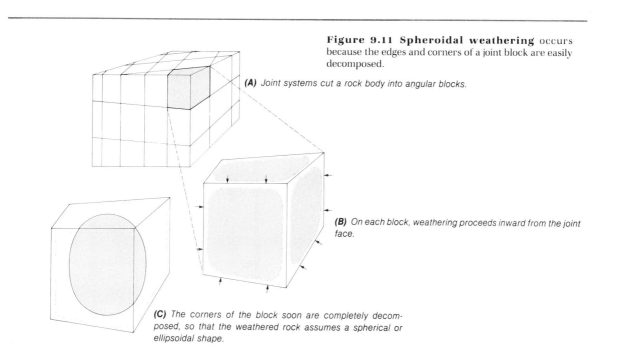

Figure 9.11 Spheroidal weathering occurs because the edges and corners of a joint block are easily decomposed.

(A) Joint systems cut a rock body into angular blocks.

(B) On each block, weathering proceeds inward from the joint face.

(C) The corners of the block soon are completely decomposed, so that the weathered rock assumes a spherical or ellipsoidal shape.

Figure 9.12 Spheroidal weathering is clearly displayed in the building blocks of the Parthenon, in Athens, Greece. The original blocks were rectangular and fit closely together, like the restored section shown on the left. Weathering, proceeding inward from the sides of each block, has produced spheroidal forms.

Figure 9.13 Exfoliation domes are well developed in the Sierra Nevada, in California. These mountains are composed of massive granite cut by joints that separate the rock into large blocks. Exfoliation domes developed as huge slabs of rock spalled off. The joints separating the rock slabs probably resulted from expansion as erosion removed the overlying rock.

CLIMATE AND WEATHERING

Statement

Climate is the single most important factor influencing weathering. More than anything else, it determines the type and rate of weathering in a given region. The products of weathering, therefore, usually directly reflect the climate. Water is the most important agent in most types of weathering. The total amount of precipitation is significant, but the intensity of rain, seasonal variations, infiltration, runoff, and the rate of evaporation all combine to influence weathering and weathered products in a given region. Temperature is important, because it greatly influences the rate of chemical reaction. An increase of only 10 °C commonly doubles reaction rates. Other thermal factors include average temperature, temperature ranges, and fluctuations of temperature around the freezing point.

Discussion

Perhaps the best way to appreciate the influence of climate on weathering is to consider variations in the types and thicknesses of soils from the equator to the poles. The diagram in *figure 9.14* summarizes the relationships between the amount of chemical weathering and variations in precipitation and temperature.

In the humid tropical climates, extreme chemical weathering rapidly develops thick soils to depths greater than 70 m. Under such conditions, the feldspars in granites and related rocks are completely altered to clays, and all soluble minerals are leached out. Only the most insoluble materials (such as silica, aluminum, and iron) remain in the thick, deep soil, with the result that the soil commonly is infertile. The high temperatures in tropical zones speed chemical reactions, so that chemical decomposition is very rapid. Frost action, of course, is essentially nonexistent in the tropics, except on the tops of high mountains.

In the low-latitude deserts north and south of the tropical rain forests, chemical weathering is minimal because of the lack of precipitation. Soil is practically nonexistent, and exposures of fresh, unaltered bedrock are abundant. Mechanical weathering is evident, however, in the fresh, angular rock debris that litters most slopes.

In the temperate regions of the higher latitudes, the climate ranges from subhumid to subarid, and temperatures range from cool to warm. Both chemical and mechanical processes operate, and the soil and regolith develop to depths of several meters. In savanna climates, where alternating wet and dry seasons occur, the type of weathering is somewhat different from weathering in regions where precipitation is more evenly distributed throughout the year.

In the polar climates, weathering is largely mechanical. Temperatures are too low for much chemical weathering, so that the soil typically is thin and composed mostly of angular, unaltered rock fragments. In permafrost zones, unique polygonal ground patterns result from thermal contractions and differential thawing and freezing in areas of swamp, lakes, and dry ground.

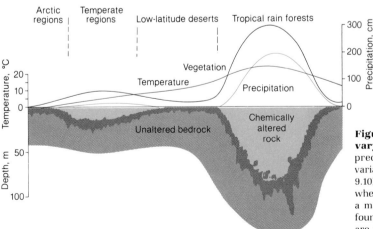

Figure 9.14 The type and extent of weathering vary with climate, because of the combined effects of precipitation, temperature, and vegetation. (Other variables are also involved, such as those shown in figure 9.10). Weathering is most pronounced in the tropics, where precipitation, temperature, and vegetation reach a maximum. Conversely, a minimum of weathering is found in deserts and polar regions, where these factors are minimal.

RATES OF WEATHERING

Statement

The rate at which weathering processes decompose and break down a solid rock body varies greatly. It depends on three main factors:

1. The susceptibility of the constituent minerals to weathering
2. The climate
3. The amount of surface exposed to the atmosphere

Minerals vary widely in their ability to resist weathering. Although many details are not completely understood, minerals that crystallize at a high temperature are most susceptible to weathering. Thus, olivine and Ca-plagioclase weather most rapidly, followed by pyroxene and amphibole and then by Na-plagioclase and K-feldspar. Quartz is one of the most highly resistant minerals, because it resists alteration by chemical decomposition and alteration by mechanical processes.

Discussion

In warm, humid climates, the rate of weathering is very rapid. This fact is demonstrated by some of the recent lava flows on Hawaii, which within a few years have decomposed sufficiently to support some vegetation. By contrast, in colder climates, the same rock types have remained fresh and essentially unaltered since the ice age.

Figure 9.15 The products of weathering can be observed on the Great Pyramids of Egypt. The rectangular building blocks have been modified into ellipsoidal forms by spheroidal weathering, and talus debris has accumulated on each step.

Figure 9.16 Deep weathering is apparent on the ancient stepped pyramids of Egypt. Each step is completely covered with talus, and many individual blocks are weathered into spherical form.

In spite of the complexity of the weathering process, we can get some idea of weathering rates by measuring the amount of decay on rock surfaces of known age. Tombstones, ancient buildings, and monuments, for example, provide rock surfaces that were fresh and unaltered when they were constructed but have since suffered weathering over a known period of time.

The Egyptian pyramids are an interesting example. The Great Pyramid of Cheops, near Cairo, was originally faced with polished, well-fitted blocks of limestone. These protected the rock of the core from weathering until the outer layers were removed, about 1000 years ago, to build mosques in Cairo. Since then, weathering has attacked all four main rock types used in the construction of the pyramid. The most durable (least weathered) rock in the pyramid is a granite, which remains essentially unweathered. Also resistant is a hard, gray limestone, which still retains marks of the quarry tools. The shaly limestone used for other blocks, however, has weathered rapidly. Many of these blocks have a zone of decay as deep as 20 cm. Most of the weathered debris remains as a talus on individual tiers and around the base of the pyramid (*figure 9.15*). The volume of weathered debris produced from the pyramid during the last 1000 years has been calculated to be 50,000 m³, or an average of 50 m³ per year. This is a loss of approximately 3 mm per year over the entire surface of the pyramid. Some of the older stepped pyramids, built 4500 years ago, show much greater weathering. Many of their large blocks are almost completely decayed or reduced to small spherical boulders. In addition, high piles of debris have accumulated on each major terrace (*figure 9.16*).

SUMMARY

Weathering processes break rocks into small pieces and decompose their unstable minerals. Most rock bodies are fractured. The fractures greatly affect weathering, because they increase the surface area of exposed rock and provide channels through which weathering agents (air and water) gain access to rock below the surface. As weathering proceeds, two products develop: (1) a regolith and (2) spheroidal fragments. Climate exerts a profound influence on the amount and type of weathering.

Weathering processes do not erode and transport the fragmented and decomposed rock materials. They only produce them. Other agents (such as running water) pick up the weathered rock fragments and transport them as sediment toward the ocean. The process of erosion by running water is the subject of the next three chapters.

Key Words

weathering	leach
mechanical weathering	joint
disintegration	bedrock
ice wedging	regolith
frost wedging	soil
sheeting	soil profile
talus	*A* horizon
chemical weathering	*B* horizon
hydrolysis	*C* horizon
dissolution	spheroidal weathering
oxidation	exfoliation

Review Questions

1. List 10 ways in which the effects of weathering are expressed in natural outcrops of bedrock.
2. Discuss the processes involved in the most important types of mechanical weathering.
3. Which rock types are most susceptible to chemical weathering?
4. Discuss the chemical reactions involved in hydrolysis.
5. Explain why joints are important in weathering processes.
6. How does soil originate?
7. Explain how climate, rock type, topography, and time influence the types of soil produced by weathering.
8. What is spheroidal weathering?
9. Draw a schematic diagram showing variations in weathering from the Arctic to the tropics.
10. How do the Great Pyramids of Egypt provide information on rates of weathering?

Additional Readings

Carroll, D. 1970. *Rock weathering*. New York: Plenum Press.

Goldich, S. S. 1938. A study in rock weathering. *Journal of Geology* 46:17–58.

Keller, W. D. 1957. *Principles of chemical weathering*. Columbia, Mo.: Lucas Brothers.

Loughnan, F. C. 1969. *Chemical weathering of the silicate minerals*. New York: American Elsevier.

Ollier, C. D. 1969. *Weathering*. New York: American Elsevier.

Reiche, P. 1950. *A survey of weathering processes and products*. University of New Mexico Publications in Geology, no. 3, pp. 1–95. Albuquerque: University of New Mexico.

RIVER SYSTEMS

A river system is a network of connecting channels through which water precipitated on the surface is collected and funneled back to the ocean. At any given time, about 1300 km³ of water flows in the world's rivers. As it flows, it picks up weathered rock debris and carries it to the ocean. Erosion by running water continually modifies the surfaces of the continents and makes river systems one of the most important geologic agents on the earth.

Measurements of the volume of sediment in motion within a river system can be used to measure the rates of denudation. Yet, how can we prove that streams have formed the valleys and canyons through which they flow? How could the Colorado River have cut the Grand Canyon more than 2 km deep in solid bedrock? Since the erosion of a canyon cannot be observed from start to finish, we must approach the problem of erosion and the development of landforms indirectly. Two methods have been used. First, geologists have analyzed drainage networks and have found that they constitute orderly systems that are susceptible to mathematical analysis. Second, geologists have studied the processes operating in drainage basins today. These processes seem infinitely slow, but over a period of several million years, they are capable of eroding the earth's surface to its present form.

In this chapter, we will consider river systems and the processes by which they change the earth's surface.

MAJOR CONCEPTS

1. Running water is part of the earth's hydrologic system and is the most important agent of erosion. Stream valleys are the most abundant and widespread landforms on the planet.
2. A river system consists of a main channel and all of the tributaries that flow into it. It can be subdivided into three subsystems: (a) tributaries, which collect and funnel water and sediment into the main stream, (b) a main trunk, which is largely a transporting system, and (c) a dispersing system, which is found at the river's mouth.
3. There is a high degree of order among the various elements of a river system.
4. Sediment is transported if the velocity of stream exceeds the settling velocity of the sediment.
5. The capacity of running water to erode and transport sediment is largely dependent on stream velocity, although characteristics of the channel are also significant.
6. There is a universal tendency for a river system to establish equilibrium among the various factors influencing stream flow (velocity, water volume, stream gradient, and volume of sediment).
7. We can manipulate river systems by constructing dams, flood control projects, and cities. Such construction can result in many unforeseen and unwanted side effects.

THE GEOLOGIC IMPORTANCE OF RUNNING WATER

Statement

Running water is by far the most important agent of erosion. It is responsible for the configuration of the landscape of most continental surfaces (see *plates 7* and *9*). Other agents, such as ground water, glaciers, and wind, are locally dominant, but these affect only limited parts of the surface. Generally, they operate in an area only for a limited interval of geologic time. Even where other agents are especially active, running water is still likely to be a significant geologic agent.

Discussion

An attempt to appreciate the significance of streams and stream valleys in the regional landscape of the earth is a problem of perspective, much like trying to appreciate the abundance of craters on the moon from viewpoints on its surface. To an astronaut on the moon, the surface appears to be an irregular, broken landscape cluttered with rock debris. The crater systems and terrain patterns, so striking when viewed from space, are not at all apparent from vantage points on the moon's surface. Indeed, crater rims may appear only as rounded hills. Without the aid of maps or space photographs, it is difficult to recognize some of the larger craters as circular forms when they are seen only from the Apollo landing sites.

Viewed from the ground, the earth's stream valleys may appear only as relatively insignificant, irregular depressions between rolling hills, mountain peaks, and broad plains. But, viewed from space, stream valleys can be seen to dominate the landscape of the earth, much as craters dominate the landscape of the moon.

The ubiquitous stream valleys on the earth's surface and the importance of running water as the major agent of **erosion** can best be appreciated by considering a broad, regional view of the continents and their major river systems, as seen through high-altitude photography. As the photographs of the southwestern United States in *figure 10.1* show, throughout broad regions of the continents, the surface is little more than a complex of valleys created by stream erosion. Even in the desert, where it sometimes does not rain for decades, the network of stream valleys commonly is the dominant landform. No other landform on the earth is as abundant and significant.

Figure 10.1 Erosion by running water is the dominant process in the formation of the landscape. The *Skylab* photograph was taken from an altitude of 650 km. The aerial photographs were taken from 12 km and 3 km.

(A) A Skylab *photograph of an area in the arid southwestern United States shows the regional patterns of a river system and its valleys.*

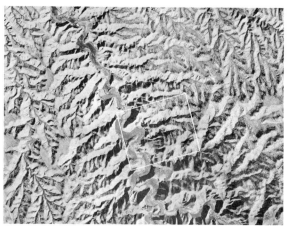

(B) A high-altitude aerial photograph of a portion of the area shown in **A** reveals an intricate network of streams and valleys within the tributary regions of the larger streams.

(C) A low-altitude aerial photograph of part of the area shown in **B** reveals many smaller streams and valleys in the drainage system.

THE MAJOR FEATURES OF A RIVER SYSTEM

Statement

A river system consists of a main channel and all of the tributaries that flow into it. Each river system (often referred to as a drainage basin) is bounded by a divide (ridge), beyond which water is drained by another system. The surface of the ground slopes toward the network of tributaries, so that the drainage system acts as a funneling mechanism for removing precipitation and weathered rock debris. A typical river system can be divided into three subsystems:

1. A collecting system
2. A transporting system
3. A dispersing system

Discussion

An idealized map of a typical **river system** is shown in *figure 10.2*. Although the boundaries between the three subsystems are somewhat gradational, the distinguishing characteristics of each subsystem are readily apparent.

The Collecting System

The collecting system of a river, consisting of a network of **tributaries** in the headwater region, collects and funnels water and sediment to the main stream. It commonly has a **dendritic** (treelike) pattern, with numerous branches, which extend upslope toward the **divide.** Indeed, one of the most remarkable characteristics of the collecting system is the intricate network of tributaries. By plotting all of the tributaries longer than 1 km in the drainage system shown in *figure 10.2*, we obtain the intricate pattern shown in the enlargement. But this is not all. Each of the smallest tributaries shown in *figure 10.2* has its own system of smaller and smaller tributaries, so that the total number becomes astronomical. From the details in *figure 10.2*, it is apparent that most of the land's surface is part of some **drainage basin.**

The Transporting System

The transporting system is the main trunk stream, which functions as a channelway through which water and sediment move from the collecting area toward the ocean. Although the major process is transportation, this subsystem collects additional water and sediment; also, some deposition occurs, where the channel meanders back and forth, and at times when the river overflows its bank during its flood stage. Thus, erosion, deposition, and transportation occur in the transporting subsystem of a river.

The Dispersing System

The dispersing system consists of a network of **distributaries** at the mouth of a river, where sediment and water are dispersed into the ocean, a lake, or a dry basin. The major processes are the deposition of the coarse sediment load and the dispersal of fine-grained material and river waters into the basin.

Figure 10.2 The major parts of a river system are characterized by different geologic processes. The tributaries in the headwaters constitute a subsystem that collects water and sediment and funnels them into a main trunk stream. Erosion is dominant in this area. The main trunk stream is a transporting subsystem. Both erosion and deposition can occur in this area. The lower end of the river is a dispersing subsystem, where most sediment is deposited in a delta or an alluvial fan and water is dispersed into the ocean. Deposition is the dominant process in this part of the river.

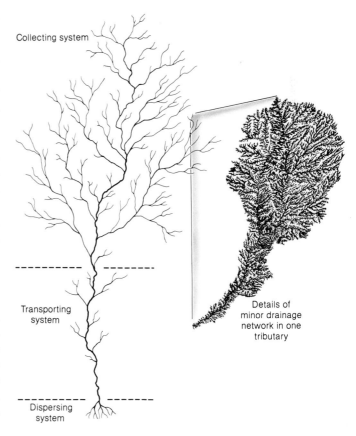

Collecting system

Transporting system

Dispersing system

Details of minor drainage network in one tributary

ORDER IN STREAM SYSTEMS

Statement

Studies of drainage systems show that, where a stream system develops freely on a homogenous surface, there are definite mathematical ratios that characterize the relationships between the tributaries and the size and gradient of the stream and the stream valley. Some of the more important relationships are the following:

1. The number of stream segments (tributaries) decreases downstream in a geometric progression.
2. The length of tributaries becomes progressively greater downstream.
3. The gradient, or slope, of tributaries decreases exponentially downstream.
4. The stream channels become progressively deeper and wider downstream.
5. The size of the valley is proportional to the size of the stream and increases downstream.

These relationships constitute the basis for the assumption that streams erode the valleys through which they flow.

Discussion

The first meaningful observations about the relationship of stream systems to their valleys were made in 1802 by John Playfair (1748–1819), an English geologist. He stated a principle that has become known as Playfair's law:

Every river appears to consist of a main trunk, fed from a variety of branches, each running in a valley proportional to its size, and all of them together forming a system of valleys connecting with one another, and having such a nice adjustment of their declivities that none of them join the principal valley at either too high or too low a level; a circumstance which would be infinitely improbable if each of these valleys were not the work of the stream which flows in it.

To appreciate **stream order** and system in a drainage basin, carefully study *figure 10.3*, which is a graphic model of a typical stream. The decrease in the **gradient** of the tributaries is represented by the curve on the side of the block diagram. The curve can be expressed by a mathematical formula. The stream channel becomes progressively deeper and wider downstream, as is shown in the diagram. Similarly, the sizes of the valleys increase downstream, in proportion to the sizes of the streams that flow through them. The length of the tributaries also increases downstream. Numerous statistical studies support these observations, showing that there are a number of definite mathematical relationships between a stream and the valley through which it flows.

If valleys were ready-made by some process other than stream erosion, such as faulting or other earth movements, these relationships would be "infinitely improbable." You can easily confirm the high degree of order in streams by studying the aerial photographs in *figure 10.1*. Does each tributary have a steeper gradient than the stream into which it flows? Does each tributary flow smoothly into a larger stream without an abrupt change in gradient? Are the tributary valleys smaller than the valleys into which they drain?

Stream erosion has been studied in great detail over the last 100 years, and geologists have been able to observe and measure many aspects of stream development and erosion by running water. The origin of valleys by erosion is well established, and it is clear that running water is the most significant agent of erosion on the earth's surface.

Figure 10.3 The characteristics of a river change systematically downstream. The gradient decreases downstream, and the channel becomes larger. Other downstream changes include an increase in the volume of water and an increase in the size of the valley through which the stream flows.

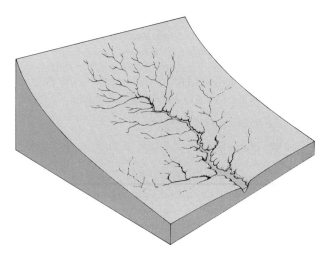

THE FLOW OF WATER IN NATURAL STREAMS

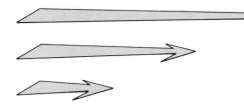

(A) *In laminar flow, water particles move in parallel lines.*

Statement

Water in natural streams moves by turbulent flow, which is characterized by numerous secondary eddies and swirls, in addition to the main downstream movement. Laminar flow, in which water moves in straight lines, is restricted to a thin film a few millimeters thick on the bottom and walls of the stream channel. The flow of water in a stream is complex and is influenced by a number of variables. The most important are the following:

1. Volume of water
2. Velocity
3. Shape and size of channel
4. Gradient (slope of the stream channel)
5. Base level (the lowest level to which a stream can erode)
6. Load (the material carried by the flowing water)

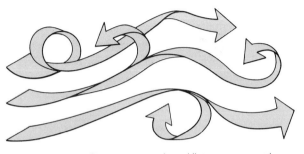

(B) *In turbulent flow, many secondary eddies are superposed on the main stream flow.*

Figure 10.4 Types of flow are illustrated by arrows.

Discussion

Laminar Flow and Turbulent Flow

The nature of flow in rivers can be studied by injecting dye into a stream and observing its movement. In the laboratory, water moving through a glass tube at very low velocities exhibits **laminar flow,** with parallel layers of water shearing over one another. Theoretically, there is no mixing. Laminar flow is rarely found in natural streams, except in a very thin layer along the bottom and banks of the channel. It is common, however, in ground-water flow.

Turbulent flow is a type of irregular movement, with eddies superposed on the main downstream flow (*figure 10.4*). Dye injected into turbulent flow moves downward, sideways, and upward, showing that water can lift loose sediment from the river bed and transport it downstream.

Discharge

Discharge is the amount of water passing a given point during a specific interval of time. It is usually measured in cubic meters per second. The discharges of most of the world's major drainage systems have been monitored by gauging stations for years. The information collected provides the basis for water resources utilization planning, flood control, estimates of erosion rates, and so on. An essential point to remember is that the discharge increases downstream because of the addition of water from each tributary (*figure 10.3*). Small tributaries discharge only 5 or 10 m³/sec. Larger streams (such as the Mississippi River) discharge from 40,500 to 337,500 m³/sec. Rates of 1,500,000 m³/sec have been measured on the Amazon River, which has the greatest discharge of any river in the world. This huge discharge is due to the great amount of precipitation in the river's collecting system, which is located in the rain forest of the tropics of South America.

The water for a river system comes from both surface runoff and seepage of ground water into the stream channels. Ground-water seepage is important because it can maintain the flow of water throughout the year. Continual seepage establishes **permanent streams.** If the supply of ground water is depleted seasonally, streams temporarily become dry. Such streams are called **intermittent streams.**

Velocity

The velocity of flowing water is not uniform throughout the stream channel. It depends on the shape and roughness of the channel and on the stream pattern. The velocity usually is greatest near the center of the channel and above the deepest part, away from the frictional drag of the channel walls and floor (*figure 10.5*). As the channel curves around a **meander,** however, the zone of maximum velocity shifts to the outside of the bend, and a zone of minimum velocity forms on the inside of the curve (*figure 10.6*). This flow pattern is important in the lateral erosion of stream channels and the migration of stream patterns.

The velocity of flowing water is proportional to the gradient of the stream channel. Steep gradients produce rapid flow, which commonly occurs in high mountain streams. Where slopes are very steep, waterfalls and rapids develop, and the velocity approaches that of free fall. Low gradients produce slow, sluggish flow, and where a stream enters a lake or the ocean, the velocity is soon reduced to zero. The velocity of flowing water in a given channel also depends on the volume. The greater the volume, the faster the flow.

Channel Shape

The cross-sectional shape of a stream channel greatly influences the nature of the flow and the velocity at which a given volume of water moves down a given slope. Velocity is greatest in channels that offer the least resistance to flow. A channel whose cross section approaches an ideal semicircle (*figure 10.7C*) has the least surface area per unit volume of water, and hence it offers the least

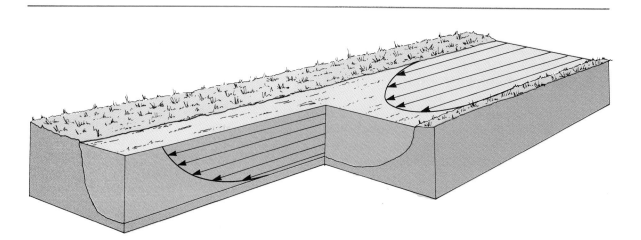

Figure 10.5 Variations in the velocity of flow in natural stream channels occur both horizontally and vertically. Friction reduces the velocity along the base and sides of the channels. The maximum velocity in a straight channel is near the top and center of the channel.

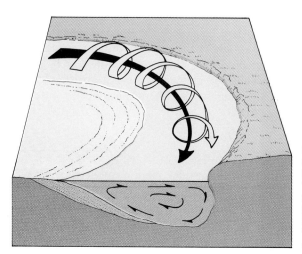

Figure 10.6 Flow in a curved channel follows a corkscrew pattern. Water on the outside of the bend is forced to flow faster than that on the inside of the curve. This difference in velocity, together with normal frictional drag on the channel walls, produces a corkscrew flow pattern. As a result, erosion occurs on the outer bank, and deposition occurs on the inside of the bend. These processes produce an asymmetrical channel, which slowly migrates laterally.

resistance to flow. Deep, narrow channels (*figure 10.7A*) and wide, shallow channels (*figure 10.7B*) both present greater surface area per unit volume of water and thus retard the flow. Streams flowing through narrow channels crowd and erode their banks, so that the channel becomes wider. In flat channels, the low velocities cause sediment depositions, and the stream channel becomes more narrow. In both cases, the channel shape approaches that of a semicircle, to produce a balance, or equilibrium, between the flowing water and the area through which it moves.

A rough channel, strewn with boulders and other obstructions, retards flow. A smooth, clay-lined channel offers the least resistance and permits greater velocity. Roughness is difficult to measure, but its influence is obvious.

Stream Gradient

The gradient of a stream, or the slope of the stream bed, is steepest in the headwaters and decreases downslope. The **longitudinal profile** (a cross section of a stream from its headwaters to its mouth) is a smooth, concave upward curve that becomes very flat at the lower end of the stream. The gradient usually is expressed in the number of meters the stream descends for each kilometer of flow. The headwater streams that drain the Rockies can have gradients of over 50 m/km; the lower reaches of the Mississippi River have a gradient of only 1 or 2 cm/km.

Base Level

The **base level** of a stream is the lowest level to which the stream can erode its channel. It is a key feature in the study of stream activity. The base level is, in effect, the elevation of the stream's mouth where it enters the ocean, a lake, or another stream. A tributary cannot erode lower than the level of the stream into which it flows. Similarly, a lake controls the level of erosion for the entire course of the river that drains into it. The levels of tributary junctions and lakes are *temporary* base levels, because lakes can be filled with sediment or be drained, and streams can be established across the former lakebed. For all practical purposes, the **ultimate base level** is sea level, because the energy of a river is quickly reduced to zero as it enters the ocean. (Exceptions are fault valleys like Death Valley, which is faulted below sea level.) Even sea level can change, however. As base level or sea level changes, the longitudinal profile of the river changes, and the stream adjusts to the new condition.

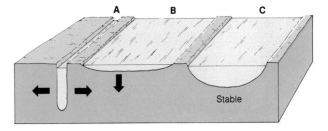

Figure 10.7 Stream channels adjust their shapes so as to minimize resistance to flow. **(A)** A deep channel presents a large surface area per unit volume of water. Its banks tend to erode, so that the channel becomes wider. **(B)** A flat channel also presents a large surface area, which reduces the velocity of the water and causes some of the load to be deposited. The channel therefore tends to become narrower. **(C)** A semicircular channel, which presents a minimum surface area per unit volume of water, is the optimum channel shape.

Stream Load

Flowing water in natural streams provides a fluid medium by which loose, disaggregated regolith is picked up and transported to the ocean. The capacity of a stream to transport sediment increases to a third or fourth power of the velocity. That is, if the velocity is doubled, the stream can move from 8 to 16 times as much sediment. A stream's **load** (the amount of material it transports at a given time) is usually less than its **capacity** (the total amount the stream can carry under given conditions). The maximum particle size a stream can transport is its **competence.** In many streams, the competence is enormous, especially in those with a relatively steep gradient and a high discharge. Flash floods in the Wasatch Range in 1923 carried boulders weighing from 80 to 90 metric tons for nearly 2 km. If a dam breaks, the large volume of water can transport blocks weighing as much as 10,000 metric tons.

By far the greatest amount of material carried by a stream is derived from the regolith in the collecting system of the drainage basin. This fact is fundamental to understanding the erosion of the land. Weathering produces a blanket of loose, unconsolidated rock (the regolith), which is easily washed downslope and carried away by the river system. This erosion exposes fresh bedrock, which is then weathered to form new regolith, and the cycle continues. The transportation of sediment and erosion of the land are discussed in subsequent sections.

THE TRANSPORTATION OF SEDIMENT BY STREAMS

Statement

Running water is the major agent of erosion, not only because it can abrade and erode its channel, but because of its enormous power to transport loose sediment produced by weathering. Within a stream system, sediment is transported in three ways:

1. Fine particles are moved in suspension (suspended load).
2. Coarse particles are moved by traction along the stream bed (bed load).
3. Dissolved material is carried in solution (dissolved load).

Discussion

Suspended Load

The **suspended load** is the most obvious, and generally the largest, fraction of the material moved by a river. It consists of mud, silt, and fine sand and can be seen in most parts of the transporting segment of a river system. Particles in suspension move if the velocity of the upward currents in turbulent flow exceeds the velocity at which the sediment particles would sink in still water. In essentially all major streams, most silt and clay-size particles remain in suspension most of the time and move downstream at the velocity of the flowing water, to be deposited in the ocean, in a lake, or on the flood plain.

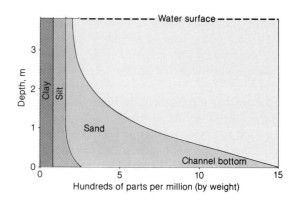

Figure 10.8 The distribution of the sediment load varies from the surface to the base of a river. This example, the Missouri River at Kansas City, shows that clay and silt are transported in suspension and are nearly equally distributed throughout the channel. The number of sand-size particles increases conspicuously toward the bottom of the channel, because they are transported mostly by saltation and traction.

Samples of the suspended load of the Missouri River at Kansas City show that clay-size particles are distributed equally from top to bottom, but sand-size particles occur in larger amounts near the bottom of the channel (*figure 10.8*).

Bed Load

Particles of sediment too large to remain in suspension collect on the stream bottom and form **bed load,** or traction load. These particles move by sliding, rolling, and **saltation** (short leaps) (*figure 10.9*). The bed load moves only if there is sufficient velocity to move the large particles. It thus differs

Figure 10.9 Movement of the bed load in a stream is accomplished in a variety of ways. Particles that are too large to remain in suspension are moved by sliding, rolling, and saltation. Increases in discharge, due to heavy rainfall or spring snowmelt, can flush out all of the loose sand and gravel, so that the bedrock is eroded by abrasion.

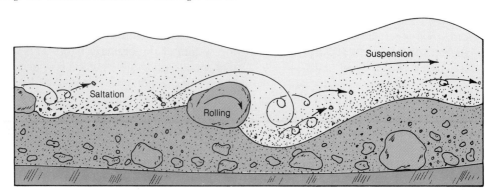

fundamentally from the suspended load, which moves constantly. The distinction between the largest particles of the suspended load and the smallest particles of the bed load is not always sharp, because the velocity of a stream fluctuates constantly, so that part of the bed load can suddenly move in suspension, and part of the suspended load can settle.

Bed load can constitute 50% of the total load in some rivers, but it usually ranges from 7% to 10% of the total sediment load. The movement of the bed load is one of the major tools of stream erosion, because as the sand and gravel move, they abrade (wear away) the sides and bottom of the stream channel.

Dissolved Load

Dissolved load is matter transported in the form of chemical ions and is essentially invisible. All streams carry some dissolved material, which is derived principally from ground water emerging from seeps and springs along the river banks. The most abundant materials in solution are calcium and bicarbonate (HCO_3^-) ions, but sodium, magnesium, chloride, ferric, and sulfate ions are also common. Various amounts of organic matter are present, and some streams are brown with organic acids derived from the decay of plant material.

Flow velocity, which is so important to the transportation of the suspended and traction load, has little effect on a river's capacity to carry dissolved material. Once mineral matter is dissolved, it remains in solution, regardless of velocity, and is precipitated and deposited only if the chemistry of the water changes.

Chemical analysis shows that most rivers carry a dissolved load of less than a thousand parts per million. Some streams in arid regions, however, carry several thousand parts per million and are distinctly salty in taste. Although these amounts of dissolved material seem small, they are far from trivial. Sampling indicates that, in some rivers, when rainfall is abundant, more than half of all the material carried to the ocean is in solution. In areas of low relief, such as the Atlantic and Gulf Coast states, runoff is slow and solution activity lowers the surface at a rate of 52 metric tons per square kilometer per year, or about 1 m in 25,000 years. Dissolved loads are somewhat smaller in mountainous terrain (from 1 m in 100,000 years to 1 m in 3 million years), but they are by no means negligible.

Sediment load can be seen and measured in the reservoir behind a dam. A dam effectively traps all of the sediment, and the reservoir eventually fills with mud and silt (*figure 10.10*). Unless this sediment is removed by dredging or other means, it terminates the useful life of the reservoir.

Figure 10.10 The volume of sediment transported by a stream is illustrated by the Mono Reservoir, California, which has been completely filled. This reservoir was built as a debris basin to protect the Gebaltar Reservoir, downstream. It was built in 1935 after a severe fire burned off much of the vegetation in the watershed area. Very heavy rains occurred during the following two winter seasons, before vegetation could be reestablished. The accelerated erosion provided enough silt and sand to completely fill the Mono Reservoir and to partially fill the lower Gebaltar Reservoir. This example serves as a vivid reminder that a river system is a system of moving water and sediment, not just water alone.

THE RELATION OF VELOCITY TO EROSION, TRANSPORTATION, AND DEPOSITION

Statement

The ability of running water to erode and transport sediment is largely dependent on stream velocity. If velocity is doubled, the size of the particles that can be transported increases by a factor of four. There are two reasons for this exponential increase: (1) each particle of water hits a particle of sediment with twice as much force, and (2) twice as much water strikes the face of a fragment during a given interval of time.

Discussion

The mechanics of sediment transportation can be illustrated by dropping a handful of soil into a container of water. The large, coarse particles sink rapidly, while the smaller particles stay in suspension for hours. Each size of particle has its own **settling velocity,** the constant velocity at which the particle falls through a still fluid. Settling velocity is determined by a particle's size, shape, and specific gravity, as well as by the density of the fluid. If a stream's velocity exceeds the settling velocity, the sediment is transported downstream with the flowing water.

The results of experimental studies of water's capacity to erode, transport, and deposit sediments are summarized in *figure 10.11*. The velocity at which a particle of a given size is picked up and moved is shown on the upper curve. This curve is a zone on the graph instead of a line, because the erosional velocity varies according to the characteristics of the water and the shape and density of the grain to be moved (shape has a definite effect on the suspension of flat mica flakes, for example). Moreover, erosional velocities vary with the depth and density of the water. The lower curve (dashed line) shows the velocity at which a particle settles out of suspension. Small particles, such as fine silt and clay, require a relatively high velocity to lift them into suspension, because they tend to stick together while they are deposited. Once in suspension, however, fine particles remain suspended with a minimum velocity.

The graph in *figure 10.11* also shows that sediment particles are deposited according to size if the velocity of the water decreases below their critical settling velocity. Thus, on gentle slopes, where the stream's velocity is reduced, a significant part of the sediment load is deposited along the channel or on the flood plains. Most of the remaining sediment is deposited where the river enters a lake or the ocean, because velocity is reduced there.

The period of greatest erosion and transportation of sediment is, of course, during floods. The increase in discharge results in an increase in both the maximum size of transported material and the total load. An exceptional flood can cause a large amount of erosion and transportation during a brief period of time. In arid regions, many streams carry little or no water throughout the year, but during a cloudburst they can move great quantities of sediment.

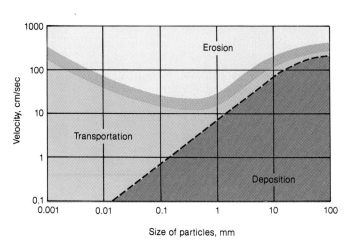

Figure 10.11 The threshold velocity for sediment transport provides important insight into processes of stream erosion, transportation, and deposition. The upper curve shows the range of minimum velocities at which a stream can pick up and move a particle of a given size. This threshold velocity is represented by a zone on the graph, not a line, because of variations resulting from stream depth, particle shape and density, and other factors. The lower curve indicates the velocity at which a particle of a given size settles out and is deposited. Note that fine particles stay in suspension at velocities much lower than those required to lift them from the surface of the stream bed.

EQUILIBRIUM IN RIVER SYSTEMS

Statement

Streams are not separate, independent entities. Each is a part of a drainage system, in which every segment is intimately related to the others. Every stream has tributaries, and every tributary has smaller tributaries, extending down to the smallest gully. A river system functions as a unified whole, and any change in one part of the system affects the other parts. The major factors that determine stream flow (discharge, velocity, gradient, base level, and load) constantly change toward a dynamic balance, or equilibrium. Adjustments within a river system are continually being made, so that, eventually, the gradient of a stream is altered to accommodate the volume of water available, the channel's characteristics, and the velocity necessary to transport the sediment load. A change in any of these factors causes compensating adjustments to restore equilibrium within the entire drainage system.

A river is in equilibrium if its channel form and gradient are balanced so that neither erosion nor deposition occurs. Rivers are constantly adjusting toward this ideal condition. This adjustment is important to an understanding of the natural evolution of the landscape. It also has practical consequences, because if we continually modify rivers to suit our needs, we should know how river systems respond to artificial modifications.

Discussion

The concept of equilibrium in a river system can best be appreciated by considering a hypothetical stream in which equilibrium has been established. In diagram *A* of *figure 10.12*, the variables in the stream system (discharge, velocity, gradient, base level, and load) are in balance, so that neither downcutting nor sedimentation occurs along the stream's profile. *There is just enough water to transport the available sediment down the existing slope.* Such a stream is known as a **graded stream.** The energy of the stream system is in equilibrium, and the stream's profile is a smooth curve. In diagram *B*, the stream's profile is displaced by a fault that creates a waterfall. The increased gradient across the falls greatly increases the stream's velocity at that point, so that rapid erosion occurs, and the waterfall (or the rapid) migrates upstream. The eroded sediment added to the stream segment on the dropped fault block is more than the stream

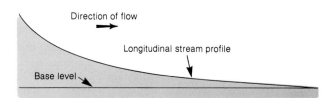

(A) *Initially, when the stream profile is at equilibrium, the velocity, load, gradient, and volume of water are in balance. Neither erosion nor deposition occurs.*

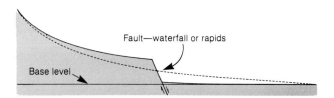

(B) *Faulting disrupts equilibrium by decreasing the gradient downstream and increasing the gradient at the fault line.*

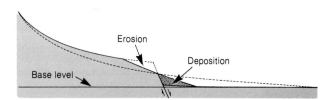

(C) *Erosion proceeds upstream from the fault, and deposition occurs downstream.*

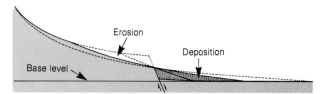

(D) *Erosion and deposition continue to develop a new stream profile.*

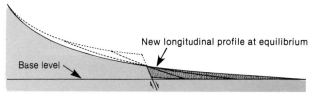

(E) *A new profile of equilibrium, in which neither erosion nor deposition occurs, is eventually reestablished.*

Figure 10.12 Adjustments of a stream to reestablish equilibrium are illustrated by profile changes after disruption by faulting.

can transport, because the system already was in equilibrium before faulting occurred. Therefore, the river deposits part of its load, building up the channel gradient (the shaded area in diagrams *C*, *D*, and *E*) until a new profile of equilibrium is established (diagram *E*).

An example of the adjustments described above occurred in Cabin Creek, a small tributary of the Madison River, north of Hebgen Dam, in Montana. In 1959, a 3-m fault scarp formed across the creek during the Hebgen Lake earthquake. By June 1960, erosion by Cabin Creek had erased the waterfall at the cliff formed by the fault, and only a small rapid was left. By 1965, the rapid was completely removed, and equilibrium was reestablished.

The tendency for river systems to establish a profile of equilibrium is also illustrated by the results of dam construction. In the reservoir behind a dam, the gradient is reduced to zero. Hence, where the stream enters the reservoir, its sediment load is deposited as a delta and as layers of silt and mud over the reservoir floor (*figure 10.10*).

Since most sediment is trapped in the reservoir, the water released downstream has practically no sediment load. Therefore, it is capable of much more erosion than the previous river, which carried a sediment load adjusted to its gradient. Extensive scour and erosion commonly result downstream from a new dam.

Another important effect of a dam is that the sediment supply to the ocean is cut off or reduced drastically. The river's delta ceases to grow, and it is vigorously eroded by wave action, resulting in significant changes in the shoreline.

Other examples of adjustments toward equilibrium can be found in canal systems, where sediment is deposited along the canal banks as the system adjusts to the new gradient and channel patterns (*figure 10.13*).

Figure 10.13 Sedimentation in a canal results in part from the universal tendency of streams to develop a meandering pattern. The flow pattern causes variations in velocity, which are accentuated on the inside of a bend. Deposition occurs, and the stream develops meanders.

THE MANIPULATION OF RIVER SYSTEMS

Statement

As population has grown, the artificial modification of the natural water systems of the continents has greatly increased, largely through the construction of dams and canals and through urbanization. Dams are built to store water for industrial and irrigational use, to control floods, and to produce electrical power. Yet, when a dam is built, the balance of the river system, established over thousands or millions of years, is instantly upset. Many unforeseen, long-term effects occur both upstream and downstream from the construction.

Discussion

The Aswân High Dam of the Nile provides a good example of the many consequences of modifying a river system. For centuries, the Nile River has been the sole source of life in Egypt. The principal headwaters of the Nile are located in the high plateaus of Ethiopia. Once a year, for approximately a month, the Nile normally rises to flood stage and covers much of the fertile farmland in the Nile Delta area. The Aswân High Dam, completed in the summer of 1970, was to provide Egypt with water for irrigating 1 million acres of arid land, generate 10 billion kilowatts of power, double the national income, and permit industrialization. The dam destroyed the equilibrium of the Nile, and many adjustments in the river resulted. This is what happened.

The Nile is not only the source of water for the delta, it is also the source of sediment. When the dam was finished and began to trap sediment in Lake Nasser, it destroyed the physical and biological balance in the delta area. Without the annual "gift of the Nile," the delta coastline is now exposed to the full force of marine currents, and wave erosion is eating away at the delta front. Some parts of the delta are receding several meters a year.

The sediment previously carried by the Nile was an important link in the aquatic food chain nourishing marine life in front of the delta. The lack of Nile sediment has reduced plankton and organic carbon to a third of their former levels. This change either killed off or drove away sardines, mackerel, clams, and crustaceans. The annual harvest of 16,000 metric tons of sardines and a fifth of the fish catch have been lost.

The sediment of the Nile also naturally fertilized the flood plain. Without this annual addition of soil nutrients, Egypt's 1 million cultivated acres need artificial fertilizer.

The water discharged from the reservoir is clear, free of most of its sediment load. Without its load, the discharged water flows downstream much faster and is vigorously eroding the channel bank. This scouring process has already destroyed three old barrier dams and 550 bridges built since 1953. Ten new barrier dams between Aswân and the ocean must be built at a cost equal to one-fourth the cost of the Aswân dam itself.

The annual flood of the Nile was also important to the ecology of the area, because it washed away salts that formed in the soil. Soil salinity has already increased, not only in the delta, but throughout the middle and upper Nile. Unless corrective measures are taken (at a cost of over $1 billion), millions of acres will revert to desert within a decade.

The change in the river system has permitted double cropping, but there are no periods of dryness. The dry seasons previously helped limit the population of bilharzia, a blood parasite carried by snails, which infects the intestinal and urinary tracks of humans. One out of every two Egyptians now has the infection, and it causes a tenth of the deaths in the country.

Problems also occur in the lake behind the dam. The lake was to have reached a maximum level in 1970, but it might actually take 200 years to fill. More than 15,000,000 m³ of water annually escapes underground into the porous Nubian Sandstone, which lines 480 km of the lake's western bank. It is capable of absorbing an almost unlimited quantity of water. Moreover, the lake is located in one of the hottest and dryest places on the earth, and the rate of evaporation is staggering. A high rate was expected, but additional losses from transpiration by plants growing along the lakeshore and increased evaporation from high winds have brought the total loss of water from the lake to nearly double the expected rate. This loss equals half the total amount of water that once was "wasted" while flowing unused to the ocean.

The Colorado River, like many others in the United States, has been altered by a series of dams. These caused environmental alterations similar to those experienced on the Nile, but the economy and population of the United States are not tied completely to one river, so that many changes go unnoticed. In the case of the Colorado River, erosion downstream from Hoover Dam has extended all the way to Yuma, Arizona, 560 km below the dam. The channel there has been lowered by approximately 3 m, and the drop has eliminated gravity flow irrigation systems in some areas. This has necessitated pumping to lift the water to the level of irrigation intakes.

Another way in which society has modified river systems is urbanization. The construction of cities may at first seem unrelated to the modification of river systems, but a city significantly changes the surface runoff, and the resulting changes in river dynamics are becoming serious and costly. Water that falls to the earth as precipitation usually follows several paths in the hydrologic system. Generally, from 54% to 97% returns into the air directly by evaporation and transpiration; from 2% to 27% collects in stream systems as surface runoff; and from 1% to 20% infiltrates the ground and moves slowly through the subsurface toward the ocean. Urbanization disrupts the normal hydrologic system. It changes the nature of the terrain and consequently affects the rates and percentages of runoff and infiltration. Roads, sidewalks, and roofs of buildings render a large percentage of the surface impervious to infiltration. Not only does the volume of surface runoff increase, but runoff is much faster, because water is channeled through gutters, storm drains, and sewers. As a result, flooding increases in intensity and frequency.

The effect of urbanization on natural runoff can be understood by considering the characteristics of stream flow shown in *figures 10.14* and *10.15*. The hydrograph in *figure 10.14* shows variations in flow with time. The shape of the curve depends on the nature of the stream system. It is influenced by such factors as relief, vegetative cover, permeability of the ground, and number of tributaries. The low, flat part of the curve represents the portion of stream flow attributable to ground water; it is referred to as base flow. The shaded area at the left side of the graph represents a period of rainfall, and the high part of the curve shows the resulting increase in discharge. An important feature is the time lag, or interval of time, between the storm and the runoff.

The effect of urbanization on runoff is shown in the hydrograph in *figure 10.15.* Lag time is materially decreased, because water runs off faster from streets and roofs than from naturally vegetated areas. As the runoff time decreases, the peak rate of runoff increases. Thus, floods are more frequent and more severe.

Figure 10.14 Variations in stream flow under normal terrain conditions are illustrated by a hydrograph. During a flood, there is a considerable time lag between the period of maximum precipitation and the period of maximum runoff. Water is held back by vegetation, seepage into the subsurface, and a slowdown of runoff due to the curves and bends in stream channels.

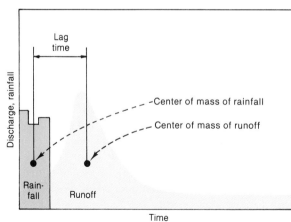

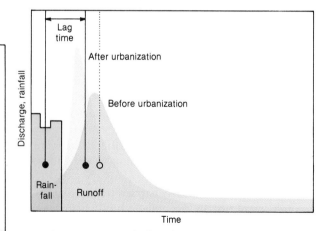

Figure 10.15 Variations in stream flow in an urban area show a tendency toward flash flooding. There is little lag time between maximum precipitation and maximum runoff, because much of the surface is made impermeable by roads, sidewalks, buildings, and parking lots. In addition, water is channeled directly through sewers and gutters for rapid runoff. As a result, the number and intensity of floods increase.

SUMMARY

A river system is a network of connecting channels that collect surface water and funnel it back to the ocean. It can be divided into three subsystems: (1) tributaries, which collect and transport water and sediment into the main stream; (2) a main trunk stream, which is largely a transporting system; and (3) a dispersing system, at the river's mouth.

Studies of river systems show that there is a high degree of order among the tributaries in relation to the size and gradients of their valleys. These relationships indicate that streams erode the valleys through which they flow.

The major variables in a river system are (1) discharge, (2) velocity, (3) gradient, (4) base level, and (5) load.

A river system is best thought of as a system of moving water and sediment, because an enormous volume of sediment is constantly being transported to the ocean by running water. The sediment is moved in solution, in suspension, and by traction. It is visible in most rivers, especially where it is being deposited as deltas near the rivers' mouths. The capacity of running water to erode and transport sediment is largely dependent on the stream's velocity.

A river system functions as a unified whole, adjusting its profile to establish equilibrium among the factors that influence flow. If one factor changes, the river system adjusts to reestablish equilibrium.

Running water is the most important eroding agent on this planet, and stream valleys are the most characteristic landforms on continental surfaces. In the next chapter, we will consider the processes by which streams erode their valleys and canyons and ultimately deposit the eroded sediment as deltas, alluvial fans, and other landforms.

Key Words

drainage system	dispersing system
river system	dendritic drainage pattern
drainage basin	divide
collecting system	tributary
transporting system	distributary

stream order	stream load
gradient	capacity
turbulent flow	competence
laminar flow	suspended load
discharge	bed load
permanent stream	dissolved load
intermittent stream	saltation
meander	settling velocity
longitudinal profile	graded stream
base level	erosion
ultimate base level	

Review Questions

1. Explain the reasons for concluding that stream action (running water) is the most important process of erosion on the earth.
2. Describe and illustrate the three major subsystems of a river.
3. Explain the concept of order in stream systems.
4. How does order in stream systems indicate that streams have eroded the valleys through which they flow?
5. Show by means of diagrams the patterns and velocities of flow in (a) a straight channel and (b) a curved channel.
6. Define *stream gradient, stream load, capacity,* and *competence.*
7. Draw a diagram showing the general nature of (a) bed-load transportation, (b) suspended load, and (c) dissolved load.
8. Explain the concept of threshold velocity in the transportation and deposition of stream sediment.
9. Explain the concept of equilibrium in river systems and cite several examples of how streams adjust to attain equilibrium.
10. How does urbanization affect surface runoff?

Additional Readings

Leopold, L. B., and W. B. Langbein. 1966. River meanders. *Scientific American* 214(6):60–70. Reprint no. 869. San Francisco: W. H. Freeman.

Leopold, L. B., M. G. Wolman, and J. P. Miller. 1964. *Fluvial processes in geomorphology.* San Francisco: W. H. Freeman.

Morisawa, M. 1968. *Streams: Their dynamics and morphology.* New York: McGraw-Hill.

PROCESSES OF STREAM EROSION AND DEPOSITION

Erosion of the land is one of the major effects of the hydrologic system. It has occurred throughout all of recorded geologic time and will continue as long as the system operates and land is exposed above sea level. Evidence for erosion is ubiquitous and varied. We see it in the development of gullies on farmlands and in the cutting of great canyons. We see it where material carried by rivers is deposited in the ocean. We see it in the thick layers of sedimentary rocks, which cover large parts of the continents and bear witness to erosion during past ages on a magnitude that is difficult to conceive.

There are six major agents of erosion: running water, mass movement, ground water, glaciers, waves and currents, and wind. Running water is by far the most important; few areas on the earth have not been modified by it.

To help you understand the various processes of stream erosion, they will be analyzed in this chapter as independent, isolated activities, but remember that each is part of the larger hydrologic system.

MAJOR CONCEPTS

1. The downcutting of stream channels is accomplished by abrasion, solution, and hydraulic action.
2. Drainage systems grow by headward erosion, which commonly results in stream piracy.
3. Slope retreat causes valleys to grow wider and is associated with downcutting and headward erosion. It results from various types of mass movement under the pull of gravity.
4. As a river develops a low gradient, it deposits part of its load on point bars, on natural levees, and across the surface of its flood plain.
5. Most of a river's sediment is deposited where it empties into lakes or the ocean. This deposition commonly builds a delta at the river's mouth. In arid regions, many streams deposit their loads as alluvial fans at the bases of steep slopes.

THE REMOVAL OF REGOLITH

Statement

One of the most important processes of erosion is the transport and removal of rock debris produced by weathering. The process is simple but important. Loose, weathered rock debris is washed downslope into the drainage system and is transported as sediment load in streams and rivers. In addition, soluble material is carried in solution. The net result is that the blanket of regolith over the earth's surface is continually being removed and transported to the sea by stream action, while it is also continually being regenerated by weathering.

Discussion

The continuous removal of regolith can be observed on any hillside during a rainstorm or during spring runoff. Small particles of silt and mud are washed downslope into the system of tributaries. The vast amount of material carried by rivers bears witness to the removal of regolith on a magnitude that is difficult to conceive. As the regolith is removed, fresh rock is exposed and weathered, so that the supply of regolith is always being replenished. By this process, and by associated downcutting of stream channels and headward erosion, high mountains and entire continents are reduced to elevations close to sea level.

Figure 11.1 Sand and gravel are the tools of stream erosion. Transported by a river, they act as powerful abrasives, cutting through the bedrock as they are moved by flowing water. The abrasive action of sand and gravel cut this vertical gorge through resistant limestone in the Grand Canyon, Arizona.

THE DOWNCUTTING OF STREAM CHANNELS

Statement

The basic process of erosion in all stream channels, whether small hillside gullies or great canyons of major rivers, is abrasion by sand, cobbles, and boulders moving along the channel floor. Other erosional processes in stream channels are dissolution and direct hydraulic action.

Discussion

Abrasion

To fully understand the effect of **abrasion** on a river channel, as gravel and sand are swept along the river bed, you must have a clear conception of river systems as not only systems of flowing water but systems of moving water and sediment. Although the sediment may be temporarily deposited along the way during low water, it is picked up and moved again during high water and floods, and as it moves, it continually abrades the surface of the channel. The process of abrasion in river beds is similar to that by which the wire saws used in quarries cut and shape large blocks of stone. An abrasive such as garnet, corundum, or quartz, dragged across a rock by a wire, can cut through a stone block with remarkable speed.

Some dramatic examples of the power of streams to cut downward are the steep, nearly vertical gorges in many canyons in the southwestern United States (*figure 11.1*). The bed load of sand and gravel on the channel floor is stationary much of the time, but during spring runoff and periodic flash floods, it moves with the flowing water. This material is an effective abrasion tool and can cut the stream channel to a profile of equilibrium in a short time.

An effective and interesting type of stream abrasion is the drilling action of pebbles and cobbles trapped in a depression and swirled around by currents. The rotational movement of the sand, gravel, and boulders acts like a drill and cuts deep holes called **potholes.** As the pebbles and cobbles are worn away, new ones take their places and continue to drill into the stream channel. They eventually develop holes that can measure several meters in diameter and more than 5 m deep (*figure 11.2*).

Abrasion also effectively undercuts stream

Figure 11.2 Potholes are eroded in the hard rock of a stream bed by sand, pebbles, and cobbles whirled around by eddies.

banks, and large blocks of rock and regolith slump off and become part of the stream load (*figure 11.3*).

As they abrade and downcut a stream channel, the pebbles and cobbles themselves are worn down from striking one another and the channel bottom. Their corners and edges are chipped off, and the particles become smaller and smooth and rounded. Thus, large boulders that fall into a stream or are transported only during a flood are slowly broken and worn down to smaller fragments. Ultimately, they are washed away as grains of sand.

Solution

Although the solution action of running water generally is not visible, chemical analysis of rivers shows that **dissolution** is an effective and important means of lowering the land surface. From measurements of the chemical content of the world's major rivers, it is estimated that each year about 3.5 billion metric tons of dissolved material is carried from the continents by running water. The effectiveness of such chemical erosion is illustrated by the weathering and erosion of limestone. In humid areas, limestone (which is the most soluble of the major rock types) forms valleys and is con-

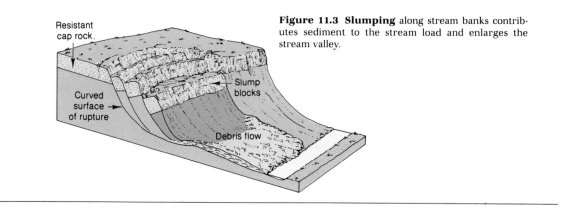

Figure 11.3 Slumping along stream banks contributes sediment to the stream load and enlarges the stream valley.

sidered a weak rock. In arid regions, it is resistant to erosion by solution and forms cliffs. Chemical erosion, especially in limestone terrains, also occurs in stream channels and in the small particles of soluble rock carried by streams.

Hydraulic Action

The force of running water alone, without the tools of abrasion, can quickly erode soft, unconsolidated material, such as soil, clay, sand, and gravel. It is especially effective if it moves at high velocities. Although directed **hydraulic** action cannot rapidly erode large quantities of solid bedrock,

the pressure exerted by moving water into cracks and bedding planes can remove slabs and blocks from the channel floor and sides.

An important factor in the downcutting of a stream channel is the upstream migration of waterfalls and rapids. Here, again, the process is simple but important. It can be appreciated by considering the erosion of Niagara Falls (*figure 11.4*). The increased velocity of the falling water sets up strong turbulence at the base of the falls, which rapidly erodes the weak, nonresistant rock layers. The cliff is gradually undermined and the falls retreat upstream.

Figure 11.4 Retreat of Niagara Falls upstream occurs as hydraulic action undercuts the weak shale below the limestone. The Niagara River originated as the last glacier receded from the area and water flowed from Lake Erie to Lake Ontario over the Niagara cliffs. Erosion causes the waterfalls to migrate upstream at an average rate of 1.3 m per year.

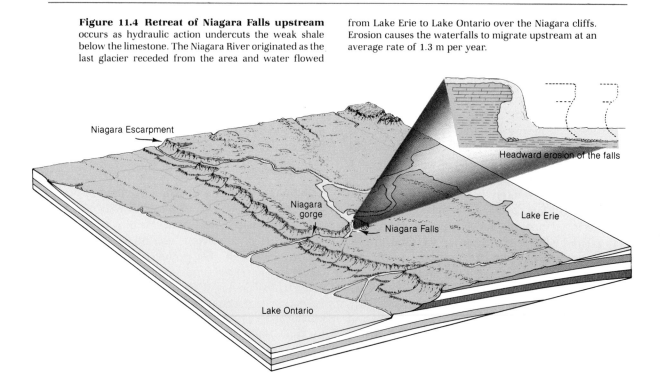

HEADWARD EROSION AND THE GROWTH OF DRAINAGE SYSTEMS UPSLOPE

Statement

In the process of stream erosion and valley evolution, there is a universal tendency for a stream to erode headward, or upslope, and increase the length of its valley until it reaches the divide. Every tributary is involved in this process until the valley extends upslope as far as the divide.

Discussion

Headward erosion can be analyzed by reference to *figure 11.5*. The reason erosion is more vigorous at the head of a valley than on its sides is apparent from the relationships between the valley and the regional slope. Above the head of a valley, water flows down the regional slope as a sheet (sheet flow), but it begins to converge to a point where a definite stream channel begins. As the water is concentrated into a channel, its velocity and erosive power increase far beyond that of the slower-moving sheet of water on the surrounding, ungullied surface. The additional volume and velocity erode the head of the valley much faster than sheet flow erodes ungullied slopes or the valley walls. Thus, the head of the valley is extended upslope. In addition, ground water moves toward the valley, so that the head of the valley is a favorable location for the development of springs and seeps. These, in turn, help to undercut overlying resistant rock and cause headward erosion to

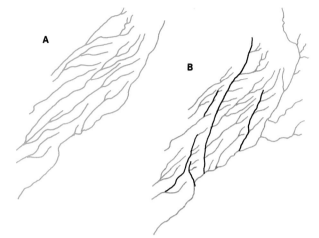

Figure 11.6 Stream piracy occurs where a tributary with a high gradient rapidly erodes headward and captures the tributary of another stream.

occur much faster than the retreat of the valley walls.

Careful study of the relationship between the drainage network and regional slope (*figure 11.5*) shows that a tributary's position is fixed at the point where it joins the next larger stream segment. Erosion along this part of the tributary is restricted to downcutting and valley widening. Thus, the drainage system can grow only by extending existing gullies up the regional slope or by developing new tributaries along the walls of the larger stream. Inevitably, the gullies extend up the slope of the undissected surface toward the gathering grounds of the waters that feed the stream. Thus, headward erosion is a basic process in the evolution of drainage systems.

Stream Piracy

With the universal tendency for headward erosion, the tributaries of one stream can extend upslope and intersect the middle course of another stream and thus divert the headwater of one stream to the other. This process, known as **stream piracy**, or stream capture, is illustrated in *figure 11.6*.

Stream piracy is most likely to occur if headward erosion of one stream is favored by a steeper gradient or by a course in more easily eroded rocks. Some of the most spectacular examples occur in the folded Appalachian Mountains, where nonresistant shale and limestone are interbedded with resistant quartzite formations. The process of stream capture and the evolution of the region's drainage system are shown in the series of dia-

Figure 11.5 Headward erosion is the basic mechanism by which a drainage system is extended upslope. Water flows as a sheet down the regional slope (away from the foreground of the figure). As it converges toward the head of a tributary valley, its velocity is greatly increased, so that its ability to erode also increases. The tributary valley is thus eroded headward, up the regional slope.

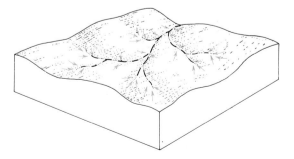

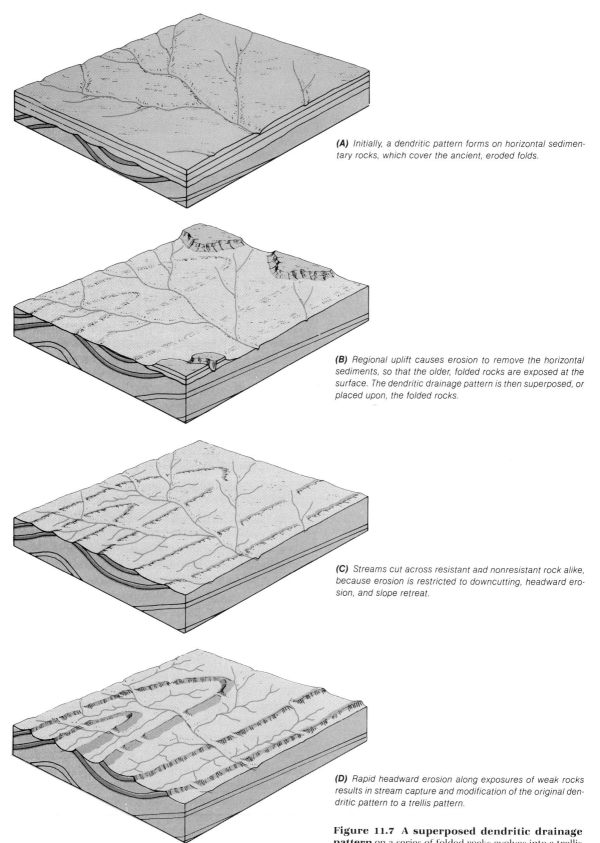

(A) Initially, a dendritic pattern forms on horizontal sedimentary rocks, which cover the ancient, eroded folds.

(B) Regional uplift causes erosion to remove the horizontal sediments, so that the older, folded rocks are exposed at the surface. The dendritic drainage pattern is then superposed, or placed upon, the folded rocks.

(C) Streams cut across resistant and nonresistant rock alike, because erosion is restricted to downcutting, headward erosion, and slope retreat.

(D) Rapid headward erosion along exposures of weak rocks results in stream capture and modification of the original dendritic pattern to a trellis pattern.

Figure 11.7 A superposed dendritic drainage pattern on a series of folded rocks evolves into a trellis pattern as headward erosion proceeds along nonresistant rock formations.

grams in *figure 11.7.* The original streams flowed in a **dendritic pattern** (a branching, treelike pattern) on horizontal sediments that once covered the folds. As uplift occurred, erosion removed the horizontal sediments, and the dendritic drainage pattern became **superposed,** or placed upon, the folded rocks beneath, so that it cuts across weak and resistant rocks alike. As the major stream cuts a valley across the folded rocks, new tributaries rapidly extend headward along the nonresistant formations. By headward erosion, these new streams progressively capture the superposed tributaries and change the dendritic drainage pattern to a **trellis pattern.**

Extensive stream piracy and development of a trellis drainage pattern can be seen almost anywhere folded rocks are exposed at the surface. In the folded Appalachian Mountains, the major streams that flow to the Atlantic (such as the Susquehanna and the Potomac) are all superposed across the folded strata. Their tributaries, however, flow along the nonresistant rocks parallel to the geologic structure and have captured many superposed streams.

Another example is the Pecos River in New Mexico. By extending itself headward to the north along the weak shale and limestone, it has captured a series of eastward-flowing streams that once extended from the Rockies across the Great Plains. The original eastward drainage (shown in *figure 11.8A)* resulted from the uplift of the Rocky Mountains. Now, the headwaters of most of the original streams have been captured by the Pecos River. Water that once would have flowed across the Llano Estacado (the High Plains of Texas) now flows down the Pecos valley (*figure 11.8B).*

In summary, headward erosion is one of the most important processes by which a drainage system evolves. It continues to extend the drainage network upslope until the tributaries reach the crest of the divide (the ridge separating two drainage systems). There, it is effectively eliminated, because no undissected slope remains. Subsequent erosion is restricted to downcutting and valley widening. Therefore, a drainage system grows in a specific way. Once sheet flow is concentrated into a channel, the future development of the drainage is somewhat predictable.

Figure 11.8 The evolution of the Pecos River occurred as headward erosion extended the drainage network northward and captured the eastward-flowing streams.

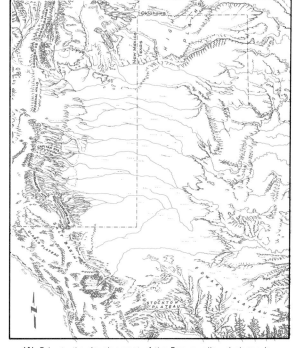

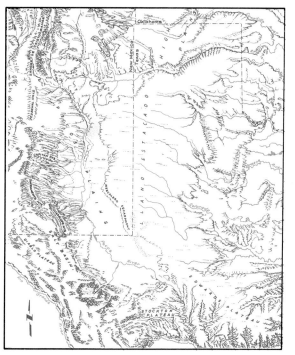

(A) *Prior to the development of the Pecos valley, drainage is believed to have been eastward from the Rocky Mountains across the Great Plains.*

(B) *Headward erosion of the Pecos River northward along the nonresistant rocks of the Pecos plains captured the headwaters of the eastward-flowing streams.*

THE EXTENSION OF DRAINAGE SYSTEMS DOWNSLOPE

Statement

In addition to downcutting and headward erosion, a drainage system can grow in length simply by extending its course downslope as sea level falls or as the landmass rises. This process is probably fundamental in determining the original course of many major streams, especially in the interior lowlands, where the oceans once covered much of the continents and then slowly withdrew. As the oceans retreated, drainage systems were extended down the newly exposed slopes. Later, they were modified by headward erosion and stream piracy.

Discussion

Tidal channels along coastal plains are an example of the beginning of a new segment of a drainage system as a result of a fall in sea level. The pattern

---

Figure 11.9 Tidal channels develop a characteristic dendritic drainage pattern between high tide and low tide, because the sediment is homogeneous and the tidal flat slopes gently seaward.

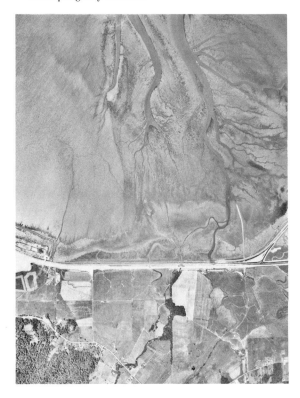

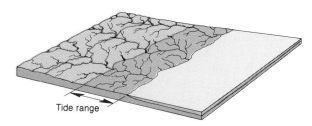

(A) *In the original position of the shore, tidal channels develop between high tide and low tide.*

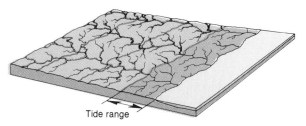

(B) *As the sea level falls and the shoreline recedes, tidal channels become part of the permanent drainage system.*

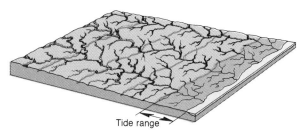

(C) *With each successive retreat of the shoreline, new tidal channels develop and drainage is extended further downslope. A dendritic drainage pattern typically is produced on the homogeneous tidal-flat material.*

Figure 11.10 Extension of a drainage system downslope occurs as a shoreline recedes. This downslope extension commonly results in a dendritic pattern.

---

of land and tidal drainage is shown in *figure 11.9.* It is characteristically dendritic, because the material upon which this drainage is established consists of recently deposited horizontal sediments. If the slope is pronounced, however, the tributaries, as well as the major streams, flow parallel for a long distance. If sea level were to drop, the streams would continue to flow downslope, following the courses established by tidal channels, as is shown in *figure 11.10.* Major streams would extend their drainage patterns over the deltas that they deposit. If sea level is falling, the youngest part of a river is near the shoreline and upslope, where headward erosion develops new channels.

MASS MOVEMENT AND SLOPE RETREAT

Statement

The valley walls associated with a stream system are subject to a variety of slope processes that cause them to recede from the river channel. These processes, called mass movement, can be rapid and devastating (such as a great landslide) or imperceptibly slow (such as creep down the gentle slope of a grass-covered field).

Gravity is the driving force behind all slope processes. Mass movement therefore is not limited to stream valleys but occurs on all slopes, including sea cliffs and fault-block mountain fronts, and on the slopes of craters on other planets. The force of gravity is continuous, of course, but it can move material only if it can overcome inertia. Any factor that tends to reduce resistance to motion aids mass movement. Such factors include saturation of the material by water, oversteepening of slopes through undercutting by streams or waves, alternating freezing and thawing, and vibrations from earthquakes. Water is an important factor in mass movement, because it lubricates the unconsolidated material and adds weight to the mass of material.

Mass movement of surface material downslope is the most universal of erosional processes. It occurs under all geologic conditions on the earth—on the steep slopes of high mountains, the gently rolling plains, the sea cliffs, and the slopes beneath the oceans—and on the moon, Mars, and other planets.

Discussion

The mechanics of **mass movement** can be very complicated. For convenience we will consider examples from two groups: (1) predominantly rapid downslope movements and (2) predominantly slow movements. Some examples of each are shown in the diagrams in *figure 11.11.*

Rockfalls—free falls of rocks from steep cliffs —are the most rapid types of mass movement. The sizes of falling rocks range from large masses that break loose from the face of a cliff to small fragments loosened by frost action and other weathering processes. In areas where rockfalls are common, the debris can be seen at the bases of cliffs as talus.

Rockslides are rapid movements of rock material along a bedding plane or another plane of structural weakness. Rockslides can move as a large block for a short distance, but, more frequently, they break up into smaller blocks and rubble. They usually occur on steep mountain fronts, but they can develop on slopes with gradients as low as 15°. Rockslides are among the most catastrophic of all forms of mass movement. Sometimes millions of metric tons of rock plunge down the side of a mountain in a few seconds (*figure 11.12*).

Debris slides are rapid movements of solid and loose rock fragments. They are composed of dry or moderately wet, loose rock fragments, moving rapidly over the underlying bedrock.

Debris flows are rapid flows of rock fragments and water. They consist of mixtures of rock fragments, mud, and water that flow downslope as a thick, viscous fluid. Movement can range from that similar to the flow of freshly mixed concrete to that of a stream of fluid mud in which rates of flow are equal to those of rivers. Many debris flows begin as slumps and continue as flows near the lower margins of the slump block.

Mudflows are a variety of debris flow consisting of a large percentage of silt and clay-size particles. They almost invariably result from an unusually heavy rain or a sudden thaw. Water content can be as much as 30%. As a result of the predominance of fine-grained particles and the high water content, mudflows typically follow stream valleys. They are common in arid and semiarid regions and typically originate in steep-sided gullies where there is abundant loose, weathered debris (*figure 11.13*). As they reach the open country at a mountain front, they spread out in the shape of a large lobe, or **fan.** Because of their great density, mudflows can transport large boulders by "floating" them over slopes as gentle as 5° and have been known to move houses and barns from their foundations. Many of the disastrous landslides in southern California are really mudflows that move rapidly down a valley for considerable distances. Mudflows vary in size and rate of flow, depending on water content, slope angle, and available debris. Many are over 100 m thick, and some can be as much as 80 km long.

Subaqueous sand flows are a type of mass movement that occurs beneath the water. They are the most common type of subaqueous mass movement, but rockfalls and slump blocks also occur underwater.

Creep is an extremely slow, almost imperceptible downslope movement of soil and rock debris. The motion is so slow that it generally is difficult to

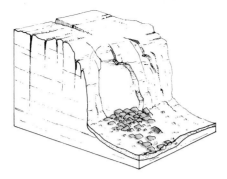

(A) **A rockfall** *is the free fall of rock from steep cliffs.*

(B) **A rockslide** *is the rapid downslope movement of rock material along a bedding plane, joint, or other plane of structural weakness.*

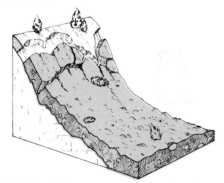

(C) **A debris slide** *is the rapid movement of soil fragments and loose bedrock. The mass can be dry or moderately wet.*

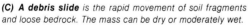
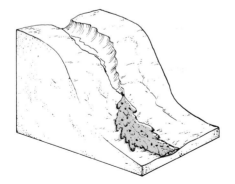

(D) **A debris flow** *is the rapid flow of a mixture of rock fragments, soil, mud, and water. The mixture generally contains a large proportion of mud and water. Mudflows are a common variety.*

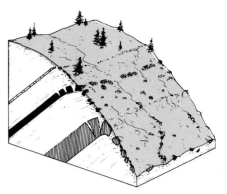

(E) **Creep** *is the slow downslope migration of soil and loose rock fragments, resulting from a variety of processes, including frost heaving.*

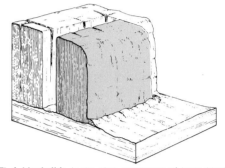

(F) **A blockslide** *is the slow movement of large blocks of material over a layer of weak, plastic material (such as clay or shale).*

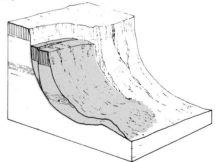

(G) **Slump** *is the slow or moderately rapid movement of a coherent body of rock along a curved rupture surface.*

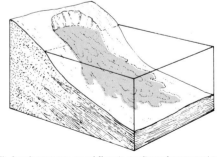

(H) **A subaqueous sand flow** *is the flow of saturated sand or silt beneath the surface of a lake or an ocean.*

Figure 11.11 Mass movement takes various forms, all of which produce slope retreat and enlarge valleys. Examples of various types of mass movement are illustrated in the diagrams.

Figure 11.12 A recent rockfall in the Swiss Alps is a good example of mass movement and slope retreat. The process will be repeated until the mountain is completely removed.

observe directly, but it is expressed in a variety of ways. On weakly consolidated, grass-covered slopes, evidence of creep can be seen as bulges or low, wavelike swells. In road cuts and stream banks, creep can be expressed by the bending of steeply dipping strata in a downslope direction, or the movement of blocks of a distinctive rock type downslope from their outcrop. Additional signs of creep include tilted trees and posts, displaced monuments, deformed roads and fence lines, and tilted retaining walls (*figure 11.14*). The slow movement of large blocks (blockslides) can be considered a type of creep.

Many factors combine to cause creep. Undoubtedly, soil moisture is important, because it weakens the soil's resistance to movement. In cold regions, creep is produced by **frost heaving.** Water percolates into the pore spaces of the rocks, freezes, and expands, lifting the ground surface at a right angle to the slope. When the ground thaws, each particle drops vertically, coming to rest slightly downslope from its original position (*figure 11.15*). Repeated freezing and thawing cause the

particles to move downslope in a series of zigzags. Wetting and drying can cause a similar motion of loose particles, since moisture causes expansion of clay minerals.

Many other factors contribute to creep. Growing plants exert a wedgelike pressure between rock particles in the soil and thus cause them to be displaced downslope. Burrowing organisms also displace particles, so that the force of gravity can move them.

Landslides comprise a wide variety of predominantly rapid earth movements, including the subtle slumping of stream banks or sea cliffs as well as the more obvious sliding (slipping) of mountainsides. Both unconsolidated regolith and solid bedrock can move in landslides. A landslide block moves as a unit (or a series of units) along a definite plane. In contrast to debris flows, which move as viscous fluids, landslides occur when gravitational stress causes a definite fracture or system of fractures to develop in a rock body. Much of the material moves as a large block, downward and outward along a curved plane, in a type of movement called **slump.** In the lower part of a slump, debris flow can occur. Slumping is common in weak shales that are semiconsolidated. In some areas, large blocks can be as much as 5 km long and 150 m thick. They can move in a matter of seconds or can gradually slip throughout a period of several weeks.

Solifluction is a downslope movement of regolith saturated with water. It is a special type of earthflow, which is common in arctic and subarctic regions, where the subsoil is permanently frozen. During the spring and early summer, the ground begins to thaw from the surface downward. Since

Figure 11.13 Mudflows typically originate on steep slopes where there is an abundance of unconsolidated shale, volcanic ash, or thick regolith. They commonly spread out in the form of a large fan, or lobe.

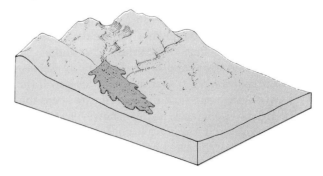

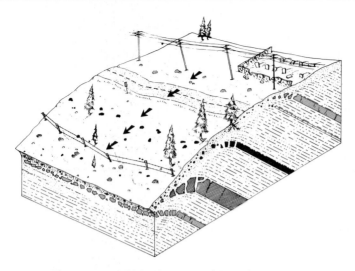

Figure 11.14 Creep is expressed in various ways, such as the downslope displacement of fence lines, roads, telephone poles, headstones in cemeteries, and rock debris from original outcrops.

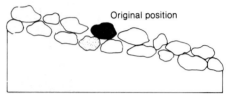

(A) Water seeps into the pore spaces between fragments of loose rock debris.

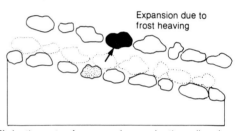

(B) As the water freezes and expands, the soil and rock fragments are lifted perpendicularly to the ground surface.

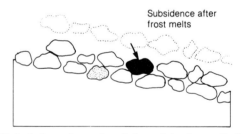

(C) As the ice melts, gravity pulls the particles down vertically, displacing them slightly downhill.

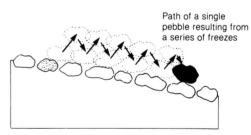

(D) Repeated freezing and thawing causes a significant net displacement downslope.

Figure 11.15 Creep can result from repeated frost heaving. With each freeze, there is a net downslope displacement of all loose material.

the meltwater cannot percolate downward into the impermeable **permafrost,** the upper zone of soil is always saturated and slowly flows down even the most gentle slopes. Solifluction can also occur in temperate regions, in nonfrozen soil, if a sufficient amount of water accumulates in the upper soil.

Rock glaciers are glacierlike bodies of rock fragments slowly flowing downslope. These tonguelike masses of angular rock debris, resembling glaciers, form in cold regions, such as Alaska, the high Rockies, and other alpine areas, and move as a body downslope at rates ranging from 3 cm per day to 1 m per year. Evidence of their flow includes concentric ridges within the body, its lobate form, and its steep front. Excavations into rock glaciers reveal ice in the pore spaces between the rock fragments. Presumably, the ice is responsible for much of the flow movement. With a continuous supply of rock fragments from above, the constantly increasing weight causes the interstitial ice to flow. Favorable conditions for the development of rock glaciers thus include steep cliffs and a cold climate. The steep cliffs supply coarse rock debris, which allows large spaces in which ice can form between fragments, and the cold climate keeps the ice frozen. Some rock glaciers, however, may be debris-covered, formerly active glaciers (no longer active because the rate of flow is less than the accepted minimum rate for active glaciers).

SLOPE SYSTEMS

Statement

The previous section described some ways in which rock debris is transported downslope by gravity. Now we will consider slopes as systems in which the effects of weathering, mass movement, and headward erosion of small gully tributaries combine to transport material down to a main stream. The net result is slope retreat. The diagram in *figure 11.16* illustrates the major components of a slope system and shows how they function.

The resistant rock units break up into blocks and fragments, which accumulate at the base of the cliff. This coarse talus then moves further downslope, chiefly by rockslides and by rolling. Downslope, the debris moves slowly by creep. In the process, the rock particles are weathered and broken into smaller fragments. The fine-grained rock debris then gradually enters a river, where it is carried away by running water.

Discussion

The talus slope can be thought of as a local open system, with an input of coarse rock debris at the base of the cliff and an output of fine rock fragments into a stream. As the system operates, the valley slope gradually retreats from the stream channel, although the shape of the profile can remain essentially constant.

The diagram in *figure 11.16* illustrates how mass movement is involved in **slope retreat.** Remember that headward erosion is also active. Numerous tributaries and small gullies operate on essentially all slopes, as they collect and funnel debris to the main stream. These small tributaries work headward, undercutting the resistant cliff, and thus make it more susceptible to rockfalls and slumping. During heavy storms, much of the loose rock debris on the slope is moved in the tributary system. Debris flows commonly follow the tributary stream channels. Therefore, in the slope system, mass movement and headward erosion of minor tributaries operate together. Each facilitates and reinforces the other in transporting debris and causing slope retreat.

To appreciate the importance of mass movement and headward erosion of minor tributaries in the process of valley development, try to visualize what a valley would look like without these processes. If downcutting of the stream channel were the only process in operation, a deep, vertical-walled canyon would result (*figure 11.17A*). This does occur where rocks are strong enough to resist the force of gravity, such as in the deep canyons cut in resistant sandstones and limestones in the Colorado Plateau (*figure 11.1*). More commonly, slope retreat occurs contemporaneously with downcutting, to produce sloping valley walls (*figure 11.17B*).

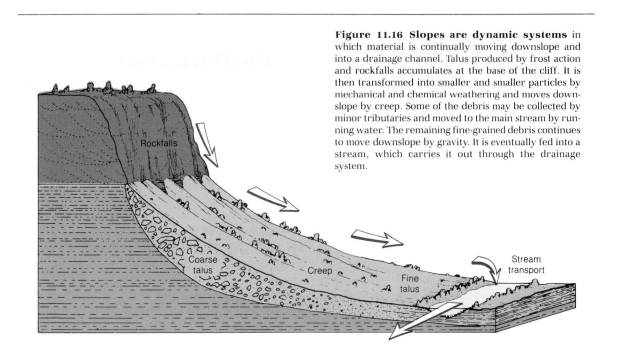

Figure 11.16 Slopes are dynamic systems in which material is continually moving downslope and into a drainage channel. Talus produced by frost action and rockfalls accumulates at the base of the cliff. It is then transformed into smaller and smaller particles by mechanical and chemical weathering and moves downslope by creep. Some of the debris may be collected by minor tributaries and moved to the main stream by running water. The remaining fine-grained debris continues to move downslope by gravity. It is eventually fed into a stream, which carries it out through the drainage system.

Rockfalls

Coarse talus

Creep

Fine talus

Stream transport

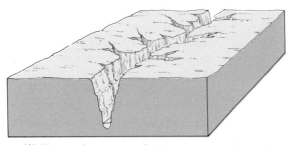

(A) *Where rocks are very resistant, mass movement cannot keep pace with downcutting, so that a vertical-walled canyon develops.*

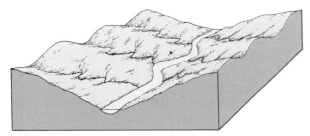

(B). *If slope processes can keep pace with downcuttting, gently sloping valley walls develop.*

Figure 11.17 Slope processes affect the development of landscapes in important ways.

One of the most important features of slopes is that, although most of them may appear to be stable and static, they actually are dynamic systems, in which surface material is constantly moving downslope in a variety of ways. Large, catastrophic landslides and debris flows are the most hazardous, but imperceptible creep and mass movements of moderate magnitude probably move the most material. Downslope movements are natural events that occur everywhere. The artifical modification of slopes, however, by the construction of dams, roads, and buildings can cause an increase in mass movement or, where stabilizing structures are built, a temporary reduction. Whether catastrophic or slow-moving, the mass movement of slope material represents a major geologic hazard and a continual challenge for effective land-use planning.

Effects of Artificial Modification of Slope Systems

Many catastrophic landslides are the natural result of high, steep slopes with weak surface material and occur regardless of human activity. They are commonly triggered by exceptionally heavy snowfalls or rainfalls or by earthquakes. For example, during the earthquake in Peru on May 31, 1970, 4000 m³ of debris roared down the slope of Mount Huascarán at a speed of more than 300 km per hour. The debris killed approximately 40,000 people as it covered the town of Yungay with as much as 14 m of mud and rock. Similar types of mass movement can be expected to occur on the high mountainous areas in the future.

Many landslides, however, are actually caused by artificial manipulation of the landscape, though the event may be initiated by excess moisture. An example is the slide that occurred at Vaiont Dam, in northern Italy. This, the worst dam disaster in history, resulted from a huge landslide into the Vaiont Reservoir on October 9, 1963. The landslide began as a relatively slow creep over a 3-year period. The rate of creep ranged to as much as 7 cm per week until a month before the catastrophe, when it increased to 25 cm per day. On October 1, animals grazing on the slopes sensed the danger and moved away. Finally, on the day before the slide, the rate of creep was about 40 cm per day. Engineers expected a small landslide but did not realize until the day before the disaster that a large area of the mountain slope was moving en masse at a uniform rate.

When the slide broke loose, more than 240,000,000 m³ of rock rushed down the hillslope and splashed into the reservoir. It produced a wave of water over 100 m high, which swept over the dam and completely destroyed everything in its path for many kilometers downstream. The entire catastrophic event, including the slide and flood, lasted only 7 minutes, but it took approximately 2600 lives and caused untold property damage.

Several adverse geologic conditions contributed to the slide. The rocks consisted of weak limestone with interbedded thin layers of clay. The beds dipped steeply toward the reservoir, so that strong driving (gravitational) forces were created. But it was the rising of water level in the reservoir that reduced the slope's resistance to shear and caused the slide.

The well-known and frequent landslide problems in southern California are commonly associated with uncontrolled development of hillsides underlain by poorly consolidated rock material. Landslides would naturally be common in the area, but as slopes are artificially modified for building sites and roads, the magnitude and frequency of mass movement increase greatly. This results in the loss of millions of dollars' worth of property each year. Careful geological investigations and proper land-use planning could greatly reduce these losses.

MODELS OF STREAM EROSION

Statement

The evolution of a drainage system may require tens of millions of years, so that the origin of stream valleys can be studied only indirectly. One approach is to study the nature of drainage systems in newly exposed areas, such as coastal plains and former lakebeds. Another is to use computer models to analyze relationships among the major variables—downcutting, headward erosion, slope retreat, and uplift. The results of these studies indicate that a river system evolves through a series of predictable stages. Careful observation in the field can reveal the major trends of landscape development in a given area.

Discussion

Drainage Systems on Former Lakebeds

The development of a drainage system by downcutting, headward erosion, slope retreat, and extension of the drainage net downslope can be seen in the photograph in *figure 11.18*. This area was once occupied by a prehistoric lake, whose remnants can be seen in the background. The well-defined **terrace** midway up the mountain front marks its highest level, and lower terraces formed as the lake receded.

The lake levels provide reference lines with respect to the growth of the drainage network. Above the highest lake level, stream erosion has continued without interruption. The drainage system was established before the lake was created, and downcutting, headward erosion, and slope retreat have produced the dissected mountain front. Note that the valleys above the lake level are much wider than those below, because they have been subjected to a much longer period of slope retreat.

When the lake came into existence, the drainage system ended at the upper shoreline, and sediment carried by the streams was deposited in the lake. These deposits were redistributed by waves to form the prominent terrace that appears in the middle of the photograph. As the lake level fell, the major streams extended downslope, and new, sub-parallel streams formed on the newly exposed lakebed. Many of these terminated at the lower shoreline, because the lake level remained stationary for some time at this elevation. The lowest zones have been exposed only since the lake receded to its present level. This surface is marked by the extension of the major drainage lines downslope.

Within this small area, all of the major processes of drainage development can be observed, including headward erosion, downcutting, slope retreat, and extension of the drainage system downslope as the shoreline retreats.

Figure 11.18 The shoreline of ancient Lake Bonneville illustrates all of the major processes involved in the dissection of the land by running water. Downcutting and slope retreat occur in all segments of the streams. Headward erosion is extending the drainage network upslope in the mountains above the shoreline. As the shore of the ancient lake retreated, the drainage was also extended downslope.

Computer Models of the Erosion of the Grand Canyon

Geologists have studied the interaction of downcutting and slope retreat by means of a computer model of the Colorado River's erosion of the rock sequence in the Grand Canyon area. Variables such as rates of downcutting and slope retreat affecting various rock formations were analyzed. This study produced a series of hundreds of computer-calculated profiles of the Grand Canyon, showing changes between the time the Colorado River began cutting through the Colorado Plateau and the present. Several of these profiles are shown in *figure 11.19*. Erosion of the model canyon was not uniformly fast or slow but occurred in a series of pulses. Downcutting of the main stream was extremely rapid and was largely a function of the rates of uplift. In areas of little uplift, erosion was slow. In areas of rapid uplift, erosion was correspondingly fast, and the stream maintained a smooth, gently curving longitudinal profile.

On minor intermittent streams, downcutting was slow where resistant rocks were encountered, and rapid erosion occurred on nonresistant strata. The main river, however, cut through resistant and nonresistant rocks with almost equal ease. The rate of slope retreat was shown to be intimately related to the rate of downcutting.

Although this model cannot be verified directly, you can get a glimpse of the stages of canyon development by studying the canyon longitudinally (*figure 11.20*). Upstream, near Lees Ferry, the river flows on top of the Kaibab Limestone. Here, the entire sequence of strata exposed farther downstream in the Grand Canyon is below the surface. The canyon cut into the resistant formation is a narrow gorge in the Kaibab Limestone. Farther downstream (in the foreground of the figure), uplift permitted the river to cut much deeper into the rock sequence, and the profile that has formed is similar to the one developed by the computer model. Thus, evidence of the evolution of slope morphology from the canyon itself supports the findings of the computer model.

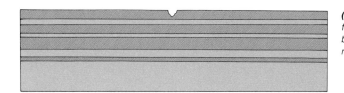

(A) *The original undissected surface* is a flat, stepped surface formed on horizontal sedimentary rock. The underlying bedrock contains a sequence of alternating resistant and nonresistant rock types.

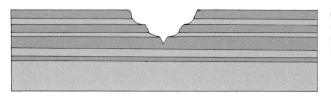

(B) *Initial dissection and slope retreat* occur as a result of uplift. Downcutting by the stream is slow on resistant rock units. Slope retreat causes nonresistant rocks to recede from the river, so that a terrace is left on the resistant rock layers.

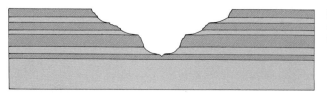

(C) *Downcutting accelerates* as the river breaks through a resistant formation and rapidly erodes the weaker underlying rocks.

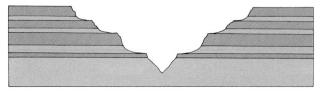

(D) *Downcutting continues* as differential slope retreat produces alternating cliffs and terraces.

Figure 11.19 A series of profiles representing the evolution of the Grand Canyon has been constructed from a computer model.

Figure 11.20 The effects of erosion of the eastern Grand Canyon are seen in this high-altitude photograph. At Lees Ferry, in the eastern Grand Canyon, the river is just beginning to cut through the rock sequence and has produced a profile like the one in figure 11.19B. Downstream, uplift has permitted the river to cut deeper, so that it has produced profiles similar to the ones in figures 11.19C and 11.19D.

FLOOD-PLAIN DEPOSITS

Statement

Erosion is the dominant process in the high headwaters of a drainage system. In the transporting subsystem of a river system (which commonly occurs across a stable platform or a shield), the surface of the land slopes gently toward the ocean, and the stream gradient is very low. There, the river system is unable to transport all of its sediment load, and some deposition occurs. The low gradient in the transporting segment causes the development of a number of landforms that are not found in parts of the system with higher gradients. Foremost among these are

1. Flood plains
2. Meanders
3. Point bars
4. Natural levees
5. Backswamps

Discussion

A schematic diagram showing features commonly developed on a river **flood plain** is shown in *figure 11.21*. It serves as a simple graphic model of flood-plain sedimentation.

Meanders and Point Bars

All rivers naturally tend to flow in a sinuous pattern, even if the slope is relatively steep. This is because water flow is turbulent, and any bend or irregularity in the channel deflects the flow of water to the opposite bank. The force of the water striking the stream bank causes erosion and undercutting, which initiate a small bend in the river channel. In time, as the current continues to impinge on the outside of the channel, the bend grows larger and is accentuated, and a small bend ultimately grows into a large **meander** bend (*figure 11.22*; see also *plate 9*). On the inside of the meander, velocity is at a minimum, so that some of the sediment load is deposited. These deposits occur on the point of the meander bend and are called a **point bar.** These two major processes, erosion on the outside of the meander and deposition on the inside, cause meander loops to migrate laterally.

Because the valley surface slopes toward the ocean, erosion is more effective on the downstream side of the meander bend. Thus, the meander also migrates slowly down the valley (*figure 11.22*). As a meander bend becomes accentuated, it develops an almost complete circle. Eventually, the river channel cuts across the meander loop and follows a more direct course to the ocean.

Figure 11.21   The major features of a flood plain include meanders, point bars, oxbow lakes, natural levees, backswamps, and yazoo streams. A stream flowing around a meander bend erodes the outside curve and deposits sediment on the inside, to form a point bar. The meander bend migrates and is ultimately cut off, to form an oxbow lake. Natural levees build up the banks of the stream, and backswamps develop on the lower surfaces of the flood plain. Yazoo tributaries find it difficult to enter the main stream and flow parallel to it for considerable distances. Slope retreat continues to widen the low valley, which is partly filled with river sediment.

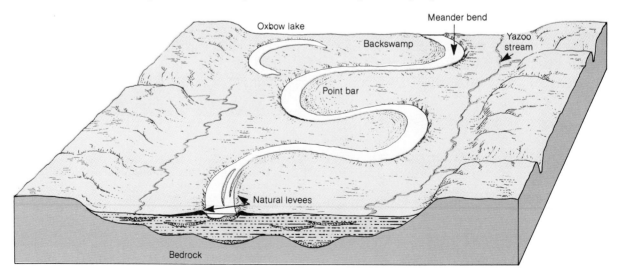

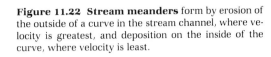

Figure 11.22 Stream meanders form by erosion of the outside of a curve in the stream channel, where velocity is greatest, and deposition on the inside of the curve, where velocity is least.

(A) *Patterns of stream flow are deflected by any irregularity and move to the opposite bank, where erosion begins.*

(B) *Once the bend begins to form, the flow of water continues to impinge on its outside curve, so that a meander loop develops. At the same time, deposition occurs on the inside of the bend as a result of the lower stream velocities in that area.*

(C) *The meander is enlarged and migrates laterally, with the contemporaneous growth of a point bar. There is a general downslope migration of meanders as they grow larger and ultimately cut themselves off to form oxbow lakes.*

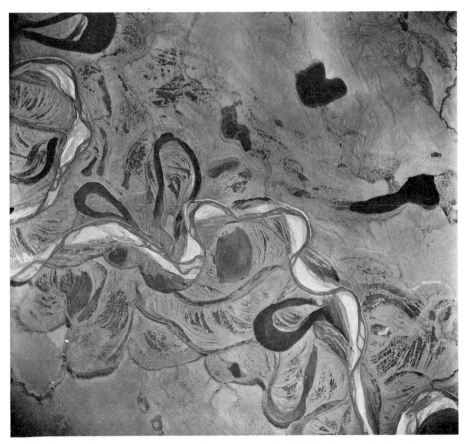

Figure 11.23 Flood-plain features in Canada were photographed from a height of 6 km. Note the meanders, cutoffs, oxbow lakes, and old meander scars.

The meander cutoff forms a short but sharp increase in stream gradient, so that the river completely abandons the old meander loop. It remains as a crescent-shaped lake called an **oxbow lake** (*figure 11.23*).

Natural Levees

Another key process operating on a flood plain is the development of natural **levees.** If a river overflows its banks during flood stage, the water is no longer confined to a channel but flows over the surface as a broad sheet. This flow pattern significantly reduces the water's velocity, and some of the sediment settles out. The coarsest material is deposited close to the channel, where it builds up an embankment known as a natural levee. Natural levees grow with each flood, and they soon form high embankments. Thus, a river actually can build its channel higher than the surrounding area (*figure 11.24*).

Backswamps

As a result of the growth and development of natural levees, much of the flood plain may be below river level. This area, known as the **backswamp,** is poorly drained and commonly is the site of marshes and swamps. Tributary streams are unable to flow up the slope of the natural levees, so that they are forced either to empty into the backswamp or to flow as **yazoo streams,** running parallel to the main stream for many kilometers before they connect with it. Strangely enough, then, the highest parts of the flood plain may be along the natural levees immediately adjacent to the river.

The aerial photograph in *figure 11.25* displays various flood-plain features. There are many excellent examples of flood-plain deposits, because most streams that flow across the lowland of the continental interior have a low gradient and deposit part of their sediment load. The lower Mississippi River is a well-known example. Between Cairo, Illinois, and the Gulf of Mexico, the Mississippi meanders over a broad flood plain, and much of the sediment it carries is deposited in natural levees, oxbow lakes, and backswamps. The dynamics of the river and the changes that it can bring about by deposition are illustrated by the fact that, during the period from 1765 to 1932, the river abandoned 19 meanders between Cairo and Baton Rouge, Louisiana. Now the Mississippi is controlled by dams and artificial levees, which have modified its hydrology, much as the Nile and Colorado rivers have been artificially manipulated.

Braided Streams

If streams are supplied with more sediment than they can carry, they deposit the excess materials on the channel floors as sandbars and gravel bars. This deposition forces a stream to split into two or more channels, so that it forms an interlacing network of channels and islands (*figure 11.26*), called a **braided stream.**

Braided streams have wide, shallow channels. The pattern is best developed in rivers that are heavily laden with erosional debris from a high mountain range. As such a river emerges from the mountain front, its velocity is rapidly reduced by the abrupt change in gradient, and part of the load, including all of the coarser debris, is dropped.

Melting ice caps and glaciers also produce favorable conditions for braided streams, because

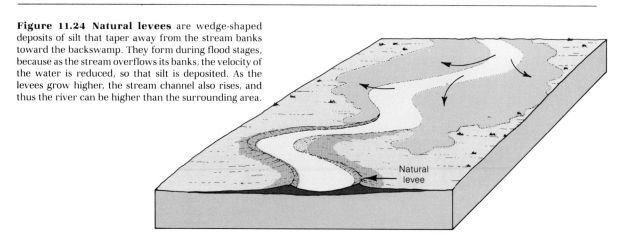

Figure 11.24 Natural levees are wedge-shaped deposits of silt that taper away from the stream banks toward the backswamp. They form during flood stages, because as the stream overflows its banks, the velocity of the water is reduced, so that silt is deposited. As the levees grow higher, the stream channel also rises, and thus the river can be higher than the surrounding area.

Natural levee

Figure 11.25 Flood-plain features include (*a*) meander bends, (*b*) point bars, (*c*) natural levees, (*d*) oxbow lakes, and (*e*) backswamps.

the streams in front of these bodies cannot transport the exceptionally large load of reworked glacial debris.

Stream Terraces

We have seen in this section how a stream can deposit much of its sediment load across a flood plain, building up a deposit of **alluvium.** Deposition can be initiated by any change that reduces a stream's capacity to transport sediment. These

changes include (1) a reduction in discharge (as a result of climatic changes or of water volume lost to stream piracy), (2) a change in gradient (caused by a rise in sea level or by regional tilting), and (3) an increase in sediment load. During the last ice age, the hydrology of most rivers changed significantly. Stream runoff was increased greatly by the melting ice, and large quantities of sediment deposited by the glaciers were reworked by the streams. In addition, the climatic changes accompa-

Figure 11.26 A braided stream pattern commonly results if a river is supplied with more sediment than it can carry. Deposition occurs, causing the river to develop new channels repeatedly.

nying the ice age caused a general increase in precipitation. As a result, many streams filled part of their valleys with sediment during the ice age, and now they are cutting through that sediment fill to form **stream terraces**.

The basic steps in the evolution of stream terraces are shown in the diagrams in *figure 11.27*. In block *A*, a stream cuts a valley by downcutting and slope retreat. In block *B*, changes such as regional tilting of the land or rise of sea level cause the stream to deposit part of its sediment load and build up a flood plain, which forms a broad, flat valley floor. In block *C*, subsequent changes (such as uplift or increased runoff) cause renewed downcutting into the easily eroded flood-plain deposits, so that a single set of terraces develops on both sides of the river. Further erosion can produce additional terraces (block *D*) by the lateral shifting of the meandering stream.

Figure 11.27 The evolution of stream terraces
involves the deposition of sediment in a stream valley
and subsequent erosion. These changes can be initiated
by various factors that change a stream's capacity to
transport sediment, such as changes in climate, changes
in base level, or regional uplift.

(A) A stream cuts a valley by normal downcutting and headward erosion processes.

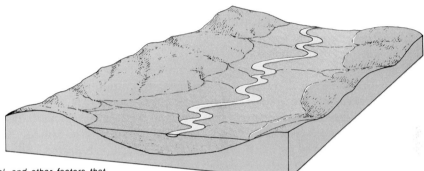

(B) Changes in climate, base level, and other factors that
reduce flow energy cause the stream to partially fill its valley
with sediments, forming a broad, flat floor.

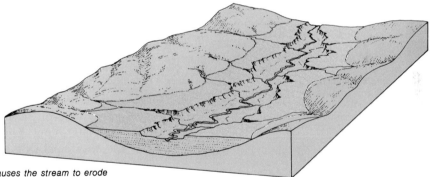

(C) An increase in flow energy causes the stream to erode
through the previously deposited alluvium. A pair of terraces
are left as remnants of the former flood plain.

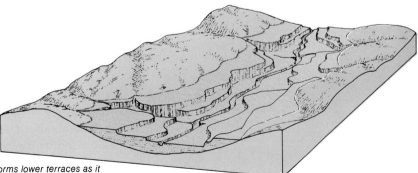

(D) The stream shifts laterally and forms lower terraces as it
erodes through the valley fill.

DELTAS

Statement

As a river enters a lake or the ocean, its velocity suddenly diminishes, and most of its sediment load is deposited to form a delta. The growth of a delta can be complex, especially for large rivers depositing huge bodies of sediments. Two major processes, however, are fundamental to the formation and growth of a delta:

1. The splitting of a stream into a distributary channel system, which extends into the open water in a branching pattern
2. The development of local breaks, or crevasses, in natural levees, through which sediment is diverted and deposited as splays in the area between the distributaries

The manner in which a delta grows depends on such factors as the rate of sediment supply, the rate of subsidence, and the removal of sediment by waves and tides. If stream deposition dominates, the delta is extended seaward. If waves dominate, the sediment delivered by the river is transported along the coast and deposited as beaches and bars. If tides are strong, sediment is transported up and down the distributary channels, forming linear

sedimentary bodies parallel to the flow of tidal currents.

Discussion

Deltaic Processes

The diagrams in *figure 11.28* illustrate the development of **distributaries,** which is one of the major processes involved in the formation of a **delta.** As a river enters the ocean (or a lake) and the flowing water is no longer confined to a channel, the currents flare out, rapidly losing velocity and flow energy. The coarse material carried by the stream is deposited in two specific areas: (1) along the margin of quiet water on either side of the main channel (these deposits build up subaqueous natural levees) and (2) in the channel at the river mouth, where there is a sudden loss of velocity (these deposits build a **bar** at the mouth of the channel). These two deposits effectively create two smaller channels (distributary channels), which extend seaward for some distance. The process is then repeated, and each new distributary is divided into two smaller distributaries. In this manner, a system of branching distributaries builds seaward in a fan-shaped pattern.

Figure 11.29 shows how the area between distributaries is filled with sediment. A local break in the levee, called a **crevasse,** forms during periods of high runoff and diverts a significant volume of water and sediment from the main stream. The escaping water spreads out and deposits its sediment to form a **splay,** which is essentially a small delta with small distributaries and systems of subsplays.

An interesting example of the growth of splays in the Mississippi Delta is documented by a series of historical maps dating back to 1838 (*figure 11.30*). The area is near the end of the major distributary just north of Head of Passes, the junction of the three largest distributaries. Before 1838, the water was confined to the distributary channels. Then a crevasse formed in the natural levee, and a splay began to form and grow. By 1903, the splay had developed a complex system of distributaries and subsplays, which filled in much of the interdistributary bay.

The sediment deposited by distributaries is vulnerable to erosion and transportation by marine waves and tides. The growth of a delta is therefore influenced by the balance between the rate of input of sediment by the river and the rate of erosion by marine processes. If waves or tides are strong,

Figure 11.28 Distributaries develop where a stream enters a lake or the sea. A bar forms at the mouth of a river channel where water that was previously confined to the channel loses velocity as it enters the ocean. The bar then diverts the water coming from the main stream into two distributary channels, which grow seaward. This process is repeated, forming branching tributaries.

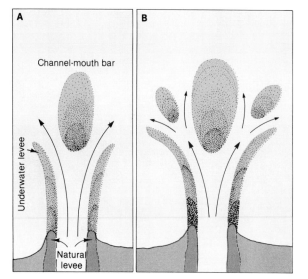

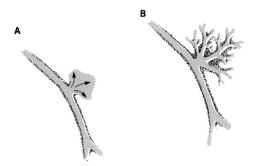
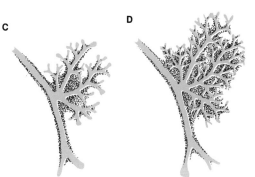
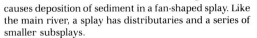

Figure 11.29 Splays form where a break in a natural levee permits part of the stream to be diverted to the backswamp. This diversion reduces the velocity and causes deposition of sediment in a fan-shaped splay. Like the main river, a splay has distributaries and a series of smaller subsplays.

(A) In 1838, a section of the main channel was bordered by a natural levee.

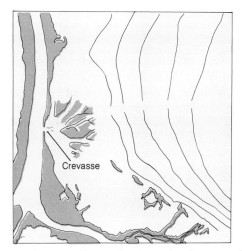

(B) In 1870, about 10 years after the opening of a crevasse in the natural levee, a splay had begun to grow, as water was diverted from the main stream and deposited its sediment on the backswamp.

(C) In 1877, about 17 years after the crevasse opened, minor splays and distributaries had been formed, extending into the area along the main channel.

Figure 11.30 The evolution of a splay in the Mississippi River is documented by a series of maps drawn between 1838 and 1903.

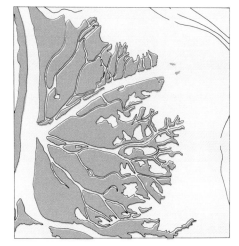

(D) In 1903, about 43 years after the crevasse opened, the splay had been extended by subsplays and minor distributaries to fill much of the area along the main channel. The splay has continued to grow by formation of subsplays and distributary channels.

the development of distributaries is limited, and the sediment is reworked into bars, beaches, and tidal flats.

A major phenomenon in the construction of deltas is the shifting of the entire course of a river. Distributaries cannot extend indefinitely into the ocean, because the gradient and capacity of the river to flow gradually decrease. The river therefore is ,eventually diverted to a new course, which has a higher gradient. This diversion generally happens during a flood. The river breaks through its natural levee, far inland from the active distributaries of the delta, and develops a new course to the ocean. The new course shifts the site of

sedimentation to a different area, and the abandoned delta is quickly attacked by wave and current action. The new, active delta builds seaward, developing distributaries and splays, until it also is abandoned, and another site of active sedimentation is formed.

Examples of Modern Deltas

Several types of deltas are illustrated in *figure 11.31*. Each shows a different balance between the forces of stream deposition and the forces reworking the sediment (waves and tides). In the Mississippi Delta, processes of river deposition dominate (*figure 11.31A*). The delta is fed by the extensive

Figure 11.31 The shape of a delta depends on the balance between fluvial and marine processes.

(A) The Mississippi Delta is dominated by fluvial processes, which produce a bird-foot extension.

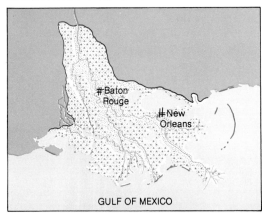

(B) The Nile Delta is dominated by wave action, which produces an arcuate delta front.

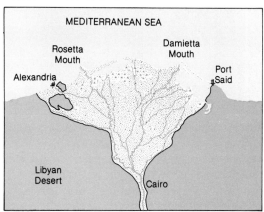

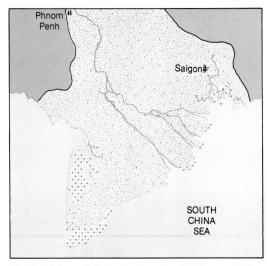

(C) The Mekong Delta is dominated by tidal forces, which produce wide distributary channels.

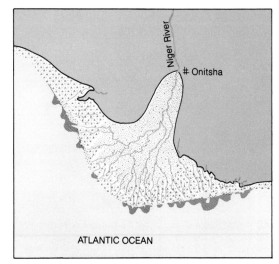

(D) The Niger Delta has formed where stream deposition, wave action, and tidal forces are about equal. An arcuate delta front and wide distributary channels are thus produced.

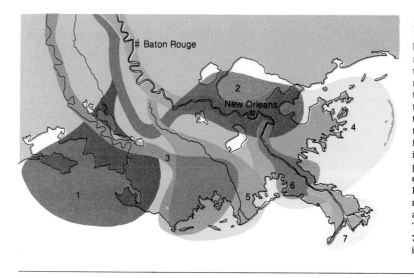

Figure 11.32 The Mississippi Delta contains seven subdeltas, built as the river shifted its course. The reason for the shifting of the channel is that the flow of the Mississippi River is relatively constant throughout most of the year, and the water is confined to the main distributary, so that deposition is concentrated in a small sector of the delta front. After one sector is built far out into the ocean, the entire flow is shifted to some other sector, and the process is repeated. Wave action then erodes back the bird-foot deltas. Previous subdeltas are indicated by numbers (1 through 6) according to age. Number 7 is the present subdelta. The active distributary system (6 and 7) has built a major bird-foot delta during the last 500 years.

Mississippi River system, which drains a large part of North America, and discharges an annual sediment load of approximately 454 million metric tons per year. The river is confined to its channel throughout most of its course, except during high floods. Therefore, most of the sediment reaches the ocean through two or three main distributary channels and has rapidly extended the river channels far into the Gulf of Mexico (see *plate 9*). This extension is known as a **bird-foot delta.**

If a bird-foot extension grows too far out to sea, the gradient of the river becomes too low for the river to flow. The river then shifts its course by breaking through its natural levee upstream, and the process is repeated. Seven major subdeltas have been constructed by the Mississippi River during the last 5000 years. These are shown in *figure 11.32.* The oldest lobe (1) was abandoned approximately 4000 years ago and since then has been eroded back and inundated. Only small remnants remain exposed today. The successive lobes, or subdeltas (2 through 7), have been modified to varying degrees. The abandoned channels of the Mississippi are well preserved and can be recognized on satellite photographs (*plate 9*). The presently active delta lobe (7) has been constructed during the last 500 years. It is apparent from river studies that the present bird-foot delta has been extended as far as the balance of natural forces permits. Without continued human intervention, the Mississippi will shift to the present course of the Atchafalaya River.

The Nile Delta differs from the Mississippi Delta in several ways. Instead of being confined to one channel, the Nile begins to split up into distrib-

utaries at Cairo, Egypt, more than 160 km inland, and fans out over the entire delta (*figure 11.31B*). Before construction of the Aswân High Dam, the Nile's annual flood briefly covered much of the delta each year and deposited a thin layer of silty mud. In the past 3000 years, 3 m of sediment has accumulated near Memphis. Two of the larger distributaries have built major lobes extending beyond the general front of the delta, but strong wave action in the Mediterranean redistributes the sediment at the delta front. The reworked sediment forms a series of arcuate barrier bars, which close off segments of the ocean to form lagoons. The lagoons form a subenvironment, which soon become filled with fine sediment. The difference between the Nile Delta and the Mississippi Delta is due largely to different balances between the influx of sediment, which builds bird-foot deltas, and the strength of wave action, which redistributes the sediment to form barrier bars.

The Mekong Delta, along the southern coast of Vietnam, is dominated by tidal currents, which redistribute sediment in the river channels and along the delta fronts (*figure 11.31C*). The distributaries branch into two main courses near Phnom Penh, about 500 km inland. Sediment carried by the river is reworked by tidal currents, forming broad distributary channels and a series of elongate shoals.

The Niger Delta is a good example of a delta in which the important energy systems are in equilibrium. Stream deposition, wave action, and tidal currents are more evenly balanced here than in the other types, so that the Niger Delta is remarkably symmetrical.

ALLUVIAL FANS

Statement

Alluvial fans are stream deposits that accumulate in dry basins at the bases of mountain fronts. Deposition results from the sudden decrease in velocity as a stream emerges from the steep slopes of the upland and flows across the adjacent basin, with its gentle gradient. They form mostly in arid regions, where streams flow intermittently. In such areas, there is usually a large quantity of loose, weathered rock debris on the surface, so that when rain falls, the streams transport huge volumes of sediment. Where a river emerges from a mountain front into the adjacent basin, its gradient is sharply reduced and deposition occurs. The channel soon becomes clogged with sediment, and the stream is forced to seek a new course. In this manner, the stream shifts from side to side and builds up an arcuate, fan-shaped deposit (*figure 11.33*). As several fans build basinward at the mouths of adjacent canyons, they eventually merge to form a broad slope of alluvium at the base of the mountain range (*figure 11.34*).

Discussion

The Basin and Range province of the western United States is an arid region with large areas of internal drainage. Most of the sediment eroded from the ranges accumulates as **alluvial fans** in the basins. The fans range from very small deposits, covering an area of less than 1 km², to large alluvial slopes on which a series of fans merge into a continuous alluvial plain along the base of the mountain front (*figure 11.34*).

Although alluvial fans and deltas are somewhat similar, they differ in mode of origin and internal structure. In deltas, stream flow is checked by standing water, and sediment is deposited in an aqueous environment. The level of the ocean or lake effectively forms the upper limit to which the delta can be built. In contrast, a fan is deposited in a dry basin, and its upper surface is not limited by water level. The coarse, unweathered, poorly sorted sands and gravels of an alluvial fan also contrast with the fine sand, silt, and mud that predominate in a delta.

Figure 11.33 Alluvial fans form in arid regions where a stream enters a dry basin and deposits its load of sediment.

Figure 11.34 Alluvial slopes develop as fans grow and merge together. This photograph of part of Nevada shows large alluvial slopes, which cover much of the dry basin.

SUMMARY

A river system is continually changing. It erodes primarily in the headwater regions by downcutting, slope retreat, and headward erosion. Some sediment is deposited in the transporting segment of a river system, forming a variety of flood-plain features. Most of the sediment load, however, is deposited as a delta near the mouth of the river or as an alluvial fan. Erosion occurs in a systematic way, and the resulting landscape evolves through a predictable sequence of stages. Although we cannot observe the evolution of a landscape from start to finish, we can construct conceptual models to explain landscape development. In the next chapter, we will consider a general model that provides insight into the changing surface of the earth, and we will look briefly at the origin of landforms in the major physiographic provinces of North America.

Key Words

erosion	frost heaving
abrasion	permafrost
pothole	slope retreat
dissolution	fan
hydraulic action	flood plain
headward erosion	meander
stream piracy	point bar
dendritic drainage pattern	oxbow lake
trellis drainage pattern	natural levee
superposed stream	backswamp
mass movement	yazoo stream
rockfall	braided stream
rockslide	alluvium
debris slide	stream terrace
debris flow	delta
mudflow	bar
subaqueous sand flow	crevasse
creep	splay
landslide	distributary
slump	bird-foot delta
solifluction	alluvial fan
rock glacier	

Review Questions

1. Explain how stream action is able to cut a valley through solid bedrock.

2. What is headward erosion? Why does it occur?
3. Explain the process of stream piracy and cite examples of how it modifies a drainage system.
4. How does superposition of a drainage pattern occur?
5. Explain how a drainage system grows by extending itself downslope.
6. Name and describe the major types of rapid downslope mass movement.
7. What causes creep?
8. Describe some of the obvious expressions of creep.
9. What geologic conditions are most favorable for the development of mudflows?
10. Annotate *figure 11.16* to explain the various types of downslope movement that cause valley slopes to retreat away from the stream channel.
11. How are talus cones constructed?
12. What is solifluction?
13. Name and describe the important landforms associated with flood-plain deposits.
14. Describe the steps involved in the growth of a stream meander and the formation of an oxbow lake.
15. How does a point bar develop?
16. Explain the origin of natural levees.
17. What conditions are conducive to the development of braided streams?
18. Describe and illustrate the steps in the development of stream terraces.
19. Explain how a delta is built where a stream enters a lake or the sea.
20. Explain the history of the Mississippi Delta.
21. Make a series of sketches to show the form of a delta in which (a) fluvial processes dominate, (b) wave processes dominate, and (c) tidal processes dominate.
22. Explain how an alluvial fan is built.

Additional Readings

Crandell, D. R., and H. H. Waldron. 1956. A recent volcanic mudflow of exceptional dimensions from Mt. Rainier, Washington. *American Journal of Science* 254: 359–62.

Morisawa, M. 1968. *Streams: Their dynamics and morphology.* New York: McGraw-Hill.

Sharpe, C. F. S. 1938. *Landslides and related phenomena.* New York: Columbia University Press.

Young, A. 1972. *Slopes.* Edinburgh: Oliver and Boyd.

12

EVOLUTION OF LANDFORMS

River systems are not simply channels through which water and sediment flow. They are the major agents by which the earth's surface is sculptured into an infinite variety of erosional and depositional landforms. Almost every landform on the continents is related in some way to a drainage system. As rivers flow endlessly to the oceans, the landscapes of drainage basins continually change and evolve. The preceding chapter described how a stream valley evolves by three main processes, (1) downcutting of the stream channel, (2) headward erosion and extension of the drainage net upslope, and (3) slope retreat (or lateral migration) of the valley walls away from the stream channel. These processes operate as a system. As a result, a landscape evolves through a predictable series of stages from the time erosion begins until it is modified or stopped. No two regions are ever exactly alike, but through the study of regions in various stages of development, geologists have constructed conceptual models of landscape evolution. These models can help you better understand the importance of river systems in the formation of the earth's surface features and the sequence of steps in landscape development.

In this chapter, we will describe a general model of landscape development. We will then explore the origins of distinctive landscapes of some of the major regions of North America.

MAJOR CONCEPTS

1. A landscape eroded by running water evolves through a predictable series of stages.
2. climate and rock structure greatly influence the evolution of a landscape.
3. Rock formations erode at different rates, so that resistant and nonresistant units are etched out in relief. This differential erosion produces alternating cliffs and slopes on horizontal rocks and alternating ridges and valleys on folded rocks.
4. The regional landforms of North America are the result of the hydrologic system's operating on the continent's major tectonic provinces. Each province—the shield, stable platform, mountain belt, rift system, and volcanic provinces—is characterized by distinctive landscapes.

A MODEL OF STREAM EROSION

Statement

Theoretically, a newly uplifted area evolves through a series of stages. In the initial stage, headward erosion and downcutting begin to dissect the area. Divides are broad and wide, but headward erosion extends the drainage net upslope. In time, the entire area becomes completely dissected into rolling hills. Slope retreat then becomes a major process, and the valley walls recede from the stream channel. Ultimately, the area is eroded to a surface of low relief lying near sea level.

Discussion

The block diagrams in *figure 12.1* illustrate the sequence of changes following uplift in an area with a humid climate. The figure depicts landscape development according to a general, idealized model. It assumes that uplift was rapid and that the surface was elevated more than 500 m above sea level. The major processes involved are headward erosion, downcutting, and slope retreat.

Initial Stage. The newly exposed surface shown in *figure 12.1A* is at first poorly drained, but eventually an integrated drainage system develops by headward erosion. Streams begin to erode V-shaped valleys into the uplifted landmass; downcutting and headward erosion are the dominant processes. The valley walls extend right down to the stream bank. Tributaries develop rapidly as the drainage net extends up the regional slope by headward erosion. New tributaries form on the slopes of the valley walls of the major rivers. Stream capture is common in the headwater region. Stream gradients are steep, and stream channels are marked by waterfalls and rapids. Broad areas of the original surface remain uncut,

---

Figure 12.1 A model of landscape development in a humid climate shows how a newly uplifted area may evolve through a definite series of stages until it is reduced to a surface of low relief near sea level.

(A) Initial stage. Large areas are undissected. Downcutting and headward erosion are the dominant processes. Streams have steep gradients, with rapids and waterfalls. Only a small percentage of the area consists of valley slopes. The streams have not developed flood plains or an extensive meandering pattern.

(B) Intermediate stage. The area is completely dissected, so that most of the surface consists of valley slopes. Relief is at a maximum, and a well-integrated drainage system is established. The main streams meander, and a flood plain begins to develop. The topography is characterized by smooth, rolling hills.

(C) Late stage. The landscape has been eroded to a nearly featureless surface near sea level. Rivers meander over a broad flood plain. Some isolated erosion remnants remain, but deposition is nearly as important as erosion.

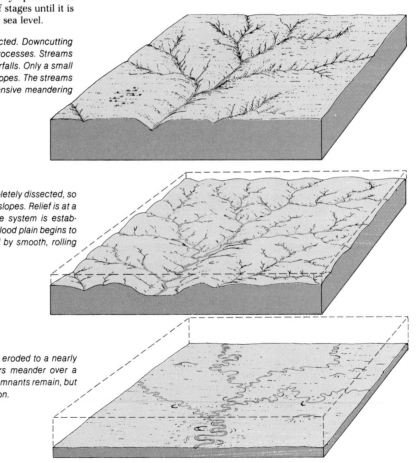

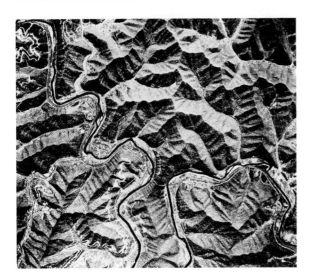

Figure 12.2 The Allegheny Plateau is an example of an area in the intermediate stage of erosional development. Valley slopes make up a large percentage of the surface; relief is at a maximum; and the drainage system is well integrated, with a large number of streams per unit area.

and marshes and lakes occupy original depressions in the newly uplifted surface.

As erosion continues, the local **relief** of the area (the difference in elevation between valley floor and divide summit) increases, and valleys are cut deeper and wider. Tributary valleys branch out from the larger streams and further dissect the surface. Slope retreat then becomes a more important process.

Intermediate Stage. As tributaries increase in number and length, the original surface is completely dissected, and the landscape becomes marked by steep slopes leading down to the deeply incised stream channels (*figure 12.1B; see also plate 7*). The drainage network is extended upslope until the entire area is well drained. Local relief steadily increases. Streams also begin to erode laterally during this stage and develop flood plains. Slope retreat becomes more prominent. Ultimately, downcutting ceases to produce greater local relief. Thereafter, erosion and deposition tend to reduce the local relief and the general elevation of the area. With the land reduced to a lower elevation, erosion proceeds more slowly.

The Allegheny Plateau, in Pennsylvania, with its adjacent areas, is an example of a region near the intermediate stage of erosion in a humid climate. Most of the area is dissected to some extent, and

relief will be diminished with continued erosion (*figure 12.2*).

Late Stage. Ultimately, the landscape is reduced to a low, nearly featureless erosion surface near sea level (*figure 12.1C*). This surface is a **peneplain,** over which the streams meander slowly, depositing almost as much sediment as they erode. The peneplain is the ultimate result of erosion in this model. It is near sea level, and no further effective stream erosion can occur upon it. Subsequent modifications, if conditions remain stable, are deep weathering and the deposition of sediment on the flood plain. Oxbow lakes, meander scars, and natural levees are common. Locally, an isolated erosional remnant of resistant rock, a **monadnock,** may protude above the peneplain surface.

Erosion of a landscape is much more complex than this model and is greatly influenced by the tectonic setting and climate. It is doubtful that a large area evolves through all three stages, but there are numerous examples of regions in various stages of dissection.

Interruptions in Landscape Development

The model described above presents a hypothesis of how stream erosion systematically sculptures the landscape by downcutting, headward erosion, and slope retreat. From the model, we would expect a newly uplifted segment of the land to evolve through a predictable series of stages. In reality, this exact sequence rarely occurs. Various forces can interrupt the development at any stage and greatly complicate the simple process of erosion. Perhaps the most significant interruptions in landscape development are due to earth movements. These can be broad, regional uplift, subsidence of the coast, or the intense folding and faulting associated with mountain building. Each results in a change in the energy of the stream system, with a corresponding change in erosion or deposition.

Uplift, for example, increases the elevation of the land, causing an increase in stream energy. As a result, many aspects of landscape development return to the initial stage, and the cycle begins again. If uplift occurs in the late stage, the gradient of a meandering stream pattern is increased, and the stream begins to cut down its channel. As a result, the meandering stream pattern, which is characteristic of the late stage in erosion, can take the form of an **entrenched meander,** cutting a deep, winding canyon. Such a river, however, has many characteristics of streams in the initial stage of erosion, such as a steep gradient and rapids.

Figure 12.3 Entrenched meanders in the drainage system of the Colorado Plateau illustrate the rejuvenation of stream erosion. The meandering pattern of the river was formed during an earlier period, when the river's gradient was lower. Recent uplift increased the gradient and the river's power to erode. Downcutting is now the dominant process, and the meanders have become entrenched.

Classic examples of entrenched meanders are found throughout the Colorado Plateau (*figure 12.3*).

Climatic changes can also disrupt the erosional processes and drastically alter the landscape. A change from humid conditions to extremely arid conditions could cause an entire region to become a desert with little or no stream erosion. Wind would then become the most important geologic agent, developing sand dunes, which would cover much of the landscape. In another situation, the humid climate might turn cold, and glaciers could develop and become the dominant geologic agents. The last ice age, from which the earth is just emerging, is an excellent example. Large ice sheets, over 3000 m thick, covered much of North America and Europe, and they completely modified the landscape. Previous drainage systems were obliterated, and deposits of glacial debris were spread over large areas. Thus, much of the surface in Canada and northern Europe shows a strong imprint of glaciation rather than erosion by streams and rivers. Also, the melting ice deposited a huge load of debris along the margins of the glaciers, so that the remaining streams became overloaded with sediment. Instead of eroding, the streams began to fill their valleys with sediment. For example, in the lower Mississippi valley, more than 60 m of sediment apparently accumulated, largely from deposition of the great load of glacial sediment. The river now flows over a broad, flat-floored valley built up by fluvial sedimentation.

Volcanic activity can also interrupt the development of a landscape. Floods of lava can completely cover a large area, like the Snake River plain and the Columbia Plateau, in the northwestern United States. As illustrated in *figure 12.4*, erosion in such an area must begin from scratch. Other volcanic terrain can produce a series of high composite volcanoes with associated ash and lava flows. These flows block drainage, so that lakes form, and they build up a distinctive constructional landscape. The Cascade Range is the best example of this type of terrain in the United States.

Figure 12.4 The Snake River plain, in southern Idaho, an area of recent volcanic activity, is experiencing the initial stage of erosion.

DIFFERENTIAL EROSION

Statement

Different types of rocks erode at vastly different rates. Factors such as the climate, the degree of compaction and cementation, and the nature of the dominant minerals cause some rocks to be hard and resistant and others to be soft and weak. The net result is that, when a sequence of rocks is exposed to the surface, some rock bodies erode much faster than others. This variation in rates is called differential erosion.

Differential erosion occurs on all scales, from steep cliffs and gentle slopes formed on alternating hard and soft rock bodies, to thin laminae within a rock etched out in delicate relief. This variation in rates of erosion, occurring on such a vast range of scales, is responsible for much of the beauty and the spectacular scenery of the earth.

Discussion

Differential erosion generally is well expressed in arid regions, where differences in rock type, jointing, and the availability of surface water and ground water combine to produce fascinating details of the landscape.

Cliffs and Slopes. Probably the most widespread examples of differential erosion are the alternating cliffs and slopes that develop on sequences of alternating hard and soft sedimentary rocks. Soft shales typically form slopes, and the more resistant sandstones and limestones produce cliffs (*figure 12.5*). The height of a cliff and the width of a slope are largely functions of the thickness of the formations involved. Similarly, if the series of resistant and nonresistant rocks are tilted, the nonresistant rocks are quickly eroded to form lowlands or valleys, leaving the resistant rocks as hills or ridges (*figure 12.6*).

Columns and Pillars. Buttes, pinnacles, and pillars are simply details of differential erosion on receding cliffs. If a massive cliff has well-developed joints, vertical columns develop, like those shown in *figure 12.7*. Where stratification produces alternating layers of hard and soft rock, additional detail can be etched out (*figure 12.8*). Jointing commonly plays an important role, for it permits weathering to attack a rock body from many sides at once. The

Figure 12.5 Alternating cliffs and slopes develop in the Colorado Plateau by differential erosion. Cliffs form on resistant sandstone and limestone formations. Slopes develop on nonresistant shale.

Figure 12.6 Differential erosion of a sequence of tilted strata produces ridges on resistant formations and long, narrow valleys on nonresistant rocks.

Figure 12.7 Rock columns and pinnacles develop from rapid weathering along joint systems.

Figure 12.8 Eroded columns provide spectacular scenery in Bryce Canyon National Park, Utah. Rapid erosion along intersecting joint systems separates the columns from the main cliff. Differential erosion, accentuating the difference between rock layers, produces the fluted columns.

famous columns and pillars in Bryce Canyon National Park, Utah, for example, result from differential weathering along a set of intersecting joints. The joints divide the rock into columns. Nonresistant shales separate the more resistant sandstone and limestone, and differential erosion forms deep recesses in the columns, producing the fascinating slopes and forms (*figure 12.9*).

Natural Arches. Natural arches are an interesting product of differential erosion. The best examples are found in the massive sandstones of the Colorado Plateau, in the western United States. The diagrams in *figure 12.10* illustrate how natural arches may be formed. The arid western parts of the United States receive little precipitation, and much of the surface water seeps into the thick, porous sandstone. Subsurface water is most abundant beneath dry stream channels, and its movement follows the general surface drainage lines.

Ground water, emerging as a seep in a cliff beneath a dry waterfall, dissolves the cement in that area. Loose sand grains are washed or blown away, so that an **alcove** soon develops at the base of the normally dry waterfall. If the sandstone is cut by joints, a large block can be separated from the cliff as the joints are enlarged by weathering processes. The alcove and joint surface continue to enlarge, and an isolated arch is eventually produced (*figure 12.11*). Weathering then proceeds inward from all surfaces until the arch is destroyed, leaving only columns standing.

The main point of these examples is that erosion is a selective process. It rapidly removes weak rock to form valleys or depressions in the landscape and leaves resistant rock bodies standing in relief as mountains, hills, and ridges. In this way, landscapes commonly reflect the structure of the rocks exposed at the surface.

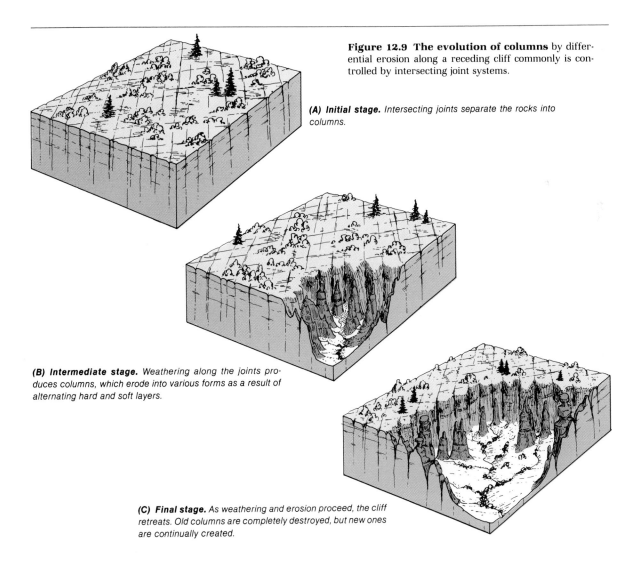

Figure 12.9 The evolution of columns by differential erosion along a receding cliff commonly is controlled by intersecting joint systems.

(A) Initial stage. Intersecting joints separate the rocks into columns.

(B) Intermediate stage. Weathering along the joints produces columns, which erode into various forms as a result of alternating hard and soft layers.

(C) Final stage. As weathering and erosion proceed, the cliff retreats. Old columns are completely destroyed, but new ones are continually created.

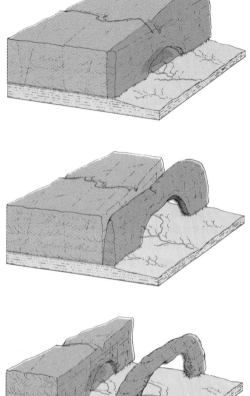

Figure 12.10 Natural arches develop in massive sandstone formations by selective solution activity.

(A) Initial stage. In arid regions, much water seeps into the subsurface below a stream channel. This water moves laterally above an impermeable layer and eventually emerges as a seep near the base of a cliff. The cement that holds the sand grains together is soon dissolved in this area of greatest moisture, and the sand grains fall away, so that a recess, or alcove, forms beneath the dry falls from the intermittent stream above.

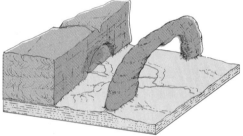

(B) Intermediate stage. If a joint system in the sandstone is roughly parallel to the cliff face, it can be enlarged by weathering, which separates a slab from the main cliff.

(C) Final stage. An arch is produced as the alcove enlarges. Weathering then proceeds inward from all surfaces until the arch collapses.

Figure 12.11 The initial stage in the development of a natural arch can be observed in Zion National Park, Utah. In this photograph, a well-developed alcove has formed beneath a dry waterfall. Weathering along a large joint will soon separate the alcove from the cliff to produce a natural arch.

THE LANDSCAPE OF NORTH AMERICA

Statement

The landscape of a continent may appear at first to be a vast chaos, but as you study the earth's surface from a variety of perspectives, you can find order in the apparent chaos and meaning in every landform. A meaningful way to "read the landscape" and understand the earth's surface features is to consider them in their tectonic setting. North America is a large continent, with symmetry, variety, and beauty. The core of the continent is the Canadian Shield, which is composed of Precambrian igneous and metamorphic rock eroded down to a surface of low relief lying near sea level. The shield is nearly surrounded by the stable platform, sometimes referred to as the interior lowlands. This region is covered with layers of sedimentary rock that have experienced little deformation since they were deposited. The landscape developed on the stable platform is mostly plains, which are remarkably flat and have little relief. The Appalachian Mountains, near the eastern margins of the continent, are built from deformed Paleozoic rocks. Erosion has etched out the resistant layers, forming the major mountain ridges. The western Cordillera (the Rocky Mountains) has a variety of tectonic features and associated landforms resulting from upwarps of the crust, rifting, volcanic activity, and plate collision.

Discussion

A complete description of each major division of North America is beyond the scope of this chapter, but we will consider some of the major landforms of each area and how they evolved. We will proceed from the more simple provinces to the more complex.

The Stable Platform—Erosion of Gently Inclined Strata

The stable platform is simply part of the shield that is covered by sedimentary strata. The thickness of the sedimentary cover rarely exceeds 2000 or 3000 m. Thus, on a regional basis, the cover of sedimentary strata is only a thin veneer over the igneous and metamorphic rocks of the shield.

Some of the landforms that typically develop on horizontal strata are illustrated in *figure 12.12*. A resistant layer, commonly sandstone or limestone, forms a protective cover, or cap rock, resulting in a nearly flat **plateau.** Erosion and slope retreat along the edges of the plateau form alternating cliffs and slopes on the alternating resistant and weak rock. As erosion eats headward, a large portion of the plateau can detach to form a **mesa,** which then shrinks by slope retreat on all sides to form groups of **buttes.** These continue to waste away into **pillars** and **pinnacles,** which ultimately disappear.

A given layer of sedimentary rocks is essentially homogeneous in all directions. Such formations have no appreciable structural control over the drainage system. All tributaries are free to grow with equal ease, so that a dendritic drainage pattern typically develops on horizontal rocks.

Examples of erosion on horizontal or gently inclined strata of the stable platform are found in the Colorado Plateau, Kansas, and the Appalachian Plateau (*figure 12.13*).

On a regional scale, the strata of the stable platform are gently inclined 1° or 2°. As stream erosion acts on the inclined strata, the weak rock layers are eroded down into low valleys, trending parallel to the **strike,** or trend, of the rocks. The

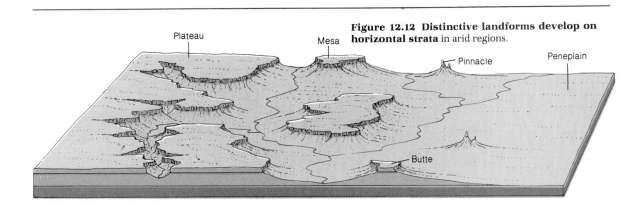

Figure 12.12 Distinctive landforms develop on horizontal strata in arid regions.

Plateau

Mesa

Pinnacle

Peneplain

Butte

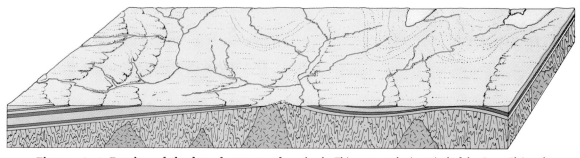

Figure 12.13 Erosion of the broad structural domes and basins in the stable platform produces a series of gently inclined cuestas with adjacent low-lands. This topography is typical of the Great Plains, the Interior Lowlands, and the Coastal Plains.

resistant layers are left standing as ridges called **cuestas.** In the coastal plains of the Atlantic and Gulf Coast states, the alternating layers of sandstone and shale deposited during the Mesozoic and Cenozoic eras have a persistent **dip** (inclination) seaward. The topography of the coastal plains therefore consists of a series of low cuestas (formed on the sandstone layers) and broad, low valleys (formed in the soft shales). The old major streams developed as the seas receded, following the direction of the initial slope. These streams are called **consequent streams.** A younger set of streams eroded headward along the weak shale formations and excavated a belt of lowlands. These are called **subsequent streams.** The resulting drainage system in the coastal plains has a trellis pattern.

In many areas of the stable platform, the sedimentary rocks are warped into broad **domes** and **basins.** On the flanks of these structures, the strata may dip at angles of 20° or 30° or more. As erosion progresses, strata are removed from the top of a dome and are eroded outward to form a series of sharp-crested cuestas, or **hogbacks,** with intervening **strike valleys** (*figure 12.14*). If the older rocks in the center of the dome are nonresistant, the center of the uplift may be eroded down into a topographic lowland bordered by inward-facing **escarpments** formed on the younger, resistant units. If the older rocks are more resistant, the center of the dome remains high, forming a dome-shaped hill or ridge. Large structural domes have inward-facing cliffs, whereas large basins have outward-facing escarpments.

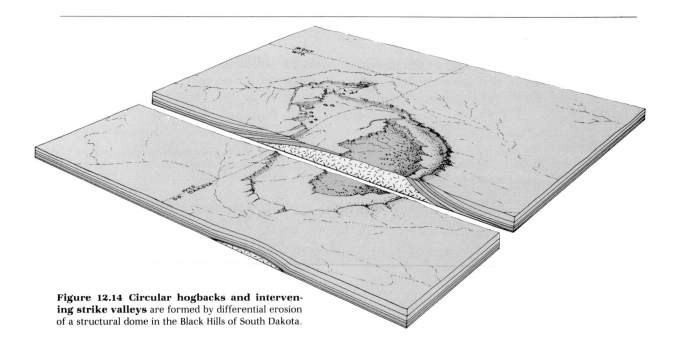

Figure 12.14 Circular hogbacks and intervening strike valleys are formed by differential erosion of a structural dome in the Black Hills of South Dakota.

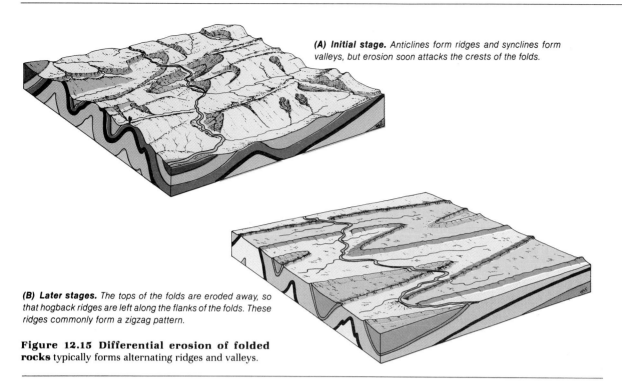

(A) Initial stage. *Anticlines form ridges and synclines form valleys, but erosion soon attacks the crests of the folds.*

(B) Later stages. *The tops of the folds are eroded away, so that hogback ridges are left along the flanks of the folds. These ridges commonly form a zigzag pattern.*

Figure 12.15 Differential erosion of folded rocks typically forms alternating ridges and valleys.

Excellent examples of domal structures in the stable platform include the Black Hills of South Dakota, the Ozark Mountains of Missouri, and the Nashville dome of Tennessee (see *plate 4*).

The Appalachian Mountains—Erosion of Folded Strata

Landforms that typically develop on folded strata are illustrated in *figure 12.15.* In the initial stage of erosion, the uparched folds, or **anticlines,** form ridges, and the downwarped folds, or **synclines,** form long valleys. Some major streams from previous cycles of erosion may be superposed across the anticlinal ridges, as is shown in *figure*

11.7. In places, parts of the resistant formations have been stripped off the crests of anticlines, so that hogback ridges have formed along the flanks of the folds.

As erosion proceeds, the crests of the anticlines are eroded away, and the surface topography bears little or no resemblance to the underlying folded structure. Differential erosion effectively removes the weak rock layers to form long strike valleys. Resistant rock bodies stand up as narrow hogback ridges. The ridges (mountains) thus mark the limbs of the folds. The pattern of topography is typically one of alternating valleys and ridges, and the drainage system generally forms a trellis pattern.

Figure 12.16 Erosion of folded strata in the Appalachian Mountains, in the eastern United States, produces a series of zigzag ridges and valleys.

The deeply eroded folds of the Appalachian Mountains, extending from Pennsylvania to central Alabama, are a classic example of this type of topography (*figure 12.16*). The deformation that produced the folds occurred during the late Paleozoic Era, and they have since been deeply eroded. Resistant ridges carved on the hard sandstone and conglomerate layers commonly rise more than 300 m above the strike valleys eroded into the weaker shale and limestone formations (see *plate 5*).

The Canadian Shield—Erosion of Complex Igneous and Metamorphic Rocks

Most of the Canadian Shield is composed of metamorphic rocks—gneisses, schists, slates, quartzites, and marbles—all of which have been intruded by granitic stocks and batholiths. The shield is the largest physiographic region of North America. These rocks are interpreted to be the roots of a series of ancient mountain belts, formed by plate collision during the Precambrian. Their erosional history is long and complex, and most of the area has been eroded down to within 200 or 300 m above sea level (*figure 12.17*). On a regional scale, the shield is a flat lowland, but differential erosion has etched out resistant rock bodies in relief of 100 m or so (*figure 12.18*). Commonly, metamorphic rocks lie in roughly parallel belts or zones, reflecting the deep structural trend that developed during mountain building. The resulting topography of discontinuous ridges and valleys in roughly parallel zones reflects the complex structure. Quartzite, gneiss, and granite generally form the resistant ridges or high areas. Slate, schist, and marble form the lowlands. The topography of the

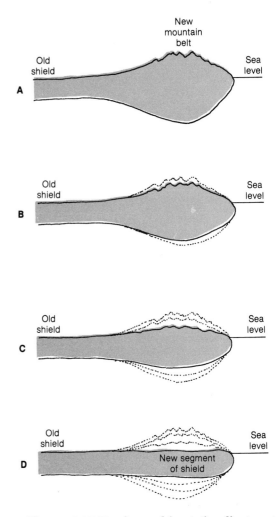

Figure 12.17 Erosion and isostatic adjustment of a mountain belt eventually produce a new segment of the shield.

Figure 12.18 Erosion of the complex igneous and metamorphic rocks of the Canadian Shield etches out the resistant rocks in relief, so that the landscape reflects the internal structure of the rocks. Resistant rock units, such as quartzite, granite, and gneiss, form ridges. Less resistant rocks are eroded down to form lowlands.

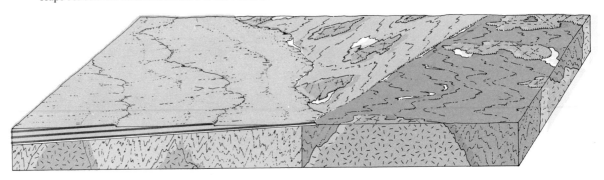

Figure 12.25 Basalt flows displaced by recurrent movement along the Hurricane Fault, in southern Utah, provide a means of measuring rates of downcutting in a stream channel. The basalt was extruded 125,000 years ago. Since then, downcutting by the Virgin River has cut a gorge 134 m deep in the upthrown block, while stream erosion on the downthrown block has only cut through the basalt to the river's original profile. This example indicates that perennial streams are capable of very rapid downcutting when uplift occurs.

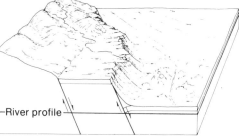

River profile

Figure 12.26 Lava cascades into the Grand Canyon provide a means of measuring rates of erosion. The remnants of basalts at the base of the cliff are 1.2 million years old. These formed a lava dam across the river 350 m high. Erosion by the Colorado River has cut down through this dam plus a more recent one. The average rate of downcutting is approximately 1 m per 1000 years.

models show how, under ideal conditions, a landscape evolves. We can see various stages of landscape evolution in the major physiographic provinces of North America. Much of the earth's scenic landscape is the product of differential erosion, in which rock units of varying resistance are etched into cliffs, slopes, columns, and natural arches. Rates of erosion are slow by human standards, but in a geologic time frame they can be remarkably rapid. Measurements of the amount of sediment transported by rivers indicate that, on an average, the continental surface of North America is being lowered at a rate of 6 cm per 1000 years.

Although running water is by far the most effective agent of erosion, water also seeps into the ground and percolates through the pore spaces of rocks. As it moves, it dissolves soluble minerals and erodes the rocks by subsurface solution activity. We will discuss the operation of ground-water systems in the next chapter.

Key Words

denudation	basin
peneplain	hogback
monadnock	dip
entrenched meander	strike
differential erosion	strike valley
butte	escarpment
pillar	anticline
pinnacle	syncline
natural arch	rift system
alcove	normal fault
plateau	horst
mesa	graben
cuesta	fault scarp
consequent stream	faceted spurs
subsequent stream	block faulting
dome	fault block

playa lake	inverted valley
bajada	volcanic neck
pediment	

Review Questions

1. Describe the model of erosion in a humid climate.
2. Explain the origin of alternating cliffs and slopes as a result of differential erosion.
3. Explain the origin of columns and pillars such as those in Bryce Canyon National Park.
4. How are natural arches formed?
5. Describe and illustrate by means of a cross section the landforms that typically develop on a stable platform.
6. Describe and illustrate by means of a cross section the landforms that typically develop by erosion of folded sedimentary rocks.
7. Describe and illustrate by means of a cross section the origin of landforms in the Canadian Shield.
8. Describe the steps involved in the development of an inverted valley.

Additional Readings

Bloom, A. L. 1969. *The surface of the earth.* Englewood Cliffs, N.J.: Prentice-Hall.

Easterbrook, D. J. 1969. *Principles of geomorphology.* New York: McGraw-Hill.

Garner, H. F. 1974. *The origin of landscapes.* New York: Oxford University Press.

Gordon, R. B. 1972. *Physics of the earth.* New York: Holt, Rinehart and Winston.

Hunt, C. B. 1973. *Natural regions of the United States and Canada.* San Francisco: W. H. Freeman.

Thornbury, W. D. 1969. *Principles of geomorphology.* 2d ed. New York: John Wiley and Sons.

Tuttle, S. D. 1970. *Landforms and landscapes.* Dubuque, Iowa: William C. Brown.

GROUND-WATER SYSTEMS

The movement of water in the pore spaces of rocks beneath the earth's surface is a geologic process that is not easily observed and therefore not readily appreciated. Yet, ground water is an integral part of the hydrologic system and a vital natural resource. Ground water is not a rare or unusual phenomenon. It is distributed everywhere beneath the surface. It occurs not only in humid areas but beneath desert regions, under the frozen Arctic, and in high mountain ranges. In many areas, the amount of water seeping into the ground equals or exceeds the surface runoff.

Ground water rarely flows as distinct underground streams. Rather, it slowly percolates through the pore spaces of the rocks, and eventually it is discharged back to the surface. As it moves, it dissolves the more soluble minerals and thus produces caves, and it is responsible for a special type of topography unlike that formed by streams and rivers.

MAJOR CONCEPTS

1. The movement of ground water is controlled largely by the porosity and permeability of the rocks through which it flows.
2. The water table is the surface below which all pore spaces in the rocks are saturated with water.
3. Ground water moves slowly (percolates) through the pore spaces in rocks by the pull of gravity. In artesian systems, it is moved by hydrostatic pressure.
4. The natural discharge of ground water is generally into streams, marshes, and lakes.
5. Artesian water is confined under pressure, like water in a pipe. It occurs in permeable beds bounded by impermeable formations.
6. Erosion by ground water produces karst topography, characterized by sinkholes, caves, and disappearing streams.

POROSITY AND PERMEABILITY

Statement

Water can infiltrate the subsurface because solid bedrock, like loose soil, sand, and gravel, contains pore spaces. The pores, or voids, within a rock can be spaces between mineral grains, cracks, solution cavities, or vesicles. Two physical properties of a rock largely control the amount and movement of ground water. One is porosity, the percentage of the total volume of the rock consisting of voids. Porosity determines how much water a rock body can hold. The second property is permeability, the capacity of a rock to transmit fluids. Permeability depends on such factors as the size of the voids and the degree to which they are interconnected. Porosity and permeability are not identical. Some very porous rocks (such as shale) have such small pore spaces that it is difficult for water to move through them. Even though they have high porosity, such rocks are impermeable.

Contrary to popular belief, underground rivers occur only in cavernous limestone and in some lava tunnels in volcanic terrain.

Discussion

Porosity

There are four main types of **pore spaces,** or voids, in rocks (*figure 13.1*): (1) spaces between mineral grains, (2) fractures, (3) solution cavities, and (4) vesicles. In sand and gravel deposits, pore space can constitute from 12% to 45% of the total volume (*figure 13.1A*). If several grain sizes are abundant or if a significant amount of cementing material is present, the **porosity** is greatly reduced. All rocks are cut by fractures, and in some dense rocks (such as granite), fractures constitute the only significant pore spaces (*figure 13.1B*). Solution activity, especially in limestone, commonly removes soluble material, forming pits and holes (*figure 13.1C*). Some limestones thus have high porosity. As water moves along joints and bedding planes in limestone, solution activity enlarges fractures in the rock and develops passageways, which may grow to become caves. In basalts and other volcanic rocks, vesicles formed by trapped gas bubbles significantly affect porosity (*figure 13.1D*). Vesicles commonly are concentrated near the top of a lava flow and form zones of very high porosity, which can be interconnected by columnar joints or by cinders and rubble at the top and base of the flow.

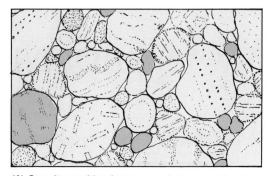

(A) Porosity resulting from spaces between grains is exemplified in sandstone and conglomerate.

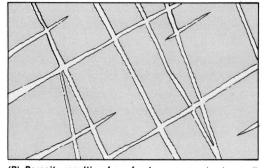

(B) Porosity resulting from fractures occurs in almost all rocks.

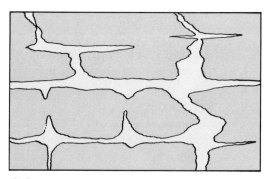

(C) Porosity resulting from solution activity is common in limestones.

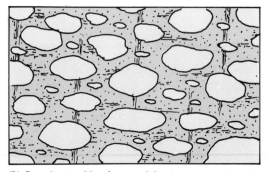

(D) Porosity resulting from vesicles is common in basalt.

Figure 13.1 Various types of pore spaces in rocks permit the flow of ground water.

Permeability

Permeability, the capacity of a rock to transmit a fluid, varies with the fluid's viscosity, **hydrostatic pressure,** the size of openings, and, particularly, the degree to which the openings are interconnected. A rock can have high porosity but low permeability.

Rocks that commonly have high permeability are conglomerates, sandstones, basalt, and certain limestones. Permeability in sandstones and conglomerates is high because of the relatively large, interconnected pore spaces between the grains. Basalt is permeable because it often is extensively fractured with columnar jointing and because the tops of most flows are vesicular. Fractured limestones are permeable, as are limestones in which solution activity has created many small solution cavities. Rocks that have low permeability are shale, unfractured granite, quartzite, and other dense, crystalline metamorphic rocks.

Water moves through the available pore spaces, twisting and turning through the tiny voids. Regardless of the degree of permeability, ground water flows extremely slowly in comparison to the turbulent flow of rivers. Whereas the flow velocity of water in rivers is measured in kilometers per hour, the flow velocity of ground water commonly ranges from 1 m per day to 1 m per year. The highest rate of percolation measured in the United States, in exceptionally permeable material, is only 250 m per day. Only in special cases, such as the flow of water in caves, does the movement of ground water approach the velocity of slow-moving surface streams.

THE WATER TABLE

Statement

As water seeps into the ground, gravity pulls it downward through two zones of soil and rock. In the upper zone, the pore spaces in the rocks are only partly saturated, and the water forms a thin film, clinging to grains by surface tension. This zone, in which pore space is filled partly with air and partly with water, is called the zone of aeration. Below a certain level, all of the openings of the rock are completely filled with water (*figure 13.2*). This area is called the zone of saturation. The water table, which is the upper surface of the zone of saturation, is an important element in the ground-water system. It can be only a meter or so deep in humid regions but might be hundreds of meters below the surface in deserts. In swamps and lakes, the water table is essentially at the surface.

Discussion

Although the **water table** (the upper surface of the **zone of saturation**) cannot be observed directly, it has been studied and mapped with data collected from wells, springs, and surface drainage. In addition, the movement of ground water has been studied by means of dyes and other tracers, so that extensive knowledge of this invisible body of ground water has been acquired.

Several important generalizations can be made about the water table and its relation to surface

Figure 13.2 The water table is the upper surface of the zone of saturation.

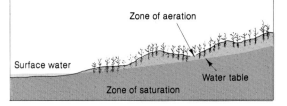

(A) Water seeps into the ground through pore spaces in the rock and soil. It passes first through the zone of aeration, in which the pore spaces are occupied by both air and water, and then into the zone of saturation, in which all of the pore spaces are filled with water. The depth of the water table varies with the climate and amount of precipitation. In most areas, it is near the surface (or at the surface, in lakes and swamps), but it can be hundreds of meters deep in desert regions.

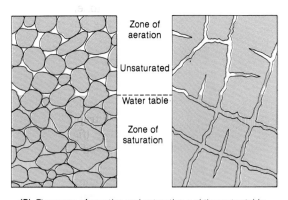

(B) The zones of aeration and saturation and the water table are shown in microscopic view.

topography and surface drainage. These are diagrammed in *figure 13.3.* In general, the water table roughly parallels the topographic surface. In flat country, the water table is flat. In areas of rolling hills, it rises and falls with the surface of the land. The reason is that ground water moves very slowly, so that the water table is maintained at a higher level in areas of high elevation. During periods of heavy rainfall, the water table rises in the areas beneath the hills, and during droughts it tends to flatten out.

Where impermeable layers (such as shale) occur within the **zone of aeration,** the ground water is trapped, forming a local **perched water table** within the zone of aeration. If a perched water table extends to the side of a valley, springs and seeps occur.

The water table is at the surface in lakes, swamps, and most streams, and water moves from the subsurface toward these areas. In arid regions, most streams lie above the water table, so that they lose much of their water through seepage into the subsurface.

The Movement of Ground Water

The difference in elevation between parts of the water table is called the **hydraulic head.** It causes the water to follow the paths illustrated in *figure 13.3.* If we could trace the path of a particle of water, we would find that gravity slowly pulls it through the zone of aeration to the water table. When a particle encounters the water table, it continues to move downward by the pull of gravity along curved paths, from areas where the water table is high toward areas where it is low (lakes, streams, and swamps). The path of ground water movement is *not* down the slope of the water table, as one first might suspect. The explanation for this seemingly indirect flow is that the water table is not a solid surface like the ground surface. It is a surface of liquid, which in some ways resembles the surface of a wave. Water at any given point below the water table beneath a hill is under greater pressure than water at the same elevation below the lower water table in a valley. Therefore, it moves directly downward toward points of lesser pressure.

Although these paths of ground-water movement may seem indirect, they conform to the laws of fluid physics and have been mapped in many areas by tracing the movement of dye injected into the system. The movement of the dye reveals that there is a continual slow circulation of ground water from infiltration at the surface to seepage into streams, rivers, and lakes.

Thus, the ground-water body is not stagnant and motionless. Rather, it is an important part of the hydrologic system and is intimately related to surface drainage. At considerable depths, all pore spaces in the rocks are closed by high pressure and there is no free water. This is the lower limit, or base, of the ground-water system.

Like other parts of the hydrologic system (rivers and glaciers), the ground-water system is open, with an input, transport, and discharge of water.

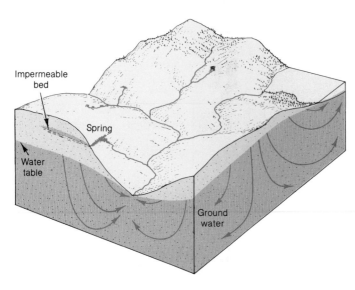

Figure 13.3 The movement of ground water is directed toward areas of least pressure. Therefore, ground water can move in any direction. In the idealized system depicted here, ground water moves downward to the water table (by the pull of gravity) and then moves toward areas of least pressure. The configuration of the water table has a strong influence on the direction of movement. In most areas, the water table is a subdued replica of the topography: a cross section shows high and low areas much like hills and valleys on the surface. Differences in the height of the water table cause differences in the pressure on water in the saturation zone at a given point. Thus, water moves downward beneath the high areas of the water table (because of the higher pressure) and upward beneath the low areas. It commonly seeps into streams, lakes, and swamps, where the water table is near the surface. In areas where the water table is low, water from streams and lakes can move downward toward the zone of saturation. A line of springs and seeps commonly occurs where an impermeable rock layer that has formed a perched water table is exposed at the surface. Ground water also interchanges with water in the unsaturated zone and with streams, lakes, and oceans.

Labels in figure: Impermeable bed; Spring; Water table; Ground water

NATURAL AND ARTIFICIAL DISCHARGE

Statement

Ground-water reservoirs discharge naturally wherever the water table intersects the surface of the ground. Generally, such places are inconspicuous, typically occuring in the channels of streams and on the floors and banks of marshes and lakes. Under various special geologic conditions, the water table intersects the ground surface and discharges as springs and seeps.

The natural discharge of ground-water reservoirs into the drainage system introduces a significant volume of water into streams. It is the major link between ground-water reservoirs and other parts of the hydrologic system. If it were not for ground-water discharge, many permanent streams would be dry during parts of the year. Most natural discharge is near or below the surface of streams and lakes and therefore usually goes unnoticed. It is detected and measured indirectly by comparing the volume of precipitation with the volume of surface runoff.

Water wells are a form of artificial discharge, made simply by drilling holes into the water table.

Discussion

Natural Discharge

Several geologic conditions that produce natural discharge in **seeps** and **springs** are shown in *figure 13.4*. If permeable beds alternate with impermeable layers (*figure 13.4A*), the ground water is forced to move laterally to the outcrop of the bed. Conditions such as this usually are found in mesas and plateaus where interbedded sandstone and shale occur. The spring line commonly is marked by a line of vegetation. *Figure 13.4B* shows a limestone terrain in which springs occur where the base of the cavernous limestone outcrops. The Mammoth Cave area, in Kentucky, is a good example. As is shown in *figure 13.4C*, lava formations outcrop along the sides of a canyon, and springs develop, because ground water migrates readily through the vesicular and jointed basalt. Note that surface drainage disappears as the water flows over the lava plain. An excellent example of this type of discharge is found in the Thousand Springs area, in Idaho, where numerous springs occur along the side of the Snake River valley. *Figure 13.4D* shows springs along a fault that produces an avenue of greater permeability. Faults frequently

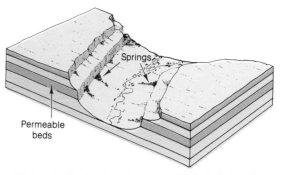

(A) *A spring line develops where permeable beds that overlie an impermeable layer are exposed in valley walls.*

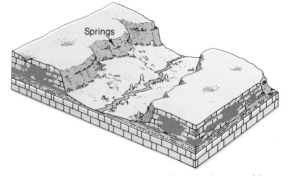

(B) *Springs form where cavernous limestone is exposed in valley walls.*

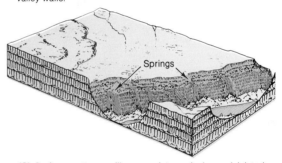

(C) *Surface water readily seeps into vesicular and jointed basalt flows. It then migrates laterally and forms springs where basalt units are exposed in canyon walls.*

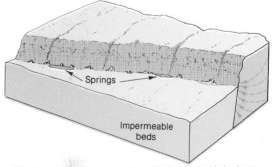

(D) *Many faults displace rocks so that impermeable beds are placed next to permeable beds. A spring line commonly results as ground water migrates along a fault line.*

Figure 13.4 Springs can be produced under a variety of geologic conditions, some of which are illustrated in the block diagrams here. They are natural discharges of the ground-water reservoir and introduce a significant volume of water to surface runoff.

displace strata so that impermeable beds block the flow from permeable layers. The water then moves up the fault plane.

Wells

Ordinary wells are made simply by digging or drilling holes through the zone of aeration into the zone of saturation, as shown in *figure 13.5*. Water then flows out of the pores into the well, filling it to the level of the water table. When a well is pumped, the water table is drawn down around the well in the shape of a cone, called the **cone of depression.** If water is withdrawn faster than it can be replenished, the cone of depression continues to grow, and the well ultimately can go dry. The cone of depression around large wells, such as those used by cities and industrial plants, can be many hundreds of meters in diameter. All wells within the cone of depression are affected (*figure 13.6*). This undesirable condition has been the cause of "water wars," fought physically as well as in the courts. Because ground water is not fixed in one place, as mineral deposits are, it is difficult to determine who owns it. Many disputes are now being arbitrated with computer models that simulate subsurface conditions such as permeability, direction of flow, and level of the water table. The models can predict what changes occur in the ground system if given amounts of water are drawn out of a well over specified periods of time.

Extensive pumping lowers the general surface of the water table. This effect has had serious consequences in some metropolitan areas of the southwestern United States, where the water table has fallen hundreds of meters. There is a limited supply of ground water. Although the ground-water reservoir is continually being replenished by precipitation, the migration of ground water is so slow that it can take hundreds of years to raise a water table to its former position of balance with the hydrologic system.

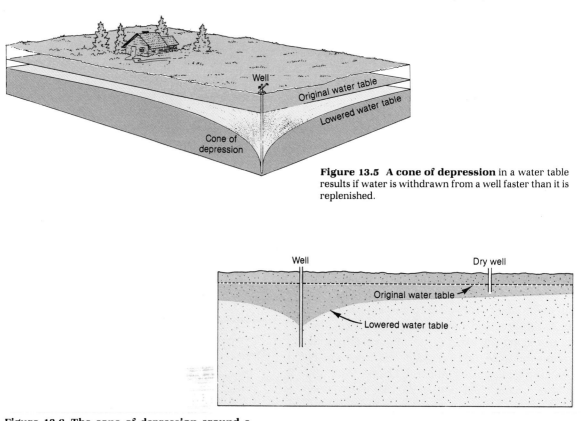

Figure 13.5 A cone of depression in a water table results if water is withdrawn from a well faster than it is replenished.

Figure 13.6 The cone of depression around a deep well can extend outward for hundreds of meters, effectively lowering the water table over a large area. Shallow wells in the area then run dry, because they lie above the water table.

ARTESIAN WATER

Statement

Artesian water is ground water that is confined so that it builds up an abnormally high hydrostatic pressure. The necessary geologic conditions for artesian water, illustrated in *figure 13.7*, include the following:

1. The rock sequence must contain interbedded permeable and impermeable strata. The sequence occurs commonly in nature as interbedded sandstone and shale. Permeable beds usually are called aquifers.
2. The rocks must be tilted and exposed in an elevated area where water can infiltrate into the aquifer.
3. Sufficient rain must fall in the outcrop area to keep the aquifer filled.

The water confined in an aquifer bed behaves much like water in a pipe. Hydrostatic pressure builds up, so that where a well or fracture intersects the bed, water rises in the opening, producing a flowing well or artesian spring.

Discussion

The height to which **artesian water** rises above the **aquifer** is shown by the dashed colored line in *figure 13.7*. The surface defined by this line is called the **artesian-pressure surface**. You might expect it to be a horizontal surface at the elevation of the water surface in the aquifer in the source area. Actually, an artesian-pressure surface slopes away from the **recharge** area, because pores in the aquifer provide resistance to flow, and because pressure is lost through fractures (leaks) in the underground plumbing system. If a well were drilled at location *A* or *C* in *figure 13.7*, water would rise in it, but it would not flow to the surface, because the artesian-pressure surface is below the

Figure 13.7 Necessary geologic conditions for an artesian system include the following: (1) a permeable bed (aquifer) must be confined between impermeable layers; (2) rocks must be tilted so that the aquifer can receive infiltration from surface waters; and (3) there must be adequate infiltration to fill the aquifer and create hydrostatic pressure. The diagram shows an idealized artesian system. All wells are artesian wells (that is, water rises in them under pressure). Flowing wells occur only where the top of the well is below the artesian-pressure surface.

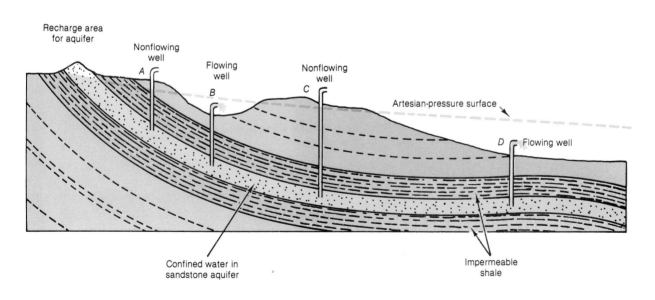

Recharge area for aquifer

Nonflowing well

A

Flowing well

B

Nonflowing well

C

Artesian-pressure surface

D Flowing well

Confined water in sandstone aquifer

Impermeable shale

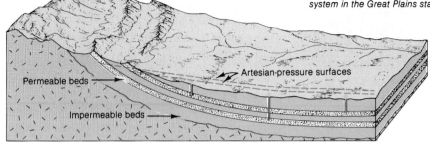

(A) The Great Plains states are characterized by slightly inclined underlying beds of permeable Cretaceous sandstones, which are warped up along the Rocky Mountains, where they receive infiltration. This structure forms a widespread artesian system in the Great Plains states.

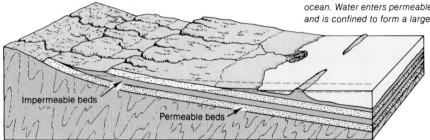

(B) The Atlantic and Gulf Coast states are characterized by Tertiary and Cretaceous rocks dipping uniformly toward the ocean. Water enters permeable beds where they are exposed and is confined to form a large artesian system.

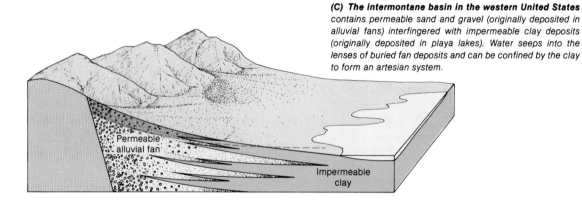

(C) The intermontane basin in the western United States contains permeable sand and gravel (originally deposited in alluvial fans) interfingered with impermeable clay deposits (originally deposited in playa lakes). Water seeps into the lenses of buried fan deposits and can be confined by the clay to form an artesian system.

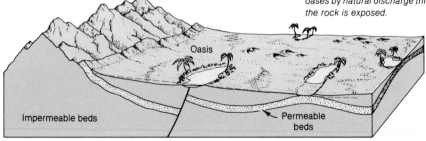

(D) The Sahara Desert is underlain mostly by gently warped permeable beds, which receive water where they are exposed at the base of the Atlas Mountains. The artesian water forms oases by natural discharge through fractures or at sites where the rock is exposed.

Figure 13.8 Artesian systems develop under a variety of geologic conditions, some of which are illustrated in these block diagrams. All of them satisfy the basic conditions necessary for an artesian system, as illustrated in figure 13.7. The main difference is in the geometry of the rock structures in each area.

ground surface. A well at location *B* or *D* would flow, because the artesian-pressure surface is above the ground surface. Nonetheless, all of these wells are artesian wells. That is, the water is under artesian pressure and rises above the top of the aquifer.

Examples of Artesian Systems

Artesian water is commonplace in most areas underlain by sedimentary rocks, because the necessary geologic conditions for an artesian system are present in sedimentary rocks in a variety of ways. Examples of well-known areas in the United States should help you appreciate how artesian systems result from different geologic settings.

One of the better-known artesian systems underlies the Great Plains states (*figure 13.8A*). The sequence of interbedded sandstones, shales, and limestones is nearly horizontal throughout most of Kansas, Nebraska, and the Dakotas, but it is upwarped along the eastern front of the Rockies and the margins of the Black Hills. Several sandstone formations are important aquifers. Water confined in them under hydrostatic pressure gives rise to an extensive artesian system in the Great Plains states. The recharge area is along the foothills of the Rockies.

Figure 13.8B illustrates a regional artesian system in the inclined strata of the Atlantic and Gulf Coast plains. The rock sequence consists of permeable sandstone and limestone beds alternating with impermeable clay. Surface water, flowing toward the coast, seeps into the beds where they are exposed at the surface. It then moves slowly down the dip of the permeable strata.

A third example is from the western states, where the arid climate makes artesian water an important resource (*figure 13.8C*). In this region, the subsurface rocks in the intermontane basin consist of sand and gravel deposited in ancient alluvial fans. Further into the basin, in the **playa,** these deposits become interbedded with layers of clay and silt. The playa deposits act as confining layers between the permeable sand and gravel. Water seeping into the fan deposits becomes confined as it moves away from the mountain front.

Artesian systems also underlie some of the world's great desert regions. Natural discharge from them is largely responsible for oases. Part of the Sahara system is shown in *figure 13.8D*. Oases occur where artesian water is brought to the surface by fractures or folds or where the desert floor is eroded down to the top of the aquifer.

THERMAL SPRINGS AND GEYSERS

Statement

In areas of recent volcanic activity, rocks associated with old magma chambers can remain hot for hundreds of thousands of years. Ground water migrating through these geothermal areas becomes heated and, when discharged to the surface, produces thermal springs and geysers.

Discussion

The three most famous regions of hot springs and **geysers** are Yellowstone National Park, Iceland, and New Zealand. All are regions of recent volcanic activity, so that rock temperatures just below the surface are quite high. Although no two geysers are alike, several conditions are necessary for their development: (1) a body of hot rocks must lie relatively close to the surface, (2) a system of irregular fractures must extend downward from the surface, and (3) a relatively large and constant supply of ground water must be present. Geyser eruptions are caused if ground-water pressure in fractures, caverns, or layers of porous rock builds up to a critical point (*figure 13.9*). Since the water at the base of the fracture is under greater pressure than the water above, it must be heated to a higher temperature before it boils. Eventually, a slight increase in temperature or a decrease in pressure (resulting from the liberation of dissolved gases) causes the deeper water to boil. The expanding steam throws water from the underground chambers high into the air. After the pressure is released, the caverns refill with water and the process is repeated.

This process accounts for the periodic eruption of many geysers. The interval between eruptions is the amount of time required for water to percolate into the fracture and be heated to the critical temperature. Geysers like Old Faithful, in Yellowstone National Park, erupt at definite intervals, because the rocks are permeable and the "plumbing system" refills rapidly. Other geysers, which require more time for water to percolate into the chambers, erupt at irregular intervals, because water supply over a longer period of time can fluctuate.

Geothermal Energy

The thermal energy of ground water, called **geothermal energy,** offers an attractive source of energy for human uses. presently, it is used in various ways in local areas of Italy, Japan, and Iceland. Recent estimates show that from 1% to 2% of our current energy needs could come from geothermal sources.

In Iceland, geothermal energy has been used successfully since 1928. The plan is simple. Wells are drilled into geothermal areas, and the steam and hot water are piped to storage tanks and then pumped to homes and municipal buildings for heating and hot water. The cost of this direct heating is only about 60% of that of fuel-oil heating and about 75% of the cost of the cheapest method of electrical heating. Steam from geothermal energy is also used to run electric generators, producing an easily transported form of energy. Corrosion is a problem, however, because most geothermal waters are acid and contain undesirable dissolved salts.

Figure 13.9 The origin of geysers is depicted in this series of diagrams. A geyser can develop only if (1) a body of hot rock lies relatively close to the surface, (2) a system of irregular fractures extends down from the surface, and (3) there is a constant supply of ground water.

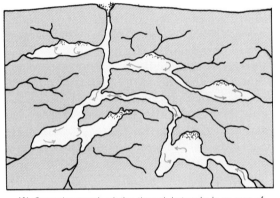

(A) Ground water circulating through hot rocks in an area of recent volcanic activity collects in caverns and fractures. As steam bubbles rise, they grow in size and number and tend to accumulate in restricted parts of the geyser tube.

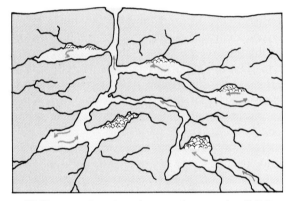

(B) The expanding steam forces water upward until it is discharged at the surface vent. The deeper part of the geyser system then becomes ready for the major eruption.

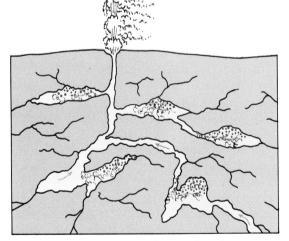

(C) The preliminary discharge of water reduces the pressure on the water at depth. Water from the side chambers and pore spaces begins to flash into steam, forcing the water in the geyser system to erupt.

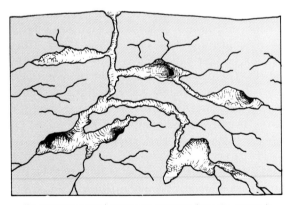

(D) Eruption ceases when the pressure from the steam is spent and the geyser tubes are empty. The system then begins to fill with water again, and the eruption cycle starts anew.

Figure 13.12 Karst topography in the limestone region of Kentucky is dominated by sinkholes and solution valleys. Note the absence of an apparent drainage system in this region.

DEPOSITION BY GROUND WATER

Statement

The mineral matter dissolved by ground water can be deposited in a variety of ways. The most spectacular deposits are commonly formed in caves as dripstone. Less obvious are the deposits in permeable rocks, such as sandstone and conglomerates, where ground water commonly deposits mineral matter as a cement between grains. The precipitation of minerals by ground water is also responsible for the formation of certain mineral deposits, such as the uranium found in the Colorado Plateau.

Discussion

Cave deposits are familiar to almost everyone, and a great variety of forms have been named. Most originate in a similar way, however, and they are referred to collectively as **dripstone.** The process of dripstone formation is shown in *figure 13.13.* As water enters the cave (usually from a fracture in the ceiling), part of it evaporates, and a small amount of calcium carbonate is left behind. Each succeeding drop adds more calcium carbonate, so that eventually a cylindrical or cone-shaped projection is built downward from the ceiling. Many beautiful and strange forms result, some of which are shown in *figures 13.14* and *13.15.* Icicle-shaped forms growing down from the ceiling are called **stalactites.** These commonly are matched by columns growing up from the floor, called **stalagmites,** as the water dripping from a stalactite precipitates additional calcium carbonate on the floor directly below. Many stalactites and stalagmites unite to form columns. Water percolating from a fracture in the roof may form **drip curtains.** Pools of water on the cave floor flow from one place to another, and as they evaporate, calcium carbonate is deposited on the floor, forming **travertine terraces.**

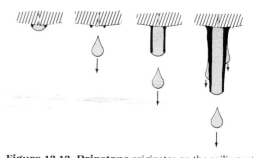

Figure 13.13 Dripstone originates on the ceilings of caves. Water seeps through a crack and deposits a small ring of calcite as it partially evaporates. The ring grows into a tube, which commonly acquires a tapering shape, as water seeps from adjacent areas and flows down its outer surface.

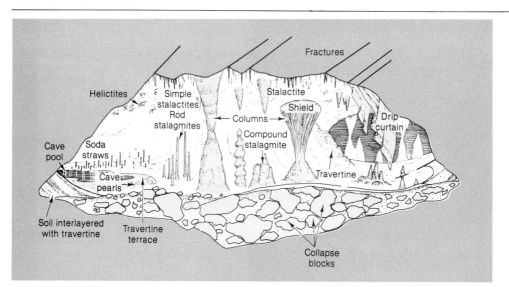

Figure 13.14 Many varieties of cave deposits have been recognized. All of them are composed of calcite. They originate from dissolved calcite deposited by water that seeps into the cave and then evaporates.

Figure 13.15 Onadaga Cave, Crawford County, Missouri, shows many of the forms of dripstone illustrated in figure 13.14.

THE ALTERATION OF GROUND-WATER SYSTEMS

Statement

Ground water is part of the hydrologic system and is intimately related to precipitation, surface drainage, and discharge. As is the case in many other natural systems, ground-water systems display a tendency towards equilibrium. For example, the level of the water table represents an equilibrium between the amount of precipitation and infiltration, surface drainage, the porosity and permeability of the rock, and the volume of ground water discharged back to the surface. If one of these factors is modified, the others respond to reestablish equilibrium.

Ground water constitutes a valuable resource, which is being exploited at an ever-increasing rate. A variety of problems result from human activities that alter the ground-water system. Some of the more important problems are the following:

1. Changes in the chemical composition of ground water
2. Salt-water encroachment
3. Changes in the position of the water table
4. Subsidence

Discussion

Changes in Composition

The composition of ground water can be changed by increases in the concentration of dissolved solids in surface water. The soil mantle of the earth is like a filtration system through which ground water moves. Obviously, any concentrations of chemicals or waste create local pockets that potentially can contaminate the ground-water reservoir. Material that is **leached** (dissolved by percolating ground water) from waste-disposal sites, for example, includes both chemical and biological contaminants. Upon entering the ground-water flow system, the contaminants move according to the hydraulics of that system. The character and strength of the **leachates** depends partly on the length of time the infiltrated water is in contact with the waste deposit and partly on the volume of infiltrated water. In humid areas where the water table is shallow and in constant contact with refuse, leaching continually produces maximum potential for pollution.

Figure 13.16 illustrates four geologic environments in which waste disposal affects the ground-water system. In the environment shown

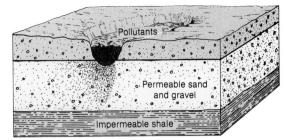

(A) *A permeable layer of sand and gravel overlying an impermeable shale creates a potential pollution problem, because contaminants are free to move with ground water.*

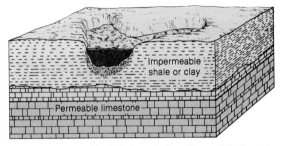

(B) *An impermeable shale (or a clay) confines pollutants and prevents significant infiltration into the ground-water system in the limestone below.*

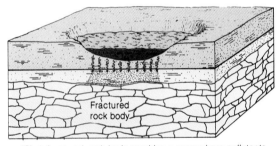

(C) *A fractured rock body provides a zone where pollutants can move readily in the general direction of ground-water flow.*

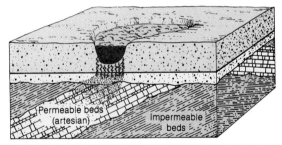

(D) *An inclined, permeable aquifer below a disposal site permits pollutants to enter an artesian system and move down the dip of the beds, so that they contaminate the system.*

Figure 13.16 The effects of solid waste disposal on a ground-water system depend on the geologic setting. In many cases, water seeping through the disposal site enters and pollutes the ground-water system.

in *figure 13.16A*, the near-surface material is permeable and essentially homogeneous. Leachates percolate downward through the zone of aeration and, upon reaching the water table, enter the ground-water flow system. The flowing leachates ultimately become part of the surface drainage system. As is shown in *figure 13.16B*, an impermeable layer of shale confines pollutants and prevents their free movement in the ground-water

Figure 13.17 The relationship between fresh water and salt water on an island or a peninsula is affected by the withdrawal of water from wells. Excessive pumping causes a cone of salt-water encroachment, which limits the usefulness of the well.

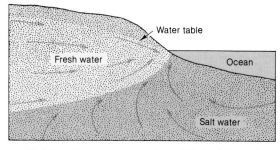

(A) A lens of fresh ground water beneath the land is buoyed up by denser salt water.

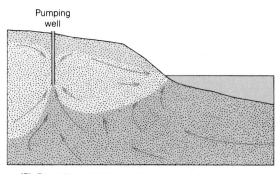

(B) Excessive pumping causes a cone of depression in the water table and a cone of salt-water encroachment at the base of the freshwater lens.

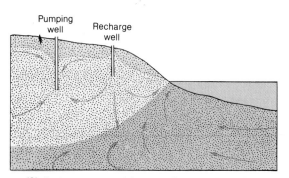

(C) Fresh water pumped down an adjacent well can raise the water table around the well and lower the interface between the fresh water and the salt water.

system. *Figure 13.16C* illustrates a disposal site above a fractured rock body. Upon reaching the fractured rock, the contaminants can move more readily in the general direction of ground-water flow. Dispersion of the contaminants is limited, however, because of the restriction of flow to the fractures. *Figure 13.16D* illustrates a critical condition in which a waste-disposal site is located in permeable sand and gravel above an inclined aquifer. Here, leachates move down past the water table and enter the aquifer as recharge. If the waste-disposal site were located directly above the aquifer, as is shown in the diagram, most of the leachates would enter the aquifer, but some would move down the gradient of the impermeable shale. If the site were located at position *B*, very little of the leachates would penetrate the aquifer. Instead, they would move downgradient over the impermeable shale.

Salt-Water Encroachment

On an island or a peninsula where permeable rocks are in contact with the ocean, a lens-shaped body of fresh ground water is buoyed up by the denser salt water below, as is illustrated in *figure 13.17A*. The fresh water literally floats on the salt water and is in a state of balance with it. If excessive pumping develops a large cone of depression in the water table, the pressure of the fresh water on the salt water below the well is decreased, and a large cone of **salt-water encroachment** develops below the well, as is shown in *figure 13.17B*. Continued excessive pumping causes the cone of salt water to extend up the well and contaminate the fresh water. It is then necessary to stop pumping for a long time to allow the water table to rise to its former position and depress the cone of salt water. Restoration of the balance between the freshwater lens and the underlying salt water can be hastened if fresh water is pumped down into an adjacent well (*figure 13.17C*).

Changes in the Position of the Water Table

The water table is intimately related to surface runoff, the configuration of the landscape, and the ecological conditions at the surface. The balance between the water table and surface conditions, established over thousands or millions of years, can be completely upset by changes in the position of the water table. Two examples illustrate some of the many potential ecological problems.

In southern Florida, water from Lake Okeechobee flowed for the past 5000 years as an almost imperceptible river, only a few centimeters deep and 64 km wide. This sheet of shallow water created the swampy Everglades. The movement of the water was not confined to channels. It flowed as a sheet in a great curving swath for more than 160 km (*figure 13.18*). The surface of the Everglades slopes southward only 2 cm per km, but this gradient was high enough that the water moved slowly to the coast, preventing salt water from invading the Everglades and the subsurface aquifers along the coast. In effect, the water table in the swamp was at the surface, and the ecology of the Everglades was in balance with the water table.

Today, many canals have been constructed to drain swamp areas for farmland, to help control flooding, and to supply fresh water to the coastal megalopolis (*figure 13.19*). The canals diverted the natural flow of water across the swamp and, in effect, lowered the water table, in some places as much as 0.5 m below sea level. This change in position of the water table produced many unforeseen and often unfortunate results. As the water table was lowered, salt-water encroachment occurred in wells all along the coast. Some cities had to move their wells far inland to obtain fresh water.

The most visible effects, however, involve the ecology of the swamp. In the past, the high water table could maintain a marsh during periods of natural drought. Now the surface is dry during

droughts. Forest fires ignite the dry organic muck, which burns like peat, smoldering long after the surface fires die out. This effectively destroys the ecology of the swamp. The lowering of the water table also caused the muck to compact, so that it subsides in places as much as 2 m. In addition, muck exposed to the air oxidizes and disappears at a rate of about 2.5 cm per year. Once the muck is gone from the swamp, it can be replaced only by nature.

Raising the water table can also modify many surface processes. An example is found in the environmental changes caused by irrigation in the Pasco basin of Washington. This area, which lies in the rain shadow of the Cascade Mountains, receives only from 15 to 25 cm of precipitation a year. The basin's surface conditions developed in response to an overall increase in aridity over the last several million years. The surface material, slope, and vegetation developed a balance with an arid climate and a low water table.

In recent years, extensive irrigation has caused the water table to rise, introducing many changes in surface conditions. Today, from 100 to 150 cm of water is applied to the ground by irrigation, which produces the effect of a simulated climatic change of considerable magnitude. The higher water table

Figure 13.18 Natural drainage of southern Florida in 1871 spread southward from Lake Okeechobee in a broad sheet only a few centimeters deep. This maintained swampy conditions in the Everglades and established a water table very close to the surface.

Figure 13.19 Modification of the natural drainage system by canals diverted the natural flow of surface water across the Everglades. The water table was lowered, the swamp was destroyed in some areas, and salt water encroached in wells along the coast.

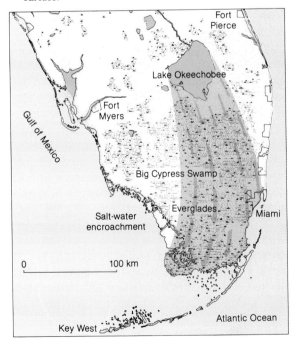

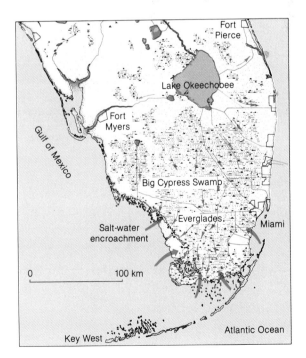

has rapidly developed large springs along the sides of river valleys. The springs are now permanent, reflecting saturation of much of the ground. Erosion is accelerated, and many farms and roads have been damaged severely. Landslides present the most serious problems. Slopes that were stable under arid conditions are now unstable, because they are partly saturated from the high water table and from the formation of perched water bodies.

In many areas, it is imperative that people modify the environment by reclaiming land or by irrigation. But unless we are careful, the detrimental effects may outweigh the advantages. Before we modify an environment seriously, we must attempt to understand the many consequences of altering the natural systems.

Subsidence

Surface **subsidence** related to ground water can result from natural earth processes, such as the development of sinks in a karst area, or from the artificial withdrawal of fluids.

An ever-present hazard in limestone terrains is the collapse of subterranean caverns and sinkhole formations. Buildings and roads have frequently been damaged by sudden collapses into previously undiscovered caverns below. In the United States, important karst regions occur in central Tennessee, Kentucky, southern Indiana, Alabama, Florida, and Texas (*figure 13.20*). The problem of collapse is difficult to solve. Important construction in karst regions should be preceded by test borings to determine whether subterranean cavernous zones are present. Concrete slurries can be pumped into

Figure 13.21 Subsidence of buildings in Mexico City resulted from compaction after ground water was pumped excessively from unconsolidated beds beneath the city. Subsidence has tilted this building more than 2 m, so that the windows shown in the background of the photograph are now at the level of the sidewalk.

solution cavities, but such remedies can be very expensive.

Compaction and subsidence present serious problems, especially in areas of newly deposited sediments. In New Orleans, for example, large areas of the city are now 4 m below sea level, a drop due largely to the pumping of ground water. As a result, the Mississippi River flows some 5 m above the city, and rainwater must be pumped out of the city at considerable cost. Also, as the earth subsides, water lines and sewers are damaged.

Where ground water, oil, or gas is withdrawn from the subsurface, significant subsidence can occur, damaging construction, water supply lines, sewers, and roads. Long Beach, California, has subsided 9 m as the result of 34 years of oil production from the Wilmington oil field. This subsidence has resulted in almost $100 million damage to wells, pipelines, transportation facilities, and harbor installations. Houston, Texas, has subsided as much as 1.5 m from the withdrawal of ground water.

Probably the most spectacular example of subsidence is Mexico City, which is built on a former lakebed. The subsurface formations are water-saturated clay, sand, and volcanic ash. The sediment compacts as ground water is pumped for domestic and industrial use, and slow subsidence is widespread. The opera house (weighing 54,000 metric tons) has settled more than 3 m, and half of the first floor is now below ground level. Other large structures are noticeably tilted, as is illustrated in *figure 13.21*.

Figure 13.20 The major areas of karst topography in the continental United States are restricted to the central and southern states, where outcrops of limestone occur in humid climatic conditions. Limestone outcrops in many areas in the western states, but in most cases the climate there is too arid to develop a typical karst topography.

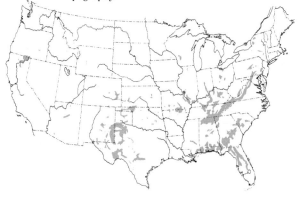

SUMMARY

Ground water is an integral part of the hydrologic system. It is intimately related to surface water. The movement of ground water is very slow, controlled largely by the porosity and permeability of the rock. At some depth below the surface, all pore spaces are filled with water. The upper surface of this saturated zone is called the water table. Movement of ground water below the water table is closely associated with surface runoff. The ground-water reservoir discharges into streams, lakes, and swamps.

Artesian water is water confined between impermeable beds under hydrostatic pressure.

Ground water erodes by dissolving soluble rocks, such as limestone, salt, and gypsum. Such erosion produces karst topography, characterized by sinkholes, solution valleys, and disappearing streams.

Under normal conditions, ground water is essentially in balance with surface runoff, surface topography, and salt water in the ocean. Alteration of the ground-water system can produce many unforeseen problems, as does alteration of any segment of the hydrologic system.

In the next chapter, we will study a natural alteration—in fact, a complete disruption—of the hydrologic system. When the earth's climate changes sufficiently to produce widespread and prolonged cold, the water in the hydrologic system freezes and glaciers form. Although an ice age is a rare event in the earth's history, the phenomenon significantly modifies and interrupts geologic processes on the earth's surface.

Key Words

ground water	cone of depression
porosity	artesian water
permeability	aquifer
hydrostatic pressure	artesian-pressure surface
zone of aeration	recharge
zone of saturation	geyser
water table	geothermal energy
perched water table	karst topography
hydraulic head	cave
spring	sinkhole
seep	solution valley
disappearing stream	travertine terrace
dripstone	leaching
stalactite	leachate
stalagmite	salt-water encroachment
drip curtain	subsidence

Review Questions

1. Define porosity and permeability.
2. Describe and illustrate the major types of pores, or voids, in rocks.
3. What rock types are impermeable or nearly impermeable?
4. Describe the zones of subsurface water and explain how water moves through each zone.
5. Explain some of the ways in which springs originate.
6. What effects are produced in the water table by excessive and rapid pumping?
7. Explain how artesian flow occurs.
8. Explain the origin of geysers.
9. What is the source of heat for hot springs and geysers?
10. Describe the evolution of a landscape in which ground water is the dominant agent of erosion.
11. Explain how stalagmites and stalactites originate.
12. Describe the forms and processes of ground-water pollution.
13. Describe the relation between salt ground water and fresh ground water beneath an island or a peninsula.
14. What undesirable effects can result from withdrawing ground water from wells close to the ocean?
15. Explain how the alteration of the natural drainage system in southern Florida has affected the Everglades.
16. How can subsidence of the land result from the withdrawal of ground water? Give examples.

Additional Readings

Davis, S. N., and R. J. M. DeWiest. 1966. *Hydrogeology.* New York: John Wiley and Sons.

Sayre, A. N. 1950. Ground water. *Scientific American* 183(5):14–19. Reprint no. 818. San Francisco: W. H. Freeman.

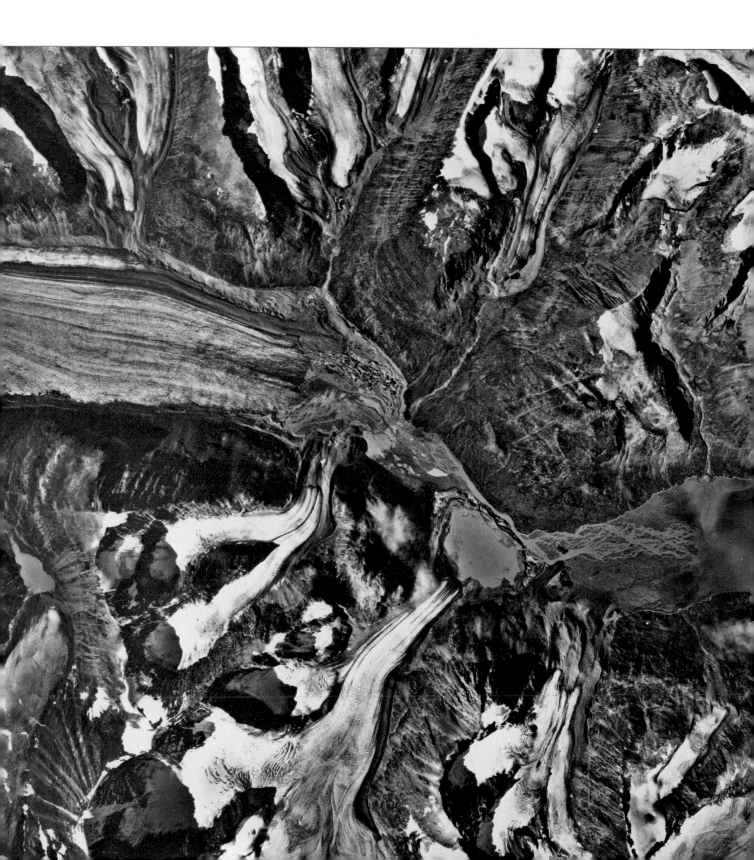

GLACIAL SYSTEMS

The earth is just emerging from an ice age—a rare event in its history, for, throughout most of geologic time, its climate has been relatively mild. When glaciation does occur, it completely disrupts the hydrologic system, and many geologic processes are interrupted or modified significantly. During an ice age, much precipitation becomes trapped in glaciers instead of flowing immediately back to the ocean. Consequently, sea level drops and the hydrology of streams is greatly altered. As the great ice sheets advance over the continents, they obliterate preexisting drainage networks. The moving ice scours and erodes the landscape and deposits the debris near its margins, covering the preexisting topography. The crust of the earth is pushed down by the weight of the ice, and meltwater commonly collects and forms lakes along the ice margins. As the glaciers melt, new drainage systems are established to accommodate the large volume of meltwater. Far beyond the margins of the glaciers, stream systems are modified by changing climatic patterns. Even in arid regions, the imprint of climatic changes associated with glaciation is seen in the development of large lakes in closed basins.

The area beneath the ice is completely depopulated, as both plants and animals are forced to migrate in front of the advancing ice. This forced migration causes great stress on animal populations, and many species become extinct because they fail to adapt.

The last ice age also had a profound effect on human populations. The history of migrations of early peoples is largely an account of responses to the changing environment of the ice age, and the rigorous glacial climate probably stimulated many amazing adaptations that otherwise might not have occurred.

The causes of an ice age remain a tantalizing, unanswered question. Will the ice age return, or will the existing glaciers melt? In either case, some climatic change is certain.

MAJOR CONCEPTS

1. Glaciers are systems of flowing ice that form where more snow accumulates each year than melts.
2. As ice flows, it erodes the surface by abrasion and plucking. Sediment is transported by the glacier and deposited where the ice melts.
3. The two major types of glaciers—continental and valley glaciers—produce distinctive erosional and depositional landforms.
4. The major effects of an ice age include the rise and fall of sea level, isostatic adjustments of land, modification of drainage systems, creation of numerous lakes, and migration and selective extinction of plant and animal species.
5. The cause of glaciation is not clearly understood, but it may be related to ocean currents and continental drift.

GLACIAL SYSTEMS

Statement

A glacier is a system of flowing ice that originates on land through the accumulation and recrystallization of snow. The necessary conditions for the development of a glacier are simple: more snow must fall each year than is lost by melting and evaporation. Under these conditions, a new layer of snow accumulates each year, and, over many years, the mass of ice eventually becomes thick enough to flow under its own weight. Perennial snowfields that do not move are not considered glaciers, nor is pack ice formed from seawater in polar latitudes.

Glaciers are open systems and have much in common with other gravity flow systems, such as rivers and ground water. Water enters the system primarily in the upper parts of the glacier, where snow accumulates and is transformed into ice. The ice then flows out of the zone of accumulation, generally moving a few centimeters per day. At the lower end (or terminus) of the glacier, ice leaves the system by melting and evaporating. There are two principal types of glaciers:

1. Valley glaciers are streams of ice that originate in the snowfields of high mountain ranges.
2. Continental glaciers are ice sheets thousands of meters thick that spread out and cover large parts of continents.

Discussion

Valley Glacial Systems

Valley glaciers, or alpine glaciers, occur in some of the most scenic mountain ranges of the world and are relatively accessible for direct observation. As a result, they have been studied for many years, and their general characteristics are well understood. *Figure 14.1* is an idealized diagram showing how a valley glacial system operates. Valley glaciers are long, narrow rivers of ice that originate in the snowfields of high mountain ranges and flow down preexisting stream valleys. They range from a few hundred meters to more than a hundred kilometers long. In many ways, they resemble river systems. They receive an input of water in the higher reaches of the mountains, they flow downslope through a system of the tributaries leading to a main trunk system, and they erode and modify the landscape over which they flow. The essential parts of the system are (1) the zone of ac-

cumulation, where there is a net gain of ice, and (2) the zone of **ablation,** where ice leaves the system by melting and evaporating.

In the zone of accumulation, snow is transformed into glacial ice through a process identical to rock metamorphism. Freshly fallen snow consists of delicate, hexagonal ice crystals or needles, with as much as 90% void space (*figure 14.2*). As snow accumulates, the ice at the points of the snowflake melts from the pressure buildup and migrates toward the center, eventually forming an elliptical granule of recrystallized ice approximately 1 mm thick. The accumulation of these particles packed together is called **firn,** or **névé.** With repeated annual deposits, the loosely packed névé granules are compressed by the weight of the overlying snow. Meltwater, which results from daily temperature fluctuations and the pressure of the overlying snow, seeps through the pore spaces between the grains and freezes, aiding in the recrystallization process. Air in the pore spaces is driven out. When the ice reaches a thickness of approximately 30 or 40 m, it can no longer support its own weight and yields to plastic flow. The upper part of a glacier is rigid and tends to fracture, but the ice beneath moves by plastic flow.

As is shown in *figure 14.1,* the boundary between the zone of accumulation and the zone of ablation is approximated by the **snowline.** Above the snowline, the surface of the glacier is smooth and white, because more snow accumulates than is lost by melting, and any irregularities are soon covered and filled with snow. Below the snowline, melting and evaporation exceed snowfall. There, the surface of the ice is rough and pitted and commonly is broken by open **crevasses.**

At the terminal margin of a glacier, the loss of ice by melting and evaporation combined exceeds the rate of accumulation. This margin is the major exit boundary of the system.

It is important for you to understand that the margins of a glacier constitute the boundaries of a system of flowing ice, much as the banks and mouth of a river constitute the boundaries of a river system. If more snow is added in the zone of accumulation than is lost by melting or evaporating at the end of the glacier, the ice mass increases, and the glacial system expands. If the accumulation of ice is less than ablation, there is a net loss of mass, and the size of the glacial system is reduced. If accumulation and ablation are in balance, the mass of ice remains constant, the size of the system remains constant, and the terminus of the ice remains stationary. *Ice within the glacier continually*

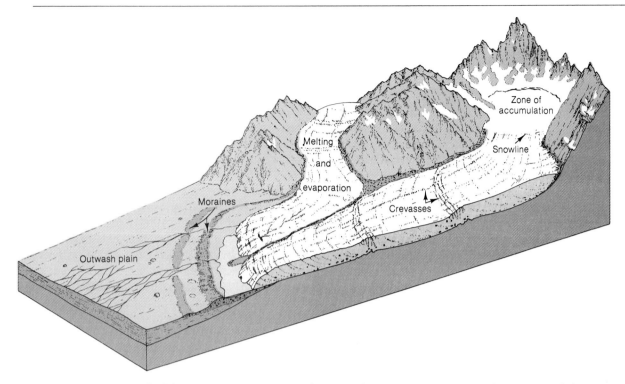

Figure 14.1 A glacial system is an open system of ice that flows under the pull of gravity. Snow enters the system by precipitation and is transformed into ice. The ice flows outward from the zone of accumulation under the pressure of its own weight. The ice leaves the system by evaporating and melting in the zone of ablation. The balance between the rate of accumulation and the rate of melting determines the size of the glacial system. The major parts of a valley glacial system are the zone of ac- cumulation, where snow enters the system, and the zone of ablation, where ice leaves the system by melting and evaporation. The boundary between these zones is approximated by the snowline. As the ice moves through the system, it erodes the preexisting stream valley and deposits sediment at the end of the glacier. Meltwater can rework much of the glacial sediment and redeposit it downstream.

flows toward the terminal margins, however, re- gardless of whether the terminal margins are ad- vancing, retreating, or stationary.

The behavior of a glacial system is determined by the balance between the rate of input and the rate of output of ice. The two major variables in this balance are temperature and precipitation. A glacier can grow or shrink at a given temperature if the amount of precipitation (rate of input) varies. Also, it can grow or shrink with a given rate of precipitation if the temperature varies enough to increase or decrease the rate of melting (rate of output). The length of a glacier in no way repre- sents the amount of ice that has moved through the system, just as the length of a river does not repre- sent the volume of water that has flowed through it. It simply shows the amount of ice currently in the system.

An example from the last ice age illustrates this point. A glacial valley 20 km long in the Rocky Mountains was eroded 600 m deeper than the orig- inal stream valley. This large amount of erosion was not accomplished by 20 km of ice moving down the valley. It was the result of many thou- sands of kilometers of ice flowing through the valley. If the ice occupied the valley during each glacial epoch and moved 0.3 m per day, a total of approximately 72,000 km of ice would have moved

Figure 14.2 The transformation of a snowflake into granular ice involves melting the arms of the snow crystal and refreezing the water near the center to produce a more compact grain of ice.

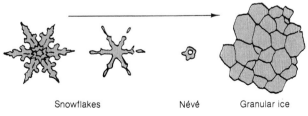

Snowflakes   Névé   Granular ice

down the valley. The enormous abrasion caused by such a long stream of ice would be adequate to wear down the valley 600 m deep.

Continental Glacial Systems

In terms of their effect on the landscape and the earth's hydrologic system, **continental glaciers** are by far the more important type of glacial system. These large **ice sheets** form in some of the most rigorous and inhospitable climates on the earth. Nonetheless, teams of scientists from various countries have used modern technology to study existing continental glaciers in Canada, Greenland, and Antarctica. From these studies, we can construct a reasonably accurate model of an idealized continental glacial system and analyze how it operates (*figure 14.3*).

The basic elements of a continental glacier are the same as those in a valley glacier. Both systems have a zone of accumulation, where there is a net gain of ice from snowfall. The ice flows out from the zone of accumulation to the zone of ablation, where it leaves the system through melting and evaporation. As shown in *figure 14.3*, a continental glacier is a roughly circular or elliptical plate of ice, rarely more than 3000 m thick. Ice does not have the strength to support the weight of an appreciably thicker accumulation. If more ice is added by increased precipitation, the glacier simply flows out from the centers of accumulation more quickly.

The weight of such a huge ice mass causes the earth's crust to subside, so that the surface of the land commonly slopes toward the glacier. Subsidence creates a lowland along the ice margin, which traps meltwater to form large lakes. If the margin of the glacier is near the coast, an arm of the sea may invade the depression, as is shown in *figure 14.3*.

Preexisting drainage systems are modified or completely obliterated. Rivers that flow toward the ice margins are impounded, forming lakes, which may overflow and develop a river channel parallel to the ice margin. Drainage systems covered by the glacier are destroyed. Thus, when the ice melts, there is no established drainage system, and numerous lakes form in the natural depressions.

The margins of continental glaciers commonly form large lobes. These develop because, where the ice is not confined to a system of valleys, it moves more rapidly into preexisting lowlands. The sediment deposited at the ice margins is therefore arcuate or lobate in map pattern.

The Barnes Ice Cap, of Baffin Island, Canada (*figure 14.4*), is the last remnant of the glacier that recently covered much of Canada and parts of the northern United States. This example illustrates the

---

Figure 14.3 A continental glacial system covers a large part of a continent and causes a number of significant changes in the regional physical setting. The weight of the ice depresses the continent, so that the land slopes toward the glacier. Consequently, glacial lakes form along the ice margins, or an arm of the ocean can invade the depression. The original drainage system is greatly modified, as some streams flow toward the ice margins and form lakes. The glacier advances more rapidly into lowlands, so that the margins typically are lobate. As the system expands and contracts, ridges of sediment are deposited along its margins, and a variety of erosional and depositional landforms develop beneath the ice.

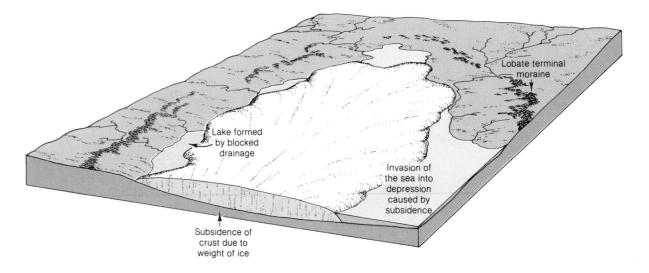

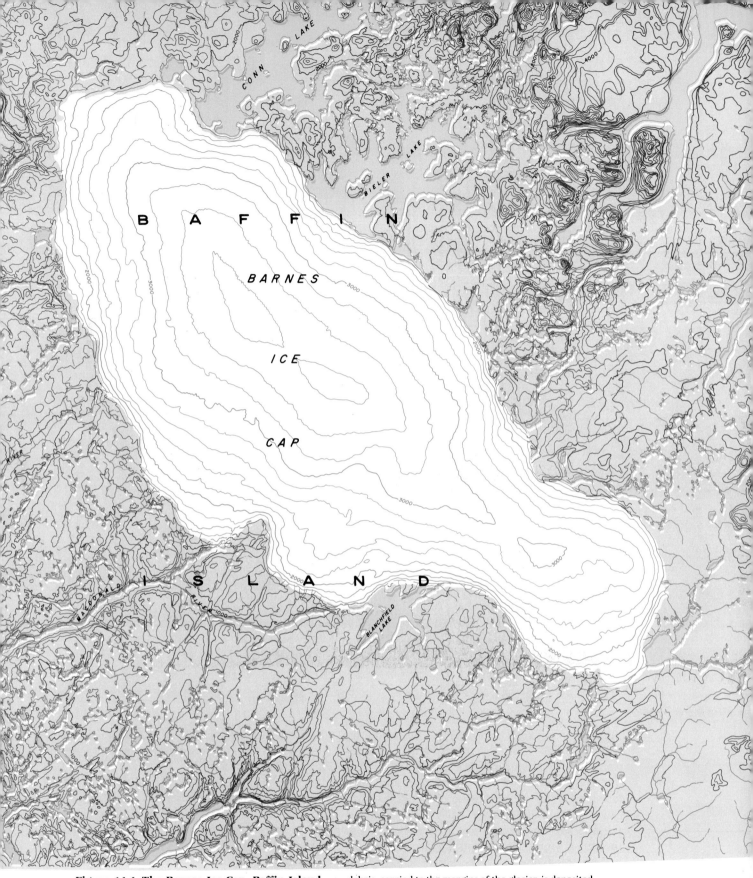

Figure 14.4 The Barnes Ice Cap, Baffin Island, Canada, shows many features produced by continental glaciation. Drainage coming from the north is blocked by the ice and forms lakes near the ice margins. Erosional debris, carried to the margins of the glacier, is deposited as moraine. Irregularities in the surface over which the ice flows cause the ice margins to be irregular or lobate.

Figure 14.5 The margins of the Barnes Ice Cap are marked by ridges of sediment deposited by the ice. The glacier's upper surface is gently arched, and meltwater has formed small meandering streams on it. The landscape of the Great Lakes region must have appeared something like this 20,000 years ago.

relationship between a continental glacier and the regional landform. As shown on the map, the glacier is elliptical, with irregular or lobate margins. The ice is thickest in the central part and thins toward the edges. The presence of the glacier has caused an isostatic subsidence of the crust, so that the land slopes toward the ice margins. In addition, the glacier completely disrupts the former drainage system. Meltwater has therefore accumulated along the ice margins, forming a group of lakes. A photograph of the southern margins of the ice cap (*figure 14.5*) shows the large, gently arched surface

of the glacier, sediment deposited along the ice contact, and stream channels formed by meltwater on the glacier's surface.

The ice cap that covers nearly 80% of Greenland is much larger than the remnant on Baffin Island. In cross section, the glacier is shaped like a drop of water on a table (*figure 14.6*). Its upper surface is a broad, almost flat-topped arch and is typically smooth and featureless. The base of the glacier is relatively flat. The Greenland glacier is over 3000 m thick in its central part, but it thins toward the margins. The zone of accumulation is in the central

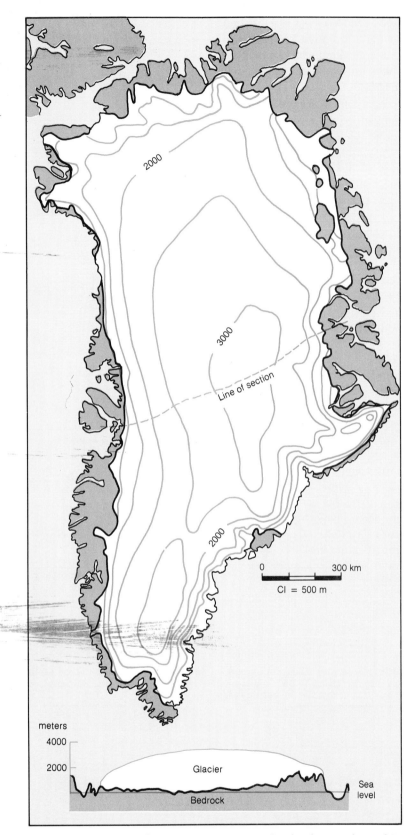

Figure 14.6 The Greenland Ice Sheet covers nearly 80% of the island. In this diagram, the thickness of the glacier is shown by contour lines. Note from the cross section that the central part of Greenland has been depressed below sea level by the weight of the ice.

part of the island, where the ice sheet is nourished by snowstorms moving from west to east. The snowline lies from 50 to 250 km inland, so that the area of ablation constitutes only a narrow belt along the glacial margins.

In rugged terrain, especially in areas close to the margins, the direction of ice movement is greatly influenced by mountain ranges, and the ice moves through mountain passes in large streams called **outlet glaciers** (*figure 14.7*). These resemble valley glaciers in that they are confined by the topography. Pressure builds up in the ice behind a mountain range and forces outlet glaciers through mountain passes at relatively high speeds. measurements in Greenland show that the main ice mass advances from approximately 10 to 30 cm per day. Outlet glaciers, however, can move as fast as 1 m per hour. In some places, you can actually watch the ice move.

The glacier of Antarctica is similar to that of Greenland in that it covers essentially the entire continent (*figure 14.8*). But Antarctica is much larger than Greenland, and its glacier contains more than 90% of the earth's ice. Much of the glacier is over 3000 m thick, and its weight has depressed large parts of the continent's surface below sea level. Parts of Antarctica (mostly near the continental margins) are mountainous, with higher peaks and ranges protruding above the ice. In the mountains, outlet glaciers funnel ice from the interior to the coast. Two large ice shelves and numerous small ones occur along the coast. These break up, in a process called **calving,** to form large icebergs in the South Atlantic Ocean.

Both Greenland and Antarctica are surrounded by water, so that there is an ample supply of moisture to feed their glaciers. In contrast, Siberia

is cold enough for glaciers to exist, but there is not sufficient precipitation for ice to accumulate.

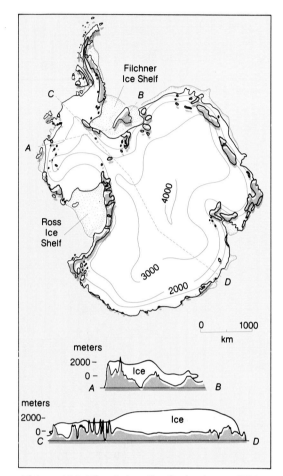

Figure 14.8 The Antarctic glacier is over 3000 m thick in many areas and has a general cross-sectional shape similar to that of the Greenland glacier.

Figure 14.7 Outlet glaciers are segments of a continental glacier that advance ahead of the main sheet through a mountain pass. A mountain range is a physical barrier to the advancement of a continental glacier, and great pressures build up in the ice behind the range. This pressure causes the ice in an outlet glacier to move very rapidly.

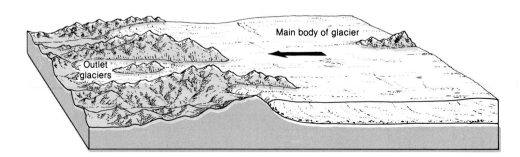

EROSION, TRANSPORTATION, AND DEPOSITION

Statement

As glacial ice flows over the land, it erodes, transports, and deposits vast amounts of rock material and greatly modifies the preexisting landscape. Glaciers erode bedrock in two ways: by glacial plucking and by abrasion (grinding). The eroded material is then carried in suspension in the ice and is deposited near the margins of the glacier, where melting dominates.

Discussion

Erosion

Glacial **plucking** is the lifting out and removal of fragments of bedrock by a glacier. It is one of the most effective ways in which a glacier erodes the land. The process involves frost wedging. Beneath the glacier, meltwater seeps into joints or fractures, where it freezes and expands, wedging blocks of rock loose. The loosened blocks freeze to the bottom of the glacier and are plucked or quarried from the bedrock, becoming incorporated into the moving ice (*figure 14.9*). The process is especially effective where the bedrock is cut by numerous joints and where the surface is unsupported on the downstream side.

Abrasion is essentially a filing process. The angular blocks plucked and quarried by the moving ice freeze firmly into the glacier, and they act as tools that grind and scrape the bedrock. Aided by the pressure of the overlying ice, the angular blocks are very effective agents of erosion, capable of wearing away large quantities of bedrock. The rock fragments incorporated into the glacial ice are themselves abraded and worn down as they grind against the walls and floor of the valley. As a result, they usually develop flat surfaces that are deeply scratched.

Evidence of the distinctive abrasive and quarrying action of glaciers can be seen on most bedrock surfaces over which glacial ice has moved. Small hills of bedrock commonly are streamlined by glacial abrasion. Their upstream side typically is rounded off, while the downstream side is made steep and rugged by glacial plucking (*figure 14.10*). Glacial **striations**, such as those illustrated in *figure 14.11*, range from hairline scratches to large furrows more than a kilometer long.

Glacial Transport

The rock fragments transported by a glacier are collectively referred to as its **load**. The manner in which a glacier transports its load, however, differs from stream transport in a very important way. The load in a glacier is carried in suspension with large blocks transported side by side with small grains, without sorting or separation of the material according to size. As a result, the deposits of a

---

Figure 14.9 Glaciers erode, transport, and deposit rock material in the manner illustrated in this diagram. Erosion by glacial plucking occurs as blocks of rock, separated by joints, freeze to the base or sides of the glacier and are lifted from the outcrop by the moving ice. These fragments, frozen in the ice, then act as abrasives and wear down the surface by grinding. Near the end of the glacier, the ice is commonly stagnant and no longer moves. Much of the sediment load is thus forced upward along shear planes and is concentrated at the surface of the glacier. As the ice melts, this sediment accumulates in an end moraine.

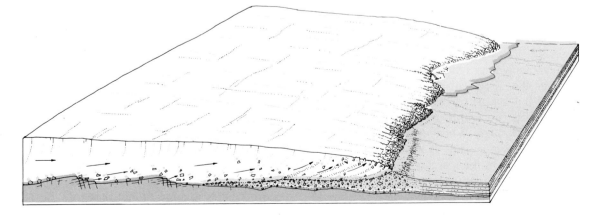

Direction of ice flow

Bedrock

Figure 14.10 Roche moutonnée is an erosional feature that forms as ice moves over bedrock, eroding it into a streamlined shape. Glacial plucking commonly produces a ragged edge on the downcurrent side.

glacier are unsorted and unstratified, and thus they differ markedly from stream deposits. The load of a glacier is concentrated near the contact between the ice and the bedrock, from which it was derived.

Most of the particles carried by a glacier are fresh and unweathered and have angular, jagged surfaces. The grinding action of the moving ice further crushes the grains to produce an abundance of fine particles known as **rock flour.**

Deposition

Most of the particles transported by a glacier are deposited near the terminus, where melting dominates. There, the ice can become stagnant. This commonly forces the ice upstream to move up and over the "dead ice" at the terminus, so that much of the load is carried upward along shear planes and is concentrated on the surface of the glacier (*figure 14.9*). When melting is completed, this material accumulates as a ridge, marking the former position of the margins of the glacier.

The sediment deposited by a glacier is commonly picked up and reworked by meltwaters and rede-

posited by streams. This sediment, whether deposited directly by the glacier or indirectly in glacial lakes and streams, is referred to as glacial **drift** (the name is held over from the time when such deposits were believed to have drifted to their present resting place during the flood of Noah). There is thus stratified and unstratified glacial drift. The term *till* is used to designate unsorted, unstratified drift, deposited directly by the glacier.

Rates of Ice Flow

Ice flow in a glacier may seem extremely slow in comparison to the flow of water in rivers, but the movement is continuous, and over the years, vast quantities of ice can move across the landscape. Measurements show that some of Switzerland's large valley glaciers move as much as 180 m per year. Smaller glaciers move from 90 to 150 m per year. Some of the most rapid rates have been measured on the outlet glaciers of Greenland, where ice is funneled through mountain passes at a speed of 8 km per year. From these and other measurements, it appears that flow rates of a few

Figure 14.11 Glacial striations result from the abrasive action of a glacier.

centimeters per day are common and that velocities of 3 m per day are exceptional. The most rapid movement occurs in outlet glaciers, where pressure built up behind a mountain range helps to force the ice through a mountain pass.

Recent close observations of the movements of valley glaciers show that, occasionally, a glacier can surge forward more than several hundred meters per day. Now that movements can be monitored by satellite photography, these surges are known to be fairly common. (In the past, only a few surges were seen, because the flow is so short-lived.) A glacier in the Himalayas, for instance, recently moved 11 km in 3 months. In 1966, the Steele Glacier, in the Yukon Territory, had a maximum surface displacement of approximately 8 km within a matter of weeks. Glacial surges apparently result from sudden slippages along the bases of glaciers, caused by the buildup of extreme stress upstream. Stagnant or slow-moving ice near the terminus can act as a

dam for the faster-moving ice upstream. If this happens, stress builds up behind the slow-moving ice, causing a sudden surge when a critical point is reached. Surges can also be caused by a sudden addition of mass to the glacier, such as a large avalanche or a landslide on its surface.

In considering the flow of ice in a glacier and the erosion it can cause, remember that a glacier is an open system. Material enters the system in the zone of accumulation, flows through, and then exits at the distal margins. Ice within a glacier flows continuously through the system, regardless of whether the terminal margin is advancing, retreating, or stationary. The length of a valley glacier is therefore no indication of the amount of erosion it can accomplish or has accomplished. Erosion by a glacier is a function of the duration of the process, the thickness of the ice, and the velocity at which it flows.

LANDFORMS DEVELOPED BY VALLEY GLACIERS

Statement

Valley glaciers are responsible for some of the most rugged and scenic mountainous terrain of the world. The Alps, the Sierra Nevada, the Rockies, and the Himalayas were all greatly modified by glaciers during the last ice age, and the shapes of their valleys, peaks, and divides retain the unmistakable imprint of erosion by ice. A valley glacier commonly fills more than half of the valley, and as it moves, it modifies the former V-shaped stream valley to a broad U-shaped, or troughlike,

form. The head of the valley is sculptured into a large amphitheater called a cirque. Where several cirques approach a summit from different directions, a sharp, pyramid-shaped peak called a horn is formed. The projecting ridges and divides between glacial valleys are subjected to rigorous ice wedging, abrasion, and mass movement. These processes produce sharp, angular crests and divides (called arêtes) instead of the rounded topography developed by stream erosion.

Part of a valley glacier's load consists of rock fragments that avalanche down the steep valley sides and accumulate along the glacier's margins. Frost action is especially active in the cold climate of valley glaciers and produces large quantities of angular rock fragments. This material is trans-

Figure 14.12 Lateral moraines are accumulations of debris derived from erosion of the valley walls. Much of the debris is produced by ice wedging on the high valley slopes. This material moves downslope by mass movement and accumulates along the margins of the glacier. It is then transported downstream by the moving ice. Where tributary glaciers merge into the main stream, the lateral moraines also merge to form a medial moraine. (See plate 10.)

ported along the surface of the glacier's margins, forming conspicuous dark bands called lateral moraines (*figure 14.12*). Where a tributary glacier enters the main valley, the lateral moraine of the tributary glacier merges with a lateral moraine of the main glacier to form a medial moraine in the central part of the main glacier. Thus, in addition to transporting the load near its base, a valley glacier acts as a conveyer belt and transports a large quantity of surface sediment to the terminus. At the terminus, ice leaves the system through melting and evaporation, and the load is deposited as an end moraine (*figure 14.13*). End moraines commonly block the ends of the valleys, so that meltwater from the ice accumulates and forms ponds and lakes. Downstream from the glacier, meltwater reworks the glacial sediments and redeposits them to form an outwash plain.

Discussion

The idealized diagrams in *figure 14.14* illustrate the major landforms resulting from valley glaciation. This figure permits a comparison and contrast of landscapes formed only by running water with those that have been modified by valley glaciers (see also *plate 10*). *Figure 14.14A* shows the typical topography of a mountain region being eroded by streams. A relatively thick mantle of soil and weathered rock debris covers the slopes. The valleys are V-shaped in cross section and have many bends at tributary junctions, so that ridges and divides between tributaries appear to overlap if you look up the valley. In *figure 14.14B*, the area is shown occupied by glaciers. The growing glaciers flow down the tributary valleys and merge to form a major glacier. During glaciation, several

Figure 14.13 Terminal moraines form at the end of a glacier, where sediment transported by the glacier is deposited as the ice melts.

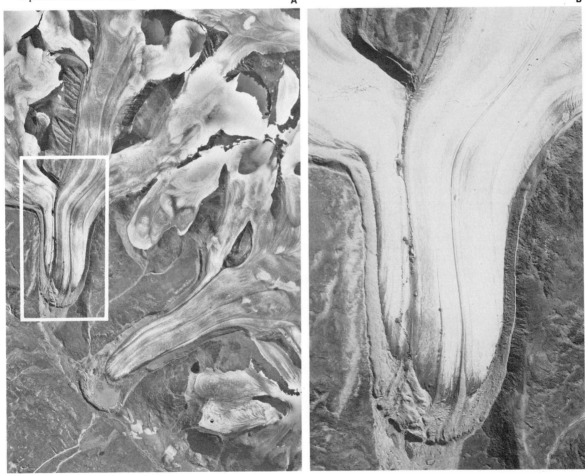

A

B

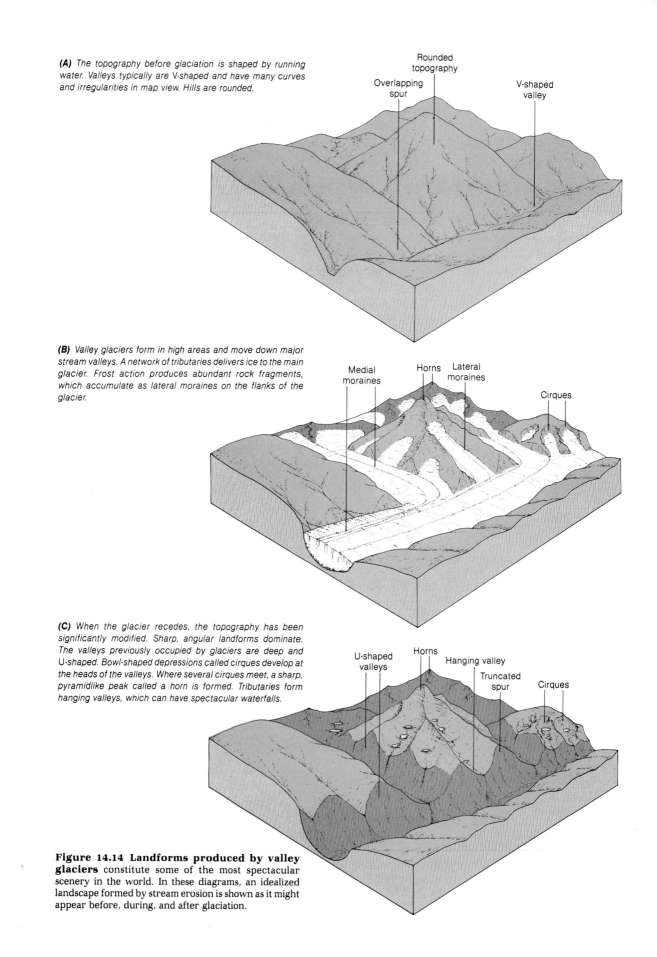

(A) The topography before glaciation is shaped by running water. Valleys typically are V-shaped and have many curves and irregularities in map view. Hills are rounded.

Rounded topography

Overlapping spur

V-shaped valley

(B) Valley glaciers form in high areas and move down major stream valleys. A network of tributaries delivers ice to the main glacier. Frost action produces abundant rock fragments, which accumulate as lateral moraines on the flanks of the glacier.

Medial moraines

Horns

Lateral moraines

Cirques

(C) When the glacier recedes, the topography has been significantly modified. Sharp, angular landforms dominate. The valleys previously occupied by glaciers are deep and U-shaped. Bowl-shaped depressions called cirques develop at the heads of the valleys. Where several cirques meet, a sharp, pyramidlike peak called a horn is formed. Tributaries form hanging valleys, which can have spectacular waterfalls.

U-shaped valleys

Horns

Hanging valley

Truncated spur

Cirques

Figure 14.14 Landforms produced by valley glaciers constitute some of the most spectacular scenery in the world. In these diagrams, an idealized landscape formed by stream erosion is shown as it might appear before, during, and after glaciation.

Figure 14.28 Pluvial lakes were formed in the closed basins of the western United States as a result of climatic changes associated with the glacial epoch. Most are now dry lakebeds because of the arid climate. Former shorelines of the pluvial lakes are well marked along the basin margins (see figure 14.29). Lake Bonneville, in western Utah, was the largest. Its present remnants are Great Salt Lake, Utah Lake, and Sevier Lake.

Figure 14.29 Shoreline features of Lake Bonneville include deltas, beaches, bars, spits, and wave-cut cliffs. The horizontal terrace shown in the photograph marks the high-water mark of the lake.

rivers entered the lake from the high Wasatch Range, to the east. They built large deltas, shoreline terraces, and other coastal features that are now high above the valley floors. The most conspicuous feature is a horizontal terrace high on the Wasatch mountain front (*figure 14.29*).

As the level of the lake rose to 300 m above the floor of the valley, it overflowed to the north into the Snake River and thence to the ocean. The outlet, established on unconsolidated alluvium, rapidly eroded down to bedrock, 100 m below the original pass. The level of the lake was stabilized, fluctuating only with the pluvial epochs associated with glaciation. Some valley glaciers from the Wasatch Range extended down to the shoreline of the old lake, and some of their moraines were carved by wave action. This wave erosion shows conclusively that glaciation was contemporaneous with the high level of the lake. As the climate became dryer, the lake evaporated, leaving faint shorelines at lower levels.

The Channeled Scablands

The continental glacier in western North America moved southward from Canada only a short distance into Washington, but it played an important role in producing a strange complex of interlaced deep channels, a type of topography found nowhere else on the earth. This area, the Channeled Scablands, covers much of eastern Washington and consists of a network of braided channels

from 15 to 30 m deep. The term *scabland* is appropriately descriptive because, viewed from the air, the surface of the area has the appearance of great wounds or scars (*figure 14.30*). Many of the channels have steep walls and dry waterfalls or cataracts. In additon, there are deposits with giant ripple marks and huge bars of sand and gravel (*figure 14.31*). These features attest to erosion by running water of exceptional degree, which would be considered catastrophic by normal standards. Yet, today, there is not enough rainfall in the area to maintain a single permanent stream.

Briefly, the scablands were eroded by the following process. A large lobe of ice advanced southward across the Columbia Plateau and temporarily blocked the Clark Fork River, one of the major northward-flowing tributaries of the Columbia River (*figure 14.32*). The impounded water backed up to form glacial Lake Missoula, a long, narrow lake extending diagonally across part of western Montana. Sediments deposited in this lake now partly fill the long, narrow valley. As the glacier receded, the ice dam failed, releasing a tremendous flood over the southwestward-sloping Columbia Plateau. The enormous discharge, barely diverted by the preexisting shallow valleys, spread over the basalt surface, scouring out channels and forming giant ripple marks, bars, and other sediment deposits. Estimates suggest that, during the flood, as much as 40 km³ of water per hour might have discharged from Lake Missoula. Since the glaciers

Figure 14.30 The Channeled Scablands of Washington consist of a complex of deep channels cut into the basalt bedrock. The scabland topography is completely different from that produced by a normal drainage system. It is believed to have been produced by catastrophic flooding.

advanced several times into the region, such catastrophic flooding probably occurred many times. Lake Missoula formed each time the ice front advanced past the Clark Fork River and then flooded the scablands with each recession of the ice and subsequent dam failure.

Biological Effects of the Ice Age

The severe climatic changes during the ice age had a drastic impact on most life forms. With each advance of the ice, large areas of the continents (the areas beneath the ice) became totally depopulated. In addition, plants and animals retreating southward in front of the advancing glacier were under tremendous stress. The most severe stresses resulted from drastic climatic changes, reduction of living space, and curtailment of total food supply. As the glaciers advanced, most species were displaced, along with their environments, across distances of approximately 3200 km. As the ice retreated, new living space became available in the deglaciated areas, but the formerly exposed continental shelves were inundated by the rising sea. During the major glacial advances, when sea level was lower, new routes of migration opened from Asia to North America, because much of Alaska and Siberia were not glaciated (see *figure 14.19*), and from Southeast Asia to the islands of Indonesia. Land plants were forced to migrate with the climatic zones in front of the glaciers. As the glaciers pushed cold-weather belts southward, displaced storm tracks and changes in precipitation affected even the tropics.

Many life forms could not cope with the repeated and overwhelming environmental changes brought about by the cycles of advancing and retreating ice. Numerous species, particularly giant mammals, became extinct. The imperial mam-

Figure 14.31 Giant ripples were formed in the surface of a gravel bar by catastrophic flooding in the Channeled Scablands.

moth, 4.2 m high at the shoulders, roamed much of North America. The saber-toothed tiger became extinct about 14,000 years ago. Fossils of the giant beaver, as large as a black bear, and the giant ground sloth, which measured 6 m tall standing on its hind legs, have been found in Pleistocene sediments. In Africa, fossil sheep 2 m tall have been found, in addition to pigs as big as a present-day rhinoceros. In Australia, giant kangaroos and other marsupials thrived during the Pleistocene.

Effects of Winds

The presence of ice over so much of the continents greatly modified patterns of atmospheric circulation. Winds near the glacial margins were strong and unusually persistent because of the abundance of dense, cold air coming off the glacier fields. These winds picked up and transported large quantities of loose, fine-grained sediment

brought down by the glaciers. This dust accumulated as **loess** (windblown silt), sometimes hundreds of meters thick, forming an irregular blanket over much of the Missouri River valley, central Europe, and northern China.

Sand dunes were much more widespread and active in many areas during the Pleistocene. A good example is the Sand Hills region, in western Nebraska, which covers an area of about 60,000 km². This region was a large, active dune field during the Pleistocene, but today the dunes are largely stabilized by a cover of grass.

The Oceans

Pleistocene glaciation affected the waters of all of the oceans to some degree. Besides changing the sea level, so that shorelines were altered and much of the continental shelves were exposed, the glacial periods cooled the ocean waters. The lower tem-

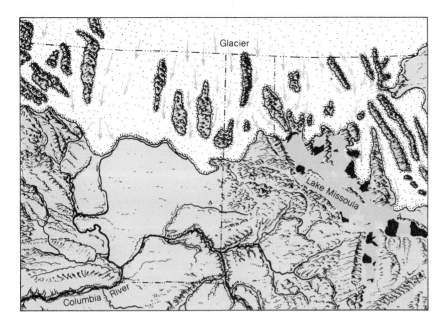

Figure 14.32 The origin of the Channeled Scablands is attributed to catastrophic flooding, on a magnitude apparently unique in the earth's history. The flood resulted when the ice dam that formed glacial Lake Missoula failed as the glacier receded.

(A) The ice sheet in northern Washington blocked the drainage of the northward-flowing Clark Fork River to form Lake Missoula, a long, deep lake in northern Idaho and western Montana.

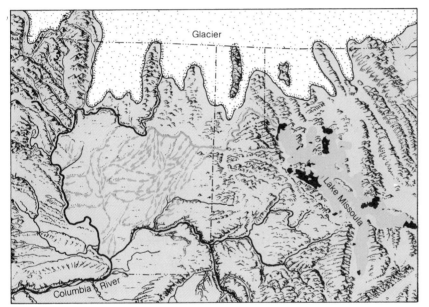

(B) As the glacier receded, the ice dam that formed Lake Missoula failed catastrophically, and water from the lake quickly drained across the scablands, eroding deep channels. Repeated advance and retreat of the glacier probably produced several ice dams that failed as the ice melted, and each caused catastrophic flooding.

peratures affected the kind and distribution of marine life and also influenced seawater chemistry. Furthermore, patterns and strengths of oceanic currents were changed. Circulation was significantly restricted by features such as the Bering Strait, extensive pack ice, and exposed shelves.

Even the deep-ocean basins did not escape the influence of glaciation. Where glaciers entered the ocean, icebergs broke off and rafted their enclosed load of sediment into the ocean. As the ice melted, debris ranging from huge boulders to fine clay settled on the deep-ocean floor, resulting in an unusual accumulation of coarse glacial boulders in fine oceanic mud. Ice-rafted sediment is most common in the Arctic, the Antarctic, the North Atlantic, and the northeastern Pacific.

In the warmer reaches of the ocean, the glacial and interglacial periods are recorded by alternating layers of red clay and small calcareous shells of microscopic organisms. The red mud accumulated during cold periods, when fewer organisms inhabited the cold water. During the warm interglacial periods, life flourished, and layers of shells mixed with mud were deposited.

RECORDS OF PRE-PLEISTOCENE GLACIATION

Statement

Glaciation has been a very rare event in the history of the earth. Prior to the last ice age, which began from 2 million to 3 million years ago, the earth's climate was typically mild and uniform for long periods of time. This climatic history is implied by the types of fossil plants and animals and the characteristics of sediments preserved in the stratigraphic record. There are, however, widespread glacial deposits—unsorted, unstratified debris containing striated and faceted cobbles and boulders—recording several major periods of ancient glaciation. These glacial deposits commonly rest upon striated and polished bedrock, and they are associated with varved shales and with sandstones and conglomerates that are typical of outwash deposits. Such evidence implies several periods of glaciation in the remote history of the earth.

Discussion

The best-documented record of pre-Pleistocene glaciation is found in late Paleozoic rocks (formed from 200 million to 300 million years ago) of South Africa, India, South America, Antarctica, and Australia. Exposures of ancient glacial deposits are numerous in these areas, many of them resting upon a striated surface of older rock (*figure 14.33*).

Deposits of even older glacial sediment exist on every continent but South America. These indicate that two other periods of very widespread glaciation occurred during late Precambrian time.

Small bodies of glacial sediment from other geologic periods have been found in local areas, but they are not nearly as well documented or as widespread. Glaciation therefore has been a relatively rare phenomenon, and it has apparently occurred in regular cycles. One feature seems to be common to all periods of glaciation: the continents appear to have been relatively high and undergoing periods of mountain building. Mountain-building episodes were numerous in the geologic past, but often without accompanying glaciation. This fact suggests that an elevated landmass is not the primary cause of glaciation, but only one of several prerequisites. Glacial epochs must require a special combination of conditions, which has occurred only a few times in the last 4 billion years.

CAUSES OF GLACIATION

Statement

Although the history of Pleistocene glaciation is well established and the many and varied effects of glaciation are clearly recognized, we do not know why glaciation takes place. For over a century, geologists and climatologists have struggled with the problem, but it remains unsolved. An adequate theory of glaciation must account for the following facts:

1. During the last ice age, which lasted from 2 million to 3 million years, repeated advances of the ice were separated by interglacial periods of warm climate. Therefore, glaciation is not related to a slow process involving long-term cooling.

2. Glaciation is an unusual event in the earth's history. There was widespread glaciation at the end of the Paleozoic Era, from 200 million to 300 million years ago, and during late Precambrian time, approximately 700 million years ago. The few other episodes of glaciation apparently were not widespread. Therefore,

---

Figure 14.33 Late Paleozoic glaciation is well documented in the southern continents by deposits of moraine, striated bedrock surfaces, and other glacial features. This map shows the areas covered by ice during the late Pennsylvanian and Permian periods.

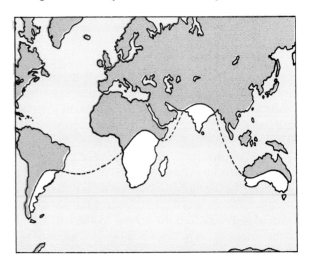

glaciation does not appear to be a result of some normal cyclic process. Rather, it seems to stem from a rare combination of conditions.

3. Throughout most of the earth's history, the climate has been milder and more uniform than it is now. A period of glaciation would require a lowering of the earth's average surface temperature by about 5 °C and, perhaps, an increase in precipitation.

4. Continental glaciers grow on elevated or polar landmasses that are situated so that storms bring moist, cold air to them. Glaciers can move into lowlands in lower latitudes, but they originate in highlands or in high latitudes. Greenland and Antarctica provide favorable topographic conditions today, as do the Labrador Peninsula, the northern Rocky Mountains, Scandinavia, and the Andes Mountains.

5. Precipitation is critical for the growth of glaciers. A number of areas are presently cold enough to produce glaciers but have insufficient snowfall, so that glacial systems do not develop in them.

Discussion

Astronomical Hypotheses

Many hypotheses have been advanced to explain glacial periods by extraterrestrial or astronomical influences. These can be grouped into categories involving (1) variations in solar radiation and (2) variations in the earth's orbital path.

Short-term fluctuations in solar radiation of as much as 3% have been observed, and some meteorologists believe that larger changes are possible. If periods of minimum solar radiation correspond with favorable topographic conditions on the earth, a glacial period could result. Unfortunately, there is presently no way to determine past variations in solar radiation, so that this hypothesis cannot be tested.

Variations in the earth's orbit are measurable. They alter the length of the seasons, perhaps enough to produce alternating colder and warmer cycles. If the cooling produced by rotational wobble is synchronous with the cooling produced by orbital deviation and minimal radiation from the sun, the earth (at least one hemisphere) could be cooled sufficiently to initiate glaciation. These astronomical variations occur in cycles repeated every few thousand years and apparently have existed throughout geologic time. The geologic

record, however, does not seem to show a corresponding cyclic repetition of glaciation. Thus, astronomical variations apparently fail to explain the erratic occurrence of glaciation throughout geologic time.

Atmospheric Changes

Could glacial periods be caused by changes in the composition of the earth's atmosphere? A drop in air temperature of 5 °C could be sufficient to initiate an ice age. A widely supported hypothesis attributes the necessary temperature fluctuation to a decrease in atmospheric carbon dioxide. Carbon dioxide and water vapor produce an important greenhouse effect. With a high carbon dioxide content in the atmosphere, shortwave solar radiation can reach the earth, but longwave thermal radiation cannot escape. Heat from the sun is thus trapped by the atmosphere, and the earth's climate remains moderate. Extensive growth of vegetation could remove carbon dioxide from the atmosphere and bring about a drop in temperature. The drop in temperature, however, would retard plant growth and rebalance the carbon dioxide content.

Another proposal suggests that glaciation results from fine volcanic dust injected into the atmosphere, which could insulate the earth by reflecting part of the solar radiation back into space, thus causing a drop in temperature. Periods of exceptional volcanic activity do not correspond with glaciation, however, so that this hypothesis is generally discarded.

Oceanic Controls

Ocean currents are intimately involved in the temperature and moisture balance in climate. In the past, they may have played an important role in the growth and decay of ice sheets.

One hypothesis is based on the circulation of ocean currents in the Atlantic and Pacific. Before the Pleistocene, North America and South America were *not* connected by the Isthmus of Panama. The warm equatorial currents of the Atlantic moved westward into the Pacific instead of northward in the Gulf Stream. The Arctic Ocean, being frozen, could not supply the moisture needed to feed the glaciers, although the temperature was sufficiently low.

As the Isthmus of Panama grew and connected the Americas, the equatorial currents were deflected northward, providing enough heat to melt the Arctic Ocean's ice. The Arctic waters were then free to evaporate and provide moisture for the glaciers. As the ice sheets grew, sea level dropped and thus reduced the effectiveness of the Gulf Stream. The Arctic waters then froze again, and the glaciers decayed and disappeared from lack of nourishment. As the glaciers receded and the ocean rose, the Gulf Stream became more effective. Melt from the Arctic Ocean's ice again provided the precipitation necessary for advance of the ice sheets. An important aspect of this hypothesis is that it explains the repeated advance and retreat of the glaciers during the Pleistocene Epoch under an essentially stable heat budget.

Continental Drift

The theory of continental drift has recently been used to help explain the growth and decay of ice sheets during the Pleistocene and earlier glacial periods. Throughout most of geologic time, the polar regions appear to have been broad, open oceans, which allowed major ocean currents to move unrestricted. Equatorial waters were spread over the polar regions, warming them with water from more temperate latitudes. The unrestricted circulation produced mild, uniform climates, which persisted throughout most of geologic time.

The large North American and South American continental plates moved westward from the Eurasian plate throughout Tertiary time. This drift culminated in the development of the Atlantic Ocean, trending north-south, with the North Pole in the small, nearly landlocked basin of the Arctic Ocean. Meanwhile, by late Miocene time, the Antarctic continent had drifted over the South Pole, and continental glaciation began. Evidence from deep-sea cores in the southern oceans strongly suggests that glaciation in the Antarctic began long before the Pleistocene Epoch and has continued there ever since. The continental glacier on Antarctica probably caused further cooling of both hemispheres. By the beginning of Pleistocene time, the present location and configuration of the continents and ocean basins had been established. With the present belts of climatic zones, areas near the polar regions were at the glacial threshold.

According to a theory proposed by W. L. Donn and M. Ewing, precipitation derived from the ice-free Arctic Ocean initiated continental glaciation in the high latitudes of the Northern Hemisphere. The Arctic Ocean connected with the Atlantic through shallow seaways on either side of Greenland. Cold polar air, originating over the snow-covered continents, had very low humidity, but the air masses from the open oceans contained substantial amounts of water. Ice caps grew into ice sheets and spread over the northern parts of North America

and Eurasia. Sea level dropped, so that circulation through the shallow channels connecting the Atlantic and Arctic oceans was restricted. This restriction reduced the amount of warm water carried to the Arctic, and thus the region's water temperature was lowered. The glaciers themselves also lowered the average atmospheric temperature. The ice-free Arctic Ocean then became frozen over, probably by the middle of the glacial stage. This freezing essentially eliminated evaporation from the Arctic, so that the snow supply to the major centers of the ice sheets was cut off. The southern extent of glaciation was determined by the amount of moisture available from the North Atlantic and North Pacific oceans.

An interglacial period began when the temperatures of the ocean surface, particularly of the North Atlantic, dropped enough to decrease the evaporation rate. This limited precipitation over the ice sheets, and the glaciers, starved for moisture, began to decay and retreat toward their spreading centers.

As the ice sheets receded, the oceans' surface waters became warmer, and the Arctic ice eventually began to melt. This melting initiated another ice age. A time lag of about 5000 years between the melting of the Arctic ice and the warming of the water's surface permitted the ice to disappear completely before the beginning of a new glacial cycle.

Greenland and Antarctica sustain continental glaciers because they are surrounded by the open ocean, which supplies enough moisture for glaciers to exist.

This hypothesis accounts for the complete short-term cycle of glacial and interglacial stages that recurred throughout the Pleistocene. By requiring special geographic conditions involving continental plates, polar regions, and ocean currents, it correctly implies that glaciation is a rare event. In addition, it explains the glacial period in the Southern Hemisphere during late Paleozoic time. The supercontinent existing at that time was located in the area of the South Pole, surrounded by oceans that supplied the necessary moisture. The major ice age thus developed but was terminated as the continents split and drifted toward the equator.

Moreover, this theory emphasizes precipitation as the major controlling factor, whereas other theories are based on almost inexplicable changes in temperature. Preliminary exploration of Arctic sea-floor sediments, however, shows no traces of the alternating cold- and warm-water fauna that should be found if the Arctic were alternately frozen and unfrozen.

SUMMARY

Several times in the geologic past, glaciers have interrupted the normal hydrologic system over much of the world. We recognize two main types of glaciers: (1) valley glaciers and (2) continental glaciers. Each type produces distinctive landforms.

The last ice age began to terminate from 15,000 to 20,000 years ago. The period of glaciation lasted for more than 1 million years and profoundly affected many aspects of the physical environment. The most significant results of glaciation include (1) the rise and fall of sea level, (2) isostatic adjustment of the crust, (3) the modification of drainage systems, (4) the creation of numerous lakes, and (5) the migration and selective extinction of many plants and animals.

The causes of glaciation are not completely clear, but a glacial period may be related to continental drift and the modification of ocean currents.

Melting of the glaciers raised the level of the oceans and inundated many coastlines. Today, these shores are continually modified by movement of the water in the oceans. In the next chapter, we will discuss these shoreline processes.

Key Words

glacier	hanging valley
zone of accumulation	arête
zone of ablation	moraine
ablation	lateral moraine
valley glacier	medial moraine
alpine glacier	end moraine
firn	terminal end moraine
névé	recessional end moraine
snowline	ground moraine
crevasse	outwash
continental glacier	outwash plain
ice sheet	drumlin
outlet glacier	esker
calving	varve
glacial plucking	kettle
abrasion	peat
striation	eustatic change
load	of sea level
rock flour	pluvial lake
glacial drift	Channeled Scablands
till	loess
cirque	Pleistocene Epoch
horn	interglacial period

Review Questions

1. Describe the processes by which snow is transformed into glacial ice.
2. Draw a cross section of a typical valley glacier and explain how a valley glacial system operates.
3. Sketch a model of a continental glacial system and explain how it operates.
4. Explain how glaciers erode the surface over which they flow.
5. Name and describe the landforms produced by valley glaciers.
6. Make a sketch map of North America showing the extent of the ice sheet during Pleistocene time.
7. Briefly describe the major effects, both direct and indirect, of Pleistocene glaciation.
8. Explain the origin of the Channeled Scablands.
9. How did the Great Lakes originate?
10. List several hypotheses to explain the cause of glaciation.
11. Explain the origin of Hudson Bay.
12. Why did sea level change during each period of advance and retreat of the ice?
13. Explain the origin of the Missouri River.
14. Study *figure 14.20* and explain why the terminal moraines occur in a series of lobate patterns rather than a straight or gently curving line.
15. How do geologists measure isostatic adjustments of the crust that result from glaciation?
16. Why did a large number of lakes develop in the arid part of the western United States during each major advance of the ice during the Pleistocene ice age?
17. List the periods of major pre-Pleistocene glaciation that are well documented in the geologic record.

Additional Readings

Flint, R. F. 1971. *Glacial and Quaternary geology.* New York: John Wiley and Sons.

Paterson, W. S. B. 1969. *The physics of glaciers.* Elmsford, N.Y.: Pergamon Press.

Post, A., and E. R. LaChapelle. 1971. *Glacier ice.* Seattle: University of Washington Press.

Price, R. J. 1972. *Glacial and fluvioglacial landforms.* New York: Hafner.

Sharp, R. P. 1960. *Glaciers.* Eugene: University of Oregon Press.

Turekian, K. K., ed. 1971. *The Late Cenozoic glacial ages.* New Haven: Yale University Press.

SHORELINE SYSTEMS

Water in the ocean is in constant motion. It moves by wind-generated waves, tides, tsunamis (seismic sea waves), and a variety of density currents. As it moves, it constantly modifies the shores of all the continents and islands of the world, reshaping coastlines with the ceaseless activity of waves and currents. Shoreline processes can change in intensity from day to day and from season to season, but they never stop.

The present shorelines of the world, however, are not the result of present-day processes alone. Nearly all coasts were profoundly affected by the rise in sea level caused by the melting of the Pleistocene glaciers, between 15,000 and 20,000 years ago. The rising sea flooded large parts of the continents, and shorelines moved inland over landscapes formed by continental processes.

Therefore, the configuration and the other characteristics of a given shoreline may be largely the results of stream erosion or deposition, glaciation, volcanism, earth movements, marine erosion or deposition, or even the growth of organisms.

The great concentration of population on or near shorelines indicates their importance to society. To live properly in this environment, we must understand its processes and work in harmony with shoreline dynamics. In this chapter, we will consider shorelines as dynamic systems and discuss the origins of coastal features.

MAJOR CONCEPTS

1. Wind-generated waves provide most of the energy for shoreline processes.
2. Wave refraction concentrates energy on headlands and disperses it in bays.
3. Longshore currents, generated by waves advancing obliquely toward the shore, are one of the most important shoreline processes.
4. Erosion along a coast tends to develop sea cliffs by the undercutting action of waves and longshore currents. As a cliff recedes, a wave-cut platform develops, until equilibrium is established between wave energy and the configuration of the coast.
5. Sediment transported by waves and longshore currents is deposited in areas of low energy to form beaches, spits, and barrier islands.
6. Erosion and deposition along a coast tend to develop a straight or gently curving shoreline that is in equilibrium with the energy expended upon it.
7. Reefs grow in a special environment and form coasts that can evolve into atolls.
8. The worldwide rise in sea level associated with the melting of the Pleistocene glaciers drowned many coasts between 15,000 and 20,000 years ago.
9. Tides are produced by the gravitational attraction of the moon and exert a major local influence on shorelines.
10. Tsunamis are waves generated by earth movements on the sea floor.

WAVES

Statement

Most shoreline processes are directly or indirectly the result of wave action. An understanding of wave phenomena is therefore fundamental to the study of shoreline processes. All waves are a means of moving some form of energy from one place to another; this is true of sound waves, radio waves, and water waves. All waves must be generated by some source of energy. By far the most important types of ocean waves are those generated by wind.

As wind moves over the open ocean, the turbulent air distorts the surface of the water. Gusts of wind depress the surface where they move downward; as they move upward, they cause a decrease in pressure, which elevates the surface. These changes in pressure produce an irregular, wavy surface on the ocean and transfer part of the wind's energy to the water. In a stormy area, waves tend to be choppy and irregular, and wave systems of different sizes and orientations can be superposed on each other. As they move out from their place of origin, however, the shorter waves move more slowly and are left behind, and the wave patterns develop some measure of order.

As a wave approaches the shore, it collapses forward, or breaks, into a surf. The water then rushes forward to the shore and returns as backwash. It is this energy that causes erosion, transportation, and deposition along the shores of all the continents and islands of the world.

Discussion

Wave Motion in Water

Ocean waves are described in the same terms as those applied to other wave phenomena. These are illustrated in *figure 15.1.* The **wavelength** is the horizontal distance between adjacent **wave crests** or adjacent **wave troughs.** The **wave height** is the vertical distance between wave crest and wave trough. The time between the passage of two successive crests is called the **wave period,** or frequency.

Wave motion can easily be observed by watching a floating object move forward as the crest of a wave approaches and then sink back into the following trough. Viewed from the side, the object moves in a circular orbit whose diameter is equal to the wave height (*figure 15.2*). Beneath the surface,

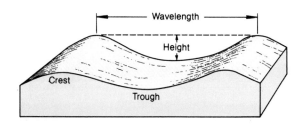

Figure 15.1 The morphology of a wave can be described in terms of its length (the distance from crest to crest), height (the vertical distance between crest and trough), and period (the time between the passage of two successive crests).

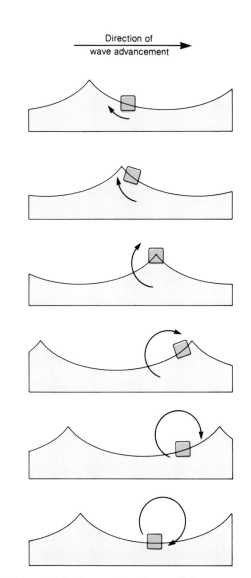

Figure 15.2 The motion of a particle as a wave advances is indicated by the movement of a floating object. As the wave advances (from left to right), the object is lifted up to the crest and then returns to the trough. The wave form advances, but the water particles move in an orbit, returning to their original position.

this orbital motion dies out rapidly, becoming negligible at a depth equal to about one-half the wavelength. This level is known as the **wave base** (*figure 15.3*). The motion of water in waves is therefore distinctly different from motion in current systems, in which water moves in a given direction and does not return to its original position.

The energy of a wave depends on its length and height. The greater the wave height, the greater the size of the orbit in which the water moves. The total energy of a wave can be represented by a column of water in orbital motion.

Breakers

Wave action produces little or no net forward motion of the water, because the water moves in an orbital path as the wave form advances. As a wave approaches shallow water, however, some important changes occur (*figure 15.4*). First, the wavelength decreases, because the wave base encounters the bottom, and the resulting friction gradually slows down the wave. Second, the wave height increases as the column of orbiting water encounters the sea floor. As the wave form becomes progressively higher and the velocity decreases, a critical point is reached, at which the forward velocity of the orbit distorts the wave form. The wave crest then extends beyond the support of the underlying column of water, and the wave collapses, or breaks. At this point, all of the water in the column moves forward, releasing its energy as a wall of moving, turbulent surf called a **breaker.**

After a breaker collapses, the **swash** (turbulent sheet of water) flows up the beach slope. The swash is a powerful surge, which causes a landward movement of sand and gravel on the beach. As the force of the swash is dissipated against the slope of the beach, the water flows down the beach slope as **backwash,** but much of it seeps into the permeable sand and gravel.

In summary, waves are generated by the wind on the open ocean. The wave form moves out from the storm area, but the water itself moves in a circular orbit with little or no forward motion. As a wave approaches the shore, it breaks, and the energy of the forward-moving surf is expended on the shore, causing erosion, transportation, and deposition of sediment.

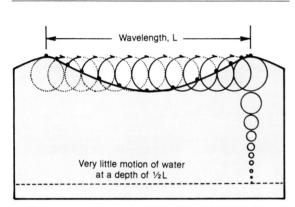

Figure 15.3 The orbital motion of water in a wave decreases with depth and dies out at a depth equal to about half the wavelength.

Figure 15.4 A wave approaching the shore undergoes several significant changes as the water in orbital motion encounters the sea floor. (1) The wavelength decreases due to frictional drag, and the waves become crowded together closer to shore. (2) The wave height increases as the column of water, moving in an orbit, stacks up on the shallow sea floor. (3) The wave becomes asymmetrical because of increasing height and frictional drag on the sea floor, and ultimately it breaks. The water then ceases to move in an orbit and rushes forward to the shore.

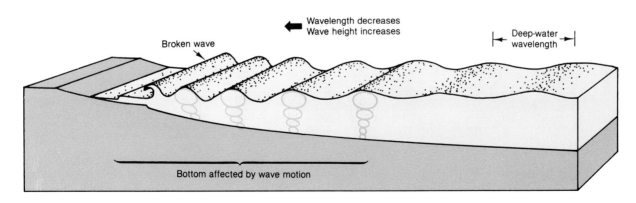

WAVE REFRACTION

Statement

As waves approach the shore, they commonly are bent, or refracted, so that their crest line tends to become parallel to the shore. Wave refraction is a key factor in shoreline processes, because it influences the distribution of energy along the shore as well as the direction in which coastal water and sediment move. It occurs because the part of a wave in shallow water begins to drag the bottom and slows down while the remainder of it, in deeper water, moves forward at normal velocity. As a result, wave refraction concentrates energy on headlands and disperses it in bays.

Discussion

To appreciate the effect of **wave refraction** on the concentration and dispersion of energy, consider the energy in a single wave. In *figure 15.5*, the unrefracted wave is divided into three equal parts (*AB*, *BC*, and *CD*), each having an equal amount of energy. As the wave moves toward the shore, segment *BC*, in front of the **headland,** meets the shallow floor first and is slowed down. Meanwhile, the rest of the wave (segments *AB* and *CD*) moves forward at normal velocity. This difference in velocities causes the crest line of the wave to bend as it advances shoreward. The wave energy between point *B* and point *C* is concentrated on a relatively short segment (*B'C'*) of the headland, whereas the equal amounts of energy between *A* and *B* and between *C* and *D* are distributed over much greater distances (*A'B'* and *C'D'*). Thus, breaking waves are powerful erosional agents on the headlands but are relatively weak in bays, where they commonly deposit sediment to form beaches. Where major wave fronts are refracted around islands and headlands, the refraction patterns are obvious from the air (*figure 15.6*).

Figure 15.5 Wave refraction concentrates energy on headlands and disperses it across bays. Each segment of the wave, *AB*, *BC*, and *CD*, has the same amount of energy. The segment *BC* encounters the sea floor sooner than *AB* or *CD* and moves more slowly. This difference in the velocities of the three segments causes the wave to bend, so that the energy contained in segment *BC* is concentrated on the headland, while the energy contained in *AB* and *CD* is dispersed along the beach.

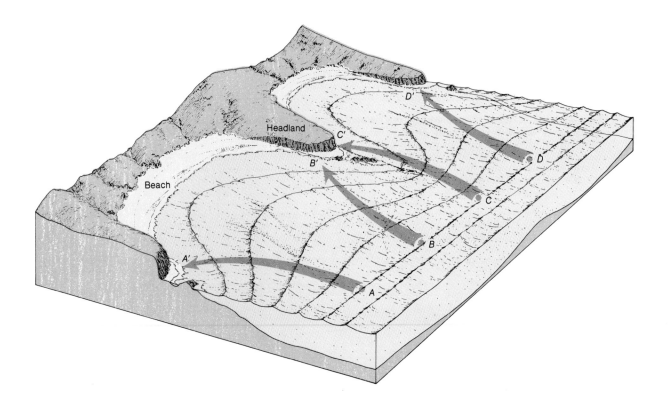

LONGSHORE DRIFT

Statement

Longshore drift is one of the most important shoreline processes. It is generated as waves advance obliquely to the shore (see *figure 15.7*). As a wave strikes the shore at an angle, water and sediment moved by the breaker are transported obliquely up the beach in the direction of the wave's advance. When the wave's energy is spent, the water and sediment return with the backwash directly down the beach, perpendicularly to the shore. The next wave moves the material obliquely up the shore again, and the backwash returns it directly down the beach slope. A single grain of sand is thus moved in an endless series of small steps, with a resulting net transport parallel to the shore. This process is known as beach drift. A similar process that develops in the breaker zone is called a longshore current. Thus, longshore movement occurs in two zones. One is along the upper limits of wave action and is related to the surge and backwash of the waves. The other is in the surf and breaker zone, where material is transported in suspension and by saltation. Both processes work together, and their combined action is called longshore drift.

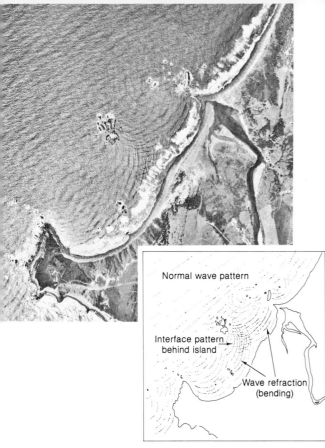

Figure 15.6 Wave refraction around headlands and islands was photographed along the coast of Oregon from an altitude of 6 km. The energy concentrated on the headlands has reduced some of them to offshore islands.

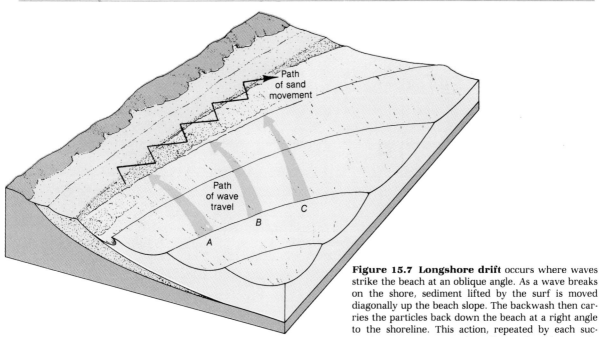

Figure 15.7 Longshore drift occurs where waves strike the beach at an oblique angle. As a wave breaks on the shore, sediment lifted by the surf is moved diagonally up the beach slope. The backwash then carries the particles back down the beach at a right angle to the shoreline. This action, repeated by each successive wave, transports the sediment along the coast in a zigzag pattern. Particles also are moved underwater in the breaker surf zone by this action.

Discussion

Longshore drift results in the movement of an enormous volume of sediment. A beach can be thought of as a river of sand, moving by the action of **beach drift**. If the wave direction is constant, longshore drift occurs in one direction. If waves approach the shore at different angles during different seasons, longshore drift is periodically reversed. **Longshore currents** can pile significant volumes of water on the beach, which return seaward through the breaker zone as a narrow **rip current**. These currents can be strong enough to be dangerous to swimmers.

One of the best ways to appreciate the process of longshore drift is to consider how it has influenced human affairs. A good example is in the area of Santa Barbara, along the southern coast of Cali-

fornia, where data have been collected over a considerable period. Santa Barbara is a picturesque coastal town at the base of the Santa Ynez Mountains. It is an important educational, agricultural, and recreational area, and the people there wanted a harbor that could accommodate deep-water vessels. Studies by the United States Army Corps of Engineers indicated that the site was unfavorable because of the strong longshore currents, which carry large volumes of sand to the south (*figure 15.8*). An important geologic system in this area transports sediment from the mountains to the deep-sea floor. Rivers draining the mountains of the coastal ranges supply new sediment to the coast at a rate of 592 m³ per day. Longshore drift continually moves the sand southward from beach to beach. The currents are so strong that boulders 0.6 m in diameter can be transported. Ultimately, the

Figure 15.8 The effect of a breakwater on long-shore drift in Santa Barbara, California, is docu-mented by a series of maps of the coast from 1925 to 1938.

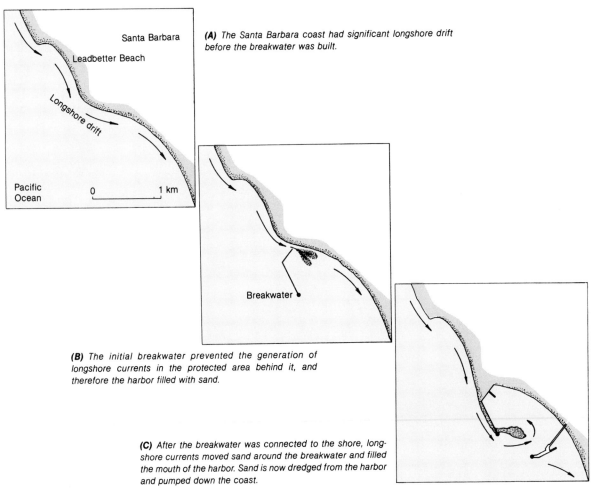

(A) The Santa Barbara coast had significant longshore drift before the breakwater was built.

(B) The initial breakwater prevented the generation of longshore currents in the protected area behind it, and therefore the harbor filled with sand.

(C) After the breakwater was connected to the shore, long-shore currents moved sand around the breakwater and filled the mouth of the harbor. Sand is now dredged from the harbor and pumped down the coast.

sand transported by longshore drift is delivered to the head of a submarine canyon and then moves down the canyon to the deep-sea floor.

In spite of reports advising against the project, a breakwater 460 m long was built and a deep-water harbor was constructed in 1925, at a cost of $750,000 (*figure 15.8B*). This breakwater was not tied to the shore, and sand, moved by longshore drift, began to pour through the gap and fill the harbor, which was protected from wave refraction and longshore currents by the breakwater (*figure 15.8B*). To stop the filling of the harbor, it was necessary to connect the breakwater to the shore. Sand then accumulated behind the breakwater, at its southern end. Soon a smooth, curving beach developed around the breakwater, and longshore drift carried sand around the breakwater and deposited it inside the harbor (*figure 15.8C*). This produced two disastrous effects. First, the harbor became so choked with sand that it could accommodate only vessels with very shallow draft. Second, the beaches downcoast were deprived of their source of sand and began to erode. Within 12 years, more than $2 million worth of damage had been done to downcoast property, as the beach in some areas was cut back 75 m. To solve the problem a dredge was installed in the Santa Barbara harbor to pump out sand and return it to the longshore drift system on the downcurrent side of the harbor. Most of the beaches have been partly replenished, but dredging costs exceed $30,000 per year.

EROSION ALONG COASTS

Statement

Coast is a general term for broad regions of land next to the ocean. These areas are sculpted into many shapes and forms, such as steep, rocky cliffs, low beaches, quiet bays, tidal flats, and marshes. The topography of a coast results from the same forces that shape other land surfaces: erosion, deposition, tectonic uplift, and subsidence.

Wave action is the major agent of erosion along coasts, and its power is awesome during storms. When a wave breaks against a sea cliff, the sheer impact of the water can exert a pressure exceeding 100 kg/m². Water is driven into every crack and crevice of the rocks, compressing the air within them. The compressed air acts as a wedge, widening the cracks and loosening the blocks.

Solution activity also takes place along the coast and is especially effective in eroding limestone. Even noncalcareous rocks can be weathered rapidly by solution activity, because the chemical action of seawater is stronger than that of fresh water.

The most effective process of erosion along coasts, however, is the abrasive action of sand and gravel moved by the waves. These tools of erosion operate like the bed load of a river. Instead of cutting a vertical channel, however, the sand and gravel moved by waves cut horizontally, forming wave-cut cliffs and wave-cut platforms.

Figure 15.9 A profile of a wave-cut platform shows the features commonly produced by wave erosion. Wave action operates like a horizontal saw cutting at the base of the cliff. The cliff is undermined and collapses. The debris is soon removed by wave action, and undercutting continues. As erosion continues, the cliff recedes, and a gently sloping wave-cut platform is left. Some sediment eroded from the shore can be deposited in deeper water to form a complementary wave-built terrace.

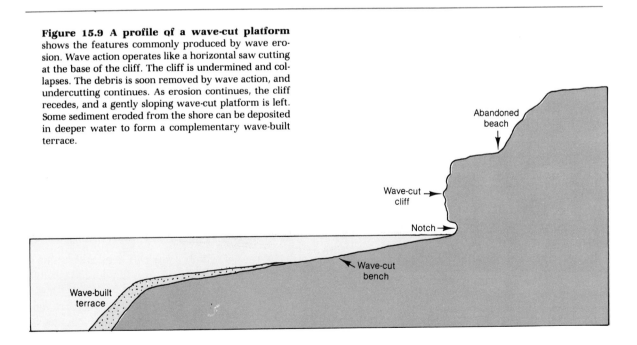

Discussion

To understand the nature of wave erosion and the principal features it forms, first consider what happens along a profile at right angles to a shore (*figure 15.9*). Where steeply sloping land descends beneath the water, waves act like a horizontal saw, cutting a notch into the bedrock at sea level (*figure 15.10*). This undercutting produces an overhanging **sea cliff**, or **wave-cut cliff**, which ultimately collapses. The fallen debris is broken up and removed by wave action, and the process is repeated on the fresh surface of the new cliff face. As the sea cliff retreats, a **wave-cut platform** is produced at its base, the upper part of which commonly is visible near shore at low tide. Sediment derived from the erosion of the cliff and transported by longshore drift can be deposited in deeper water to form a **wave-built terrace**. Stream valleys that formerly reached the coast at sea level are shortened and left as **hanging valleys** when the cliff recedes.

As the platform is enlarged, the waves break progressively farther from shore, losing much of their energy by friction as they travel across the shallow platform. Consequently, wave action on the cliff is greatly reduced. Beaches can then develop at the

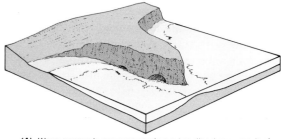

(A) Wave energy is concentrated on a headland as a result of wave refraction. Zones of weakness, such as joints, faults, and nonresistant beds, erode faster, so that sea caves develop in those areas.

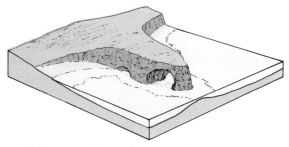

(B) Sea caves enlarge to form a sea arch.

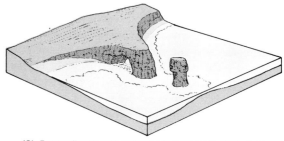

(C) Eventually, the arch collapses to form a sea stack. A new arch can develop from the remaining headland.

Figure 15.11 The evolution of sea caves, sea arches, and sea stacks is associated with differential erosion of a headland.

Figure 15.10 Wave erosion along a coast of Mexico has produced this notch and overhanging cliff. Collapse appears imminent. The process will be repeated, causing retreat of the sea cliff and development of a wave-cut platform.

base of the cliff, and the cliff face is gradually worn down, mainly by weathering and mass movement. Wave-cut platforms effectively dissipate wave energy and thus limit the size to which they can grow. Some volcanic islands, however, appear to have been truncated completely by wave action and slope retreat, so that only a flat-topped platform is left near low tide.

Sea Caves, Sea Arches, and Sea Stacks

The rate at which a sea cliff erodes depends on the durability of the rock and the degree to which the coast is exposed to direct wave attack. Zones of weakness (such as outcrops with joint systems, fault planes, and beds of shale between harder sandstones) are loci of accelerated erosion. If a joint extends across a headland, wave action can hollow out an alcove, which can later enlarge to a **sea cave.** Since the headland commonly is subjected to erosion from two sides, caves excavated along a zone of weakness can join to form a **sea arch** (*figures 15.11* and *15.12*). Eventually, the arch collapses, and an isolated pinnacle called a **sea stack** is left in front of the cliff. In the erosion of a shoreline, marine and terrestrial agents operate together to produce erosion above wave level. The seepage of ground water, frost action, wind, and mass movement all combine with the undercutting action of waves to erode the coast.

Rates of Erosion

The rate at which wave action cuts away at the shore is extremely variable. It depends on the configuration of the coast, the size and strength of the waves, and physical characteristics of the bedrock. In poorly consolidated material, the rate of cliff retreat has been measured to be from 1.5 to 2 m per year. Rates of erosion along most coasts are much slower. An interesting example of rates of coastal erosion is documented by maps made by the ancient Romans when they conquered Britain. These maps show that, in approximately 2000 years, parts of the British coast have been eroded back more than 5 km, and the sites of many villages and landmarks shown on the Roman maps have been swept away. Other examples of rapid wave erosion are found on new volcanic islands, such as Surtsey, near Iceland. The newly formed volcanic ash that makes up such islands can be completely planed off by wave action in a matter of only a few years.

The Evolution of a Wave-Cut Platform

The evolution of erosional features along a coast is shown in *figure 15.13.* In the initial stage (*figure 15.13A*), sea level rises over a landscape eroded by streams, so that an irregular shoreline is formed. Wave action develops a small notch, and abrasion of the platform begins. Continued wave erosion enlarges the platform and develops a high wave-cut cliff (*figure 15.13B*). Minor features, such as sea caves, sea arches, and sea stacks, form by differential erosion in weak places in the bedrock. These are continually being formed and destroyed as the sea cliff recedes. In the advanced stage of development (figure *15.13C*), the platform is so enlarged that it absorbs most of the wave energy. Weathering, mass movement, and stream erosion subdue the cliff, and a beach develops as a result of the low energy level along the coast. The net result is a broad wave-cut platform.

Figure 15.12 Sea stacks and a sea arch along the coast of Iceland were carved in the basalt by wave erosion.

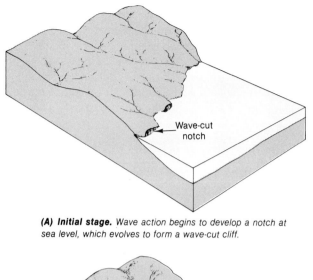

(A) Initial stage. Wave action begins to develop a notch at sea level, which evolves to form a wave-cut cliff.

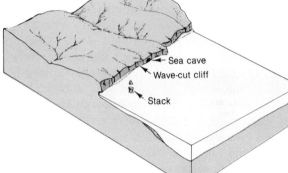

(B) Intermediate stage. Continued wave erosion causes the cliff to recede, and a wave-cut platform develops. Sea stacks, sea arches, and sea caves result from differential erosion along zones of weakness.

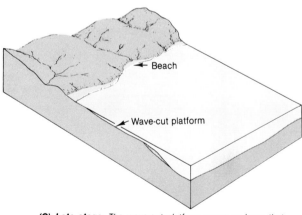

(C) Late stage. The wave-cut platform grows so large that wave energy is dissipated across it. Erosion along the shore is reduced greatly. Beaches develop, and the sea cliff retreats through mass movement.

Figure 15.13 The evolution of a shoreline involves a series of stages in which the configuration of the coast is modified by both erosion and deposition until, finally, only a minimum of energy is expended on it. The most stable form has a smooth, straight coast and a wide platform.

DEPOSITION ALONG COASTS

Statement

Sediment transported by waves and longshore currents is deposited in areas of low energy, where it builds a variety of landforms, including beaches, spits, tombolos, and barrier islands. Beaches and associated features are parts of a dynamic shoreline system. Sediment (mostly sand and gravel) is in constant motion, and the configuration of the coast is continually modified by both erosion and deposition. Changes continue until the configuration of the coast arrives at equilibrium with the available wave energy. The final configuration is usually a smooth and straight or gently curving coastline.

Discussion

Beaches

A **beach** is a shore built of unconsolidated sediment. Sand is the most common material, but some beaches are composed of cobbles and boulders and others of fine silt and clay. The physical characteristics of a beach (such as slope, composition, and shape) depend largely on wave energy, but the supply and size of available sediment particles are also important. Beaches composed of fine-grained material generally are flatter than those composed of coarse sand and gravel.

Figure 15.14 shows some of the important elements in a coastal system. The primary source of sediment for beaches and associated depositional features is the rivers that drain the continents. Sediment from the rivers is transported along the shore by longshore drift and is deposited in areas of low energy. Erosion of headlands and sea cliffs is also a source of sediment. In tropical areas, the greatest source of sand commonly is shell debris derived from wave erosion of nearshore coral reefs. Sediment can leave the system by landward migration of coastal sand dunes and by transportation into deep areas of the ocean floor, where it commonly accumulates as submarine deposits.

Spits

In areas where a straight shoreline is indented by **bays** or **estuaries,** longshore drift can extend the beach from the mainland to form a **spit.** A spit can grow far out across the bay as material is deposited at its end (*figure 15.15*). Eventually, it can extend completely across the front of the bay, forming a **baymouth bar.**

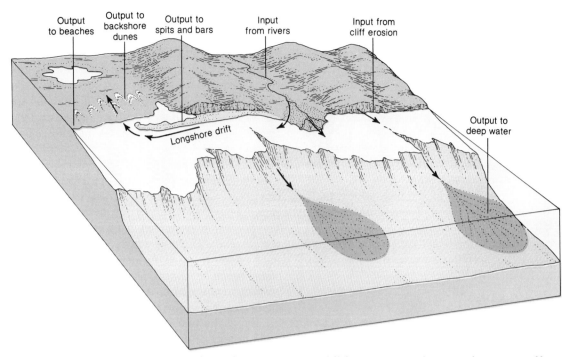

Output to beaches
Output to backshore dunes
Output to spits and bars
Input from rivers
Input from cliff erosion
Output to deep water
Longshore drift

Figure 15.14 A shoreline is a dynamic system with an input of sediment from various sources; transportation of sediment; and ultimate deposition, or output of sediment from the system. Most of the sediment in a shoreline system is supplied by rivers bringing erosional debris from the continent and by the erosion of sea cliffs by wave action. This material is transported by longshore drift and can be deposited in growing beaches, spits, and bars. Some sediment, however, leaves the system and is transported to deeper water. Some material can also leave the system by the landward migration of coastal sand dunes.

Figure 15.15 Curved spits develop as longshore drift moves sediment beyond the mainland.

Tombolos

Beach deposits can also grow outward and connect the shore with an offshore island to form a **tombolo.** This feature commonly is produced by the island's effect on wave refraction and longshore drift (*figure 15.16*). An island near shore can cause wave refraction so that little or no wave energy strikes the shore behind it. Longshore drift, which moves sediment along the coast, is not generated in this wave shadow zone. Therefore, sediment carried by longshore currents is deposited behind the island. The sediment deposit builds up and eventually connects the shore to the island, forming a tombolo. Longshore currents then move uninterrupted along the shore and around the tombolo.

Barrier Islands

Barrier islands are long, low offshore islands of sediment trending parallel to the shore (*figure 15.17*). Almost invariably, they form along shorelines that are adjacent to gently sloping coastal plains, and they typically are separated from the mainland by a lagoon. Most barrier islands are cut by one or more tidal inlets.

The origin of barrier islands has been a controversial subject for many years. One theory contends that they result from the growth of spits across irregularities in the shore, as illustrated in *figure 15.18*. Certainly, many barrier islands develop in this manner. Others may form by the shoreward migration of offshore bars. Another hypothesis contends that barrier islands formed when the worldwide rise of sea level drowned early-formed beaches and isolated them offshore. The process is illustrated in *figure 15.19*. During a period when sea level was lower, such as during the ice age, a beach formed along a low coastal plain. Beach sand was piled up in a dune ridge parallel to the shore. As sea level rose with the melting of the ice and the dune ridge was drowned, it became the nucleus for a barrier island. Later, wave action and longshore drift reworked, enlarged, and modified the dune ridge into a barrier island.

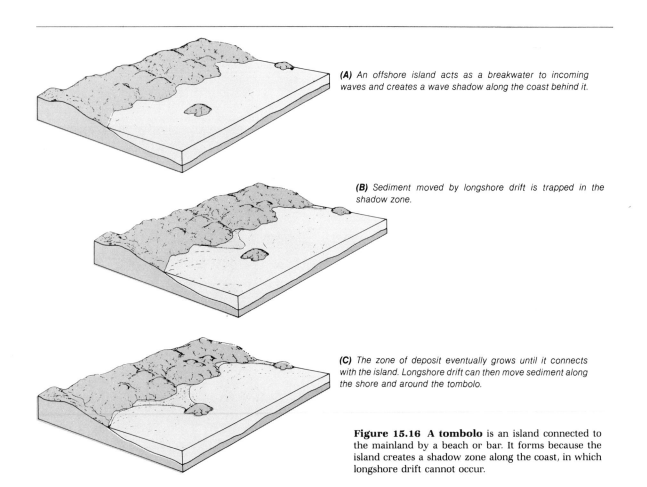

(A) An offshore island acts as a breakwater to incoming waves and creates a wave shadow along the coast behind it.

(B) Sediment moved by longshore drift is trapped in the shadow zone.

(C) The zone of deposit eventually grows until it connects with the island. Longshore drift can then move sediment along the shore and around the tombolo.

Figure 15.16 A tombolo is an island connected to the mainland by a beach or bar. It forms because the island creates a shadow zone along the coast, in which longshore drift cannot occur.

Figure 15.17 A barrier island along the Atlantic coast was photographed from an altitude of 6 km. It has a typical smooth seaward face where wave action and longshore drift actively transport sediment. A tidal inlet forms a break in the island, and sediment transported through it is deposited as a tidal delta in the lagoon.

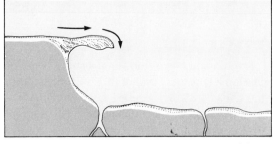

(A) Sediment moving along the shore is deposited as a spit in the deeper water near a bay.

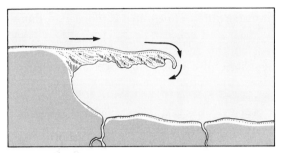

(B) The spit grows parallel to the shore by longshore drift.

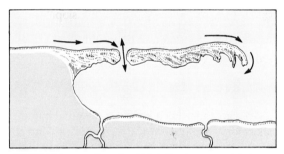

(C) Tidal inlets cut the spit, which is then long enough to be considered a barrier island.

Figure 15.18 A barrier island may be erected by longshore migration of a spit.

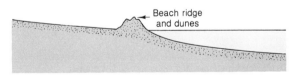

(A) On a gently inclined coast, wave action builds a beach ridge at sea level. The ridge can grow to be over 30 m high as sand is piled up by wind action.

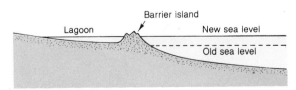

(B) A rise in sea level due to the melting of glaciers drowns the beach ridge and produces a barrier island, which is then modified by wave action.

Figure 15.19 A barrier island may also be erected as sea level rises over beach ridges.

THE EVOLUTION OF SHORELINES

Statement

Erosion and deposition along a shore combine to develop a long and straight or gently curving coastline. Headlands are eroded back, and bays and estuaries are filled with sediment. The configuration of the shoreline evolves until energy is distributed equally along the coast, and neither large-scale erosion nor deposition occurs. The energy of the waves and longshore drift is just sufficient to transport the sediment that is supplied. A shoreline with such a balance of forces is called a shoreline of equilibrium. As is the case with a stream profile of equilibrium, a delicate balance is maintained between the landform and the geologic processes operating upon it.

Very few present-day shorelines approach this state of equilibrium, however. They are all very young, having been formed during the last few thousand years by the worldwide rise in sea level associated with the melting of the Pleistocene glaciers. Most coastlines are presently being modified by erosion and deposition and are evolving toward equilibrium.

Discussion

We can construct a simple conceptual model of a shoreline's evolution toward equilibrium and show the types of changes that would be expected to occur as the processes of erosion and deposition operate. One such model is shown in *figure 15.20*. Diagram *A* shows an area originally shaped by stream erosion and subsequently partly flooded by rising sea level. River valleys are drowned to form irregular, branching bays, and some hilltops form peninsulas and islands. Next, as is shown in *figure 15.20B*, marine erosion begins to attack the shore. The islands and headlands are eroded into sea cliffs. As erosion proceeds (*figure 15.20C*), the islands and headlands are worn back, and the cliffs increase in height. A wave-cut platform develops, and sediment begins to accumulate, forming spits and beaches. The wave-cut platform enlarges, reducing wave energy, so that a beach forms at the

Figure 15.20 The evolution of a shoreline of equilibrium from an embayed coastline involves changes due to both erosion and deposition. Eventually, a smooth coastline is produced, and the forces acting on it are in equilibrium with one another, so that neither erosion nor deposition occurs on a large scale.

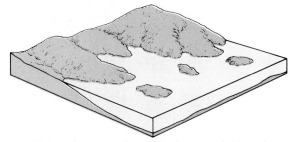

(A) A rise in sea level floods a landscape eroded by a river system and forms bays, headlands, and islands.

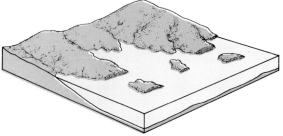

(B) Wave erosion cuts cliffs on the islands and peninsulas.

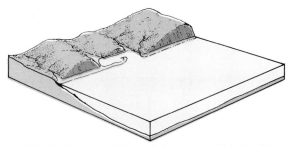

(C) Wave-cut cliffs recede and grow higher, and headlands are eroded back to a sea cliff. Sediment begins to accumulate, forming beaches and spits.

(D) Islands are eventually eroded away completely, beaches and spits enlarge, and lagoons form in the bays.

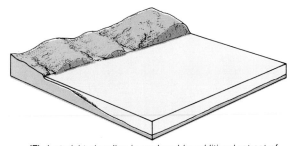

(E) A straight shoreline is produced by additional retreat of the cliffs and by sedimentation in bays and lagoons. The large wave-cut platform then limits further erosion by wave action.

base of the cliff. In a more advanced stage of development (*figure 15.20D*), the islands are completely eroded away, and bays become sealed off, partly by the growth of spits, forming lagoons. The shoreline then becomes straight and simple. In the final stages of marine development (*figure 15.20E*), the shoreline is cut back beyond the limits of the bay. Sediment moves along the coast by longshore drift, but the wave-cut platform is so wide that it effectively eliminates further erosion of the cliff by wave action. The shoreline is straight and essentially in equilibrium with the energy acting on it. Further modification of the cliffs results from weathering and stream erosion.

Naturally, the development of a shoreline is also affected by special conditions of structure and topography and by fluctuations of sea level. But the process of erosion of the headlands by wave action and the straightening of the shoreline by both erosion and deposition follows the general sequence of the idealized model. Actual shorelines rarely proceed through all of these stages, however, because fluctuations of sea level upset the previously established balance.

The development of a shoreline is interrupted in many areas by tectonic uplift, by which sea cliffs and wave-cut platforms are abruptly elevated above the level of the waves. When this happens, wave erosion begins at a new level, and the elevated marine terraces, stranded high above sea level, are attacked and eventually obliterated by weathering and stream erosion (*figure 15.21*).

Figure 15.21 A series of elevated beach terraces resulted from tectonic uplift along the southern coast of California and the offshore islands. This photograph of San Clemente Island was taken with the sun at a low angle to emphasize the sequence of terraces.

REEFS

Statement

Reefs form a unique type of coastal feature, because they are built by living organisms. Modern reefs are built by a complex community of corals, algae, sponges, and other marine invertebrates. Most grow and thrive in the warm, shallow waters of the semitropical and tropical regions. Only the upper part of a reef is organically active, because sunlight is required for vigorous growth. The living animals and plants of the reef community build new structures on the remains of dead organisms and the accumulation of shell debris. Wave action, however, continually breaks up the reef. Large bodies of calcium carbonate shell debris, originally secreted by the organisms of the reef community, accumulate along the reef flanks, adding mass to the structure. Reefs can grow upward with rising sea level, if the rate of rise is not excessive. They can also grow seaward over the flanks of reef debris.

Discussion

Reef Ecology

The marine life that forms a reef can flourish only under strict conditions of temperature, salinity, and water depth. Most modern **coral** reefs occur in warm tropical waters between the limits of 25° south latitude and 30° north latitude. Colonial corals need sunlight, and they cannot live in water deeper than about 76 m. They grow most luxuriantly just a few meters below sea level. Dirty water inhibits rapid, healthy growth, because it cuts off sunlight. Corals are therefore absent or stunted near the mouths of large, muddy rivers. They can survive only if the salinity of the water ranges from 27 to 40 parts per thousand. Thus, a reef can be killed if a flood of fresh water from the land appreciably alters the salinity. Coral reefs are remarkably flat on top, the upper surface being positioned at the level of the upper third of the tidal range. They usually are exposed at low tide but must be covered at high tide. In summary, corals thrive only in clear, warm, shallow oceans where wave action brings sufficient oxygen and food.

The fact that reefs form in such restricted environments makes them especially important as indicators of past climatic, geographic, and tectonic conditions.

Types of Reefs

The most common types of reefs in present oceans are fringing reefs, barrier reefs, and atolls (*figure 15.22*).

Fringing reefs, generally ranging from 0.5 to 1 km wide, are attached to such landmasses as the shores of volcanic islands. The corals grow seaward, toward their food supply. They are usually absent near the deltas and mouths of rivers, where the waters are muddy. Heavy sedimentation and high runoff also make some tropical coasts of continents unattractive to fringing reefs.

Barrier reefs are separated from the mainland by a lagoon, which can be more than 20 km wide. As seen from the air, the barrier reefs of islands in the South Pacific are marked by a zone of white breakers. At intervals, narrow gaps occur, through which excess shore and tidal water can exit. The finest example of this type is the Great Barrier Reef, which stretches 800 km along the northern shore of Australia, from 30 to 160 km off the Queensland coast.

Atolls are roughly circular reefs that rise from deep water, enclosing a shallow lagoon in which there is no exposed central landmass. The outer margin of an atoll is naturally the site of most vigorous coral growth. It commonly forms an overhanging rim, from which pieces of coral rock break off, accumulating as submarine talus on the slopes below. A cross-sectional view of a typical atoll shows that the lagoon floor is shallow and is composed of calcareous sand and silt with rubble derived from erosion of the outer side (see the foreground of *figure 15.22C*).

Atolls are by far the most common type of coral reef. Over 330 are known, of which all but 10 lie within the Indo-Pacific tropical area. Drilling into the coral of atolls tends to confirm the theory that atolls form on submerged volcanic islands (see the following section, "The Origin of Atolls"). In one instance, coral extends down as much as 1400 m below sea level, where it rests on a basalt platform carved on an ancient volcanic island. Since coral cannot grow at that depth, it presumably grew upward as the volcanic island sank. A reef this thick probably accumulated over 40 million or 50 million years.

Platform reefs grow in isolated oval patches in warm, shallow water on the continental shelf. They were apparently more abundant during past geologic periods of warmer climates. Modern platform reefs are somewhat randomly distributed, although some appear to be oriented in belts. The

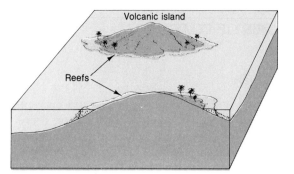

(A) *A reef begins to grow along the coast of a newly formed volcanic island.*

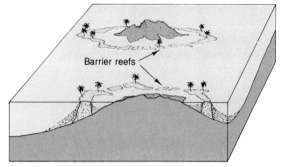

(B) *As the island subsides, the reef grows upward and develops a barrier that separates the lagoon from open water.*

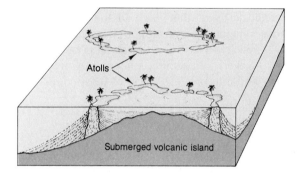
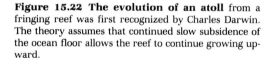

(C) *Further subsidence completely submerges the island, but the reef continues to grow upward to form an atoll if subsidence is not too rapid.*

Figure 15.22 The evolution of an atoll from a fringing reef was first recognized by Charles Darwin. The theory assumes that continued slow subsidence of the ocean floor allows the reef to continue growing upward.

latter feature suggests that they were formed on submarine topographic highs, such as drowned shorelines.

The Origin of Atolls

In 1842, Charles Darwin first proposed a theory to explain the origin of atolls. As is indicated in *figure 15.22*, the theory is based on the continued relative subsidence of a volcanic island. Darwin suggested that coral reefs are originally established as fringing reefs along the shores of new volcanic islands. As the island gradually subsides, the coral reef grows upward along its outer margins. The rate of upward growth essentially keeps pace with subsidence. With continued subsidence, the area of the island becomes smaller, and the reef becomes a barrier reef. Ultimately, the island is completely submerged, and the upward growth of the reef forms an atoll. Erosional debris from the reef fills the enclosed area of the atoll to form a shallow lagoon.

TYPES OF COASTS

Statement

Erosion and deposition by waves and currents operate to varying degrees along all coasts. Therefore, it might seem logical to classify a coast according to the stage to which it has evolved. Nearly all coastlines, however, have been profoundly affected by the worldwide rise of sea level caused by the melting of the Pleistocene glaciers. As a result, *the configuration of many coasts is not the result of marine processes at all.* Rather, it is the result of geologic processes that operated in the area before sea level rose. Therefore, coasts can be meaningfully classified on the basis of the geologic processes that are responsible for the coastal topography. Two principal subdivisions are recognized:
1. Primary coasts, shaped mainly by terrestrial processes
2. Secondary coasts, formed by marine processes
Each group can be further subdivided according to the geologic agent that has had the greatest influence. One important feature of this classification is its simplicity. With the aid of aerial photographs or maps, even a beginning student can properly classify and interpret the origin of most coasts.

Discussion

Primary Coasts

The configuration of **primary coasts** is largely the result of **subaerial** geologic agents, such as streams, glaciers, volcanism, and earth movements. These produce highly irregular coastlines characterized by bays, estuaries, fiords, headlands, peninsulas, and offshore islands. The landforms can be either erosional or depositional but are only slightly modified by marine processes. Some of the more common types are illustrated in *figure 15.23*.

Stream Erosion Coasts. If an area eroded by running water is subsequently flooded by the rise of sea level, the landscape becomes partly drowned. Stream valleys become bays or estuaries, and hills become islands. The bays extend up the tributary valley system, forming a coastline with a dendritic pattern. Chesapeake Bay is a well-known example.

Stream Deposition—Deltaic Coasts. At the mouths of major rivers, fluvial deposition builds deltas out into the ocean, which dominate the con-figuration of the coast. Deltas can assume a variety of shapes and are locally modified by marine erosion and deposition (see *plate 9*).

Glacial Erosion Coasts. Drowned glacial valleys usually are called **fiords**. They form some of the most rugged and scenic shorelines in the world. Fiords are characterized by long, troughlike bays that cut into mountainous coasts, extending

---

Figure 15.23 Primary coasts are those in which the configuration of the shoreline is produced by nonmarine processes. Rivers and glaciers are the most important processes forming this type of coast.

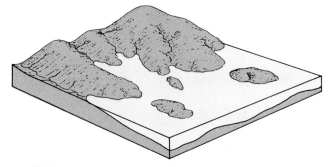

(A) *Stream erosion produces an irregular, embayed coast with offshore islands.*

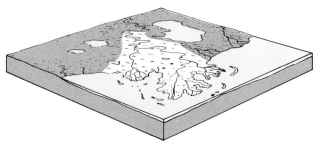

(B) *Stream deposition produces deltaic coasts.*

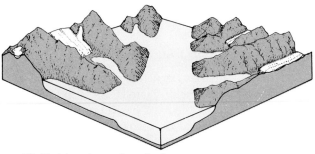

(C) *Glacial erosion produces fiords.*

inland as much as 100 km. In polar areas, glaciers still remain at the heads of many fiords. The walls of fiords are steep and straight. Hanging valleys with spectacular waterfalls are common.

Glacial Deposition Coasts. Glacial deposition dominates coasts in the northern latitudes, where continental glaciers once extended beyond the present shoreline, over the continental shelf. The ice sheets left drumlins and moraines to be drowned by the subsequent rise in sea level. Long Island, for example, is a partly submerged moraine. In Boston Harbor, partly submerged drumlins form elliptical islands.

As erosion and deposition continue, marine processes ultimately control the coastal configuration, and primary coasts become secondary coasts.

Secondary Coasts

Secondary coasts are shaped by marine erosion and deposition. They are characterized by wave-cut cliffs, beaches, barrier bars, spits, and (in some cases) sediment deposited through the action of biological agents, such as marsh grass, mangroves, and coral reefs. Marine erosion and deposition smooth out and straighten shorelines and establish a balance between the energy of the waves and the configuration of the shore (see *plate 12*). The most common types are illustrated in *figure 15.24*.

Wave Erosion Coasts. Wave erosion begins to modify primary coasts as soon as the landscape produced by other agents is submerged. Wave energy is concentrated on the headlands, and a wave-cut platform develops slightly below sea level. Ultimately, a straight cliff is formed, with hanging stream valleys and a large wave-cut platform. The cliffs of Dover, England, are a prime example.

Marine Deposition Coasts. Where abundant sediment is supplied by streams or ocean currents, marine deposits determine the characteristics of the coast. Barrier islands and beaches are the dominant features (see *plate 12*). The shoreline is modified as waves break across the barriers and transport sand inland. The barriers also increase in length and width as sand is added. The lagoons behind the barrier receive sediment and fresh water from streams. Thus, they are often capable of supporting dense marsh vegetation. Gradually, a lagoon fills with stream sediment, with sand from the barrier bar (which enters through tidal deltas), and with plant debris from swamps. The barrier coasts of the southern Atlantic and Gulf Coast states are excellent examples.

Coasts Built by Organisms. Coral reefs

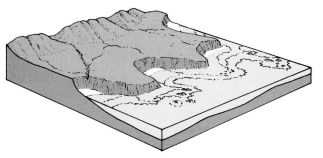

(A) *Marine erosion produces wave-cut cliffs.*

(B) *Marine deposition produces barrier bars and beaches.*

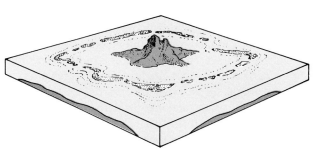

(C) *The growth of coral reefs produces barrier reefs and atolls.*

Figure 15.24 Secondary coasts are those in which the configuration of the shoreline is produced by marine processes, including erosion, deposition, and the growth of organisms.

develop a type of coast that is prominent in the islands of the southwestern Pacific. The reefs are built up to the surface by corals and algae, and they can ultimately evolve into an atoll. Another type of organic coast that is prevalent in the tropics is formed by intertwined root systems of mangrove trees, which grow in the water, particularly in shallow bays.

TIDES

Statement

Tides are produced by a combination of external forces, the principal ones being the gravitational attraction of the moon and the centrifugal force of the earth-moon system. Many variables affect the height of tides, however, including the Coriolis effect (the deflection of moving bodies due to a force arising from the earth's rotation), the position of the sun in relation to the moon and the earth, and the configuration of coastlines and the ocean floor. Tides affect coasts in two major ways:

1. By initiating a rise and fall of the water level
2. By generating currents

Discussion

The diagram in *figure 15.25* illustrates on a highly exaggerated scale the principal forces that produce tides. The gravitational force exerted by the moon tends to pull the oceans into a bulge facing the moon. Another tidal bulge, on the side of the earth opposite the moon, is caused by centrifugal force. The earth and the moon rotate around a common center of mass, which lies approximately 4500 km from the center of the earth on a line directed toward the moon. The eccentric motion of the earth as it revolves around the center of mass of the earth-moon system creates a large centrifugal force, which forms the second tidal bulge. The earth rotates beneath the bulges, so that the tides rise and fall twice every 24 hours.

The major effect of the rise and fall of tides is the transportation of sediment along the coast and over the adjacent shallow sea floor. Extremely high tides are produced in shallow seas where the rising water is funneled into bays and estuaries. For example, in the Bay of Fundy, between New Brunswick and Nova Scotia, the tide range (the difference in height between high tide and low tide) is as much as 21 m. Where fine-grained sediment is plentiful and the tide range is great, the configuration of the coast is greatly influenced by tides and tidal currents.

TSUNAMIS

Statement

Earthquakes, volcanic eruptions, and submarine landslides frequently produce an unusual wave called a seismic sea wave, or a tsunami. Tsunamis have long wavelengths and travel across the open ocean at high speeds. As they approach shore, the wavelength decreases and the wave height increases, so that tsunamis can be formidable agents of destruction along the shoreline.

Discussion

Tsunamis differ from wind waves in that energy is transferred to the water from the sea floor, so that the entire depth of water participates in the wave motion. The wave front travels from its point of origin at very high speeds, ranging from 500 to 800 km per hour, and can traverse an entire ocean. The wave height is only from 30 to 60 cm, and the wavelength ranges from 55 to 200 km. In an open ocean, therefore, a tsunami can pass unnoticed. As the wave approaches the shore, however, important changes take place. The energy distributed in the deep column of water becomes concentrated in an increasingly shorter column, resulting in a rapid increase in wave height. Waves that were less than 60 cm high in the deep ocean can build rapidly to heights exceeding 15 m in many cases and well over 30 m in rare instances. They exert an enormous force against the shore and can inflict serious damage and great loss of life.

A number of tsunamis have been well documented by seismic stations and coastal observers. For example, the tsunami that hit Hawaii on April 1, 1946, originated in the Aleutian Trench off the island of Unimak. The waves moving across the open ocean were imperceptible to ships in their path, because the wave height was only 30 cm. Moving at an average speed of 760 km per hour, they reached the Hawaiian Islands, 3200 km away, in less than 5 hours. Since the wavelength was 150 km, the wave crests arrived about 12 minutes apart. As the waves approached the island, their

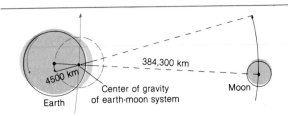

384,300 km

4500 km

Center of gravity of earth-moon system

Earth

Moon

Figure 15.25 Ocean tides are caused by the gravitational attraction of the moon and the centrifugal force of the earth-moon system. On the side of the earth facing the moon, the gravitational attraction is greater, forming a tidal bulge in the ocean's water. On the other side of the earth, the centrifugal force is greater, causing another tidal bulge.

height increased at least 17 m and thus produced an extremely destructive surf, which swept inland and demolished houses, trees, and almost everything else in its path.

SUMMARY

Shorelines are dynamic systems involving the forces of waves and currents. Wind-generated waves provide most of the energy for erosion, transportation, and deposition of sediment, but tides and tsunamis can be locally important.

Waves approaching a shore are bent, or refracted, so that energy is concentrated on headlands and dispersed in bays.

Longshore drift, one of the most important shoreline processes, is generated as waves strike the shore at an angle. Water and sediment move obliquely up the beach face but return with the backwash directly down the beach, perpendicular to the shoreline. This results in a net transport parallel to the shore.

Erosion along coasts results from the abrasive action of sand and gravel moved by the waves and currents and, to a lesser extent, from solution and hydraulic action. The undercutting action of waves and currents typically produces sea cliffs. As a sea cliff recedes, a wave-cut platform develops. Minor erosional forms of sea cliffs include sea caves, sea arches, and sea stacks.

Sediment transported by waves and longshore drift is deposited in areas of low energy. It produces a variety of landforms, including beaches, spits, tombolos, and barrier islands.

Erosion and deposition processes tend to develop a straight or gently curving coastline where neither large-scale erosion nor large-scale deposition occurs. A shoreline with such a balance of forces is called a shoreline of equilibrium.

Reefs are a special type of coastal feature, because the active process is organic. They commonly evolve through a series of types, from fringing reefs to barrier reefs to atolls.

The worldwide rise in sea level associated with the melting of the Pleistocene glaciers drowned many coasts. Coasts are classified on the basis of the process that has been most significant in developing their configurations.

Wind is the major force that causes ocean waves, which in turn modify the coasts. Wind also effectively transports and deposits loose sand and dust on the continents. It is responsible for the major

landforms in the great desert regions of the world. In the next chapter, we will discuss the geologic work of wind on the earth.

Key Words

wavelength	sea arch
wave height	sea stack
wave crest	beach
wave trough	bay
wave period	estuary
wave base	spit
breaker	baymouth bar
swash	tombolo
backwash	barrier island
wave refraction	shoreline of equilibrium
headland	elevated terrace
longshore drift	coral
beach drift	fringing reef
longshore current	barrier reef
rip current	atoll
sea cliff	platform reef
wave-cut cliff	primary coast
wave-cut platform	fiord
(terrace)	secondary coast
wave-built terrace	tide
hanging valley	tsunami
sea cave	

Review Questions

1. Describe the motion of water in a wind-generated wave.
2. Explain how wave refraction alters the form of a coastline.
3. Explain the origin of longshore drift.
4. Describe the stages in the evolution of a sea cliff and wave-cut platform.
5. Name the major depositional landforms along a coast and explain the origin of each.
6. How are marine terraces formed?
7. Explain the origin of atolls.
8. Describe six common types of shorelines.
9. Explain how ocean tides are generated.
10. Explain the origin of tsunamis.

Additional Readings

Bascom, W. 1959. Ocean waves. *Scientific American* 201(2):74–84. Reprint no. 828. San Francisco: W. H. Freeman.

———. 1964. *Waves and beaches.* New York: Doubleday, Anchor Books.

Moore, J. R., ed. 1971. *Oceanography: Readings from Scientific American.* San Francisco: W. H. Freeman.

EOLIAN SYSTEMS

Where climatic conditions are favorable, wind is the principal geologic agent by which sand and dust are transported, and a variety of sand dunes locally dominate the landscape. Foremost among these are the great "seas of sand" in large deserts, but coastal dunes are also common along many of the world's shorelines. Also, blankets of windblown dust, called loess, cover millions of square kilometers of the middle-latitude continents.

Wind is probably the least effective agent of erosion, although many curious erosional landforms are mistakenly attributed to it. Even in the desert, most erosional landforms are the products of running water, and the greatest effects of wind are shifting sand dunes. The transportation and deposition of loose, fine-grained sediment eroded by other geologic agents is the most significant effect of wind.

In this chapter, we will examine the transportation and deposition of sediment by wind and the resulting distinctive landforms.

MAJOR CONCEPTS

1. Wind is not an effective agent of erosion in most areas, but it can transport loose, unconsolidated fragments of sand and dust.
2. Wind transports sand by saltation and surface creep. Dust is transported in suspension, and it can remain high in the atmosphere for long periods of time.
3. Sand dunes migrate as sand grains are blown up and over the windward side of the dune and accumulate on the lee slope. The internal structure of a dune consists of strata inclined in a downwind direction.
4. Various types of dunes form, depending on wind velocity, sand supply, constancy of wind direction, and characteristics of the surface over which the sand migrates.
5. Windblown dust (loess) forms blanket deposits, which can mask the older landscape beneath them. The source of loess is desert dust or the fine rock debris deposited by glaciers.

WIND AS A GEOLOGIC AGENT

Statement

Wind is an effective local geologic agent, capable of lifting and transporting loose sand and dust, but its ability to erode solid rock is limited. The main effects of wind as a geologic agent are transportation and deposition in arid regions.

Discussion

Formerly, geologists thought that wind, like running water and glaciers, has great erosional power—power to abrade and wear down the earth's surface. It has become increasingly apparent, however, that few major topographical features are formed by wind erosion. Abrasion by wind-transported sand aids in eroding and shaping some landforms, but wind is incapable of producing erosional features of regional extent. Even in the desert, where water is not an obvious geologic agent, wind is not the major agent of erosion. Most erosional landforms in deserts were produced by weathering and running water in times of wetter climate. They are, in a sense, "fossil" landscapes, formed by processes that are no longer active. Even minor alcoves and niches, or wind caves, and certain topographic features called pedestal rocks, which are often thought to be caused by wind ero-

Figure 16.1 Atmospheric circulation and prevailing wind patterns are generated by solar radiation and by the earth's rotation. Heated air from the equatorial regions rises in a convection column and moves toward higher latitudes. There it is cooled, compressed, and forced to descend, forming subtropical high-pressure belts. The air then moves toward the equator as trade winds. In the Northern Hemisphere, this air is deflected by the earth's rotation to flow southwestward. In the Southern Hemisphere, it is deflected to flow northwestward. Cold polar air tends to wedge itself toward the lower latitudes and forms polar fronts.

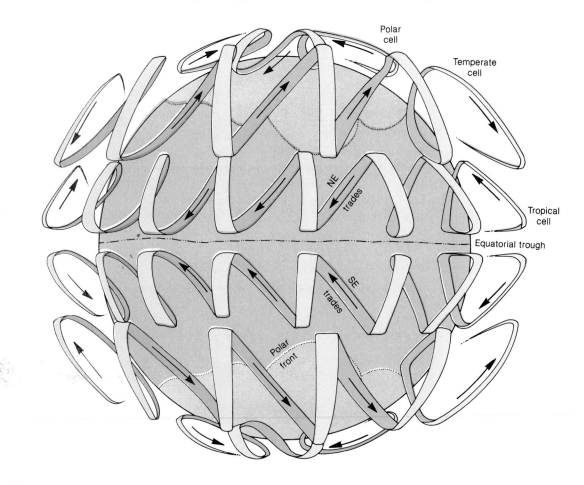

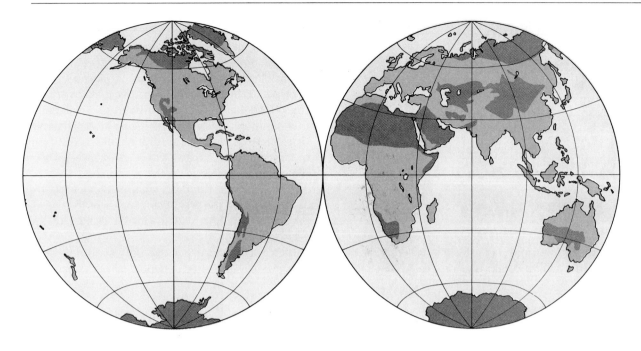

Figure 16.2 The major desert areas of the world, such as the Sahara, the Arabian, the Kalahari, and the deserts of Australia, all lie between 10° and 30° north or south of the equator. These areas are under almost constant high atmospheric pressure and are characterized by subsiding dry air and low humidity. Desert and near-desert areas cover nearly one-third of the land surface.

sion, are actually produced by differential weathering, not by wind.

Although it is relatively insignificant as an erosional agent, wind is effective in transporting loose, unconsolidated sand, silt, and dust. It is responsible for the formation of great "seas of sand" in the Sahara, the Arabian, and other deserts, as well as the blankets of windblown dust covering millions of square kilometers in China, the central United States, and parts of Europe.

Prevailing wind patterns are determined by the **Coriolis effect** (deflection due to the earth's rotation), the configuration of continents and oceans, the location of mountain ranges, and variations of solar radiation with latitude (*figure 16.1*). The world's great deserts, such as the Sahara and the deserts of Asia, are mostly located in low-latitude belts (*figure 16.2*). Equatorial air, heated by solar radiation, rises in convection cells, which descend again about 30° or 35° north and south of the equator. As the air descends, it is heated by compression. It is heated further as it moves along the ground toward the equator. The rise in temperature of the air increases its ability to hold water vapor. As a result, evaporation of the surface moisture, rather than precipitation, occurs in the low latitudes where the convection cells descend. When the air reaches the equator, it again rises and cools, releasing this moisture as rain.

Other deserts lie in the **wind shadow** behind high mountain ranges, which intercept the moisture-laden air. As the air is forced to rise over a mountain range, it cools and precipitates its moisture. Here, too, the dry descending air is heated by compression. A good example of an arid region in the rain shadow of a high mountain range is in Nevada and Utah, which lie in the rain shadow of the Sierra Nevada.

Wind action is most significant in desert areas, but it is not confined to them. Most coasts are modified by winds that pick up loose sand on the beach and transport it inland.

It is interesting to note the significance of wind action on the planet Mars, which was dramatically recorded by the space probes of 1971, 1972, and 1976 (*figure 24.32*). Because of the absence of running water and vegetation, wind action is probably the dominant process on the surface of Mars at the present time. In former geologic periods here on the earth, when climatic conditions were much different than they are now, the work of wind may have been more significant in various regions.

WIND EROSION

Statement

Wind erosion acts in two ways:
1. By deflation, the lifting and removal of loose sand and dust particles from the earth's surface
2. By abrasion, the sandblast action of windblown sand

Discussion

Deflation

Deflation causes depressions called **blowouts** in areas where weak, unconsolidated sediment is exposed at the surface. These **deflation basins** commonly develop where calcium carbonate cement in the bedrock is dissolved by ground water, so that loose material is left to be picked up and transported by the wind. In the Great Plains area of the United States, tens of thousands of small deflation basins dot the landscape (*figure 16.3*). They can be shallow depressions several meters in diameter or large ones more than 50 m deep and several kilometers across. Large deflation basins, covering areas of several hundred square kilometers, are associated with the great desert areas of the world, particularly in North Africa in the vicinity of the Nile Delta.

Generally, wind can move only sand and dust-size particles, so that deflation leaves concentra-

Figure 16.3 Deflation basins in the Great Plains are produced where solution activity dissolves the cement that binds the sand grains together in horizontal rocks. The loose sand is removed by the wind, so that a basin is formed. Water trapped in the basin dissolves more cement, and the basin is enlarged.

tions of coarser material called **lag deposits,** or **desert pavements** (*figure 16.4*). These striking desert features stand out in contrast to deposits in dune fields and playa lakes.

Deflation occurs only where unconsolidated material is exposed at the surface; it does not occur where there are thick covers of vegetation or layers of gravel. Therefore, the process is limited to areas such as deserts, beaches, and barren fields.

Abrasion

The effects of wind abrasion can be seen on the surface of bedrock in most desert regions. In areas where soft, poorly consolidated rock is exposed, wind erosion can be both spectacular and distinc-

tive. Wind abrasion is essentially the same process as the artificial sandblasting used to clean building stone. Some pebbles, called **ventifacts,** are shaped and polished by the wind (*figure 16.5*). Such pebbles are distinguished by two or more flat faces that meet at sharp ridges. Generally, they are well polished, and sometimes they have surface irregularities and grooving aligned with the wind direction.

A striking, but somewhat rare, effect of wind erosion is elongate ridges and grooves parallel to the prevailing wind direction. These features, called **yardangs,** commonly occur in groups where relatively unconsolidated rock is exposed at the surface.

Figure 16.4 Wind selectively transports sand and fine sediment, leaving the coarser gravels to form a lag deposit called desert pavement. The protective cover of lag gravels limits future deflation.

Figure 16.5 Ventifacts are pebbles shaped and polished by wind action. They commonly have three well-polished sides, which are formed as the sandblasting action of the wind reshapes one side of the pebble and then another.

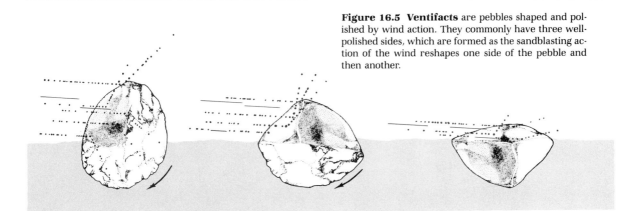

THE TRANSPORTATION OF SEDIMENT BY WIND

Statement

Field observations and wind-tunnel experiments indicate that windblown sand grains move by skipping or bounding into the air (a process called saltation) and by rolling or sliding along the surface (called surface creep). Fine silt and dust are carried in suspension over great distances and settle back to the ground only after the turbulent wind ceases.

Discussion

Movement of Sand

Although both wind and water transport sand by **saltation,** the mechanics of motion involved are different, because the viscosity of water is so much greater than that of air. In water, the process involves a hydraulic lift. In air, saltation results from impact and elastic bounces (*figure 16.6*). The energy that initially lifts sand grains into the air comes from collision with other grains. If the wind reaches a critical velocity, the grains begin to roll or slide. As one grain moves along, it strikes others. The impact can cause one or more of the colliding grains to bounce into the air, where they are driven forward by the stronger wind above the ground surface. Gravity soon pulls them back, and they strike the ground at an angle generally ranging from 10° to 16°. If the sand is moving over solid rock, the grains bounce back into the air. If the surface is loose sand, the impact of a falling grain can knock other grains into the air, setting up a chain reaction in which the entire sand surface is eventually set in motion.

Some sand grains, too large to be ejected into the air, move by **surface creep** (rolling and sliding). They are moved by the impact of saltating grains, not directly by the wind. Between approximately one-fifth and one-fourth of the sand moved by a sandstorm travels by rolling and sliding. Particles whose diameter is greater than 1 cm are rarely moved by wind.

Movement of Dust

Dust-size particles are light enough to be lifted by eddies in the air, but it is difficult for turbulent air alone to lift them off the surface. Initially, dust particles must be ejected into the air by the impact of saltating sand grains or by other disturbances. Once airborne, they can be blown high into the atmosphere and transported many kilometers.

Samples taken from many parts of the world show that large quantities of material are transported great distances by wind. For example, some dust falls in the eastern United States were composed of particles that originated 3200 km to the west, in Texas and Oklahoma. During the dust-bowl years of the 1930s in the United States, a dust cloud extending 3650 m above the ground near Wichita, Kansas, was estimated to contain 50,000 metric tons of dust per cubic kilometer.

Figure 16.6 The transportation of sediment by wind is accomplished by surface creep, saltation, and suspension. Coarse grains move by impact from other grains and slide or roll (surface creep). Medium grains move by skipping or bouncing (saltation). Fine silt and dust move in suspension.

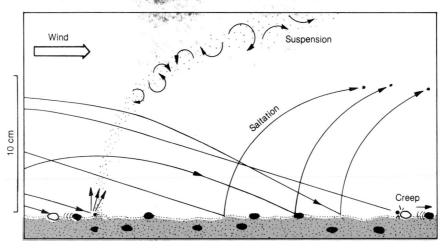

THE MIGRATION OF SAND DUNES

Statement

Wind commonly deposits sand in the form of dunes (mounds or ridges), which generally migrate downwind. In many respects, dunes are similar to ripple marks (formed either by air or water) and large sand waves or sandbars found in many streams and in shallow marine water.

Discussion

Many **dunes** originate where an obstacle, such as a large rock, a clump of vegetation, or a fence post, creates a zone of quieter air behind it (*figures 16.7* and *16.8*). As sand is blown up or around the obstruction into the protected area (the wind shadow), its velocity is reduced and deposition occurs. Once a small dune is formed, it acts as a barrier itself, disrupting the flow of air and causing continued deposition downwind (*figure 16.8*).

Dunes range from a few meters high to as much as 200 m high and 1 km wide.

The movement of sand in a typical dune is diagrammed in *figure 16.9*. Dunes are asymmetrical, with a gently inclined windward slope and a steeper downwind slope, called the **lee slope**, or **slip face**. The steep slip face of the dune indicates the direction of the prevailing wind. Dunes migrate grain by grain as wind transports sand by saltation and surface creep up the windward slope. The wind continues upward past the crest of the dune, creating divergent airflow and eddies just over the lee slope. Beyond the crest, the sand drops out of the wind stream and accumulates on the slip face. As more sand is transported from the windward slope and accumulates on the lee slope, the dune migrates downwind. The internal structure of a migrating dune consists of cross-bedding. Strata form on the lee slope and are therefore inclined in a downwind direction. Geologists map the directions of ancient eolian systems by measuring the directions in which the cross-strata of windblown sandstone are inclined (*figure 16.10*).

Figure 16.7 Sand dunes commonly originate in wind shadows. Any obstacle that diverts the wind, such as a bush or a fence post, creates eddies and reduces wind velocity. Windblown sand is deposited in protected areas, and eventually enough sand accumulates to form a dune. The dune itself then acts as a barrier, making its own wind shadow, and thus causes additional accumulation of sand.

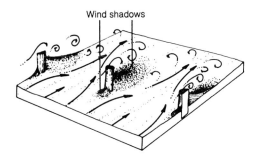

Figure 16.8 The wind shadow of a fence post caused this small dune to form on the Great Salt Lake Desert.

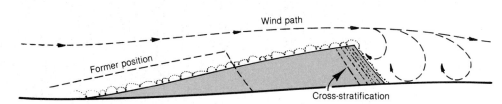

Figure 16.9 A sand dune migrates as sand grains move up the slope of the dune and accumulate in a protected area on the downwind face. The dune slowly moves downwind, grain by grain. As the grains accumulate on the downwind slope, they produce a series of layers (cross-beds) inclined in a downwind direction.

Figure 16.10 Cross-bedding in the Navajo Sandstone in Zion National Park, Utah, is excellent evidence that the rock formed in an ancient desert. The inclination of the strata shows that the wind blew from north to south (from left to right, in the photograph) during most of the time when this formation was deposited.

TYPES OF SAND DUNES

Statement

Sand dunes can assume a variety of fascinating shapes and patterns, depending on such factors as
1. Sand supply
2. Wind velocity
3. Variability of wind direction
4. Characteristics of the surface over which the sand moves

The most common varieties are transverse dunes, barchan dunes, longitudinal dunes, star dunes, and parabolic dunes (*figure 16.11*).

Discussion

Transverse Dunes

Transverse dunes typically develop where there is a large supply of sand and a constant wind direction (*figure 16.11A*). These dunes cover large areas and develop a wavelike form, with sinuous ridges and troughs perpendicular to the prevailing wind. Transverse dunes commonly form in desert regions where exposed ancient sandstone formations provide an ample supply of sand. They commonly cover large areas known as sand seas, so called because the wavelike dunes produce a surface resembling a stormy sea.

Barchan Dunes

Where the supply of sand is limited and winds of moderate velocity blow in a constant direction, crescent-shaped **barchan dunes** tend to develop (*figure 16.11B*). Typically, they are small, isolated dunes from 1 to 50 m high. The tips (or horns) of a barchan point downwind, and sand grains are swept around them as well as up and over the crest. With a constant wind direction, beautifully symmetrical crescents form. With shifts in wind direction, however, one horn can become larger than the other. Although barchans typically are isolated dunes, they often are arranged in a chain-like fashion extending downwind from the source of sand. They can migrate from 7 to 15 m per year.

Longitudinal Dunes

Longitudinal dunes, also called **seif dunes** (Arabic, "sword"), are long, parallel ridges of sand, elongate in a direction parallel to the vector resulting from two slightly different wind directions (*figure 16.11C; see also plate 13*). They develop where strong prevailing winds converge and blow in a constant direction over an area having a limited supply of sand. Many longitudinal dunes are less than 4 m high, but they can extend downwind for several kilometers. In larger desert areas, they can grow to 100 m high and 120 km long, and they are usually spaced from 0.5 to 3 km apart. Longitudinal dunes occupy a vast area of central Australia called the Sand Ridge Desert. They are especially well developed in some desert regions of North Africa and the Arabian Peninsula.

Star Dunes

A **star dune** is a mound of sand having a high central point from which three or four arms, or ridges, radiate (*figure 16.11D*). This type is typical of parts of North Africa and Saudi Arabia. The internal structure of these dunes suggests that they were formed by winds blowing in three or more directions.

Parabolic (Blowout) Dunes

Parabolic dunes typically develop along coastlines where vegetation partly covers the sand (*figure 16.11E*). Where vegetation is absent, small deflation basins are produced by strong onshore winds. These blowout depressions grow larger as more sand is exposed and removed. Usually, the sand piles up on the lee slope of the shallow deflation hollow, forming a crescent-shaped ridge. In map view, a parabolic dune is similar to a barchan, but the tips of the parabolic dune point upwind. Because of their form, parabolic dunes are also called hairpin dunes.

Figure 16.11 Each of the major types of sand dunes represents a unique balance between sand supply, wind velocity, and variability in wind direction.

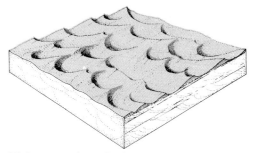

(A) Transverse dunes develop where the wind direction is constant and the sand supply is large.

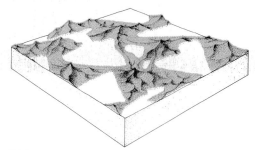

(B) Barchan dunes develop where the wind direction is constant but the sand supply is limited.

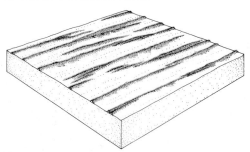

(C) Longitudinal dunes are formed by converging winds in an area with a limited sand supply.

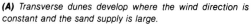

(D) Star dunes develop where the wind direction is variable.

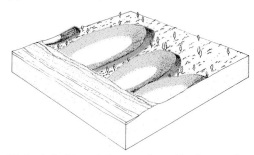

(E) Parabolic (blowout) dunes are formed by strong onshore winds.

LOESS

Statement

Loess is a deposit of windblown dust, which slowly blankets large areas, often masking preexisting landforms. It may cover as much as one-tenth of the world's land surface and is particularly widespread in semiarid regions along the margins of great deserts (*figure 16.12*). The equatorial tropics and the areas formerly covered by continental glaciers are free of loess.

Loess is a distinctive sedimentary deposit. Its particles are very similar, if not identical, to the dust in the air at the present time. Small, angular fragments of quartz predominate, with lesser amounts of feldspar, mica, and calcite. Normally, loess deposits lack stratification and erode into vertical cliffs. Large loess deposits are derived from

1. Desert regions
2. Glacial deposits

Discussion

Desert Loess

In northern China, extensive **loess** deposits consist primarily of disintegrated rock material, brought by the prevailing westerly winds from the Gobi Desert, in central Mongolia. The yellow-colored loess, which reaches a thickness of more than 60 m, blankets a large area. It is easily eroded and transported in suspension by running water and is responsible for the characteristic color of the Hwang Ho (the Yellow River) and the Hwang Hai (the Yellow Sea).

Loess in the eastern Sudan, in North Africa, probably originated from a desert, most likely the Sahara, to the west. Similarly, the loess deposits of Argentina are derived from arid regions to the west, rather than from glacial deposits to the south.

Glacial Loess

Much of the loess of North America and Europe appears to have originated from rock debris pulverized by glaciers and deposited as outwash. This sediment, commonly called **rock flour,** is picked up by the wind, transported, and deposited as a blanket of loess. In the central United States, most loess occurs in the bluffs and uplands of the Mississippi valley, the major drainage system for the meltwaters of the glaciers. Near rivers where floodplain muds provide a ready source of dust, the loess is 30 m or more thick, but it thins out away from the river channels. The greatest accumulations are east of the flood plains.

In many areas, loess lies directly upon glacial or glaciofluvial deposits. As the glaciers melted, fine rock flour apparently was deposited as glacial outwash so rapidly that a protective plant cover could not be established—a situation typical of many outwash plains. Winds then picked up the dust-size particles and deposited them to the east (downwind). In Europe, loess deposits also appear to have been derived from the adjacent glaciated areas.

Figure 16.12 Major loess deposits cover large areas of the world. They are particularly widespread in semiarid regions that are adjacent to deserts, such as those in China and South America. The loess deposits of Europe and the United States are derived from sediments deposited along the margins of continental glaciers. The tropics and areas formerly covered by glaciers are free from loess.

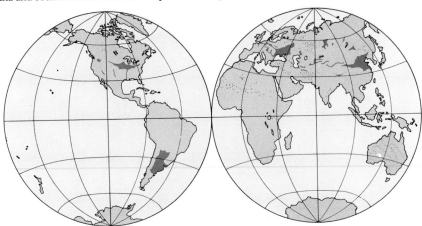

SUMMARY

Wind can transport huge quantities of sand and dust, which are deposited in distinctive landforms (sand dunes) and thick layers of loess. Wind is an important agent of sediment transport and deposition in the arid regions of the world, but it is not a major cause of abrasion of solid bedrock.

Wind transports sand by saltation and surface creep. Dust is carried in suspension. Windblown sand typically accumulates in dunes. These migrate downwind as sand is transported particle by particle up the windward slope and accumulates in the relatively quiet area of the lee slope. This process produces cross-bedding that is inclined downwind.

A variety of dune types result from variations in sand supply, wind direction and velocity, and characteristics of the desert surface. The most significant include (1) transverse dunes, (2) barchan dunes, (3) longitudinal dunes, (4) star dunes, and (5) parabolic dunes.

Loess accumulates as a blanket deposit, which can completely cover the preexisting surface. The dust is derived from rock flour near glacial margins or from desert regions.

Key Words

Coriolis effect	lag deposit
wind shadow	desert pavement
deflation	ventifact
blowout	yardang
deflation basin	saltation
surface creep	parabolic (blowout) dune
dune	lee slope
barchan dune	slip face
longitudinal (seif) dune	loess
transverse dune	rock flour
star dune	

Review Questions

1. Describe the processes involved in wind erosion.
2. What landforms are produced by deflation?
3. What controls the distribution of the major desert regions of the earth?
4. Explain the origin of ventifacts.
5. Explain the origin of desert pavement.
6. Draw a simple diagram showing how sand is transported by the wind.
7. Discuss and illustrate the migration of a dune.
8. How does cross-bedding originate?
9. List the five major types of dunes and state the conditions under which each type forms (wind direction and velocity, sand supply, and the characteristics of the surface over which the sand moves).
10. What is the origin of loess?
11. Where are the major areas of loess deposits in the world today?

Additional Reading

Bagnold, R. A. 1941. *The physics of blown sand and desert dunes*. New York: Methuen.

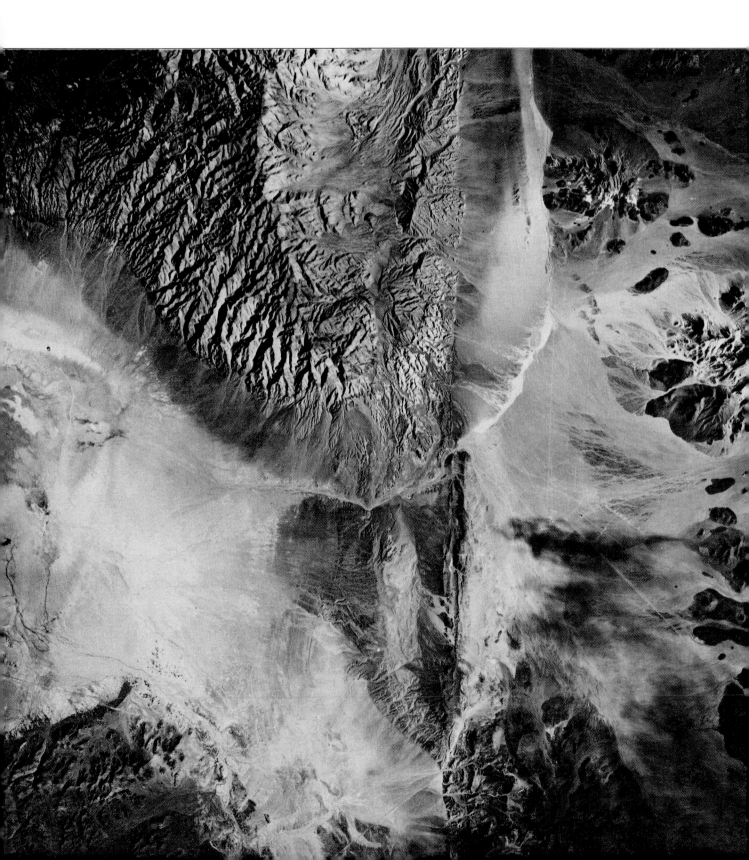

PLATE TECTONICS

The theory of plate tectonics is accepted by most geologists today. Yet, only a few years ago, the continents and the ocean basins were considered to be permanent features, which have existed since the beginning of the earth's history. We now believe that the lithosphere is in constant motion and that the continents have drifted thousands of kilometers across the earth's surface. The movement of lithospheric plates causes earthquakes, mountain building, metamorphism, and igneous activity. What is the nature of the mobile plates? What causes them to move? If the lithosphere is in motion, is it possible to measure rates and directions of movement? Have the plates always been in motion? Will they continue to move in the future? Why have we discarded the old concept of permanent continents and ocean basins? These are some of the questions we will examine in this and subsequent chapters.

Although the hypothesis of continental drift was proposed more than 50 years ago, the theory of plate tectonics was not developed until the early 1960s, when oceanographic surveys had provided enough data to make accurate regional topographic maps of the ocean floors. These data show that the ocean floors are not flat, featureless areas covered with sediment, nor are they like the continents. Instead, there is a worldwide rift system along the crest of the oceanic ridge and a system of deep trenches along some of the oceanic margins. These two systems are the most seismically active areas in the world.

In this chapter, we will briefly review the development of the theory of plate tectonics and the evidence upon which it is based. Then we will consider the nature of the lithospheric plates and their boundaries and the relative motion of the plates which produces the tectonic features of the earth.

MAJOR CONCEPTS

1. The theory of continental drift was proposed in the early 1900s, supported by a variety of geologic evidence. Without a knowledge of the nature of the oceanic crust, however, a complete theory of earth dynamics could not be developed.
2. A major breakthrough in the development of the plate tectonic theory occurred in the early 1960s, when the topography of the ocean floors was mapped, and magnetic and seismic characteristics were determined.
3. The lithosphere is divided into plates, bounded by the oceanic ridge, trenches, mountain ranges, and transform faults.
4. Tectonic plates move apart where the convecting mantle rises and spreads laterally beneath the oceanic ridge. The plates descend into the mantle beneath trenches and are consumed.
5. The energy for the plate tectonic system is internal heat, probably generated by radioactivity in the asthenosphere.

CONTINENTAL DRIFT

Statement

The theory of plate tectonics wrought a sweeping change in our understanding of the earth and the forces that shape it. Some scientists consider this conceptual change as profound as those that occurred when Darwin reorganized biology in the nineteenth century or when Copernicus in the sixteenth century determined that the earth is not the center of the universe. Yet, the concept of continental drift is an old idea.

Soon after the first reliable world maps were made, scientists noted that the continents, particularly Africa and South America, would fit together like a jigsaw puzzle if they could be moved. Antonio Snider-Pellegrini, a Frenchman, was one of the first to study the idea in some depth. In his book *Creation and its mysteries revealed* (1858), he showed how the continents looked before they separated.

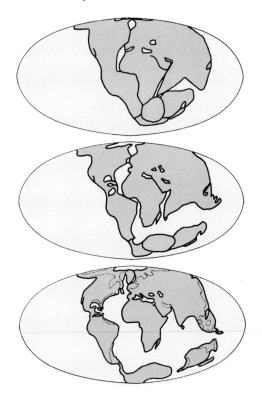

Figure 17.1 Continental drift was illustrated by Wegener in 1915 by this series of maps. Wegener called the original landmass Pangaea ("all lands") and believed that the continents somehow plowed through the oceanic crust as they drifted.

He cited fossil evidence in North America and Europe but based his reasoning on the catastrophe of Noah's flood. The theory of continental drift was not considered seriously until 1908, when an American geologist, Frank B. Taylor, pointed out a number of geologic facts that could be explained by it.

The first comprehensive presentation of the theory of continental drift was put forth by Alfred Wegener, a German meteorologist, in *The origin of continents and oceans* (1915). Wegener based his theory not only on the shape of the continents but on geologic evidence such as similarities in the fossils found in Brazil and Africa. He drew a series of maps showing three stages in the drift process, beginning with an original large landmass, which he called Pangaea (meaning "all lands"). Wegener believed that the continents, composed of light granitic rock, somehow plowed through the denser basalts of the ocean floor, driven by forces related to the rotation of the earth (*figure 17.1*).

Most geologists and geophysicists rejected Wegener's theory, although many scientific observations supporting it were known at the time. A few noted scholars, however, seriously considered the theory. Arthur Holmes, of England, developed it in his textbook *Principles of physical geology* (1944). Alexander L. du Toit, from South Africa, compared the landforms and fossils of Africa and South America and further expounded the theory in his book *Our wandering continents* (1937).

Discussion

The early arguments for the theory of **continental drift** were supported by some important and imposing evidence, most of which resulted from regional geologic studies.

Paleontologic Evidence

The striking similarity of certain fossils found on the continents on both sides of the Atlantic is difficult to explain unless the continents were once connected. The fossil record indicates that new species appear at one point and disperse outward from there. Floating and swimming organisms could migrate in the ocean from the shore of one continent to another, but the Atlantic Ocean would present an insurmountable obstacle for migration of land-dwelling animals, such as reptiles and insects, and certain land plants. Consider two examples.

Fossils of the seed fern *Glossopteris* have been found in rocks of the same age from South America, South Africa, Australia, and India and within

480 km of the South Pole, in Antarctica. Mature seeds of this plant were several millimeters in diameter, too large to have been dispersed across the ocean by winds. The simultaneous presence of *Glossopteris* on all of the southern continents, therefore, is strong supporting evidence for the theory of continental drift.

The distribution of Paleozoic and Mesozoic reptiles provides similar evidence, as fossils of several species have been found in the now separated southern continents. An example is a mammallike reptile belonging to the genus *Lystrosaurus.* This creature was strictly a land dweller. Its fossils are found in abundance in South Africa, South America, and Asia, and in 1969 a United States expedition discovered them in Antarctica. Thus, this genus once inhabited all of the southern continents. Clearly, these reptiles could not have swum thousands of kilometers across the Atlantic and Antarctic oceans, so that some previous connection must be postulated. A former land bridge, similar to present-day Central America, would explain the presence of *Lystrosaurus* in distant parts of the world, but surveys of the ocean floor show no evidence of such a submerged land bridge.

Evidence from Structure and Rock Type

A number of geologic features end abruptly at the coast of one continent and reappear on the facing continent across the Atlantic (*figure 17.2*). Folded mountain ranges at the Cape of Good Hope, at the southern tip of Africa, trend from east to west and terminate sharply at the coast. An equivalent structure, of the same age and style of deformation, appears near Buenos Aires, Argentina.

The folded Appalachian Mountains are another example. The deformed structures of the mountain belt extend northeastward across the eastern United States and through Newfoundland and terminate abruptly at the ocean. They reappear at the coast of Ireland and Brittany.

Other examples could be cited, but the important point is that the continents on both sides of the Atlantic fit together, not only in outline, but in rock type and structure. They are related much like matching pieces of a torn newspaper (*figure 17.3*).

Figure 17.2 South America and Africa fit together, not only in outline, but in rock types and geologic structure. The dark brown areas represent the shields of metamorphic and igneous rocks, formed at least 2 billion years ago. The light brown areas represent younger rock, much of which has been deformed by mountain building. Structural trends such as fold axes are shown by dashed lines. Most of the deformation occurred from 450 million to 650 million years ago. Several fragments of the African shield are stranded along the coast of Brazil. Black dots represent rocks that are more than 2 billion years old. Brown dots represent younger Precambrian rocks.

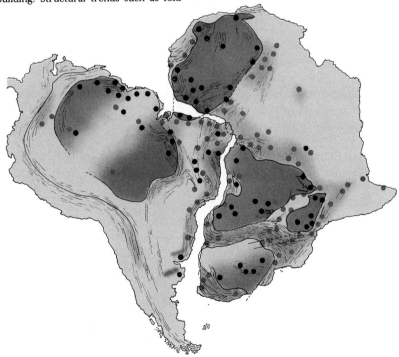

Figure 17.3 The continents fit together like a jigsaw puzzle or pieces of a torn newspaper. Not only do the outlines of the torn pieces fit together, but the printing on them (analogous to the ages and structural features of the continents) also matches across the edges of the separate pieces.

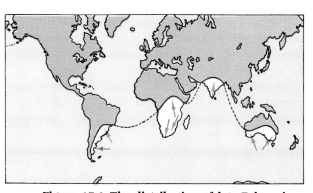

Figure 17.4 The distribution of late Paleozoic glacial deposits is restricted to the Southern Hemisphere (except for India). Arrows show the direction of ice movement. These areas are now close to the tropics. The present-day cold latitudes in the Northern Hemisphere show no evidence of glaciation during the late Paleozoic.

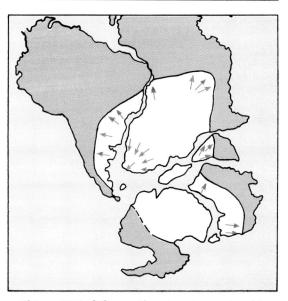

Figure 17.5 If the continents were restored to their former positons according to Wegener's theory of continental drift and if the former South Pole were located approximately where South Africa and Antarctica meet, the location of late Paleozoic glacial deposits and the directions in which the ice flowed would be explained nicely.

The jagged edges fit, and the printed lines (structure and rock types) join together into a coherent unit. One important point needs emphasis. The geologic similarities on opposite sides of the Atlantic are found only in rocks older than the Cretaceous Period, which began about 137 million years ago. The continents are believed to have split and begun to drift apart in Jurassic time, about 200 million years ago.

Evidence from Glaciation

During the latter part of the Paleozoic Era (about 300 million years ago), glaciers covered large portions of the continents in the Southern Hemisphere. The deposits left by these ancient glaciers can readily be recognized, and striations and grooves on the underlying rocks show the direction in which the ice moved (*figure 17.4*). Except for Antarctica, all of the continents in the Southern Hemisphere now lie close to the equator. In contrast, the continents in the Northern Hemisphere show no trace of glaciation during this time. In fact, fossil plants indicate a tropical climate in that area. This evidence is difficult to explain in the context of fixed continents and the climatic belts that are determined by latitude.

Even more difficult to explain is the direction in which the glaciers moved. Regional mapping of striations and grooves indicate that, in South America, India, and Australia, the ice moved inland from the oceans. Such movement would be impos-

sible unless there was a landmass where the oceans now exist.

If the continents were grouped together as Wegener proposed, the glaciated areas would have been grouped in a neat package near the South Pole (*figure 17.5*), and Paleozoic glaciation could be explained nicely. The pattern of glaciation was considered strong evidence of continental drift, and many geologists who worked in the Southern

Hemisphere became ardent supporters of the theory, because they could see the evidence with their own eyes.

Evidence from Paleoclimates

Other evidence of striking climatic changes tends to support the drift theory. Great coal deposits in Antarctica show that abundant plant life once flourished on this continent, now mostly covered with ice.

On the other continents, salt deposits and formations of windblown sandstone provide additional clues that permit us to reconstruct the climatic zones of the past. The paleoclimatic patterns are baffling with the continents in their present position, but if they are grouped together in their pre-drift position, the patterns are easily explained (*figure 17.6*).

The evidence for the theory of continental drift was considered and debated for years. Although it was convincing to many, most geophysicists objected to the theory because it simply is impossible for continents to drift through the solid rocks of the oceanic crust. In the absence of a reasonable mechanism for drift, there was little further development of the theory until after World War II. Then an explosion of knowledge not only provided renewed support for the drift hypothesis but also led to the discovery of a possible mechanism.

Figure 17.6 Paleoclimatic evidence for continental drift includes deposits of coal (C), desert sandstone (D), rock salt (S), windblown sand (dotted area), gypsum (GY), and glacial deposits (GL). Each indicates a specific climatic condition at the time of its formation. The distribution of these deposits is best explained if we assume that the continents were once grouped together as shown in this diagram.

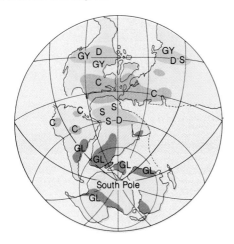

THE DEVELOPMENT OF THE THEORY OF PLATE TECTONICS

Statement

Although the theory of continental drift was supported by some convincing evidence, the data upon which it was originally based came only from the continents, because, prior to the 1950s, there was no effective means of studying the ocean floor. Before 1950, therefore, geologists faced an almost total absence of data about the geology of three-fourths of the earth's surface. Then, in the 1950s and 1960s, an outburst of new data and new ideas resulted from research efforts in two areas:

1. The topography and geology of the ocean floor
2. Rock magnetism

Discussion

The Geology of the Ocean Floor

In the 1950s and 1960s, newly developed echo-sounding devices enabled marine geologists and geophysicists to map the topography of the ocean floor in considerable detail (see chapter 21). When the results of these studies were compiled, they revealed that the ocean basins are divided by a great ridge approximately 65,000 km long. Moreover, at the crest of the ridge is a central valley, from 1000 to 3000 m deep. This feature appears to be a **rift valley,** which is splitting apart under tension. Other evidence shows that ocean basins are relatively young. Seismic studies have established that the oceanic crust (composed largely of basalt) has a completely different composition and is much thinner than the continental crust. Furthermore, the oceanic crust is not deformed into folded mountain structures and apparently is not subjected to strong compressional forces.

In 1960, H. H. Hess, a noted geologist from Princeton University, proposed a theory of sea-floor spreading, which accounts for the new data and also suggests a possible mechanism for continental drift. Hess suggested that the ocean floors are spreading apart and moving symmetrically away from the oceanic ridge, propelled by **convection currents** in the mantle. According to his theory, this continuous spreading produces fractures in the rift valley, and magma from the mantle is injected into these fractures to become new oceanic crust. The convection currents in the mantle carry the continents away from the oceanic ridge and toward deep-sea trenches. There, the oceanic crust descends into the mantle, with the descending convection current, and is reabsorbed. In this way, the entire ocean floor is completely regenerated in 200 million or 300 million years.

Hess elaborated on the theory of continental drift in the light of fresh knowledge and redefined it in the scheme of sea-floor spreading. A test of his ideas, utilizing new studies in paleomagnetism, was soon to follow.

Rock Magnetism

The study of rock magnetism developed during the 1950s with the perfection of new, highly sensitive magnetometers. Certain rocks, such as basalt, are fairly rich in iron and become weakly magnetized by the earth's magnetic field as they cool. In a sense, the mineral grains in the rock become "fossil" magnets, oriented with respect to the earth's magnetic field at the time when the rock was formed, and thus they preserve a record of **paleomagnetism.** Similarly, red sandstone contains enough iron that its grains become oriented in the earth's magnetic field as the sediment is

deposited, so that it also can indicate the orientation of the paleomagnetic fields. These rocks therefore retain an imprint of the earth's magnetic field at the time of their formation.

The magnetic field of the earth resembles that of a simple bar magnet with its axis inclined 11° from the earth's geographic axis (*figure 17.7A*). The mantle and core of the earth, however, are far too hot to retain a permanent magnetic field. The earth's magnetism, therefore, must be generated electromagnetically. The electromagnetic, or dynamo, theory postulates that the outer core of liquid iron slowly rotates with respect to the surrounding mantle. Such motion would generate strong electrical currents, which would establish a magnetic field (*figure 17.7B*).

Polar Wandering. Studies of paleomagnetism in European rocks of widely different ages demonstrate that the earth's north magnetic pole apparently has steadily changed its position with time. As is shown in *figure 17.8*, the pole has slowly migrated northward and westward to its present position. The change in position was systematic,

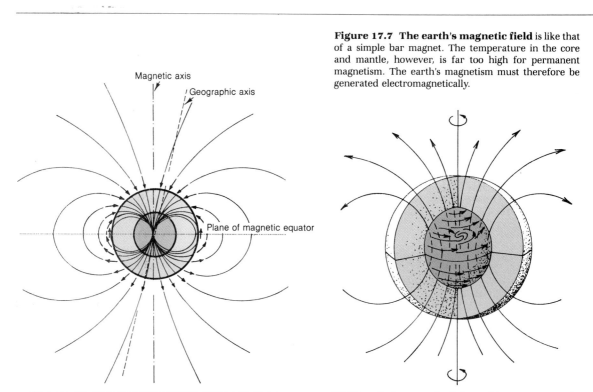

Figure 17.7 The earth's magnetic field is like that of a simple bar magnet. The temperature in the core and mantle, however, is far too high for permanent magnetism. The earth's magnetism must therefore be generated electromagnetically.

(A) Lines of force in the earth's magnetic field are shown by arrows. If a magnetic needle were free to move in space, it would be deflected by the earth's magnetic field. Close to the equator, the needle would be horizontal and would point toward the poles. At the magnetic poles, the needle would be vertical.

(B) Theoretically, convection in the earth's core can generate an electrical current (in a manner similar to the operation of a dynamo), which produces a magnetic field.

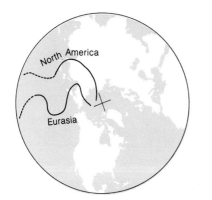

(A) *The magnetic properties of rocks in North America suggest that the north magnetic pole has migrated in a sinuous path over the last several hundred million years. Evidence from other continents shows similar migration, but along different paths. How could different continents show different paths of polar migration? The paleomagnetic evidence implies that, if the continents had remained fixed, different continents would have had different magnetic poles at the same time. But that would be impossible.*

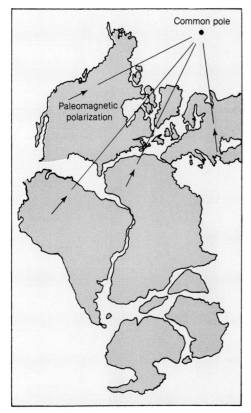

(B) *The question can be answered if the pole has remained fixed while the continents have drifted. If, for example, Europe and North America were previously joined, the paleomagnetic field preserved in their rocks would indicate a single pole location until they drifted apart. The sequence of rocks on each continent would then show the pole taking a different path of migration to its present position.*

Figure 17.8 Changes in the position of the magnetic poles during the geologic past are indicated by paleomagnetic studies of rocks.

not random. A similar migration of the magnetic pole was found from paleomagnetic work in North America, and, although the path of migration was systematically different, it paralleled that of the European shift. These observations could be explained nicely by drifting continents, and students of paleomagnetism became leading proponents of the theory of continental drift. Soon, results collected from the southern continents were reported. Again, a systematic change in the position of the magnetic pole through time was reported— but with different paths for different continents.

It is impossible that there were numerous magnetic poles that migrated systematically and eventually merged. The most logical explanation is that there has always been only one pole, which has remained fixed while the continents moved with respect to it. The results of paleomagnetic studies make sense if the continents were once arranged as shown in *figure 17.8B* and then drifted to their present positions. This discovery brought renewed interest in the theory of continental drift and lent support to the conclusion that the Atlantic Ocean opened relatively recently.

Magnetic Reversals. Recent studies of the magnetic properties of numerous samples of basalt from many parts of the world demonstrate that the earth's magnetic field has been reversed many times throughout the last 70 million or 80 million years. Epochs of normal polarity (that is, periods when the magnetic poles were close to their present location), lasting from 1 million to 3 million years, have been followed by similar periods during which the north magnetic pole and the south magnetic pole were reversed. At least nine magnetic reversals have occurred in the last 4.5 million years. The present period of normal polarity began about 700,000 years ago. It was preceded by a period of reversed polarity, which began 2.5 million years ago and contained two short periods of normal polarity.

The major intervals of alternating polarity (about 1 million years apart) are termed **polarity epochs,** and intervals of shorter duration are termed **polarity events.** The pattern of alternating polarities has been clearly defined, and evidence of the occurrence of polarity epochs has been found in widely separated parts of the earth. From the sequence of **magnetic anomalies** and their radiometric ages, a reliable chronology of magnetic reversals has been established for the most recent 4 million years (*figure 17.9*). In addition, extrapolation as far back as 76 million years reveals a sequence of at least 171 reversals.

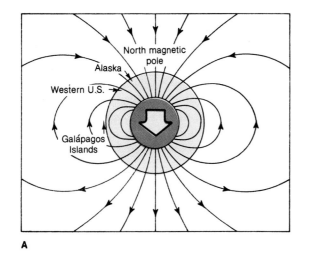

A

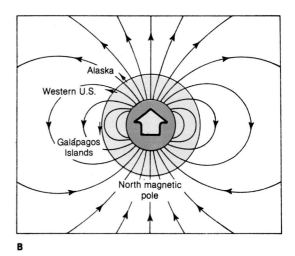

B

C

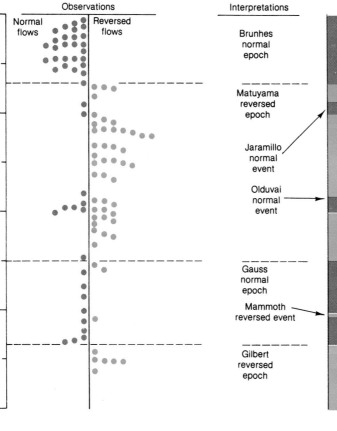

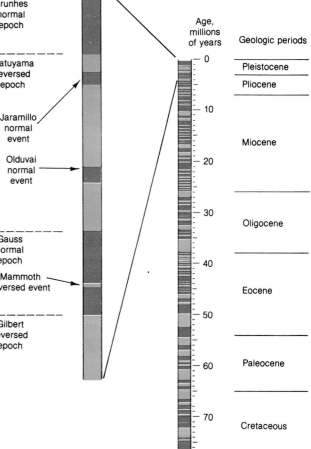

Figure 17.9 Reversals of lines of force in the earth's magnetic field are documented by paleomagnetic studies of numerous rock samples from throughout the world. Lines of force with normal polarity are shown in diagram **A**. With reverse polarity (diagram **B**), the lines of force are oriented in the op- posite direction. Diagram **C** shows the patterns of changing polarity with time. The pattern of change during a period of 1 million or 2 million years is distinctive, and it can be used to help establish the age of a rock sequence.

In 1963, Fred Vine and D. H. Matthews saw a way to test the idea of sea-floor spreading, put forth by Hess. If sea-floor spreading has occurred, they suggested, it should be recorded in the magnetism of the basalts on the ocean's crust. (The same idea was developed independently by L. W. Morley.) If the earth's magnetic field reversed inter-mittently, new basalt forming at the crest of the oceanic ridge would be magnetized according to the polarity at the time when it cooled. As the ocean floor spreads, a symmetrical series of magnetic stripes, with alternating normal and reversed polarities, would be produced in the crust along either side of the oceanic ridge. Subsequent

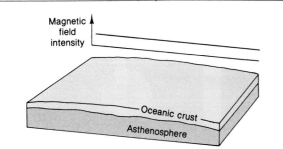

(A) The oceanic ridge during the Gauss normal polarity epoch, 2.75 million years ago. *As magma cooled and solidified along the ridge in dikes and flows, it became magnetized in the direction of the magnetic field existing at that time (normal polarity).*

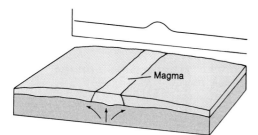

(B) The oceanic ridge during the Matuyama reversed polarity epoch, 2.25 million years ago. *As sea-floor spreading continued, the magnetized crust formed during the Gauss epoch separated into two blocks. Each was transported laterally away from the ridge, as though on a conveyor belt. New crust, formed at the ridge, became magnetized in the opposite direction.*

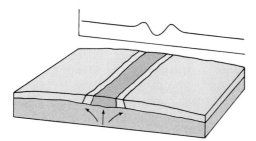

(C) The oceanic ridge at the present time, showing polarity reversals of the past 3 million years. *The alternating directions of magnetic polarity in the sea floor produced a symmetrical sequence of positive and negative magnetic anomalies on each side of the ridge. Note that the pattern of magnetic reversals away from the ridge is the same as the pattern found in a sequence of basalt flows on the continents. This permits continental rocks to be correlated with rocks of equivalent age on the sea floor.*

Figure 17.10 Specific patterns of magnetism are preserved in the newly formed crust, which is generated at the oceanic ridge as the lithosphere moves laterally. The patterns of magnetic reversals away from the ridge are identical to the patterns of magnetic reversals in a vertical sequence of rocks on the continents.

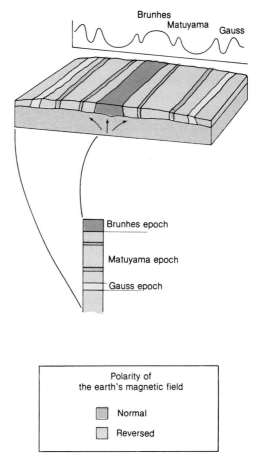

Polarity of the earth's magnetic field	
▨	Normal
□	Reversed

(D) Patterns of magnetic reversals in a vertical sequence of basalts on the continents. *The youngest (upper) continental rocks correlate with the youngest oceanic crust (at the center of the oceanic ridge).*

investigations have conclusively proved the theory proposed by Vine and Matthews and by Morley.

To better understand the origin of these magnetic patterns, consider how the sea floor could have evolved during the last few million years. *Figure 17.10A* shows the sea floor as it is considered to have been about 2.75 million years ago, during the Gauss normal polarity epoch (named for the German mathematician Karl Friedrich Gauss). Basalt, injected into the fractures of the oceanic ridge, formed dikes or was extruded over the sea floor as submarine flows. As it solidified, it became magnetized in the direction of the existing (normal) magnetic field. Thus, basalt extruded along the oceanic ridge formed a zone of new crust with normal magnetic polarity. As the sea floor spread, this zone of crust migrated away from the ridge but remained parallel to it. About 2.5 million years ago, the earth's magnetic polarity was reversed. New crust generated at the oceanic ridge was magnetized in the opposite direction (*figure 17.10B*), producing a zone of crust with reverse polarity. When the polarity changed to normal again, new crust was magnetized in a normal direction. In this way, the sequence of polarity reversals became imprinted as magnetic stripes on the oceanic crust.

Note that the patterns of magnetic stripes on the ocean floor on either side of the ridge match the patterns found in a sequence of recent basalts on the continents (*figure 17.10C*; see also *figure 17.9C*). That is, the crest of the ridge shows normal polarity and is flanked on either side by a broad stripe of rocks with reversed polarity (formed during a reversed epoch) containing two narrow bands of rocks with normal polarity (formed during normal events). Then follows a stripe with normal polarity containing one narrow band with reversed polarity, and so on. In brief, the patterns of magnetic reversals away from the crest of the ridge are the same as those found in a vertical sequence of rocks on the continents, from youngest to oldest. These data provide compelling evidence that the sea floor is spreading.

An important aspect of these reversal patterns is that they enable us to determine rates of plate movement. Magnetic reversals in rock sequences on the continents have been radiometrically dated. These studies show that the present normal polarity has existed for the last 700,000 years and was preceded by the pattern shown in *figure 17.9C*. Since the same pattern exists in the oceanic crust, we can assign provisional ages to the magnetic anomalies on the ocean floor.

Magnetic surveys have now determined patterns of magnetic reversal for much of the ocean floor (*figure 17.11*). These studies show that most of the deep-sea floor was formed during Cenozoic time (during the last 65 million years). It now seems probable that very little of the present ocean basin was formed before the Jurassic. From the patterns of magnetic reversals, the rate of sea-floor spreading appears to range from 1 to 16 cm per year.

Iceland, because it is essentially a large exposure of the **mid-Atlantic ridge,** offers a unique opportunity to study the physical mechanism of sea-floor spreading. Geologic studies there show that the island is being pulled apart by the spreading sea floor beneath it. The tension causes normal faults and fissures parallel to the axis of the ridge. Volcanic eruptions occur through these fissures, and swarms of parallel dikes are injected into them with each increment of crustal extension. The aggregate width of these dikes is about 400 km, which corresponds to the total amount of crustal extension since the beginning of Tertiary time, about 65 million years ago. A geologic map of Iceland (*figure 17.12*) shows that the oldest rocks are in the extreme eastern and western ends of the island. Rocks become progressively younger toward the center, where most of the island's present-day volcanism occurs.

Evidence from Sediment on the Ocean Floor

To many geologists, some of the most convincing evidence for the plate tectonic theory comes from recent drilling into the sediment on the ocean floor. The deep-sea drilling project is a truly remarkable expedition in scientific exploration. It began in 1968 with the *Glomar Challenger,* a special ship designed by a California offshore drilling company. The *Challenger* can lower more than 6100 m of drilling pipe into the open ocean, bore a hole in the sea floor, and bring up bottom cores and samples. The project was funded by the National Science Foundation and was planned by JOIDES—Joint Oceanographic Institutions for Deep Earth Sampling—under the direction of the Scripps Institution of Oceanography. Since 1968, the *Challenger* has drilled more than 400 holes in the sea floor and has provided much data in support of the theory of plate tectonics.

Deep-sea drilling confirms the conclusions drawn from paleomagnetic studies by checking the age of the fossils that first accumulated on different portions of the ocean floor. As is predicted by the plate tectonic theory, the youngest sediment is found near the oceanic ridge, where new crust is

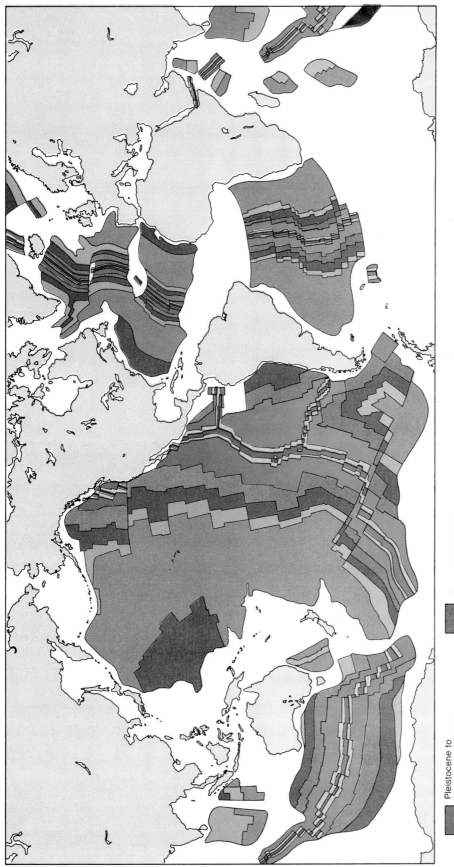

Figure 17.11 Magnetic patterns on the sea floor are symmetric with respect to the oceanic ridge. The youngest crust is along the crest of the ridge. Away from the ridge, the crust is progressively older.

Pleistocene to
Holocene (0–2 MY)

Pliocene (2–5 MY)

Miocene (5–23 MY)

Oligocene (23–38 MY)

Eocene (38–53 MY)

Paleocene (53–65 MY)

Cretaceous (65–135 MY)

Jurassic (135–190 MY)

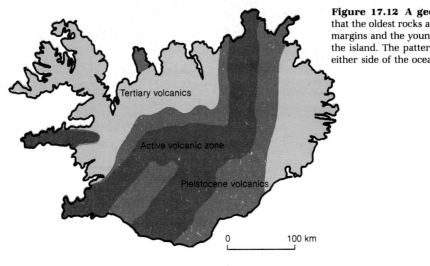

Figure 17.12 **A geologic map of Iceland** shows that the oldest rocks are along the eastern and western margins and the youngest rocks are near the center of the island. The pattern is identical to that of rocks on either side of the oceanic ridge.

being created. Away from the ridge, the sediments that lie directly above the basalt become progressively older, with the oldest sediment near the continental borders.

Measurements of rates of sedimentation in the open ocean show that between 0.9 and 1.2 cm of red clay and organic **ooze** accumulates every 1000 years. If the present ocean basins had existed since Cambrian time, for example, the sediments would be at least 5 km thick (*figure 17.13A*). But the greatest thickness of deep-ocean sediments measured to date is only 300 m, suggesting that the ocean basins are young geologic features indeed (*figure 17.13B*). In fact, the oldest sediments yet found on any ocean floor are only 160 million years old. In contrast, the metamorphic rocks of the continental shields are as much as 3.8 billion years old.

Not only do the thickness and age of sediments increase away from the crest of the oceanic ridge, but certain types of sediment also indicate sea-floor spreading. For example, plankton thrive in the upwelling, warm, nutrient-rich water of the Pacific equatorial zone. As the creatures die, their tiny skeletons rain down unceasingly to build a layer of soft, white chalk on the sea floor. The chalk can form only in the equatorial belt, as plankton do not flourish in the colder waters of higher latitudes. Yet, drilling by the *Glomar Challenger* has shown that the chalk line in the Pacific extends north of today's equator. The only logical conclusion is that the Pacific sea floor has been migrating northward for at least 100 million years.

The theory of plate tectonics has been firmly established and accepted as the fundamental theory of the earth's dynamics. Its first achieve-

ment was to explain the meaning of features on the ocean floor. Now the emphasis has switched to the continents, and most previous geologic observations of the continents are being reexamined in light of this theory.

Figure 17.13 **The thickness of sediment and magnetic reversals on the oceanic ridge** confirm the theory of sea-floor spreading.

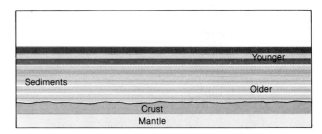

(A) With no sea-floor spreading, the entire ocean floor would be covered with a thick sequence of oceanic sediment, with alternating polarity preserving a record of the earth's magnetism since the Precambrian.

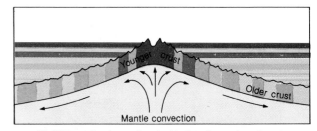

(B) With sea-floor spreading, the blanket of oceanic sediment thins progressively toward the crest of the oceanic ridge and is almost nonexistent in the rift valley. The edge of each layer of magnetized sediment lies upon basaltic crust, which was generated at the spreading center during the same time interval when the sediment was deposited.

PLATE GEOGRAPHY

Statement

The shorelines of the continents are major geographic features but have little significance from the standpoint of the earth's tectonics. Plate boundaries are the most significant structural elements of the planet, and to understand plate tectonics, you must learn a new geography—the geography of plate boundaries. This should not be difficult, because plate boundaries generally are marked by major topographic features. Understanding requires only that you focus your attention on the structural features of the earth, rather than on the boundaries between land and ocean.

Discussion

The new geography of tectonic **plates** is illustrated in *figure 17.14*. The outer, rigid layer of the earth—the lithosphere—is divided into a mosaic of seven major plates and a number of smaller subplates. The major plates are outlined by oceanic ridges, trenches, and young mountain systems. These include the Eurasian, North American, South

American, Pacific, African, Australian, and Antarctic plates. The largest is the Pacific plate, which is composed entirely of oceanic crust and covers about one-fifth of the earth's surface. The other large plates contain both continental crust and oceanic crust, but there are no major plates composed entirely of continental crust. Smaller plates include the China, Philippine, Arabian, Iran, Nazca, Cocos, Caribbean, and Scotia plates, plus a number of others that have not yet been precisely defined. The smaller plates appear to form near convergent boundaries of major plates where there is collision between continents or between a continent and an island arc. Thus, the smaller plates are characterized by rapid and complex movement.

Plates range in thickness from approximately 70 km beneath oceanic areas to 150 km beneath continents.

Individual plates are not permanent features. They are in constant motion and continually change in size and shape. Plates that do not contain continental crust can be completely consumed in a subduction zone. A plate can change its shape by splitting along new lines or by welding itself to another plate. In addition, plate margins are not fixed. They move and modify the size and shape of the plates themselves.

Figure 17.14 The major tectonic plates are delineated by the major tectonic features of the globe: (1) the oceanic ridge, (2) deep-sea trenches, and (3) young mountain belts. Plate boundaries are outlined by earthquake belts and volcanic activity. Most plates (such as the North American, African, and Australian) contain both continental and oceanic crust, but the Pacific, Cocos, and Nazca plates contain only oceanic crust.

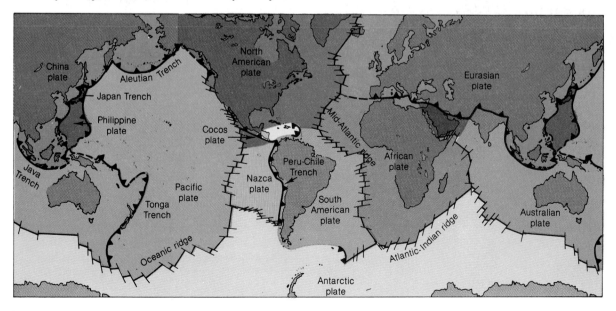

PLATE BOUNDARIES

Statement

Each tectonic plate is rigid and moves as a single mechanical unit; that is, if one part moves, the entire plate moves. It can be warped or flexed slightly as it moves, but relatively little change occurs in the middle of a plate. Nearly all major tectonic activity occurs along the plate boundaries, and thus geologists and students of geology focus their attention on the plate margins.

Three kinds of plate boundaries are recognized. These define three fundamental kinds of deformation and geologic activity (*figure 17.15*).

1. Divergent plate boundaries (also called spreading edges or spreading centers) are zones of tension, where plates split and spread apart.
2. Convergent plate boundaries (also called subduction zones or edges of consumption) are zones of compression, where plates collide and one plate moves down into the mantle.
3. Transform fault boundaries (also called passive plate margins) are zones of shearing, where plates slide past each other without diverging or converging.

Discussion

Processes at Divergent Plate Boundaries

Divergent plate boundaries, or **spreading centers,** form where a plate splits and is pulled apart. Where a zone of spreading intersects a continent, rifting occurs and the continent splits (*figure*

Figure 17.16 Stages of continental rifting are shown in this series of diagrams. The major geologic processes at divergent plate boundaries are tensional stress, block faulting, and basaltic volcanism.

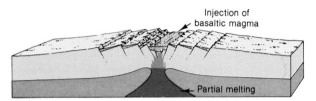

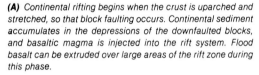

(A) *Continental rifting begins when the crust is uparched and stretched, so that block faulting occurs. Continental sediment accumulates in the depressions of the downfaulted blocks, and basaltic magma is injected into the rift system. Flood basalt can be extruded over large areas of the rift zone during this phase.*

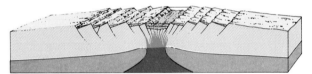

(B) *Rifting continues, and the continents separate enough for a narrow arm of the ocean to invade the rift zone. The injection of basaltic magma continues and begins to develop new oceanic crust.*

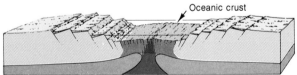

(C) *As the continents separate, new oceanic crust and new lithosphere are formed in the rift zone, and the ocean basin becomes wider. Remnants of continental sediment can be preserved in the downdropped blocks of the new continental margins.*

(D) *As spreading continues, the ocean basin grows larger. The continents move off from the uparched spreading zone, and parts of the continental crust can be covered by the ocean.*

Figure 17.15 Types of plate margins are depicted in this idealized diagram. Constructive margins (divergent plate boundaries) occur along the oceanic ridge, where plates move apart. Destructive margins (convergent plate boundaries) occur along the deep trenches. Passive plate margins coincide with fracture zones.

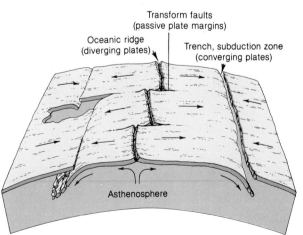

17.16). The separate continental fragments drift apart with the diverging plates, so that a new and continually enlarging ocean basin is formed in the site of the initial rift zone. Thus, divergent plate boundaries are characterized by tensional stresses that produce block faulting, fractures, and open fissures along the margins of the separating plates. Basaltic magma derived from the partial melting of the mantle is injected into the fissures, where it cools and becomes part of the moving plates. Divergent plate boundaries are some of the most active volcanic areas on the earth, generally characterized by unspectacular, quiet fissure eruptions, most of which are concealed beneath the sea.

Examples of continental rifting in various stages are shown in *figure 17.17.* The initial stage is represented by the system of great rift valleys in East Africa. The long, linear valleys, partly occupied by lakes, are huge, downdropped fault blocks, which result from the initial tensional stress. Magma rising in the rift zone produces volcanism, exemplified by the great volcanoes of Mount Kenya and Mount Kilimanjaro. The Red Sea illustrates a more advanced stage of rifting. The Arabian Peninsula has been completely separated from Africa, and a new linear ocean basin is just beginning to develop. The Atlantic Ocean represents a still more advanced stage of continental drift and sea-floor spreading. The American continents have become separated from Africa and Europe by thousands of kilometers. The mid-Atlantic ridge is the boundary between the diverging plates, with the American plates moving westward in relation to Africa and Europe.

Processes at Convergent Plate Boundaries

Convergent plate boundaries, or **subduction zones,** are areas of complicated geologic processes, including igneous activity, crustal deformation, and mountain building. The specific processes that are active along a convergent plate boundary depend on the types of crust involved in the collision of the converging plates.

If both plates at a convergent boundary contain oceanic crust, one is thrust under the margin of the other, in a process called **subduction.** The subducting plate descends into the asthenosphere, where it is heated and ultimately absorbed into the mantle. If one plate contains a continent, the lighter continental crust always resists subduction and overrides the oceanic plate. If both converging plates contain continental crust, neither can subside into the mantle, although one can override the other for a short distance. Both plates are compressed, and the continents are ultimately "fused" or "welded" together into a single continental block, with a mountain range marking the line of suture.

The zone of collision between two plates is a zone of deformation, mountain building, and metamorphism. If the overriding plate contains continental crust, compression deforms the margins into a folded mountain belt, and the deep roots of the mountains are metamorphosed.

The major processes and geologic phenomena that characterize convergent plate margins are shown in *figure 17.18.* A subduction zone (or zone of underthrusting) usually is marked by a deep-sea

Figure 17.17 Examples of continental rifting in various stages can be found today in different parts of the world.

(A) *The African rift valleys represent the initial stage of rifting, which is characterized by regional uparching of the crust; long, narrow fault blocks; and volcanic activity.*

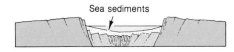

(B) *The Red Sea represents a more advanced stage of rifting, with the generation of new oceanic crust.*

(C) *The Atlantic Ocean is in an advanced stage of rifting. The symmetry of the rift is still preserved, although the continents are thousands of miles apart. New oceanic crust is continually generated along the rift valley.*

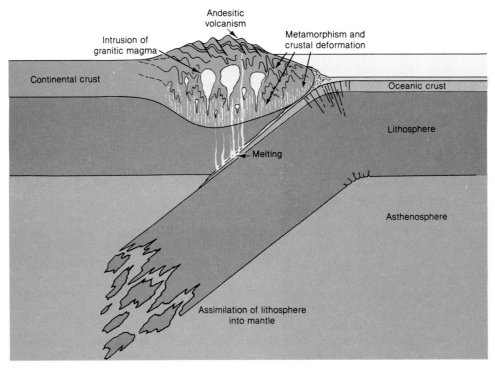

Figure 17.18 The major geologic processes at convergent plate boundaries include the deformation of continental margins into folded mountain belts; metamorphism due to high temperatures and high pressures in the mountain roots; and partial melting of the descending plate, which produces granitic intrusions and andesitic volcanism.

Figure 17.19 Examples of the main types of convergent plate boundaries can be found today in various parts of the world.

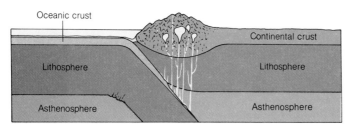

(A) The Japanese islands represent the convergence of two oceanic plates.

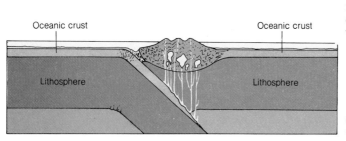

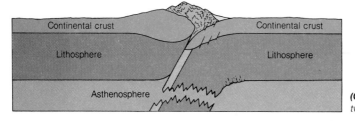

(B) South America represents the convergence of an oceanic plate and a continental plate.

(C) The Himalaya Mountains represent the convergence of two continental plates.

trench, and the movement of the descending plate generates an inclined zone of seismic activity. As the subducting plate moves down into the hot asthenosphere, partial melting of the oceanic crust generates silica-rich magma, which (being less dense than the surrounding material) moves upward. Some of this magma is extruded at the surface as lava and forms an island arc or a chain of volcanoes in the mountain belt of the overriding plate. Usually, most of it intrudes into the deformed mountain belt to produce batholiths. Both extrusion and intrusion add new material to the continental plate, so that continents grow by accretion. This is an important mechanism in the differentiation of the earth, whereby lighter material is concentrated in the upper layers of the planet.

Examples of the three types of convergent plate boundaries are illustrated in *figure 17.19*. Along the west coast of South America, the convergence of the Nazca and South American plates has produced the South American Trench and the Andes Mountains. The convergence of the Pacific and China plates has formed the island arc of Japan. The collision of the Australian plate with Asia has formed the Himalaya Mountains.

It was emphasized in chapter 2 that continents are composed of light granitic material and ride passively on the moving plates. One of their most important characteristics is that they cannot sink. They are simply not heavy enough to submerge in the denser material of the earth's mantle. The continents are therefore much older than the ocean basins. Because continents are never consumed into the mantle, they preserve records of plate movements in the early history of the earth—records in the form of ancient faults, old mountain belts, granitic batholiths, and sediments deposited along ancient continental margins.

Processes at Transform Fault Boundaries

The third type of plate boundary occurs where adjacent plates slide horizontally past each other with neither creation nor destruction of lithosphere. Thus, these boundaries are sometimes called **passive plate margins.** They occur along a special type of fault called a **transform fault,** which is simply a **strike-slip fault** between plates (that is, movement along it is horizontal and parallel to the fault). The term *transform* is used because the relative motion between plates can be changed—transformed—at different points along the fault. For example, the diverging motion between plates at an oceanic ridge can be transformed along the fault to the converging motion between plates at a subduction zone (*figure 17.20*).

Transform faults connect convergent and divergent plate boundaries in various combinations (*figure 17.20*). Where segments of the oceanic ridge have been offset, a transform fault connects the two divergent plate boundaries and creates a major topographic feature called a **fracture zone.** Fracture zones, however, are not what they might seem to be at first sight. The apparent offset of the oceanic ridge may suggest a simple strike-slip fault with displacements of thousands of kilometers. But one must keep in mind the relative motion of the plates at the spreading center. Careful study of *figure 17.21* shows that relative motion between the plates occurs only in the area between the offset segments of the ridge. This zone is the only place

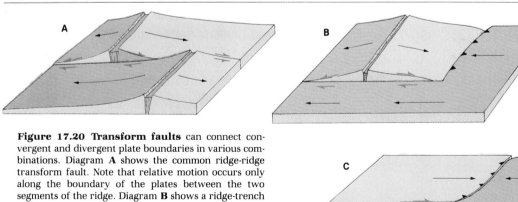
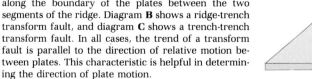
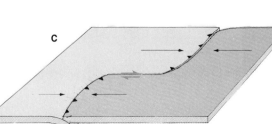

Figure 17.20 Transform faults can connect convergent and divergent plate boundaries in various combinations. Diagram **A** shows the common ridge-ridge transform fault. Note that relative motion occurs only along the boundary of the plates between the two segments of the ridge. Diagram **B** shows a ridge-trench transform fault, and diagram **C** shows a trench-trench transform fault. In all cases, the trend of a transform fault is parallel to the direction of relative motion between plates. This characteristic is helpful in determining the direction of plate motion.

where the fault forms a boundary between the plates. Beyond this zone, the plates on either side of the fracture are moving in the same direction and at the same rate and can be considered to be linked together. Note that the oceanic ridge is *not* offset by motion along the transform fault. It was offset previously and probably represents an old line of weakness. Relative movement and seismic activity occur only in the part of the fracture connecting the two segments of the oceanic ridge. Beyond this region, the fracture is inactive. Moreover, no volcanic activity is associated with transform faults. As the plates slide past each other, however, their boundaries are fractured and broken. This fracturing produces parallel ridges and troughs along the fault zone.

Transform faults can also join ridges to trenches and trenches to trenches. In all cases, transform faults are parallel to the direction of relative plate motion, so that neither divergence nor convergence occurs along this type of boundary. As a result, no rocks are generated, volcanism does not occur, and plates are neither enlarged nor destroyed. The plates slide passively along the fracture system, and their movement produces only fracturing and seismic activity.

Figure 17.21 The relative movement of plates at a ridge-ridge transform fault changes along the trend of the fracture zone. The plates are moving away from the ridge, but the relative motion between the plates along the transform faults depends on the position of the spreading center. Along an active fault, between two segments of the ridge, plates on opposite sides of the fault move in opposite directions. Beyond the spreading centers, however, the plates move in the same direction on both sides of the fault, with no relative motion along the fault plane.

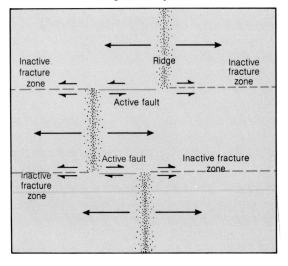

PLATE MOTION

Statement

The motion of a series of rigid plates on a sphere can be complex. Tectonic plates move as independent units, in different directions and at different velocities. Each plate is like a curved piece of shell moving on a sphere. Because of the curvature of the sphere, different parts of a plate move at different velocities. The motion of each plate can be described in terms of a pole of rotation.

Discussion

The geometry of a curved plate moving on a sphere was worked out 200 years ago by the Swiss mathematician Leonhard Euler (1707–1783). The basic analysis of this type of motion is illustrated in *figure 17.22*. In the figure, the motion of plate 1 with respect to plate 2 is a rotation around the axis A_R (called the **spreading axis**), one pole of which is the point P (the **spreading pole**). Note that the pole of the plate rotation is completely independent of the earth's axis and has no relation to the magnetic poles.

Several important facts about plate motion are immediately apparent from *figure 17.22*. First, different parts of a plate move with different velocities. Maximum velocity occurs at the equator of rotation, and minimum velocity at the poles. This may best be understood by considering a plate so large that it covers an entire hemisphere. All motion occurs around the axis of rotation. The pole of rotation has zero velocity, because it is a fixed point around which the hemispheric shell moves. Points Q, R, and S have progressively higher velocities, with a maximum velocity at point T, which lies on the equator of rotation.

Also, note that transform faults lie on lines of latitude perpendicular to the axis of rotation and that the length of a fracture zone is longest at the equator and shorter toward the poles. These conditions essentially hold for most transform faults in nature, as can be seen on a physiographic map of the Atlantic (see the endpapers of this book). Thus, we can use the geometric relationship between transform faults and spreading poles to locate the pole of rotation for each plate.

The direction of movement of the major plates in relation to their neighbors can be determined in several ways. As we have seen, the trends of the oceanic ridge and the associated fracture zones are related to the orientation of the spreading poles. In-

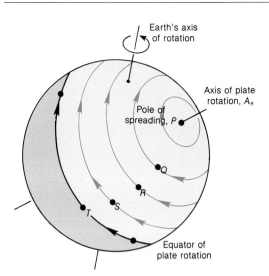

(A) *Plate motion can be easily understood by considering a plate that covers an entire hemisphere. Each point on the plate would move along lines of latitude with respect to the pole of spreading (P).*

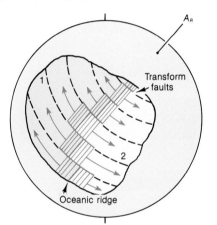

(B) *The motion of plate 1 with respect to plate 2 also can be described as rotation around some imaginary axis extending through the earth. A segment of the oceanic ridge lies on a line of longitude that passes through the pole of that axis, and transform faults lie on lines of latitude with respect to that pole. The amount and rate of spreading are at a maximum at the equatorial line and are zero at the pole.*

Figure 17.22 Plate motion on a sphere requires that the plates rotate around an axis of spreading, whose pole is called a pole of spreading. Plates always move parallel to the fracture zones and along circles of latitude perpendicular to the spreading axis.

dications of movement are also drawn from seismic data (see page 347), from relative ages of different regions of the sea floor, and from the ages of chains of volcanic islands and seamounts (see pages 365–66). From these data, geologists have determined the motion of the present tectonic plates. This motion is summarized in *figure 17.14.*

The Pacific plate is moving in a generally northwesterly direction from the eastern Pacific rise toward the system of trenches in the western Pacific. It is bordered by several plates along the subduction zone, so that the relative motion at each of the trenches differs from the general trend. The American plates are moving westward from the mid-Atlantic ridge, converging with the Pacific, Cocos, and Nazca plates. The Australian plate is moving northward.

Africa and Antarctica, however, present a different situation. Both are nearly surrounded by spreading centers, and they have no associated subduction zones to accommodate the new lithosphere generated along the spreading centers. There is some northward movement of Africa toward the convergent boundary in the Mediterranean area, but this does not accommodate the east-west spreading from the Atlantic and Indian ridges. Apparently, the African and Antarctic plates are being enlarged as new lithosphere is generated at their margins. Without subduction zones, the spreading centers surrounding these plates must be moving outward in relation to Africa and Antarctica. The African and Antarctic plates illustrate a very important point: *plate margins are not fixed but can move as much as the plates themselves can.* As is shown in *figure 17.23*, if two divergent plate margins are not separated by a subduction zone, new lithosphere is formed at each spreading center, but none is destroyed between them. The plate between the spreading centers is continually enlarged, so that the spreading centers themselves must move apart.

As a spreading center moves toward a subduction zone with the migrating plate, both the spreading center and the subduction zone are ultimately destroyed, and a single plate is formed where once there were three (*figure 17.23*).

Subduction zones also are temporary features, as an active zone can be abandoned and a new one created in a different position (*figure 17.24*). This process has apparently occurred several times in the western Pacific, where double trenches are common.

Another important change is in the lengths of plate margins. An oceanic ridge is essentially a fracture in the lithosphere. Besides growing larger with time, it can grow longer. A good example is the spreading center in the Atlantic Ocean. It has grown considerably since spreading began to separate South America from Africa.

From the foregoing discussion, it is apparent that plate margins are not permanent but can be displaced and migrate to different positions. As a plate margin moves, the shape and configuration of the plate can be changed, and in many cases, the plate itself is ultimately destroyed.

Figure 17.23 The migration of a spreading center can result in the ultimate destruction of both the spreading center and a subduction zone. If a spreading center descends into a subduction zone, it is consumed with the descending plate. The remaining plates are sutured together to form a single plate where once there were three.

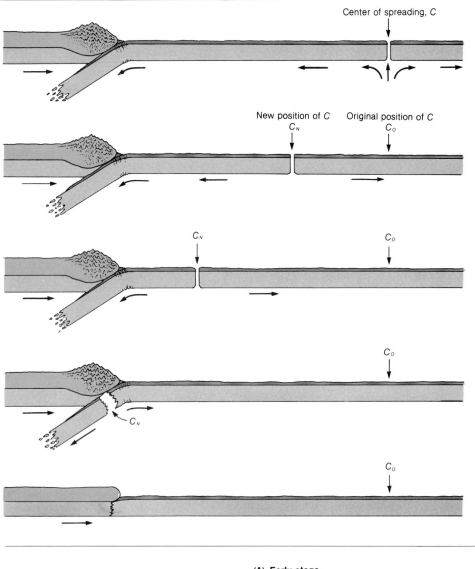

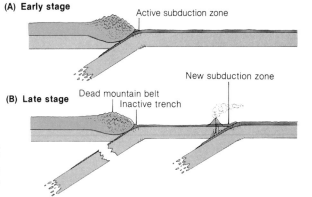

(A) Early stage

(B) Late stage

Figure 17.24 A subduction zone can be abandoned and a new one created elsewhere as a result of variations in stress in the lithosphere. This would leave an inactive trench, a dead mountain belt, and an isolated fragment of the descending lithosphere.

CONVECTION CURRENTS IN THE MANTLE

Statement

Various sources of energy have been proposed to explain convection in the mantle. Some geologists believe that the energy is associated with either the earth's rotation, the earth's tides, or heat from the core. Most advocates of plate tectonic theory, however, believe that currents are driven by radiogenic heat produced in the asthenosphere.

Discussion

Little is known about the actual driving force of plate tectonics. The problem is that, although we see the effects of the moving plates at the surface, the ultimate cause of motion lies in the mantle and is not accessible for direct measurement. The source of energy is basically thermal, and the mechanism is convection. For most geologists, the debate is not about whether convection occurs in the mantle, but about its scale and periodicity. Do **convection currents** involve the entire mantle or just the upper part? Is the process continuous, involving cycle after cycle, or does it occur only for a few hundred million years when temperature builds up to a critical point? Do **convection cells** grow and die? Do they shift positions? We do not know the answers to these questions, but some models have been proposed to explain the observed facts. Some of these models are shown in *figure 17.25*.

In *figure 17.25A*, the entire thickness of the mantle is considered to be involved in a series of convection cells. Where the currents rise and move laterally apart, they initiate horizontal motion of the lithosphere, which splits and moves away from the oceanic ridge. Where convection cells descend, they drag the lithosphere down into the mantle.

Figure 17.25B shows a similar convection system limited to the asthenosphere. According to this model, the heavy elements are concentrated in the core, and less dense elements are concentrated in the silicate minerals of the mantle. It is believed that, during the process of differentiation, radioactive elements were incorporated in the minerals of the asthenosphere. As these elements decay, they release **radiogenic heat.** Rocks within the asthenosphere begin to melt and yield to plastic flow. The less dense, partially melted material rises in a convection system. Thus, the internal energy is car-

ried to the surface and escapes. The lithosphere moves with convection cells in the manner already described.

Whatever type of convection system may exist, it is undoubtedly much more complex than those represented by the simple diagrams in *figure 17.25*. It involves melting, upwelling, horizontal movement, and sinking. Secondary eddies probably exist. Local plumes possibly rise from hot spots. Motion is probably jerky as a result of friction along plate boundaries.

The energy of this system was inherited from the early phases of the formation of the earth and is steadily diminishing. The system is irreversible. Energy within the earth, like that in the sun, will ultimately be spent, and the lithospheric plates will cease to move. When this happens, in the far-distant future, mountains will no longer form, and volcanic activity and earthquakes will end. The earth will then be a dead planet, changed only by energy from external sources.

Figure 17.25 Possible convection systems in the mantle include large, deep convection cells involving the entire mantle (diagram **A**) and shallow convection cells largely restricted to the asthenosphere (diagram **B**).

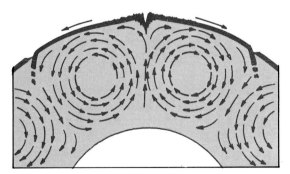

(A) Deep convection cells may involve the entire mantle.

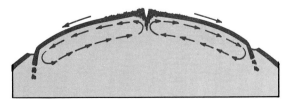

(B) Shallow convection cells may involve only the asthenosphere.

SUMMARY

The idea of drifting continents was proposed in the early 1900s. It was best developed by Alfred Wegener in his book *The origin of continents and oceans*. The theory of continental drift was supported by various types of geologic evidence, including (1) matching geologic features in the several continents, (2) paleontologic findings, (3) Paleozoic glaciation, and (4) paleoclimatic evidence. Without an understanding of the oceanic crust, however, a complete theory of the earth's dynamics could not be developed.

During the 1960s, new information about paleomagnetism and the topography of the ocean floor led to the development of the plate tectonic theory. Since then, the theory has been supported by a wide variety of geologic and geophysical data from both the continents and the oceans. Paleomagnetic studies show that the continents have changed their positions in relation to the magnetic poles and that each continent has followed a different path with respect to the poles throughout geologic time. These findings indicate that the continents have moved with respect to each other. Paleomagnetic reversals in the rocks of the sea floor occur in symmetrically matching sets on both sides of the oceanic ridge, a fact explained by sea-floor spreading.

The theory of plate tectonics explains the earth's crustal dynamics by the movement of rigid lithospheric plates due to convection in the asthenosphere. New crust is created at oceanic ridges, where plates split and spread apart and basaltic magma wells up in the rift. Oceanic crust is consumed where a plate descends into the mantle at a subduction zone.

The major structural features of the earth are formed along plate boundaries. At divergent plate boundaries, the lithosphere is under tension, and the major geologic processes are (1) rifting and block faulting, (2) the generation of basaltic magma and the formation of new oceanic crust, and (3) rifting of continents. At convergent plate boundaries, the important geologic processes include (1) subduction, (2) partial melting of the descending plate and the generation of granitic magma, and (3) compression and mountain building. Transform fault boundaries are zones where plates slide passively past each other.

Each plate is part of a curved shell on a sphere, so that different parts of the plate move at different velocities. The direction of plate motion can be determined from (1) trends of spreading centers and fracture zones, (2) seismic data, and (3) ages of the sea floor on either side of the ridge and ages of island chains. Plates move from 1 to 16 cm per year.

The driving force for plate motion is believed to be slow convection in the asthenosphere. As plates collide, diverge, and slip past each other, they generate earthquakes, and it is the study of earthquakes that tells us most about present-day plate motion. This subject is discussed in the following chapter.

Key Words

tectonics	tectonic plate
plate tectonics	divergent plate boundary
continental drift	spreading center
Pangaea	convergent plate boundary
mid-Atlantic ridge	subduction zone
rift valley	subduction
sea-floor spreading	transform fault
paleomagnetism	passive plate margin
magnetic anomaly	fracture zone
magnetic reversal	spreading axis
normal polarity	spreading pole
reversed polarity	convection current
polarity epoch	convection cell
polarity event	radiogenic heat

Review Questions

1. Discuss the paleontologic evidence supporting the theory of continental drift.
2. How do structural features and rock types along continental margins support the theory of continental drift?
3. Explain how late Paleozoic glaciation in the southern continents supports the theory of continental drift.
4. How does paleoclimatic evidence support the theory of continental drift?
5. Explain how permanent magnetism comes to be locked into lavas.
6. How does rock magnetism support the theory of continental drift?
7. Explain the origin of magnetic anomalies on the sea floor.
8. Make a simple sketch map showing magnetic patterns on the sea floor.
9. How do the age and distribution of sediment on the sea floor support the theory of plate tectonics?
10. Make a sketch map showing the boundaries of the major plates.

11. Describe the types of plate boundaries and give an example of each.
12. List the major processes that occur along each type of plate boundary.
13. Make a simple sketch map of a part of the oceanic ridge and draw arrows to show the relative motion along ridge-to-ridge transform faults.
14. Describe the geometry of lithospheric plate motion over the globe.
15. How is the pattern of a series of transform faults governed by the pole of spreading?
16. Explain how plate margins can migrate as well as the plate itself can.
17. Make a series of sketches showing how a subduction zone can be terminated.

Additional Readings

Anderson, D. L. 1971. The San Andreas Fault. *Scientific American* 225(5):52–68. Reprint no. 896. San Francisco: W. H. Freeman.

Barazangi, M., and J. Dorman. 1969. World seismicity maps compiled from ESSA, Coast and Geodetic Survey, epicenter data, 1961–1967. *Bulletin of the Seismological Society of America* 59(1):369–80. (Precise plots of earthquake activity.)

Bullard, E. C., J. E. Everett, and A. G. Smith. 1965. The fit of the continents around the Atlantic. *Philosophical Transactions of the Royal Society of London,* ser. A, Mathematical and Physical Sciences, 258:27–41.

Calder, N. 1973. *The restless earth.* New York: Viking Press.

Condie, K. C. 1976. *Plate tectonics and crustal evolution.* Elmsford, N.Y.: Pergamon Press.

Cox, A., ed. 1973. *Plate tectonics and geomagnetic reversals.* San Francisco: W. H. Freeman.

Cox, A., G. B. Dalrymple, and R. R. Doell. 1967. Reversals of the earth's magnetic field. *Scientific American* 216(2):44–54.

Dewey, J. F. 1972. Plate tectonics. *Scientific American* 226(5):56–68. Reprint no. 900. San Francisco: W. H. Freeman.

Dewey, J. F., and J. M. Bird. 1970. Mountain belts and the new global tectonics. *Journal of Geophysical Research* 75(14):2625–47.

Dickinson, W. R. 1971. Plate tectonics in geologic history. *Science* 174:107–13.

Dietz, R. S., and J. C. Holden. 1970. The breakup of Pangaea. *Scientific American* 223(4):30–41.

Hallam, A. 1973. *A revolution in the earth sciences.* Oxford: Clarendon Press.

Heirtzler, J. R. 1968. Sea-floor spreading. *Scientific American* 219(6):60–70. Reprint no. 875. San Francisco: W. H. Freeman.

Hess, H. H. 1962. History of ocean basins. In *Petrologic studies: A volume in honor of A. F. Buddington,* ed. A. E. J. Engel et al., pp. 599–620. Boulder, Colo.: Geological Society of America.

Hurley, P. M. 1968. The confirmation of continental drift. *Scientific American* 218(4):52–64. Reprint no. 874. San Francisco: W. H. Freeman.

Isacks, B., J. Oliver, and L. R. Sykes. 1968. Seismology and the new global tectonics. *Journal of Geophysical Research* 73(18):5855–99.

Le Pichon, X. 1968. Sea-floor spreading and continental drift. *Journal of Geophysical Research* 73(12):3661–97.

McKenzie, D. P. 1972. Plate tectonics and sea floor spreading. *American Scientist* 60:425–35.

Marvin, U. B. 1973. *Continental drift.* Washington, D.C.: Smithsonian Institution Press.

Sullivan, W. 1974. *Continents in motion: The new earth debate.* New York: McGraw-Hill.

Takeuchi, H., S. Uyeda, and H. Kanamori. 1970. *Debate about the earth.* Rev. ed. San Francisco: Freeman, Cooper.

Vine, F. J. 1966. Spreading of the ocean floor: New evidence. *Science* 154(3744):1405–15.

Vine, F. J., and D. H. Matthews. 1963. Magnetic anomalies over oceanic ridges. *Nature* 199:947–49.

Wilson, J. T., compiler. 1970. *Continents adrift: Readings from* Scientific American. San Francisco: W. H. Freeman.

Wyllie, P. J. 1976. *The way the earth works.* New York: John Wiley and Sons.

York, D. 1975. *Planet Earth.* New York: McGraw-Hill.

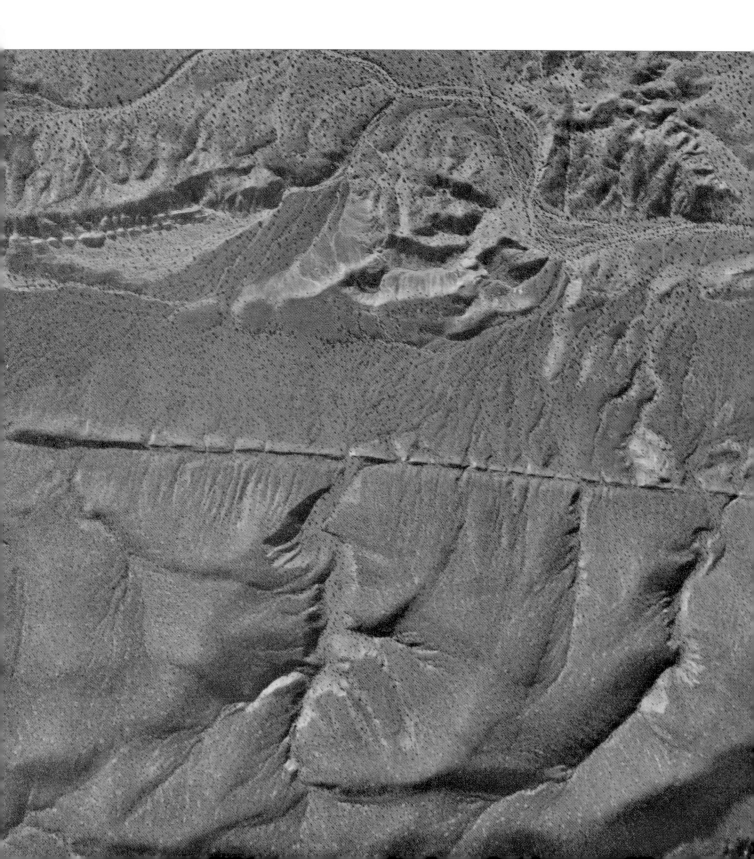

THE EARTH'S SEISMICITY

Perhaps more than any other phenomenon, earthquakes demonstrate that the earth continues to be a dynamic planet, changing each day by internal tectonic forces. Every year, more than 150,000 earthquakes are recorded by the worldwide network of seismic stations and are analyzed with the aid of computers at the earthquake data center in Boulder, Colorado. With this network, the exact location, depth, and magnitude of all detectable earthquakes are plotted on regional maps. Information about the direction of fault movement associated with the shock is also determined. As a result, we can literally monitor the details of present plate motion. Indeed, seismology provides some of the most convincing evidence supporting the theory of plate tectonics. But that is not all. Seismic waves also provide our most effective probe of the earth's interior, and they constitute the basic method of collecting data upon which our present concepts of the internal structure of the earth are based.

MAJOR CONCEPTS

1. Seismic waves are vibrations in the earth caused by the rupture and sudden movement of rocks strained beyond their elastic limits.
2. Three types of seismic waves are produced by an earthquake shock: (a) P waves, (b) S waves, and (c) surface waves.
3. The primary effect of an earthquake is ground motion. Secondary effects include (a) landslides, (b) tsunamis, and (c) regional or local uplift and subsidence.
4. Most earthquakes occur along plate boundaries. Divergent plate boundaries produce shallow-focus earthquakes. Convergent plate boundaries produce a zone of shallow-focus, intermediate-focus, and deep-focus earthquakes.
5. Infrequent shallow-focus earthquakes occur within plates, beyond the plate boundaries.
6. The velocities at which P waves and S waves travel through the earth indicate that the earth has a solid inner core, a liquid outer core, a thick mantle, a soft asthenosphere, and a rigid lithosphere. The most striking seismic discontinuity within the earth is between the core and the mantle.

CHARACTERISTICS OF EARTHQUAKES

Statement

Earthquakes are vibrations of the earth caused by the rupture and sudden movement of rocks that have been strained beyond their elastic limits. If a strained rock breaks, it then snaps into a new position and, in the process of rebounding, generates vibrations called seismic waves. These vibrations are somewhat analogous to those produced by a pebble dropped into a pool of water. The stone introduces energy into the system, and concentric waves spread out in all directions from the point of disturbance.

Land waves produced by an earthquake have been reported with heights of more than 0.5 m and wavelengths of 8 m. The period between the passage of wave crests can be as much as 10 seconds. The vibrations can continue for as long as an hour before the wave dies out. Three types of seismic waves are generated by an earthquake shock:

1. Primary waves (*P* waves), in which particles move back and forth in the direction in which the wave travels
2. Secondary waves (*S* waves), in which particles move back and forth at right angles to the direction in which the wave travels
3. Surface waves, which travel only in the outer layer of the earth and are similar to waves in water

Surface waves cause the most damage during an earthquake. Together with secondary effects from associated landslides, tsunamis, and fires, they result in the loss of approximately 10,000 lives and $100 million each year.

Discussion

Elastic-Rebound Theory

The origin of an **earthquake** can be illustrated by a simple experiment. Bend a stick until it snaps. Energy is stored in the elastic bending and is released if rupture occurs, causing the fractured ends to vibrate and send out sound waves. Detailed studies of active faults show that this model, known as the **elastic-rebound theory**, appears to hold for all major earthquakes (*figure 18.1*). Precision surveys across the San Andreas Fault in California show that railroads, fence lines, and streets are slowly deformed at first, as strain builds

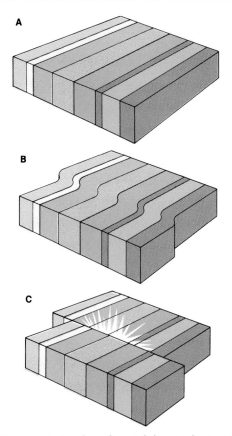

Figure 18.1 Earthquakes originate where rocks are strained beyond their elastic limits and rupture. **(A)** Rocks remain undeformed in seismically inactive areas. **(B)** Strain builds up in rocks in seismically active areas until the rocks rupture or move along preexisting fractures. **(C)** Energy is released when the rocks rupture, and seismic waves move out from the point of rupture.

up, and are offset when movement occurs along the fault, releasing the elastic strain. The San Andreas Fault is the boundary between the Pacific and North American plates. Its movement is horizontal, with the Pacific plate moving toward the northwest. On a regional basis, the plates move quite steadily at a rate of roughly 5 cm per year. Along the fault, movement can be smooth or it can occur in a series of jerks, because sections of the fault can be "locked" together until enough strain accumulates to exceed the **elastic limit** of the rock and cause displacement. The elastic-rebound theory explains earthquakes as the result of either rupture and sudden movement along a fault or recurrent movement along existing fractures.

The point within the earth where the initial slippage generates earthquake energy is called the **focus.** The point on the earth's surface directly

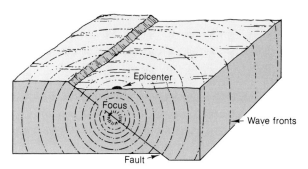

Figure 18.2 The relationship between an earthquake's focus, its epicenter, and seismic wave fronts is depicted in this diagram. The focus is the point of initial movement on the fault. Seismic waves radiate from the focus. The epicenter is the point on the earth's surface directly above the focus.

depth of 70 km. They occur in all seismic belts and produce the largest percentage of earthquakes. **Intermediate-focus earthquakes** occur between 70 and 300 km below the surface, and **deep-focus earthquakes** between 300 and 700 km. Both intermediate-focus and deep-focus earthquakes are limited in number and distribution. Generally, they are confined to convergent plate margins. The maximum energy released by an earthquake tends to become progressively smaller as the depth of focus

Figure 18.3 Motion produced by the various types of seismic waves can be illustrated by the distortions they produce in a straight fence line.

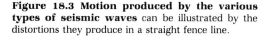

(A) Prior to seismic disturbance. *A straight fence line provides a good reference marker for future movement.*

above the focus is called the **epicenter** (*figure 18.2*).

Types of Seismic Waves

Three major types of **seismic waves** are generated by an earthquake shock (*figure 18.3*). Each type travels through the earth at a different speed. Therefore, each arrives at a **seismograph** hundreds of kilometers away at a different time. The first waves to arrive are called **primary waves (P waves)**. These are a kind of **longitudinal wave**, identical in character to sound waves passing through a liquid or gas. The particles involved in these waves move forward and backward in the direction of wave travel, causing relatively small displacements. The next waves to arrive are called **secondary waves (S waves)**. In these, particles oscillate back and forth at right angles to the direction of wave travel. *S* waves cause strong movements to be recorded on a seismograph. The last waves to arrive are **surface waves,** which travel relatively slowly over the earth's surface. Particles involved in surface waves move in an orbit similar to that of particles in water waves.

Occurrence

The location of an earthquake focus is important in the study of plate tectonics, because it indicates the depth at which rupture and movement occur. Although movement of material within the earth occurs throughout the entire mantle and core, earthquakes are concentrated in the upper 700 km.

Within the 700-km range, earthquakes can be grouped according to depth of focus. **Shallow-focus earthquakes** occur from the surface to a

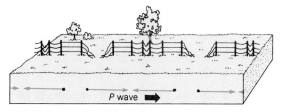

(B) Motion produced by a P wave. *Particles are compressed and then are expanded in the line of wave progression. P waves can travel through any earth material.*

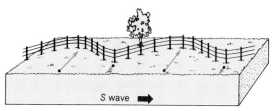

(C) Motion produced by an S wave. *Particles move back and forth at right angles to the line of wave progression. S waves travel only through solids.*

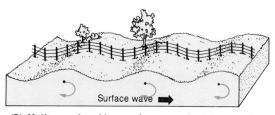

(D) Motion produced by a surface wave. *Particles move in a circular path at the surface. The motion diminishes with depth, like that produced by surface waves in the ocean (see figure 15.3).*

increases. Also, seismic energy from a source deeper than 70 km is largely dissipated by the time it reaches the surface. Most large earthquakes have a shallow focus, originating in the crust. The location of the focus of an earthquake is calculated from the time that elapses between the arrivals of the three major types of seismic waves.

The method of locating the epicenter of an earthquake is relatively simple and can easily be understood by reference to *figure 18.4*. The *P* wave, traveling faster than the *S* wave, is the first to be recorded at a seismic station. The time interval between the arrival of the *P* wave and the arrival of the *S* wave is a function of the station's distance from the epicenter. By tabulating the travel times of *P* and *S* waves from earthquakes of known sources,

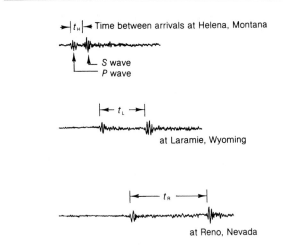

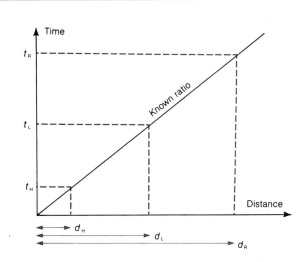

(A) *The greater the distance between the seismic event and a recording seismograph station, the more time it takes for the first waves to arrive. Also, the greater the distance, the longer the interval between the arrival of the P waves and the arrival of the S waves.*

(B) *The time between the arrival of the P waves and the arrival of the S waves is correlated with the distance between the seismic event and the recording station. For example, time at Helena, t_H yields distance from Helena, d_H.*

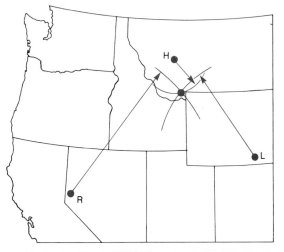

(C) *The direction of the event from any single station is not known, but by simply plotting the intersection of three arcs, whose radii are the respective distances from three stations, a common point is found. That point lies at the epicenter of the seismic event.*

Figure 18.4 Locating the epicenter of an earthquake is accomplished by comparing the arrival times of *P* waves and *S* waves at three seismic stations.

Table 18.1 Scale of Earthquake Intensity

I Not felt except by very few people under special conditions. Detected mostly by instruments.

II Felt by a few people, especially those on upper floors of buildings. Suspended objects may swing.

III Felt noticeably indoors. Standing automobiles may rock slightly.

IV Felt by many people indoors, by a few outdoors. At night, some are awakened. Dishes, windows, and doors rattle.

V Felt by nearly everyone. Many are awakened. Some dishes and windows are broken. Unstable objects are overturned.

VI Felt by everyone. Many people become frightened and run outdoors. Some heavy furniture is moved. Some plaster falls.

VII Most people are in alarm and run outside. Damage is negligible in buildings of good construction, considerable in buildings of poor construction.

VIII Damage is slight in specially designed structures, considerable in ordinary buildings, great in poorly built structures. Heavy furniture is overturned.

IX Damage is considerable in specially designed structures. Buildings shift from their foundations and partly collapse. Underground pipes are broken.

X Some well-built wooden structures are destroyed. Most masonry structures are destroyed. The ground is badly cracked. Considerable landslides occur on steep slopes.

XI Few, if any, masonry structures remain standing. Rails are bent. Broad fissures appear in the ground.

XII Virtually total destruction. Waves are seen on the ground surface. Objects are thrown into the air.

caused to buildings, dams, bridges, and other structures, as reported by witnesses. An intensity scale commonly used in the United States and a description of some of its criteria are presented in *table 18.1.*

The intensity of an earthquake depends on a number of factors. Foremost among these are (1) the total amount of energy released, (2) the distance from the epicenter, and (3) the type of rock and degree of consolidation. In general, wave amplitude and destruction are greater in soft, unconsolidated material than in dense, crystalline rock. The intensity is greatest close to the epicenter.

Magnitude

The magnitude of an earthquake is a measure of the amount of energy released. It is a much more precise measure than intensity. Earthquake magnitudes are based on direct measurements of the size (amplitude) of seismic waves, made with recording instruments, rather than on subjective observations of destruction. The total energy released by an earthquake can be calculated from the amplitude of the waves and the distance from the epicenter. Seismologists express magnitudes of earthquakes using the Richter scale, which arbitrarily assigns 0 to the lower limit of detection. Each step on the scale represents an increase in amplitude by a factor of 10 (*table 18.2*). Thus, the vibrations of an earthquake with a magnitude of 2 are 10 times greater in amplitude than those of an earthquake with a magnitude of 1, and the vibrations of an earthquake with a magnitude of 8 are 1 million times greater in amplitude than those of an earthquake with a magnitude of 2.

The largest earthquake ever recorded had a magnitude of approximately 8.5 on the Richter

seismologists have constructed time-distance graphs that can be used to determine the distance to the epicenter of a new quake. The seismic records indicate the distance, but not the direction, to the epicenter. Therefore, records from at least three stations are necessary to determine the precise location of the epicenter.

Intensity

The intensity, or destructive power, of an earthquake is an evaluation of the severity of ground motion at a given location. It is measured in relation to the effects of the earthquake on human life. Generally, it is described in terms of the destruction

Table 18.2 Richter Scale of Earthquake Magnitude

Magnitude	Approximate Number per Year
1	700,000
2	300,000
3	300,000
4	50,000
5	6,000
6	800
7	120
8	20
>8	1 every few years

scale. Significantly larger earthquakes are not likely to occur, because rocks are not strong enough to accumulate more energy.

Effects of Earthquakes

The primary effect of earthquakes is the violent ground motion accompanying movement along a fracture. This motion can shear and collapse buildings, dams, tunnels, and other rigid structures. Secondary effects include landslides, tsunamis, and regional or local submergence of the land. The following are a few examples of well-documented earthquakes in historical times.

The earthquake that shook Peru on May 31, 1970, had a magnitude of 7.8. Claiming the lives of 50,000 people, it was the deadliest earthquake in Latin American history. Much of the damage to towns and villages was caused by the collapse of adobe buildings, which are easily destroyed by ground motion. Eighty percent of the adobe houses in an area of 65,000 km² were destroyed. Vibrations caused most of the destruction to buildings, but the massive landslides triggered by the quake in the Andes were a second major cause of fatalities. A huge debris flow buried 90% of the resort town of Yungay, with a population of 20,000, in a matter of a few minutes. Similar earthquake-triggered debris flows killed 41,000 in Ecuador and Peru in 1797 and, in 1939, killed 40,000 in Chile.

In 1960, an earthquake in Chile caused extensive damage from ground motion, landslides, and flooding. It also produced a spectacular tsunami, which devastated seaports with a series of waves 7 m high. This tsunami crossed the Pacific at approximately 1000 km per hour and built up to 11 m high at Hilo, Hawaii. Twenty-two hours after the quake, it reached Japan, causing damage to property worth $70 million.

The earthquake that devastated southern Alaska late in the afternoon of March 27, 1964, was one of the greatest tectonic events of modern times. Its magnitude was between 8.30 and 8.75, and its duration ranged from 3 to 4 minutes at the epicenter (by comparison, the San Francisco earthquake of 1906 lasted about 1 minute). The Alaskan earthquake was important to geologists because it had such marked effects on the earth's surface and because perhaps no other earthquake has been better documented. Despite its magnitude and severe tectonic effects, this quake caused far less property damage and loss of life than other national disasters have (114 lives were lost and property worth $311 million was damaged), because, fortunately, much of the affected area was uninhabited. The crustal deformation associated with the Alaskan earthquake was the most extensive ever documented. The level of the land was changed in a zone 1000 km long and 500 km wide (an area of 500,000 km² was elevated or depressed). Submarine and terrestrial landslides triggered by the earthquake caused spectacular damage to communities, and the shaking spontaneously liquefied deltaic materials along the coast, causing slumping of the waterfronts of Valdez and Seward.

The San Francisco earthquake of 1906 is one of the most famous in history. It lasted only a minute but had a magnitude of 8.3. The fire that followed caused most of the destruction (an estimated cost of $400 million in damage and a reported loss of 700 lives). From a geological point of view, the earthquake was important because of the visible effects it produced along the San Andreas Fault. Horizontal displacement occurred over a distance of about 400 km and offset tiny roads, fences, and buildings by as much as 7 m.

The Tangshan quake, which shook China on July 28, 1976, was probably the second most devastating earthquake in history. (The most destructive also occurred in China, in 1556.) The enormous shock registered 8.2 on the Richter scale. Although Chinese authorities have withheld many details about this disaster, scientists from Mexico and other countries were recently allowed to investigate the affected area. They say that Tangshan looked like Hiroshima after the explosion of the atomic bomb. The devastation of the city of 1 million was literally complete. The Tangshan quake killed 750,000 people.

In addition to the descriptions of awesome damage, there have been some interesting reports about the phenomenon itself. Just before the earth began to shudder, residents were awakened by a brilliant incandescent light, which lit the early morning sky for hundreds of kilometers around. The glow reportedly was predominently red and white. When the quake struck, many people were catapulted into the air, some as high as 2 m, by what was described as a violent, hammerlike subterranean blow. Trees and crops were blown over as if by a colossal steamroller, and leaves were burned to a crisp. In the quake's aftermath, many bushes were scorched on one side only. Chinese scientists were aware of some preliminary warning signs and announced that an earthquake would occur that year in either Tangshan or Peking. Subsequent data were so conflicting and confusing, however, that no final specific prediction was made.

EARTHQUAKE PREDICTION

Statement

The preceding examples show how vulnerable the works of man are to the earth's vibrations. They provide a vivid reminder of the threat earthquakes pose to the world's heavily populated areas. The usefulness of effective earthquake prediction is obvious. Yet pinpoint earthquake prediction, which seemed so close at hand during the 1960s, is proving to be an elusive goal.

Discussion

Chinese scientists claim to have been successful in predicting about 15 earthquakes in recent years. They rely heavily on the centuries-old idea that animals sense various underground changes prior to an earthquake and behave abnormally. Most of the bizarre behavior is simply increased restlessness. Cattle, sheep, and horses refuse to enter their corrals. Rats leave their hideouts and march fearlessly through houses. Shrimp crawl on dry land. Ants pick up their eggs and migrate en masse. Fish jump above the surface of the water, and rabbits hop aimlessly about.

Chinese scientists successfully predicted the Haicheng earthquake, in February 1975, by means of unusual animal behavior. The most intriguingly bizarre behavior occurred in mid-December, when snakes came out of hibernation and froze to death on the icy ground, and groups of rats appeared and scurried about in the cold winter weather. These events were followed by a swarm of small earthquakes at the end of December. During January, Chinese scientists received thousands of reports of unusual behavior, especially in larger animals in the area that proved to be the quake epicenter.

One hypothesis offered to explain abnormal animal behavior before an earthquake is that certain animals are sensitive to small variations in the earth's magnetic field or sounds produced by microfractures prior to the larger event. Animal sensors that detect light, sound, order, touch, and temperature are well known, and they may have the ability to detect subtle changes in other physical phenomena.

Much of the work on earthquake prediction in the United States has been based on the dilatancy

Figure 18.5 Seismic gaps are important in earthquake forecasting. Areas along plate margins that are not seismically active are believed to be building up stress and may be sites of significant seismic activity in the future.

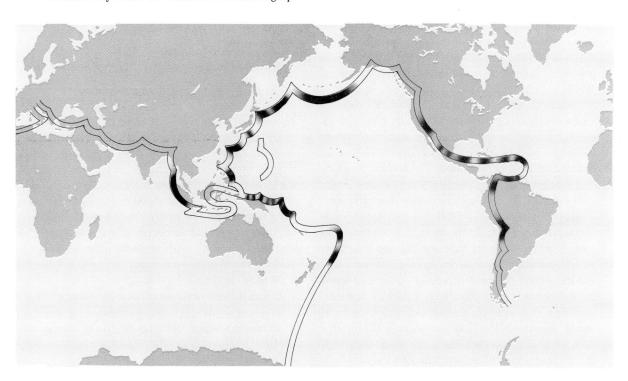

theory. Laboratory and field studies in recent years indicate that a rock subjected to stress swells just before it ruptures. This dilation is caused by the opening and extension of numerous tiny cracks, and it begins at levels of stress that are about half as great as those needed to break the rock. As rock dilates, changes occur in certain physical characteristics, such as electrical resistance, seismic wave velocities, and magnetic properties. Geologists therefore attempt to monitor uplift and tilting of the ground, electrical resistivity, the number of seismic events, and ground-water pressure.

For a prediction to benefit the populace, it must specify the time, the location, and the magnitude of the coming quake. Such accuracy is proving to be difficult to achieve. The problem is not so much that the dilatancy theory is wrong but rather that it is inadequate. Different kinds of earthquakes apparently have different kinds of precursors. Instead of attempting to predict the time, place, and magnitude of an expected earthquake, geologists are now concentrating on the more modest goal of forecasting what areas of the world may be most susceptible to significant quakes. A major contribution to forecasting has been the compilation of a map showing the seismic potential of the world's major tectonic plate boundaries (*figure 18.5*). This map essentially plots locations along the plate boundaries of the Pacific where major quakes are most likely to occur in the near future. Along the plate margins are several gaps in seismic activity, where stress may be building up to a critical level. The most susceptible areas are those where major tremors have occurred in the past, but not within the last 100 years. These include such heavily populated areas as southern California, central Japan, central Chile, Taiwan, and the west coast of Sumatra. These areas appear likely to experience a major earthquake (magnitude of 7.0 or greater) in the next few decades. One such gap along the western coast of Mexico was filled by the major quake that struck the area on November 29, 1979.

If a reliable earthquake prediction system could give from 1 to 10 years' advance warning of a "killer" quake, what would be the appropriate social response? Would the usual flow of mortgage money be terminated? New earthquake insurance would certainly become unavailable. What about fire insurance, business expansion, unemployment, tax revenues, and demands on local government? The fear that false alarms would lead to adverse social response has led some to call for the withholding of predictions until prediction techniques are perfected to absolute certainty.

EARTHQUAKES AND PLATE TECTONICS

Statement

In 1966, a worldwide network of sensitive seismic stations was established to monitor nuclear testing. As a by-product, it has enabled seismologists to compile an amazing amount of information concerning earthquakes and plate tectonics. The data are analyzed with the aid of computers at the seismic data center in Boulder, Colorado. There, the location and magnitude of the thousands of earthquakes occurring each year are plotted on regional maps. Other information, such as the direction of displacement on faults where earthquakes have occurred, is also collected. The result is a new and important insight into the details of current plate motion. This is what has been found:

1. The distribution of earthquakes delineates plate boundaries.
2. Shallow-focus earthquakes coincide with the crest of the oceanic ridge and with transform faults between ridge segments.
3. Earthquakes at convergent plate margins occur in a zone inclined downward beneath the adjacent continent or island arc.
4. Fault motion associated with earthquakes along plate margins shows the present direction of plate motion.

Discussion

Global Patterns of the Earth's Seismicity

Tens of thousands of earthquakes have been recorded since the establishment of a worldwide network of seismic observation stations. Their locations and depths are summarized in the seismicity map in *figure 18.6*. From the standpoint of the earth's dynamics, this map is an extremely significant compilation, because it shows where and how the crust of the earth is moving at the present time. Two facts about the earth's seismicity are obvious: (1) most earthquakes occur in narrow belts that coincide precisely with tectonic plate boundaries, and (2) the type of seismicity is governed by the type of plate boundary.

If a sufficient number of seismic stations record a quake, seismologists can determine the direction of movement along the fault when the earthquake occurred. This information greatly enhances our knowledge of plate motion.

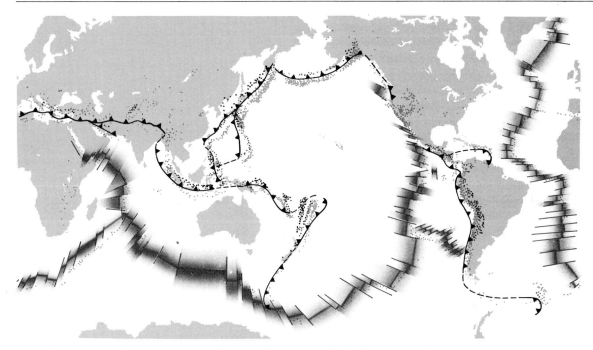

Figure 18.6 The earth's seismicity is clearly related to plate margins. This map shows the location of tens of thousands of earthquakes that occurred between 1961 and 1967. Shallow-focus earthquakes occur at both divergent and convergent plate margins, whereas intermediate-focus and deep-focus earthquakes are restricted to the subduction zones of converging plates.

Seismicity at Divergent Plate Boundaries. The global patterns of the earth's seismicity show a narrow belt of shallow-focus earthquakes coinciding almost precisely with the crest of the oceanic ridge and marking the boundaries between diverging plates. This zone is remarkably narrow in comparison to the zone of seismicity that follows the trends of young mountain belts and island arcs. The shallow earthquakes along divergent plate boundaries are less than 70 km deep and typically are small in magnitude. Although the zone looks like a nearly continuous line on regional maps, two types of seismic boundaries can be distinguished on the basis of fault motion. These are (1) spreading centers and (2) transform faults (*figure 18.7*). Earthquakes associated with the crest of the oceanic ridge occur within or near the rift valley. They appear to be associated with normal faulting and intrusions of basaltic magmas. Locally, earthquakes occur in swarms. Detailed studies indicate that earthquakes associated with the ridge crest are produced by vertical faulting, a process that appears to be responsible for the ridge topography (see pages 393–94).

Shallow-focus earthquakes also follow the transform faults that connect offset segments of the ridge crest, but they generally are not associated with volcanic activity. Studies of fault motion indicate horizontal displacement in a direction away

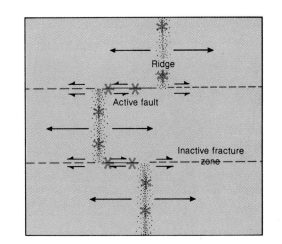

Figure 18.7 The distribution of earthquakes along divergent plate boundaries shows that seismicity along spreading centers results from normal faulting. Seismicity along transform faults results from strike-slip movement.

from the ridge crest. Moreover, as is predicted by the plate tectonic theory, earthquakes are restricted to the active transform fault zone—the area between ridge axes—and do not occur in inactive fracture zones.

Seismicity at Convergent Plate Boundaries.

The most widespread and intense earthquake activity on the earth occurs along subduction zones, at convergent plate boundaries. This belt of seismic activity is immediately apparent from the world seismicity map (*figure 18.6*), which shows a strong concentration of shallow, intermediate, and deep earthquakes coinciding with the subduction zones of the Pacific Ocean. The three-dimensional distribution of earthquakes in this belt defines a seismic zone that is inclined at moderate to steep angles from the trenches and extends down under the adjacent island arcs or continental borders. This distribution is well illustrated in the Tonga Trench, in the South Pacific. The zone of seismicity there forms a nearly planar surface that plunges into the mantle to a depth of more than 600 km (see *figure 18.8*).

Studies of fault motions from seismic waves generated in this zone indicate that the type of faulting varies with depth. Near the walls of the trench, normal faulting is typical, resulting from tensional stresses generated by the initial bending of the plate. In the zone of the shallow earthquakes, thrust faulting dominates as the descending lithosphere slides beneath the upper plate. At intermediate depths, extension or compression can occur, depending on the specific characteristics of the subduction zone. Extension and normal faulting can occur when a descending slab, which is denser than the surrounding mantle, sinks under its own weight. Compression results if the mantle resists the downward motion of the descending plate. The zone of deep earthquakes shows compression within the descending slab of lithosphere, indicating that the mantle material at that depth resists the movement of the descending plate.

Intraplate Seismicity.

Although most of the world's seismicity occurs along plate boundaries, the continental platforms also experience infrequent and scattered shallow-focus earthquakes. The zones of seismicity of East Africa and the western United States are most striking. They are probably associated with spreading centers, which can be projected into those regions. The minor shallow earthquakes in the eastern United States and Australia are more difficult to explain. Ap-

Figure 18.8 Earthquake foci in the Tonga region of the South Pacific occur in a zone inclined from the Tonga Trench toward the Fiji Islands. The map shows the areal distribution of foci, with focal depths represented by different colored dots. The cross section shows how the seismic zone is inclined away from the trench. This seismic zone accurately marks the boundary of the descending plate in the subduction zone.

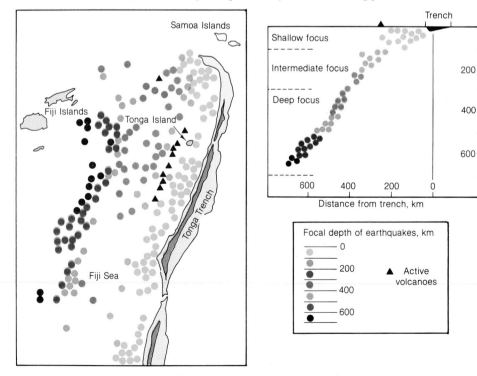

parently, lateral motion of a plate across the asthenosphere involves slight vertical movement. Built-up stress can exceed the strength of the rocks within the lithospheric plate, causing infrequent faulting and seismicity. In contrast, the ocean floors beyond the spreading centers are seismically inactive, except for isolated earthquakes associated with oceanic volcanoes.

Plate Motion as Determined from Seismicity

The data points in *figure 18.6* outline the seven major lithospheric plates, and their present directions of movement can also be deduced. The Pacific plate, consisting entirely of oceanic crust, is moving northwestward, away from the spreading center along the eastern Pacific rise. The direction of movement along the convergent plate margins varies locally, because the Pacific plate is bordered by several different plates. The width of the deep-focus earthquake zone indicates that the plate is descending into the mantle at an angle of roughly 45°. The slabs of lithosphere extending to depths of 700 km at convergent boundaries indicate that the present plate motion has continued long enough for approximately 1000 km of new crust to have been generated at the oceanic ridge.

The American plates move westward from the mid-Atlantic ridge, encountering the Pacific and adjacent plates along the trench on the west coast.

The African plate is moving north, toward the convergent boundary in the Mediterranean region, driven by spreading from the ridge, which essentially surrounds it in the Atlantic and Indian oceans. There are no subduction zones between the spreading centers of the Atlantic and Indian oceans. Therefore, the plate boundaries surrounding Africa are apparently moving in relation to the African plate and to each other. The same is probably true of the Antarctic plate, which is completely surrounded by the spreading center of the oceanic ridge.

The Himalayas and the Tibetan Plateau define a wide belt of shallow earthquakes. In this area, the convergence of plates has produced a collision of two continents. India moved in from the south and collided with Asia, which rode up and over the Indian plate to form a double thickness of continental rocks in the area. This convergence produced the wide zone of exceptionally high topography in the Himalayas and the Tibetan Plateau.

Seismicity defines the present movement of the lithosphere and provides some of the most convincing support for the theory of plate tectonics.

SEISMIC WAVES AS PROBES OF THE EARTH'S INTERIOR

Statement

Speculations about the interior of the earth have stimulated the imagination for centuries, but it was not until we learned how to use seismic waves to obtain an "X-ray" picture of the earth's interior that we have been able to probe the deep interior of the earth and formulate models of its structure and composition. Seismic waves—both *P* waves and *S* waves—travel faster through rigid material than through soft or plastic material. Therefore, the velocities of these waves through a specific part of the earth give an indication of the type of rock there. Abrupt changes in seismic wave velocities indicate significant changes in the earth's interior.

Seismic waves passing through the earth are refracted in ways that show distinct discontinuities and provide the basis for the belief that the earth has a solid inner core, a liquid outer core, a thick mantle, a soft asthenosphere, and a rigid lithosphere.

Discussion

Seismic waves are similar in many respects to light waves, and their paths are governed by laws similar to those of optics. Both seismic vibrations and light vibrations move in a straight line through a homogeneous body. If they encounter a boundary between different substances, however, they are reflected or refracted. Familiar examples are light waves reflected from a mirror or refracted (bent) as they pass from air to water.

If the earth were a homogeneous solid, seismic waves would travel through it at a constant speed. A **seismic ray** (a line perpendicular to the wave front) would then be a straight line like those shown in *figure 18.9*. Early investigations, however, found that seismic waves arrive progressively sooner than was expected at stations successively farther from the earthquake's source. As is shown in *figure 18.9*, the rays arriving at a distant station travel deeper through the earth than those reaching stations close to the epicenter. Obviously, then, if the travel times of long-distance waves are progressively shortened as they go deeper into the earth, they must travel more rapidly at depth than they do near the surface. The significant conclusion from these studies is that the earth is not a

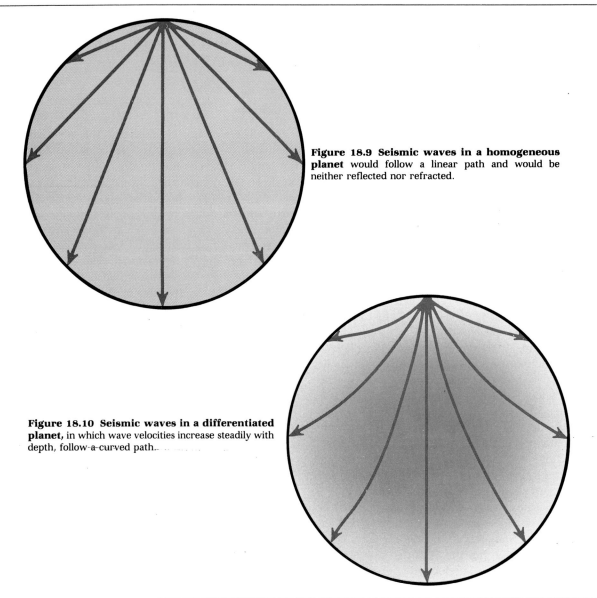

Figure 18.9 Seismic waves in a homogeneous planet would follow a linear path and would be neither reflected nor refracted.

Figure 18.10 Seismic waves in a differentiated planet, in which wave velocities increase steadily with depth, follow a curved path.

homogeneous, uniform mass but has physical properties that change with depth. As a result, seismic rays are believed to follow curved paths through the earth (*figure 18.10*).

In 1906, it was recognized that, whenever an earthquake occurs, there is a large region on the opposite side of the planet where the seismic waves are not detectable. To better understand the nature and significance of this **shadow zone,** refer to *figure 18.11*. For an earthquake at a particular spot (labeled 0°), the shadow zone for *P* waves invariably exists between 103° and 143° from the earthquake's focus. Evidently, something deflects the waves from a linear path. The best explanation for this shadow zone is that the earth has a central

core through which *P* waves travel relatively slowly. Seismic rays traveling through the mantle follow a curved path from the earthquake's focus and emerge at the surface between 0° and 103° from the focus (slightly more than a quarter of the distance around the earth). In *figure 18.11*, ray 1 just misses the core and is received by a station located 103° from the focus. Ray 2, however, being steeper than ray 1, encounters the core's boundary, where it is refracted. It travels through the core, is refracted again at the core's boundary, and is finally received at a station on the opposite side of the earth. Ray 3 is similarly refracted and emerges on the opposite side, 143° from the focus. Other rays that are steeper than ray 1 are also refracted

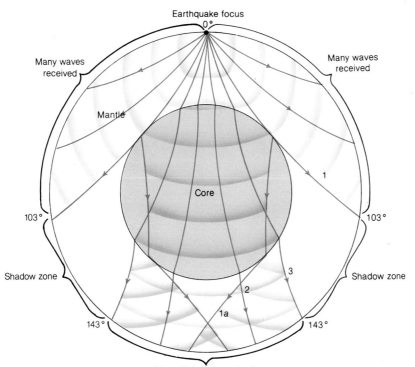

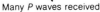

Figure 18.11 A *P*-wave shadow zone, between 103° and 143° from an earthquake's focus, results from the refraction of *P* waves passing through the earth. The best way to explain the *P*-wave shadow zone is to postulate that the earth has a central core through which *P* waves travel relatively slowly. Ray 1 just misses the core and is received at a station located 103° from the earthquake's focus. A steeper ray, such as ray 2, en- counters the boundary of the core and is refracted. It travels through the core, is refracted again at the core's boundary, and is received at a station less than 180° from the focus. Similarly, ray 3 is refracted and emerges at the surface 143° from the focus. Other rays that are steeper than ray 1 are severely bent by the core, so that no *P* waves are directly received in the shadow zone.

through the core and emerge between 143° and 180° from the focus. Thus, refraction at the boundary between the core and the mantle causes a *P*-wave shadow zone over part of the earth's surface. *Figure 18.11* is a cross section through the earth. The true nature of the shadow zone is a band across the planet, as shown in *figure 18.12.*

In 1914, Beno Gutenberg, a German seismologist, calculated that the depth to the surface of the core is 2900 km. Later analysis of more numerous and more reliable seismic data showed that Gutenberg's original estimate was remarkably accurate, with a probable error of less than two-thirds of 1%. More recent studies of the *P*-wave shadow zone show that some weak *P* waves of low amplitude are received in this zone. This suggests the presence of a solid inner core, which deflects the deep, penetrating *P* waves in the manner shown in *figure 18.13.*

The surface of the core has an even more pro-

Figure 18.12 The *P*-wave shadow zone of an earthquake whose focus is in Japan is a broad band around the earth in which no *P* waves are directly received. From shadow zones, seismologists calculate that the boundary of the core is 2900 km below the surface.

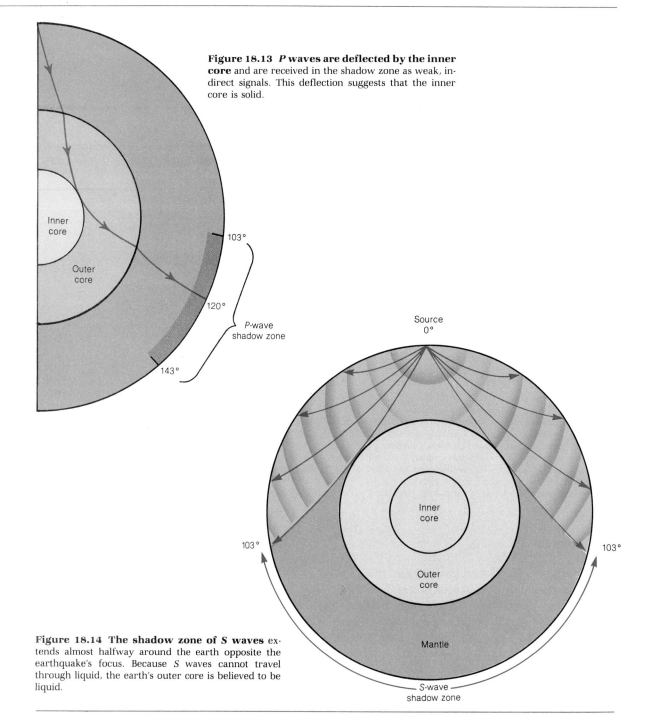

Figure 18.13 _P_ waves are deflected by the inner core and are received in the shadow zone as weak, indirect signals. This deflection suggests that the inner core is solid.

Inner core

Outer core

103°

120°

143°

_P_-wave shadow zone

Source 0°

Inner core

Outer core

Mantle

103°

103°

_S_-wave shadow zone

Figure 18.14 The shadow zone of _S_ waves extends almost halfway around the earth opposite the earthquake's focus. Because _S_ waves cannot travel through liquid, the earth's outer core is believed to be liquid.

nounced effect on _S_ waves, but this effect cannot be explained by reflection or refraction. _S_ waves simply do not pass through the core at all. They produce a huge shadow zone extending almost halfway around the earth opposite the earthquake's focus (_figure 18.14_). One difference between _P_ waves and _S_ waves is particularly significant. _P_ waves pass through any substance—solid,

liquid, or gas. _S_ waves, however, are transmitted only through solids that have enough elastic strength to return to their former shape after being distorted by the wave motion. They cannot be transmitted through a liquid. The _S_-wave shadow zone, therefore, is generally taken as evidence that the outer core is liquid.

With the present worldwide network of record-

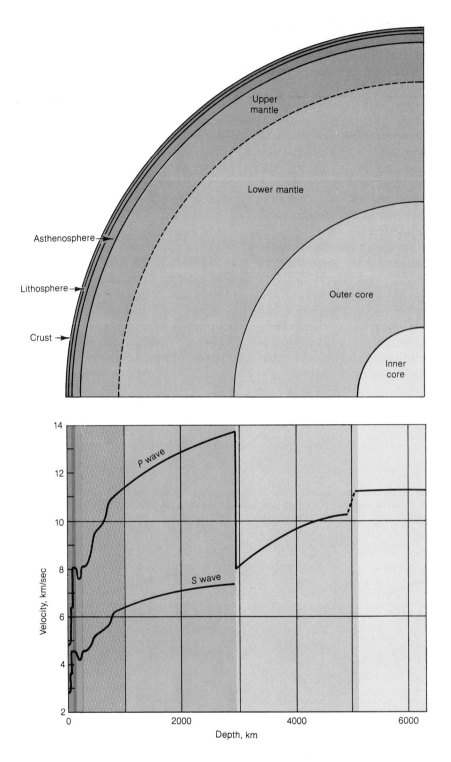

Figure 18.15 The internal structure of the earth is deduced from variations in the velocity of seismic waves at depth. The velocities of both *P* waves and *S* waves increase until they reach a depth of approximately 100 km. There the waves abruptly slow down, and their velocities continue to decrease until they have traveled to a depth of 200 km. This low-velocity layer is called the asthenosphere (see figure 18.16). Below 200 km beneath the surface, the velocities of *P* waves and *S* waves increase, until the waves reach a depth of about 3000 km, where both change abruptly. *S* waves do not travel through the central part of the earth, and the velocity of the *P* waves decreases drastically. This variation is the most striking discontinuity in the earth, and it is considered to signify the boundary between the core and the mantle. Another discontinuity in *P*-wave velocities, at a depth of 5000 km, indicates the surface of the inner core.

ing stations, even minor variations in seismic wave velocities, or **seismic discontinuities,** can be determined with considerable accuracy. These data, which can be summarized graphically in velocity-depth curves like the one in *figure 18.15,* provide additional information about the earth's interior. The most striking variation occurs at the core's boundary, at a depth of 2900 km. There, *S* waves stop and the velocity of *P* waves is drastically reduced.

Other discontinuities are apparent but are less striking. The first occurs between 5 and 70 km below the surface. This is called the **Mohorovičić discontinuity** (or simply Moho), after Andrija Mohorovičić, the Yugoslavian seismologist who first recognized it. It is considered to represent the base of the crust.

Perhaps the most significant discontinuity, however, is the low-velocity zone from 100 to 200 km below the surface (*figure 18.16*). Gutenberg recognized this zone in the 1920s, but the discovery was viewed with skepticism by most other seismologists at that time. The normal trend is for seismic wave velocities to increase with depth. In this low-velocity zone, however, the trend is reversed, and seismic waves travel about 6% slower than they do in adjacent regions. More recent seismic data again confirm Gutenberg's observations. Most seismologists are now convinced that this zone is more plastic than the areas above and below it and that it is responsible for the dynamics of the rigid lithosphere above it.

The generally accepted explanation for the low seismic wave velocities in this zone is that temperature and pressure there cause part of the material, perhaps from 1% to 10%, to melt, so that a crystal-liquid mixture is produced. A small amount of liquid film around the mineral grains serves as a lubricant, increasing the plastic nature of the material. This low-velocity zone, as we will see in later chapters, plays a key role in the theories of motion of material at the earth's surface.

Other rapid changes in seismic wave velocities occur at depths of about 350 km and 700 km. These are interpreted as the results of phase changes in the minerals, in which the atomic packing is rearranged into denser and more compact units.

Figure 18.16 The rapid decrease in velocities of both *S* waves and *P* waves suggests a zone of low strength in the upper mantle, at depths between 100 and 200 km below the surface. This zone, the asthenosphere, is commonly referred to as the low-velocity layer of the upper mantle.

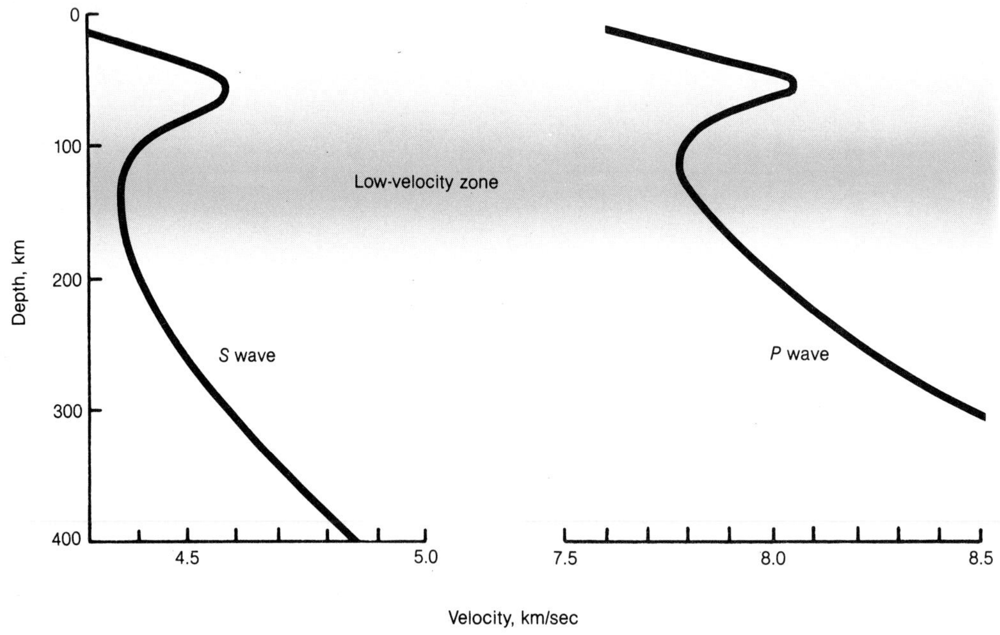

SUMMARY

Earthquakes are vibrations in the earth caused by the sudden rupture of rocks that have been strained beyond their elastic limits. Three types of seismic waves are generated by an earthquake shock: *P* waves, *S* waves, and surface waves.

The intensity of an earthquake is a measure of its destructive power. It depends on the total energy released, the distance from the quake's epicenter, and the nature of rocks in the crust. The magnitude of an earthquake is a measure of the total energy released. The primary effect of earthquakes is ground motion. Secondary effects include (1) landslides, (2) tsunamis, and (3) regional uplift and subsidence.

The establishment in 1966 of a worldwide network of sensitive seismic stations has enabled seismologists to monitor the thousands of earthquakes that occur each year. This new insight into the earth's seismicity confirms, in a most remarkable way, the theory of plate tectonics, because it shows the present-day movement of plates. The distribution pattern of earthquakes dramatically outlines the plate margins. Shallow earthquakes develop in a narrow zone along divergent plate margins, where the mantle rises and pulls the plate apart. Where plates collide and one is thrust under the other, a zone of shallow, intermediate, and deep earthquakes is produced as the oceanic plate moves down into the mantle. In the central parts of the plates, there is little differential movement. Few earthquakes occur in these stable areas.

The velocities of *P* waves and *S* waves through the earth indicate that the earth has a solid inner core, a liquid outer core, a thick mantle, a soft asthenosphere, and a rigid lithosphere.

Key Words

earthquake
seismology
seismograph
seismic wave
P wave (primary wave)
longitudinal wave
S wave (secondary wave)
surface wave
elastic-rebound theory
elastic limit
epicenter
focus
shallow-focus earthquake
intermediate-focus earthquake
deep-focus earthquake
earthquake magnitude
earthquake intensity
seismic ray
seismic discontinuity
Moho (Mohorovičić discontinuity)
shadow zone

Review Questions

1. Explain the elastic-rebound theory of the origin of earthquakes.
2. Describe the motion and velocity of the three major types of seismic waves.
3. Explain how the location of the epicenter of an earthquake is determined.
4. Describe the global pattern of earthquakes.
5. How does the depth of earthquakes indicate (*a*) convergent plate margins and (*b*) divergent plate margins?
6. Draw a diagram showing the paths that would be followed by seismic rays through the earth if the core were only half the size shown in *figure 18.11*.
7. What do seismic velocity-depth diagrams (see *figure 18.16*) tell us about the internal structure of the earth?

Additional Readings

Eckel, E. B. 1970. *The Alaskan earthquake, March 27, 1964: Lessons and conclusions.* U.S. Geological Survey Professional Paper 546. Reston, Va.: U.S. Geological Survey.

Iacopi, R. 1964. *Earthquake country.* Menlo Park, Calif.: Lane.

Nicholas, T. C. 1974. Global summary of human response to natural hazards: Earthquakes. In *Natural hazards,* ed. G. F. White, pp. 274–84. New York: Oxford University Press.

Press, F. 1975. Earthquake prediction. *Scientific American* 232(5):14–23.

Wyllie, P. J. 1976. *The way the earth works.* New York: John Wiley and Sons.

19

SUMMARY

Most volcanic activity occurs in active seismic zones and is clearly associated with plate boundaries. The type of volcanism depends on the type of plate boundary. At divergent plate margins, basaltic magma is generated by the partial melting of the asthenosphere, and it is extruded largely as fissure eruptions. Pillow lavas form on the sea floor, and plateau basalts are extruded where continents overlie rift zones.

At convergent plate margins, andesitic or granitic magma is generated in the subduction zone by the partial melting of the oceanic crust. It is intruded in the upper plate as batholiths or is extruded as composite volcanoes in island arcs or mountain belts. Andesitic magma is viscous and erupts violently, commonly in the form of an ash flow.

Minor intraplate volcanic activity probably indicates local hot spots in the mantle. The greatest intraplate volcanism occurs in the Pacific, where numerous shield volcanoes form islands and seamounts.

Key Words

pillow lava	pumice
fissure	island arc
sheet dikes	volcanic front
flood basalt	arc-trench gap
plateau basalt	mantle plume
dike swarm	hot spot

Review Questions

1. Compare the location of recently active volcanoes with the global patterns of seismicity. What appears to control the location of volcanic activity?

2. What is the Ring of Fire?

3. What volcanic features were observed by Project FAMOUS?

4. Describe the type of volcanic activity in Iceland. What is the significance of sheet dikes?

5. What does the study of plateau basalts tell us about volcanism along divergent plate margins?

6. Summarize the characteristics of volcanism along divergent plate margins.

7. Explain how magma is generated along divergent plate margins.

8. Describe the type of volcanic activity along convergent plate margins and cite several examples.

9. How does the composition of lava influence the style of volcanic eruption?

10. Why is there a gap between the trench and the volcanic front in a subduction zone?

11. What type of volcanic activity occurs within the central parts of tectonic plates, beyond the active margins?

12. Explain the origin of chains of volcanic islands and seamounts.

Additional Readings

Bullard, F. 1962. *Volcanoes: In history, in theory, in eruption.* Austin: University of Texas Press.

Francis, P. 1976. *Volcanoes.* New York: Penguin.

Green, J., and N. M. Short. 1971. *Volcanic landforms and surface features: A photographic atlas and glossary.* New York: Springer-Verlag.

Macdonald, G. A. 1972. *Volcanoes.* Englewood Cliffs, N.J.: Prentice-Hall.

Stearns, H. T. 1946. Geology of the Hawaiian Islands. *Hawaii Division of Hydrography Bulletin,* no. 8.

Tazieff, H. 1974. *The making of the earth: Volcanoes and continental drift.* New York: Saxon House.

Williams, H. 1951. Volcanoes. *Scientific American* 185(6):45–53.

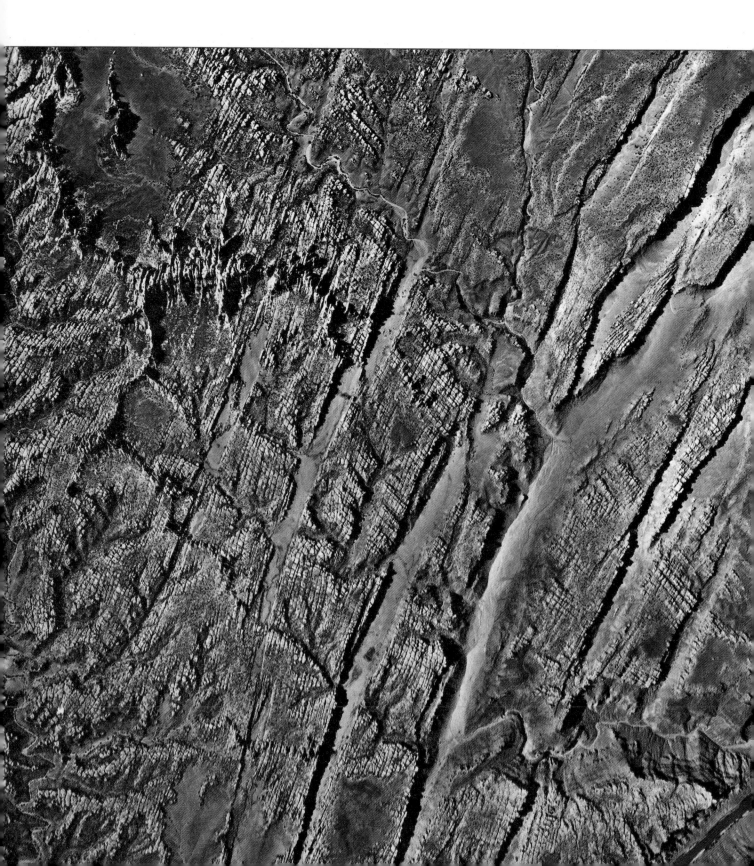

CRUSTAL DEFORMATION

Since the first geologic studies, more than 150 years ago, geologists have found that rocks in certain parts of the continents have been folded and fractured on a gigantic scale. Deformation of the crust is most intense in the great mountain belts of the world, where sedimentary rocks, which were originally horizontal, are now folded, contorted, fractured, and in some places completely overturned. In some mountains, large bodies of rock have been thrust several tens of kilometers over younger strata. In the deeper parts of mountain belts, deformation (and elevated temperature) are great enough to cause metamorphism (see chapter 7).

The extensive deformation of rock bodies in the shields and mountain belts shows that the earth's tectonic system has operated throughout geologic time, with shifting plates constantly deforming the crust at their convergent margins. As we will see in chapter 24, the tectonic activity of the earth makes it unique among the planets of the solar system. In this chapter, we will consider how rocks respond to tectonic forces by folding and fracturing and how these structures are expressed at the earth's surface.

MAJOR CONCEPTS

1. Deformation of the earth's crust is well documented by historical movement along faults, raised beach terraces, and deformed rock bodies.
2. Folds in rock strata range in size from microscopic wrinkles to large structures hundreds of kilometers long. The major types are (a) domes and basins, (b) plunging anticlines and synclines, and (c) complex flexures.
3. Faults are fractures in the crust along which slippage or displacement has occurred. The three basic types are (a) normal faults, (b) thrust faults, and (c) strike-slip faults.
4. Joints are fractures in rocks without displacement.
5. Unconformities are major discontinuities in rock sequence, which indicate interruptions in rock-forming processes because of crustal movement.

EVIDENCE OF CRUSTAL DEFORMATION

Statement

Although a casual observer may think that the crust of the earth is permanent and fixed, a great deal of evidence, both direct and indirect, indicates that the crust is in continuous motion and that it has moved on a vast scale throughout all of geologic time.

Discussion

Earthquakes are perhaps the most convincing evidence that the crust is moving. Certainly, those who experience an earthquake are convinced. During earthquakes, the crust not only vibrates, but segments of it are fractured and displaced (*figure 20.1*). One impressive example is the movement along the San Andreas Fault during the 1906 San Francisco earthquake, which offset fences and roads by as much as 7 m. Another is the 1899 earthquake at Yakutat Bay, Alaska, during which a beach was uplifted 15 m above sea level. Similar displacements, well documented with photographs, occurred during the Good Friday earthquake in Alaska in 1964, the Hebgen Lake earthquake in Montana in 1959, and the Dixie Valley earthquake in Nevada in 1954.

Various topographic features also testify to earth movements in prehistoric times. A striking example

Figure 20.1 Evidence of crustal deformation produced during historic earthquakes is seen most vividly in the disruption of man-made structures. These photographs show destruction by recent earthquakes.

(A) Anchorage, Alaska, 1964

(B) Kern County, California, 1952

(C) San Fernando, California, 1971

(D) Near Woodville, California, 1906

is the raised beach terraces along the coast of southern California. There, ancient wave-cut cliffs and terraces, containing remnants of beaches, with barnacles, shells, and sand, rise in a series of step-like forms above the present shore (*figure 20.2*).

Many rock formations contain obvious evidence of deformation on a much larger scale. Practically every mountain range exposes deformed sedimentary strata that were originally deposited horizontally below sea level. These twisted and contorted rock layers testify to the continuing motion of the lithosphere and the deformation it produces (*figure 20.3*).

Orogenesis. The most intense deformation in the earth's crust results from the convergence of lithospheric plates. Where plates converge along continental margins, sedimentary rocks (originally deposited in the ocean surrounding the continental platform) are deformed into an elongated mountain belt of strongly folded and faulted rocks. This type of deformation is called **orogenic,** and the mountain-building event is referred to as an **orogeny** (Greek, "mountain genesis"). The folded rocks in the world's major mountain belts (the Appalachians, the Rockies, the Himalayas, the Urals, and the Alps) all exemplify this type of deformation (*figure 20.4*).

Another type of deformation, found on the stable platforms of the continents, is restricted largely to broad crustal **upwarps** and **downwarps.** This type of deformation is known as **crustal warping.** Coal beds now being mined far below the surface are evidence of crustal subsidence. The bedding planes of coal, now buried thousands of meters deep, were once at the surface, where vegetation could grow. Similarly, sedimentary rocks, now buried thousands of meters below the surface, show evidence of slow subsidence during sedimentation. In contrast, in such areas such as the Colorado Plateau, marine sedimentary rocks exist 1000 m above sea level, indicating significant uplift of the crust after the sediment was deposited. All of the lines of evidence point to the same conclusions: the crust of the earth is in motion and has been in motion throughout geologic time.

Figure 20.2 Recent uplift along the southern coast of California is well documented by a series of elevated wave-cut platforms, such as these, on the Palos Verdes Hills.

Figure 20.3 The internal structure of a mountain commonly consists of highly deformed layers of sedimentary rock, such as those shown in this photograph of the Canadian Rockies. The degree of compression and deformation can be better appreciated by studying the diagram, which traces the major beds. Note that much of the rock in the upper parts of the original folds has been removed by erosion.

Figure 20.4 Folded rocks of a mountain belt can be seen in this space photograph of the southern part of Iran, taken from an elevation of approximately 1000 km.

DIP AND STRIKE

Statement

Many structural features of the crust are too large to be seen from any point on the ground. They are recognized only after the geometry of the rock bodies are determined from geologic mapping. At an outcrop, two fundamental observations—dip and strike—describe the orientations of bedding planes, fault planes, joints, and other planar features in the rock. The dip of a plane is the angle and direction of its inclination from the horizontal. The strike is the direction, or trend, of a planar feature, such as a bedding plane or a fault plane. More precisely, it is the compass bearing of a horizontal line on the planar feature. These two measurements together define the orientation of the planar surface in space.

Discussion

The concept of **dip** and **strike** can easily be understood by referring to *figure 20.5*, which shows an outcrop of tilted beds along the coast. The water provides a necessary reference to a horizontal plane. The trend of the waterline along the bedding plane of the rocks is the direction of strike. The angle between the water surface and the bedding plane is the angle of dip. The photograph in *figure 20.6* shows a sequence of beds striking south (to the top of the picture) and dipping 40° to the east (to the left of the picture). Another way to visualize dip and strike is to think of the roof of a building. The dip is the direction and amount of inclination of the roof, and the strike is the trend of the ridge.

Dip and strike are measured in the field with a geological compass, which is designed to measure both direction and angle. An example of the symbols used for recording the dip and strike of bedding planes on a map is $\measuredangle_{30}$. The long crossbar shows the strike, the short line perpendicular to it shows the direction of dip, and the number represents the angle of dip. The symbol shown here represents beds striking N 45° E and dipping 30° to the southeast.

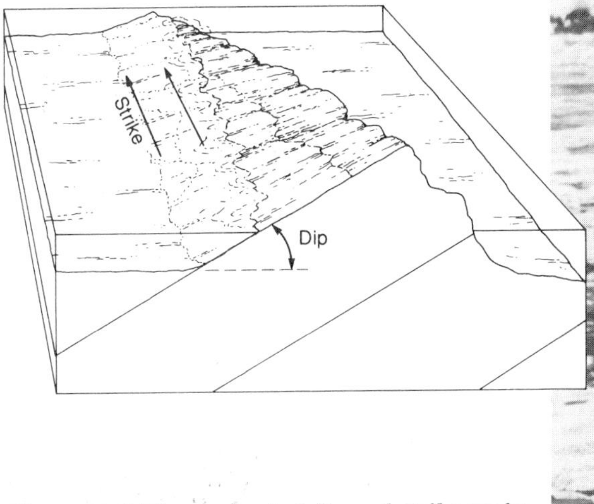

Figure 20.5 The concept of dip and strike can be understood by studying rock layers such as the ones shown in this photograph. The strike of a bed is the compass bearing of a horizontal line drawn on the bedding plane. It can readily be established by reference to the horizontal waterline in this example. The dip is the angle and direction of inclination of the bed, measured at right angles to the strike.

Figure 20.6 A sequence of inclined beds striking south (toward the background) and dipping 40° to the east is shown in this photograph of the San Rafael Swell, in Utah. The diagram shows the configuration of the flexure and the upper beds, which have been removed by erosion.

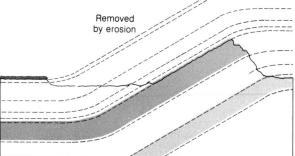

Removed by erosion

FOLDS

Statement

Folds are warps (wavelike contortions) in rock bodies. They are three-dimensional structures ranging in size from microscopic crinkles to large domes and basins hundreds of kilometers wide. Small flexures are abundant in sedimentary rocks and can be seen in mountainsides and road cuts and even in hand specimens. Large folds cover thousands of square kilometers, and they can best be recognized from aerial or space photographs or from extensive geologic mapping.

Folds indicate that rocks of the crust, like other solid bodies, can be deformed by plastic flow. The complexity of folds is proportionate to the degree of deformation. Broad, open folds form in the stable interiors of continents, where the rocks are only mildly warped. Complex folds develop in mountain belts, where deformation is more intense.

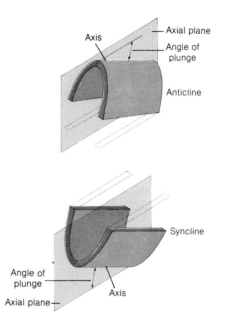

Figure 20.8 The axial plane of a fold is an imaginary plane that divides the fold into equal parts. The line formed by the intersection of the axial plane and a bedding plane is called the axis. The downward inclination of the axis is called the plunge.

Discussion

Fold Nomenclature. To characterize and interpret the significance of **folds,** you must first become familiar with the terms used to describe them. Three general types of folds are illustrated in *figure 20.7.* **Monoclines** are folds in which horizontal or gently dipping beds are modified by simple steplike bends. **Anticlines,** in their simplest form, are uparched strata, with the **limbs** (sides) of the fold dipping away from the crest. Rocks in an eroded anticline are progressively *older* toward the interior of the fold. **Synclines,** in their simplest form, are downfolds, or troughs, with the limbs dipping toward the center. Rocks in an eroded

syncline are progressively *younger* toward the center of the fold.

For purposes of description and analysis, it is useful to divide folds into two equal parts by an imaginary plane called the **axial plane.** The line formed by the intersection of the axial plane and a bedding plane is called the **axis,** and the downward inclination of the axis is referred to as the **plunge** (*figure 20.8*). In *figure 20.11,* the folds plunge toward the background of the block diagram at an angle of about 20° (see also *figure 12.15*).

Figure 20.7 The nomenclature of folds is based on the three-dimensional geometry of the structure, although most exposures show only a cross section or map view.

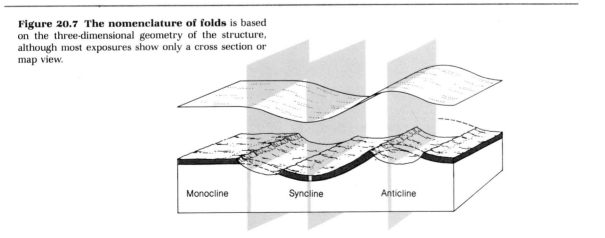

Domes and Basins

The sedimentary rocks covering much of the continental interiors have been mildly warped into broad **domes** and **basins** many kilometers in diameter. One large basin covers practically all of the state of Michigan. Another underlies the state of Illinois. An elongate dome underlies central Tennessee, central Kentucky, and southwestern Ohio. Although these flexures in the sedimentary strata are extremely large, the configuration of the folds is known from geologic mapping and from information gained through drilling. The nature of these flexures and their topographic expression are diagrammed in *figure 20.9* (see also *plate 4*).

The configuration of a single bed warped into broad domes and basins is shown in perspective in *figure 20.9A*: If erosion cuts down the tops of the domes, the layer looks like the one shown in *figure 20.9B*. The exposed rocks of both domes and basins typically have a circular or elliptical outcrop pattern. The oldest rocks are exposed in the central parts of domes, and the youngest rocks are exposed in the centers of basins.

A classic example of a broad fold in the continental interior is the large dome that forms the Black Hills of South Dakota (see *figure 12.14*). Resistant rock units form ridges that can be traced completely around the core of the dome, and nonresistant formations make up the intervening valleys. A small dome in southern Wyoming, similar to that of the Black Hills, is shown in the photograph in *figure 20.10*. It clearly illustrates the typical elliptical outcrop pattern of domes, expressed by alternating ridges and valleys that encircle the structure.

Figure 20.9 The geometry and surface expression of domes and basins are basically simple, involving broad upwarps and downwarps eroded into circular or elliptical outcrop patterns.

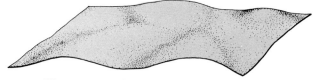

(A) *A single gently folded bed is warped into a configuration of broad domes and basins.*

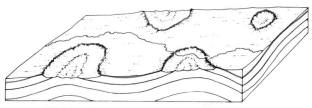

(B) *As erosion proceeds, the tops of the domes are eroded first. The outcrop pattern of eroded domes and basins typically is circular or elliptical.*

Figure 20.10 The surface expression of an elongate structural dome in western Wyoming was photographed from an altitude of approximately 6 km. The resistant layers of the structure form ridges, and the nonresistant layers are eroded into elongate valleys. Note that the oldest rocks are at the center of the structure.

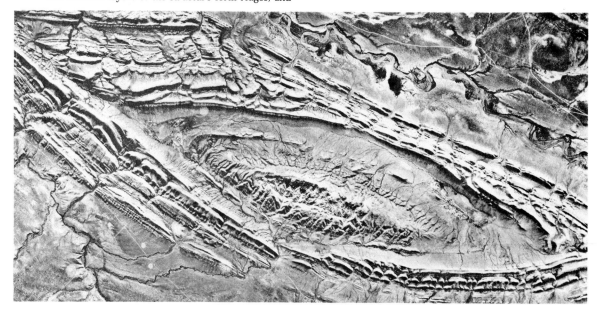

Plunging Folds

Where deformation is more intense (such as, typically, in mountain belts), the rock layers are deformed into a series of tight folds (see *plate 5*). Fold geometry is not exceedingly complex. In many ways, folds resemble the wrinkles in a rug. The diagrams in *figure 20.11* illustrate the general configuration of **plunging folds** and their surface expressions after they have been eroded. The basic form is shown in *figure 20.11A*. Diagram *B* shows a fold system after the upper part has been removed by erosion. The outcrop of the eroded plunging anticlines and synclines forms a characteristic zigzag pattern. An anticline forms a V-shaped pattern that points in the direction of plunge, and the oldest rocks are in the center of the fold. A plunging syncline forms a V-shaped pattern that opens in the direction of plunge, and the youngest rocks are in the center of the fold. Together, the outcrop pattern and the relative ages of the rocks in the center of the fold make it possible to determine the structure's subsurface configuration (*figure 20.12*).

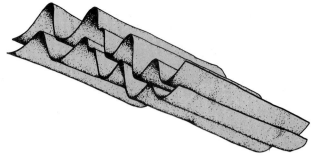

(A) The basic form of folded strata is similar to that of a wrinkled rug. In this diagram, the strata are compressed and plunge toward the background.

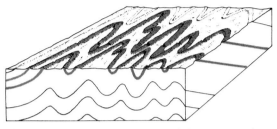

(B) If the tops of the folded strata are eroded away, a map of the individual layers shows a zigzag pattern at the surface. Rock units that are resistant to erosion form ridges, and nonresistant layers are eroded into linear valleys. (Compare this diagram with plate 5.) In a plunging anticline, like the one shown here, the surface map pattern of the beds forms a V pointing in the direction of plunge.

Figure 20.11 The geometry and surface expression of plunging folds involve rock strata deformed into tight folds and eroded into zigzag outcrop patterns.

Figure 20.12 The surface expression of a plunging fold commonly is a series of alternating ridges and valleys. In a plunging anticline, such as the one shown here, the surface trace of the beds forms a V pointing in the direction of plunge, with the oldest rocks in the center of the fold. The subsurface configuration of a fold can be determined by careful study of the surface layers.

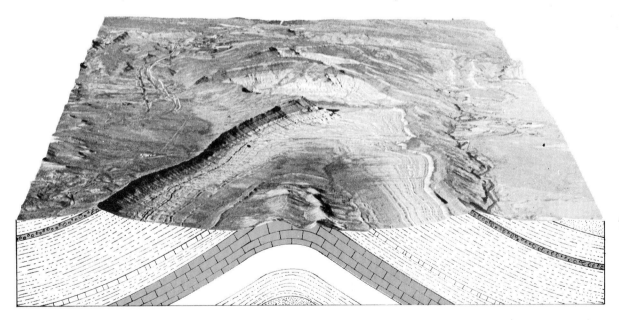

Complex Folds

Intense deformation in some mountain ranges produces complex folding, as illustrated in *figure 20.13*. Such structures commonly exceed 100 km in width, so that complex folds can extend through a large part of a mountain range. Details of such intensely deformed structures are extremely difficult to work out because of the complexity of the outcrop patterns.

Figure 20.13 illustrates the geometry and surface expression of complex folds. Diagram *A* is a perspective drawing of a single bed in a typical complex fold. This **overturned fold** is a huge anticlinal structure with numerous minor anticlines and synclines forming digits on the larger fold. Diagram *B* shows the fold after it has been subjected to considerable erosion, which has removed most of the upper limb. Note the cross section of the structure on the mountain front and the outcrop pattern (to the top of the diagram). The topographic expression of complex folds is variable. They usually are expressed in a series of mountains (*figure 20.13C*). Complex folds are common in the Swiss Alps, but they were recognized only after more than half a century of detailed geologic studies. They are also common in the roots of ancient mountain systems and thus are exposed in many areas of the shields (*figure 20.14*).

(A) *The configuration of this single bed in a large, tightly overturned fold, with minor folds on the limbs, is typical of intensely deformed strata in some mountain ranges.*

(B) *The surface outcrop of the same fold, after erosion has removed the upper surface, shows great complexity, so that its details are difficult to recognize.*

(C) *The topographic expression of complex folds can be a series of mountains or a flat, eroded surface.*

Figure 20.13 The geometry and surface expression of complex folds involve intensely deformed strata and complex outcrop patterns.

Figure 20.14 Complex folds are exposed in many areas of the Canadian Shield and commonly consist of overturned limbs, intricate interdigitation of minor folds, and associated joints and faults. Here the structure is outlined by differential erosion along resistant and nonresistant layers.

FAULTS

Statement

Faults are fractures in the earth's crust along which slippage or displacement has occurred. In a road cut or in the walls of a canyon, a fault plane may be obvious, and the displaced or offset beds can easily be seen (*figure 20.15*). Elsewhere, the surface expression of a fault can be very subtle. Detailed geologic mapping may be necessary before the precise location of the fault can be established.

Faults grow by a series of small displacements, which occur as built-up stress is suddenly released. As we have seen, the sudden release of energy produces an earthquake. Displacement can also occur by imperceptibly slow movement called tectonic creep. Three basic types of faults are recognized:

1. Normal faults
2. Thrust faults
3. Strike-slip faults

Discussion

The three basic types of **faults** are shown in *figure 20.16*.

Normal Faults

In **normal faults,** movement is predominantly vertical, and the rocks above the fault plane (the **hanging wall**) move downward in relation to those beneath the fault plane (the **footwall**). Most

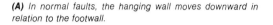
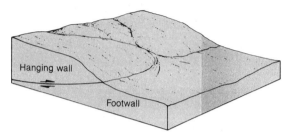
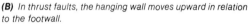

(A) In normal faults, the hanging wall moves downward in relation to the footwall.

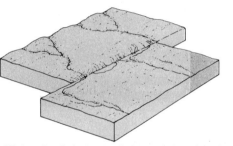

(B) In thrust faults, the hanging wall moves upward in relation to the footwall.

(C) In strike-slip faults, the displacement is horizontal.

Figure 20.16 The three major types of faults are distinguished by the direction of their relative displacement.

Figure 20.15 The displacement of beds in a fault is often well expressed on the side of a valley.

normal faults are steeply inclined, usually between 65° and 90°. Their predominantly vertical movement commonly produces a cliff, or **scarp.**

Normal faults rarely are isolated fractures. Typically, a group of parallel normal faults develops a steplike arrangement, or a series of **fault blocks.** A narrow block dropped down between two normal faults is called a **graben,** and an upraised block is a **horst** (*figure 20.17*). A graben typically forms a conspicuous fault valley or basin marked by relatively straight, parallel walls. Horsts form blocklike plateaus.

Large-scale normal faulting is the result of major tensional stresses, which stretch and pull apart the crust. This type of stress occurs on a global basis along divergent plate margins, so that normal faults are the dominant structure along the oceanic ridge and in the several continental rift systems.

In the Basin and Range province of western North America, normal faulting produced a series of fault blocks trending north and south and extending from central Mexico to Oregon and Idaho. The horsts form mountain ranges from 2000 to 4000 m high, which are considerably dissected by erosion. Grabens form topographic basins, which are partly filled with erosional debris from the adja-

Figure 20.17 Horsts and grabens in Canyonlands National Park, Utah, were photographed from an altitude of 600 m. Grabens (downdropped blocks) form elongate valleys, which are partly covered with a veneer of sediment. Horsts (upraised blocks) form elongate ridges. Relative movement along the major faults is shown in the idealized diagram.

Figure 20.18 Fault blocks of the Basin and Range province, in the western United States, are modified by erosion and deposition. Most mountain ranges in the area are horsts or tilted fault blocks. The basins are grabens that are partly filled with sediment derived from erosion of the ranges.

cent ranges (*figure 20.18*). Between the Wasatch Range, in central Utah, and the Sierra Nevada, on the Nevada-California border, the earth's crust has probably been extended some 60 km during the last 15 million years. Displacement of alluvial fans and surface soils indicates that many of the faults are still active.

The great rift valleys of Africa are another example of large-scale normal faulting produced by a zone of tension in the earth's crust (see *plate 14*).

There, large grabens have been formed by a system of normal faults extending from the Zambesi River to northern Ethiopia, a distance of 2900 km. That distance is almost doubled if the rift system of the Red Sea and the Jordan valley (which continues northward into Syria) is added. The rift valleys of Africa are remarkably uniform in width, ranging from 30 to 45 km. They are about the same size as the grabens of the Basin and Range province and the rift valleys of other continents.

Thrust Faults

Thrust faults are low-angle faults in which the hanging wall has moved up and over the footwall. Some geologists apply the term *thrust* to these faults only if they dip at angles less than 45°, and they refer to high-angle thrusts as **reverse faults.** Movement on a thrust is predominantly horizontal, and displacement can be more than 50 km.

Thrust faults result from crustal shortening. They generally are associated with intense folding, caused by powerful horizontal compression in the earth's crust. They are prominent in all of the world's major folded mountain regions, commonly evolving from folds in the manner diagrammed in *figure 20.19.* Where resistant rocks are thrust over nonresistant strata, a scarp is eroded on the upper plate. The scarp is not straight or smooth, as cliffs produced by normal faulting are. Rather, the outcrop of the fault surface typically is irregular in map view.

Figure 20.19 The evolution of thrust faults from folds is depicted in this sequence of diagrams. Diagram **A** shows the fault plane and the progressive development of folds into a thrust fault. Diagram **B** shows how the structure might be expressed at the surface.

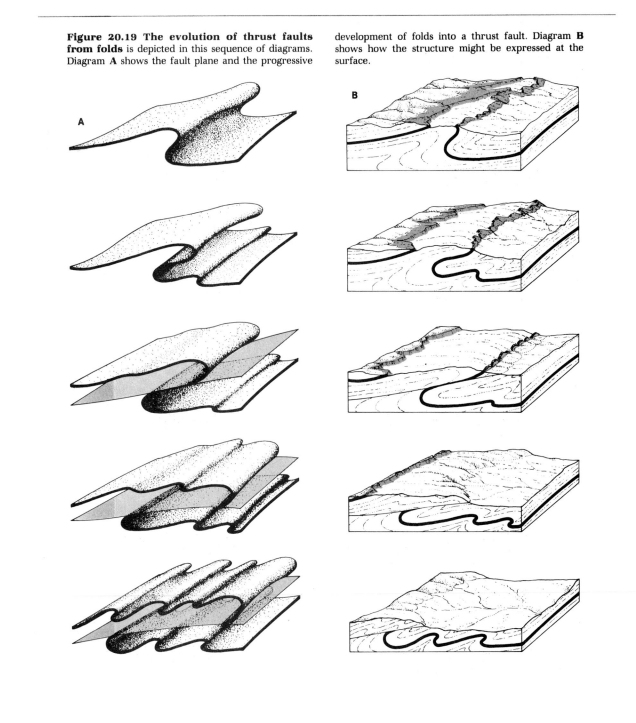

Strike-Slip Faults

Strike-slip faults are high-angle fractures in which displacement is horizontal, parallel to the strike of the fault plane. There is little or no vertical movement, so that high cliffs do not form along strike-slip faults. Instead, these faults are expressed topographically by a straight, low ridge extending across the surface, which commonly marks a discontinuity in types of landscape (see *plate 15*).

Some of the topographic features produced by strike-slip faulting and subsequent erosion are shown in *figure 20.20*. One of the more obvious is the offset of the drainage pattern. The relative movement is often shown by abrupt right-angle bends in streams at the fault line: a stream follows the fault for a short distance and then turns abruptly and continues down the regional slope. As the blocks move, some parts may be depressed to form **sag ponds.** Others buckle into low, linear ridges.

Strike-slip movement results in the juxtaposition of blocks with contrasting structures, rock types, and topographic forms (*figure 20.21*). Therefore, the fault line often marks the boundary between distinctly different surface features and rock types. This distinction is expressed in *figure 20.20* by the contrast in the degree of dissection on the fault blocks. Also, faults disrupt patterns of groundwater movement, as is reflected by contrasts in vegetation, soils, and springs along the fault trace.

Strike-slip faults result from shear stresses in the crust. They commonly are produced where one tectonic plate slides past another at a transform fault boundary.

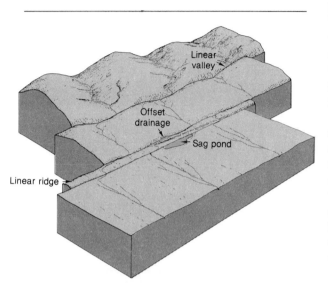

Figure 20.20 Strike-slip faults produce distinctive landforms. Streams are offset by recurrent movement, linear ridges and valleys form, and local sag ponds develop along the fault line.

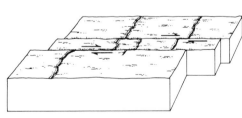

Figure 20.21 The San Andreas Fault, in California, is a major strike-slip fault. It is delineated by prominent, straight ridges and valleys. Recent movement along the fault has offset the drainage patterns on either side of the fault. Relative movement between the fault blocks is evident from the direction in which the drainage is offset.

Observed Movement on Faults

The movement along faults during earthquakes rarely exceeds a few meters. In the great San Francisco earthquake of 1906, the crust slipped horizontally as much as 7 m along the San Andreas Fault, so that roads, fence lines, and orchards were offset. Recent vertical faults in Nevada and Utah have produced fresh scarps from 3 to 6 m high (*figure 20.22*). The Good Friday earthquake in Alaska in 1964 was accompanied by a 13-m uplift near Montague Island. The largest well-authenticated displacement during an earthquake appears to have occurred in 1899 near Yakutat Bay, in Alaska, where beaches were raised as much as 15 m above sea level.

Movement along faults is not restricted to uplift during earthquakes, however. Precise surveys along the San Andreas Fault show slow shifting along the fault plane, at an average rate of 4 cm per year. Such movements, called **tectonic creep,** break buildings constructed across the fault line and eventually result in considerable displacement.

The important point is that displacement on a fault does not occur in a single violent event. Rather, it is the result of numerous periods of displacement and slow tectonic creep, commonly separated by periods of tectonic stability.

Figure 20.22 Recent displacement along the Wasatch Fault, in central Utah, has produced the fresh cliff at the base of the mountain front. Cumulative movement on the fault produced the mountain range.

JOINTS

Statement

Joints are fractures in rocks along which there has been no appreciable displacement. They are very common and are found in almost every exposure. Usually, they occur as two sets of fractures, which intersect at angles ranging from 45° to 90°. Thus, they divide the rock body into large, roughly rectangular blocks. Joints can be related to major faults, and they can form remarkably persistent patterns extending over hundreds of square kilometers. Other joints are related to broad upwarps of the crust.

Discussion

The best areas in which to study **joints** are those where brittle rocks, such as thick sandstone, have been fractured and their joint planes accentuated by erosion. The massive sandstones of the Colorado Plateau are an excellent example. Joints are expressed by deep, parallel cracks, which are most impressive as seen from the air (*figure 20.23*; see also *figure 20.17*). In places, joints control the development of stream courses, especially secondary tributaries and areas of solution activity.

Most joints result from broad regional upwarps or from compression or tension associated with faults and folds. Some tensional joints result from erosional unloading and expansion of the rock. Columnar joints in volcanic rocks are produced by stresses that are set up as the lava cools and contracts.

Joints can have economic importance. They can provide the necessary porosity for ground water and for the migration and accumulation of petroleum. Therefore, analyses of jointing have been important in both the exploration and the development of these resources. Joints also control the deposition of copper, lead, zinc, mercury, silver, gold, and tungsten ores. Hydrothermal solutions associated with igneous intrusions migrate along joint systems and precipitate minerals along the joint walls, thus forming mineral veins. Modern prospecting techniques therefore include detailed fracture analyses.

Major construction projects (such as dams) are especially affected by jointing systems within rocks, and allowances must be made for them in the project planning. For example, the Flaming Gorge Dam, on the Green River in northeastern Utah, was constructed in a small bend of the river nearly parallel to the major vertical joint system. Stresses caused by water storage therefore tend to close the fractures. Had the dam site been selected in the part of the river that trends from east to west, the joint planes would have presented major structural weaknesses in the bedrock foundation.

Joint systems can be either an asset or an obstacle to quarrying operations. Closely spaced joints severely limit the sizes of blocks that can be removed. If a quarry follows the orientation of intersecting joints, however, the expense of removing building blocks is greatly reduced, and waste is held to a minimum.

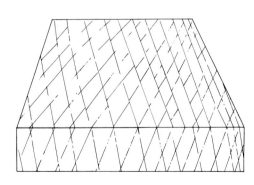

Figure 20.23 Joint systems in resistant sandstones in Arches National Park, Utah, have been enlarged by weathering so that they form deep, narrow crevasses. The set of intersecting joints reflects the orientation of stress that deformed the rock body.

UNCONFORMITIES

Statement

A record of ancient earth movements is also contained in unconformities in a rock sequence. An unconformity is a surface between two rock bodies that represents a substantial break, or hiatus, in the geologic record. It represents a period of time when deposition stopped, erosion removed some of the rock, and then deposition resumed.

Unconformities are classified into three main categories, based on the structural relationships between the underlying and overlying rocks:

1. Angular unconformities
2. Disconformities
3. Nonconformities

Each records three important geologic events:

1. The formation of the older rocks
2. Uplift and erosion
3. Subsidence and burial of this ancient erosion surface beneath younger strata

Discussion

Angular Unconformities

The relationships among the rock bodies shown in *figure 20.24* exemplify an **angular unconformity**. To appreciate its significance, consider what the angular discordance implies. The older strata were deposited first, then folded or tilted, and then eroded to an essentially horizontal surface. Subsidence followed, and the overlying beds were deposited.

Disconformities

A **disconformity** is an erosion surface carved on horizontal rocks without tilting or folding (*figure 20.25*). The rock strata above and below the erosion surface are parallel. Erosion can strip off the top of the older sequence and cut channels into the older beds, but there is no structural discordance between the two rock bodies.

Disconformities are produced by broad, gentle upwarps in which the layered sediments on the sea

Figure 20.24 An angular unconformity exposed in the wall of the Grand Canyon represents a major structural discontinuity in the rock sequence. The lower sequence of strata was tilted and eroded prior to the deposition of the younger horizontal sequence, which makes up most of the canyon wall. This deformation and erosion occurred more than 700 million years ago.

Figure 20.25 Disconformities do not show angular discordance, but an erosion surface separates the two rock bodies. The channel in the central part of this exposure reveals that the lower shale units were deposited and then eroded before the upper units were deposited.

floor are elevated above sea level without folding. The rocks are eroded, and then subsidence lowers the area below sea level again without significant tilting. A new sequence of beds is then deposited essentially parallel to the stratification of the older sequence. Generally, disconformities are more difficult to recognize than angular unconformities, because the beds above and below the erosion surface are parallel.

Nonconformities

A **nonconformity** is an unconformity separating intrusive igneous or metamorphic rocks from overlying sedimentary strata. A structure of this sort is shown in *figure 20.26.* Clearly, the light-colored granite dike in the center of the photograph did not invade the overlying horizontal sand-

stone. Moreover, the basal layers of the sandstone contain pebbles of granite and coarse sand of quartz and feldspar. These were derived from the weathering and disintegration of the granite below, indicating a significant period of erosion.

Why does sandstone deposited on a granite or a metamorphic rock indicate crustal mobility and a major discontinuity in the rock-forming processes? The answer becomes apparent if we consider the environment in which intrusive igneous and metamorphic rocks form. Both originate deep within the earth's crust: the granite cools slowly at great depth, and metamorphic rock recrystallizes at high temperatures and pressures far below the surface. Before these rocks can be weathered and eroded, the overlying cover of rock must be removed. Uplift must occur in order for these rocks to be uncovered and exposed. Thus, sedimentary rocks deposited upon granitic or metamorphic rocks imply three major events: (1) the formation of an ancient sequence of rocks, (2) uplift and erosion to remove the cover and expose the granites or metamorphic rocks to the surface, and (3) subsidence and deposition of younger sedimentary rocks on the eroded surface.

Studies of unconformities show the close relationship among crustal movements, erosion, and sedimentation, and thus provide important insight into crustal movements during the earth's history.

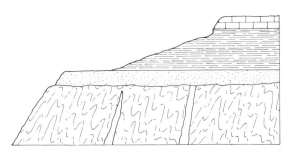

Figure 20.26 A nonconformity is an unconformity in which sedimentary rocks were deposited on the eroded surfaces of metamorphic or intrusive igneous rocks. The metamorphic rocks and the igneous dikes shown in this photograph were formed at great depths in the crust. The rocks that covered them when they formed were subsequently eroded away, so that the igneous and metamorphic rocks were exposed at the surface. The horizontal sedimentary rocks, lying above them, mark the nonconformity.

SUMMARY

Deformation of the earth's crust is well documented by historical movements along faults, raised beach terraces, and contorted rock layers. The major types of rock structures are folds, faults, and joints. Folds in stratified rocks range in size from microscopic wrinkles to large flexures hundreds of kilometers long. In the stable platform, sedimentary rocks are mildly warped into broad domes and basins. More intense deformation occurs in mountain belts, where rocks are folded into tight, plunging anticlines and synclines. Intense deformation in parts of some mountain ranges produces complex folding.

Faults are fractures in the earth's crust along which slippage or displacement has occurred. Three basic types of faults are recognized: (1) normal faults, resulting from tension; (2) thrust faults, resulting from compression; and (3) strike-slip faults, resulting from shear stresses. Normal faults develop on a large scale in rift systems. Thrust faults generally are associated with tight folds and occur at convergent plate margins. Strike-slip faults are associated with shearing along passive plate margins.

Joints are fractures with no noticeable displacement. They are associated with all types of deformation.

Unconformities are major discontinuities in rock sequences and indicate ancient crustal deformation.

The various kinds of crustal deformation record plate movement and a history of the earth's dynamics.

Key Words

orogenesis	monocline
orogeny	anticline
crustal warping	syncline
upwarp	limb
downwarp	axial plane
strike	axis
dip	plunge
fold	dome
basin	fault block
plunging fold	horst
overturned fold	graben
fault	sag pond
hanging wall	tectonic creep
footwall	joint
normal fault	unconformity
thrust fault	angular unconformity
reverse fault	disconformity
strike-slip fault	nonconformity
scarp	

Review Questions

1. List evidence that the earth's crust is in motion and has moved throughout geologic time.
2. What is orogenesis?
3. Explain the terms *dip* and *strike*.
4. Sketch a cross section of the structure of the rocks shown in *figure 20.6*.
5. Sketch an anticline and the adjacent syncline and label the following features: (a) axial plane, (b) axis, (c) angle of plunge, and (d) limbs.
6. Sketch the outcrop pattern of a plunging fold.
7. Describe the development of complex alpine-type folds.
8. Draw a simple block diagram of a normal fault, a thrust fault, and a strike-slip fault, and show the relative movement of the rock bodies along each. List the defining characteristics of each type of fault.
9. What are horsts and grabens? What global tectonic features are found where horsts and grabens most commonly are formed?
10. What global tectonic features are found where thrust faults commonly are formed?
11. List some of the surface features that commonly are produced by strike-slip faults.
12. Explain the sequence of events implied by an angular unconformity.

Additional Readings

Billings, M. P. 1972. *Structural geology.* 3d ed. Englewood Cliffs, N.J.: Prentice-Hall.

Spencer, E. W. 1977. *Introduction to the structure of the earth.* 2d ed. New York: McGraw-Hill.

21

EVOLUTION OF THE OCEAN BASINS

During the past few decades, while space exploration has held the attention of popular science, oceanographers have been quietly at work mapping the topography of the ocean floor. Through this work, in which many nations have participated, a great amount of bathymetric data has been compiled on regional physiographic maps of the surface features of the ocean floor. Special seismic reflection techniques have been developed to show the thickness of the sediment overlying the bedrock of the oceanic crust. The results are as important as they are spectacular. A great "mountain range," called the oceanic ridge, extends as a continuous unit through all of the ocean basins, a distance of more than 65,000 km. The ridge is cut by long fracture zones, some of which can be traced for several thousand kilometers. Flanking the oceanic ridge is the abyssal floor, above which rise numerous submarine volcanoes, called seamounts. The deepest parts of the ocean are the trenches, some of which descend to depths of 11,000 m. In addition, small ocean basins, commonly referred to as seas, have also been explored. These are mostly isolated from the major oceans and originate in a variety of ways.

The new knowledge of the landforms on the ocean floor helped revolutionize geological thinking. For the first time, we have been able to consider all parts of the planet. This new knowledge contributed greatly to the development of the unifying theory of plate tectonics. No amount of research on the continents alone could reveal what we now know from studies of the ocean floor and the continents together.

In this chapter, we will discuss the geology of the ocean basins in some detail, because it provides some of the most important information we have concerning the dynamics of the earth.

MAJOR CONCEPTS

1. The major features of the ocean floor are (a) the oceanic ridge, (b) abyssal hills, (c) abyssal plains, (d) trenches, (e) islands and seamounts, and (f) continental margins. Their origin is best explained by the theory of plate tectonics.
2. The oceanic crust is composed of four major layers: (a) a surface layer of marine sediment, (b) a layer of pillow basalt, (c) a zone of sheeted dikes, and (d) a layer of gabbro.
3. Although the ocean basins originated by sea-floor spreading, they are quite different in size, shape, and topography. These differences result largely from differences in ages and stages of development.
4. During the last 200 million years, the shifting plates have produced the present continents and oceans.

SEA-FLOOR MORPHOLOGY

Statement

The surface features of the continents are familiar to everyone, but the landscape of the ocean floor remained largely a mystery until the development of modern oceanographic research. Even today, only small sections of the ocean floor can be seen through photographs and television (see *figures 19.2* and *19.3*). The regional landscape of the sea floor, however, can be studied with a variety of new instruments, including echo sounders, seismic reflection profilers, and magnetometers. These instruments provide information that has been used to make maps and charts showing relief forms on the ocean floor, in detail similar to that provided by early topographic maps of the land. Within the past two decades, our knowledge of the ocean floor has increased enormously, and we have gained our first real understanding of the rocks and landforms of an area covering three-fourths of the planet. The major landforms of the ocean basins are

1. The oceanic ridge
2. The abyssal floor
3. Trenches
4. Seamounts
5. Continental margins

Discussion

Seismic Reflection Profiles

The **seismic reflection profile** reproduced in *figure 21.1* illustrates the type of information obtained about the sea floor by modern **bathymetry**. Pulses of sound emitted from a ship are reflected back from the sea floor. Some of the energy, however, penetrates the sediment and is reflected from the hard rock surface below. Automatic electronic equipment plots a continuous profile of the sea floor and the structure of the unconsolidated sediments. An example of the resulting data is shown in *figure 21.1*. This profile is a cross-sectional view of the geologic structure of the ocean floor flanking the mid-Atlantic ridge. It shows that the oceanic sediments are relatively flat and unformed, with bedrock protruding through them in irregular mounds.

From these records, we have been able to compile accurate **bathymetric charts** of the sea floor (see the physiographic map on the endpapers of this book) and can identify a variety of new and exciting geologic features not found on the continents. This is what we have found.

The Oceanic Ridge

The **oceanic ridge** is the most pronounced tectonic feature on the earth. If the ridge were not

Figure 21.1 A seismic reflection profile provides a wealth of information about the sea floor. The profile shows the morphology of the ocean floor and the configuration of unconsolidated marine sediments resting upon the solid bedrock.

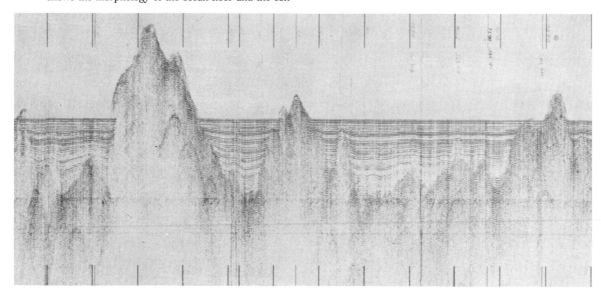

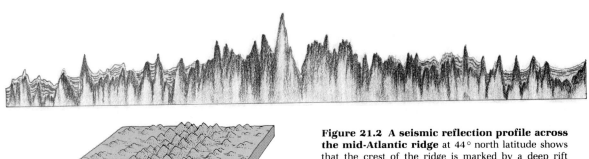

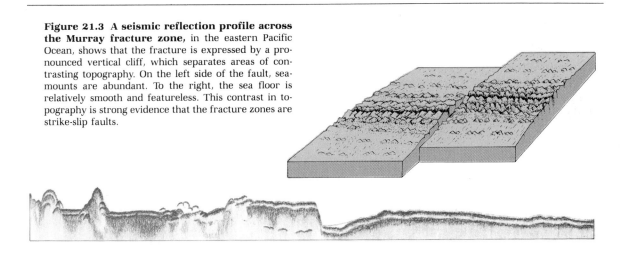

Figure 21.2 A seismic reflection profile across the mid-Atlantic ridge at 44° north latitude shows that the crest of the ridge is marked by a deep rift valley, which can be traced along the entire length of the ridge. Sediment is thickest down the flanks of the ridge, but it thins rapidly near the crest. The idealized diagram of the ridge was drawn on the basis of a series of profiles.

covered with water, it certainly would be visible from as far away as the moon. It is essentially a broad, fractured swell, generally more than 1400 km wide, with peaks rising as much as 3000 m above the surrounding ocean floor. The remarkable characteristic of the ridge is that it extends as a continuous feature around the entire globe, like the seam of a baseball. It extends from the Arctic basin, down through the center of the Atlantic, into the Indian Ocean, and across the South Pacific, terminating in the Gulf of California, a total length of more than 65,000 km. Without question, it is the greatest "mountain" system on the earth. The "mountains" of the ocean ridge, however, are nothing like the mountains of the continents, which were built largely of folded sedimentary rocks. By contrast, the ridge is composed entirely of basalt and is not deformed by folding.

Many of the characteristics of the ridge are ap-

parent in the seismic reflection profile reproduced in *figure 21.2*. On a regional basis, the ridge is a broad, uparched segment of the ocean floor, broken by numerous fault blocks, which form linear hills and valleys. The highest and most rugged topography is located along the axis, and a prominent **rift valley** marks the crest of the ridge throughout most of its length. As is shown in *figure 21.2*, oceanic sediments are thickest down the flanks of the ridge but thin rapidly toward the crest.

Throughout most of its length, the oceanic ridge is cut by a series of transform faults (*figure 21.3*). These are sites of continued seismic activity. Beyond an active transform fault, the fracture zone is expressed by an abrupt, steep cliff, which in places can be traced for several thousand kilometers.

Detailed studies of the axial spreading zone of

Figure 21.3 A seismic reflection profile across the Murray fracture zone, in the eastern Pacific Ocean, shows that the fracture is expressed by a pronounced vertical cliff, which separates areas of contrasting topography. On the left side of the fault, seamounts are abundant. To the right, the sea floor is relatively smooth and featureless. This contrast in topography is strong evidence that the fracture zones are strike-slip faults.

the mid-Atlantic ridge were made in 1974, when scientists in deep-diving vessels sampled, observed, and photographed the ridge for the first time. Without doubt, this project made some of the most remarkable submarine discoveries of modern times. The photographs show extensive pillow basalts, so recent that little or no sediment covers the fine details of their surface textures (*figures 19.2* and *19.3*). Numerous open fissures in the crust were also observed and mapped. In one small area of only 6 km², 400 open fissures were mapped, some of which are as wide as 3 m. These are considered conclusive evidence that the oceanic crust is being pulled apart. The eruption of lava from these fractures, which parallel the rift valley, would tend to create long, narrow ridges—a feature that does indeed characterize the morphology of the oceanic ridge.

The general character of the oceanic ridge seems to be a function of the rate of plate separation. Where the rate of spreading is relatively low (less than 5 cm per year), the ridge is higher and more rugged and mountainous than it is where rates of spreading are more rapid. Moreover, rift valleys on slow-spreading ridges are prominent, whereas rift valleys in areas with high rates of spreading are more subdued.

The Abyssal Floor

Vast areas of the deep ocean consist of broad, relatively smooth surfaces known as the **abyssal floor.** This type of sea-floor topography was discovered in 1947 by oceanographic expeditions surveying the mid-Atlantic ridge. Subsequently, much of it has been mapped in detail with precision depth recorders, which are capable of measuring elevations on the ocean floor with relief as small as

2 m. The abyssal floor extends from the flanks of the oceanic ridge to the continental margins, generally lying at depths of about 3000 m.

In most ocean basins, the abyssal floor can be subdivided into two sections, the abyssal hills and the abyssal plains. The **abyssal hills** are relatively small hills, rising from 75 to 900 m above the ocean floor (*figure 21.4*). They are circular or elliptical and range from 1 to 8 km wide at their bases. The hills are found along the flanks of the oceanic ridge and occur in profusion in parts of the ocean floor separated from land by trenches. In the Pacific, they cover between 80% and 85% of the ocean floor. Thus, abyssal hills can be considered the most widespread landform on the earth.

The **abyssal plains** are exceptionally flat surfaces of the ocean floor where the abyssal hills are completely buried by sediment (*figure 21.4*). Commonly, they are located near the margins of a continent, where sediment from a continental mass is transported by turbidity currents and spreads over the adjacent ocean floor.

The origin of the abyssal hills and abyssal plains can be traced to the oceanic ridge and the development of new crust at the spreading center. New lithosphere, created at the ridge crest, slowly recedes from the rift zone and is gradually modified in several important ways. The new crust cools and contracts, deepening the ocean basin as it moves away from the ridge. Fine-grained **pelagic sediments,** consisting of dust and shells of marine organisms, slowly but continually settle over all of the sea floor. Linear hills, formed by volcanic activity, intrusions of magma, and block faulting, become the foundations of the abyssal hills. As plates move away from the spreading center, the superficial features of the landforms created at the

Figure 21.4 A seismic reflection profile across the abyssal floor of the Atlantic Ocean shows abyssal hills buried with sediment, which forms the smooth abyssal plains.

ridge are gradually modified and concealed. Eventually, they can be completely buried with sediment, which forms the flat abyssal plains. The outline of the buried rock surface can be traced in seismic reflection profiles (such as *figure 21.4*).

The distribution of abyssal plains and abyssal hills substantiates this explanation of their origin. Abyssal plains occur only where the topography of the sea floor does not inhibit turbidity currents from spreading sediment from the continents over the sea floor. In the Atlantic Ocean, abyssal plains occur near the margins of the continents of North America, South America, Africa, and Europe. In the Pacific Ocean, by contrast, there are few abyssal plains, because turbidity currents cannot flow past the deep trenches that lie along most of the continental margins. The inflowing sediment accumulates in the trenches, so that most of the Pacific floor lacks abyssal plains and is covered instead with abyssal hills. The largest abyssal plains in the Pacific are found in the northeast, off the coast of Alaska and western Canada. This is the only significant segment of continental mass in the Pacific that is not bordered by deep trenches. In the North Pacific, between the Aleutian Trench and the continental shelf, the Bering abyssal plain covers most of the deep-sea floor north of the Aleutian Islands. This deep basin is underlain by an abnormally thick section of layered sediment, which in places is as thick as 2 km. As the physiographic map shows (see the endpapers of this book), the Aleutian ridge has cut off this corner of the Pacific basin, acting like a dam behind which sediments have rapidly accumulated. In the Indian Ocean, abyssal plains occur only along the margins of Africa and India. They are absent in the eastern part of the ocean, where the deep Java Trench, along the continental margins, acts as a sediment trap. Another major area of abyssal plains lies off the northern shore of Antarctica.

Large cone-shaped or fan-shaped deposits of sediment derived from the continents lie on the abyssal plains offshore from most of the world's great rivers. These are called **deep-sea fans** (see *figure 6.24*). They resemble alluvial fans and deltas on land in that they are fan-shaped accumulations of sediment located at the mouth of a river. They are different, however, because the sediment is transported and deposited primarily by turbidity currents. Most large fans are located at the base of the continental slope, with their apexes at the mouth of a submarine canyon cut into the edge of the shelf. Turbidity currents intermittently flush sediment through the canyon, building up depositional fans where they reach the lower gradient of the ocean floor.

Most, if not all, fans are marked by one or more deep-sea channels, which usually are the extension of submarine canyons cut in the continental slope. As the slope flattens, the channels develop natural levees, much like those formed by low-gradient streams on land.

The submarine fan of the Ganges River, in the northwestern Indian Ocean, is by far the largest deep-sea fan in the world. It is over 2800 km long and covers slightly more than 4,000,000 km². This accumulation represents about 70% of the debris derived from the erosion of the Himalayas. The remaining 30% goes to the flood plain and delta of the Ganges and to the Indus River, to the west.

Other large fans are the Indus fan, on the western side of the Indian peninsula; the Amazon and the Congo fans, in the South Atlantic; the Mississippi fan, in the Gulf of Mexico; and the Laurentian fan, in the North Atlantic. A number of smaller fans have been mapped off the Pacific coast of North America. There are no large fans in the Pacific, because most major rivers drain into the Atlantic and Indian oceans, and deep-marine trenches trap most of the sediment that does flow into the Pacific.

Trenches

A subduction zone, where two plates converge and one slab of lithosphere plunges down into the mantle, is expressed topographically by a **trench**. We have seen in previous chapters that a subduction zone is characterized by intense volcanic activity and seismicity. It is also marked by a large **gravity anomaly**.

Trenches, some of which reach nearly 11,000 m below sea level, are the deepest parts of the ocean. As is shown in the seismic reflection profile in *figure 21.5*, they typically are asymmetrical. A relatively steep slope lies on the landward side, along the continental landmass, and a more gentle slope lies on the side of the ocean basin. Individual trenches less than 100 km wide can form continuous features extending for several thousand kilometers across the deep-ocean floor.

The most striking examples occur in the western Pacific. A trench system there extends from the vicinity of New Zealand to Indonesia to Japan and then northeastward along the southern flank of the Aleutian Islands. Long trenches also occur along the western coast of Central America and South America, in the Indian Ocean west of Australia, in the Atlantic off the tip of South America, and in the Caribbean Sea.

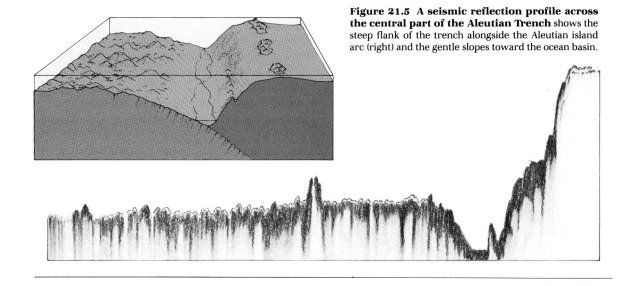

Figure 21.5 A seismic reflection profile across the central part of the Aleutian Trench shows the steep flank of the trench alongside the Aleutian island arc (right) and the gentle slopes toward the ocean basin.

Islands and Seamounts

Literally thousands of submarine volcanoes occur on the ocean floor, with the greatest concentration in the eastern Pacific. Some rise above sea level and form **islands,** but most remain submerged and are called **seamounts** (*figure 21.6*). They often occur in groups or chains, with individual volcanoes being as much as 100 km in diameter and 1000 m high. A **guyot** is a special type of seamount, whose top is a flat, mesalike surface, rather than a cone.

Chains of islands and seamounts are believed to develop as the sea floor moves over a hot spot in the upper mantle. The hot spot presumably forms from a huge column of upwelling lava known as a **mantle plume.** It lies in a fixed position under the lithosphere (see *figure 19.13*).

The origin of guyots, however, has been the subject of much debate. There is little doubt that they are ancient erosional platforms developed by stream and wave action on volcanic islands. Dredge samples show that some guyots are covered with volcanic debris that has been eroded and washed by wave action. Many are coated with coral, which can grow only in very shallow water. The big ques-

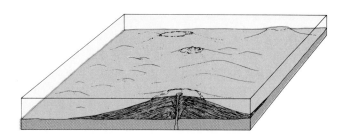

Figure 21.6 A seismic reflection profile across seamounts in the central Pacific Ocean shows the general configuration of typical seamounts rising above the ocean floor. Seamounts are submarine volcanoes, which usually occur in groups or chains. Some rise above sea level to form islands.

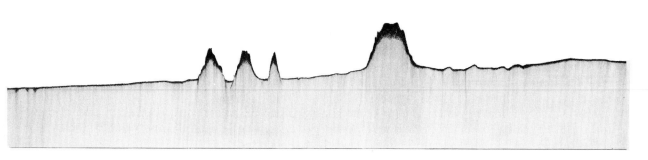

tion raised by these findings is how the guyots became submerged from 1500 to 3000 m below sea level. Also, many atolls are composed of coral reefs thousands of meters deep, all of which had to originate near sea level. These facts are accounted for by the theories of sea-floor spreading and plate tectonics. Initially, as a volcano forms near the crest of the oceanic ridge, it rises above sea level, because this area of the ocean floor is arched up by ascending convection currents. Wave action then erodes the top of the island to a flat surface, and coral reefs grow on this platform near sea level. As sea-floor spreading continues, the flat-topped island migrates off the ridge and becomes submerged. If environmental conditions are favorable, coral reefs can continue to grow upward as the seamount slowly moves off the flanks of the ridge and the sea floor sinks deeper. Thus, atolls such as Bikini Atoll could develop a reef rock over 1000 m thick as the island slowly submerged.

Continental Margins

The **continental margins** are covered by the ocean but are not geologically part of the oceanic crust. They are composed of continental crust and sediment derived from erosion of the land.

This part of the sea floor can be divided into two major sections, the continental shelf and the continental slope. The **continental shelf** is simply a submerged part of the shield or stable platform. The gently sloping platform extends from the shoreline to the area where the continental margin begins its steep descent to the ocean floor. The shelf can be as much as 1500 km wide and its depth ranges from 20 to 550 m at its outer edge. At present, the continental shelves comprise 18% of the earth's total continental area. At times in the geologic past, however, they were much larger, because the oceans spread much farther over the continental platforms.

Characteristically, the continental shelf is smooth and flat, but its topography has been influenced greatly by changes in sea level. Large areas were once exposed as dry land and thus were subjected to subaerial processes, so that the shelf topography can have features formed by both marine and nonmarine processes.

The **continental slope** descends from the outer edge of the continental shelf as a long, continuous slope to the deep-ocean basin (*figure 21.7*). It marks the edge of the continental granitic rock mass, the boundary between the continental crust and the oceanic crust. Continental slopes are found around the margins of every continent and around smaller pieces of continental crust such as Madagascar and New Zealand. Study the continental slopes shown on the physiographic map (see the endpapers of this book), especially those surrounding North America, South America, and Africa, and note that they form one of the earth's major topographic features. They are by far the steepest, longest, and highest slopes on the earth. Within this zone, from 20 to 40 km wide, the average relief above the sea floor is 4000 m. Along the marginal trenches, relief is as great as 10,000 m. In contrast to the shorelines of the continents, the edges of continental slopes are relatively straight over distances of thousands of kilometers. These long, straight margins are the

Figure 21.7 A seismic reflection profile across the western continental slope of Africa shows the profile of several submarine canyons near the upper part of the slope and the thick accumulation of undeformed sediments on the continental margin. The continental slope merges into the adjacent abyssal plains, which cover the abyssal hills.

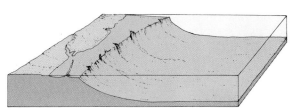

topographic expression of the geologic difference between the continental crust and the oceanic crust, reflecting a fundamental difference in structure and rock type.

Two types of continental slopes can be recognized. Each is subjected to different stresses, and thus they develop different characteristics. The slopes on the **trailing edge** of a continent are relatively passive and are not subjected to strong compressive forces. The slopes of the Red Sea are an example of a newly formed continental margin, caused by the separation of Africa from Arabia. The Atlantic continental slopes of North America and South America are similar and are considered to have originated in much the same way. The major difference is that the Atlantic slopes are much older and have been modified considerably by erosion and sedimentation.

In contrast to a continent's passive trailing edge, the **leading edge** is subjected to much more stress,

since it impinges on another moving plate. Crustal deformation results. Where a continent converges with an oceanic plate, the oceanic plate moves down and under the lighter continental crust, and a trench develops along the continental margin. This is the case along the Pacific coast of South America and Central America, where the continental slopes tend to be straight and seismically active. Volcanic activity is widespread on the adjacent continental mass.

Submarine Canyons

Submarine canyons are common along the continental slope and have been studied for many years, long before the ocean floor was mapped. As is shown on the physiographic map (see the endpapers of this book), they typically cut through the edge of the continental shelf and terminate on the deep abyssal floor, some 5000 or 6000 m below sea level. The profile in *figure 21.7* crosses three can-

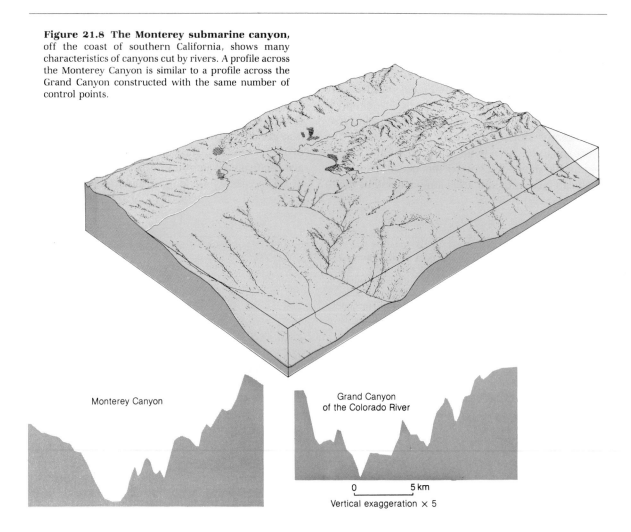

Figure 21.8 The Monterey submarine canyon, off the coast of southern California, shows many characteristics of canyons cut by rivers. A profile across the Monterey Canyon is similar to a profile across the Grand Canyon constructed with the same number of control points.

Monterey Canyon

Grand Canyon of the Colorado River

0   5 km

Vertical exaggeration × 5

yons near the upper part of the continental slope. Submarine canyons have a V-shaped profile and a system of tributaries, so that they closely resemble the great canyons cut by rivers on the continents (*figure 21.8*). Many pioneer researchers therefore suggested that submarine canyons were also cut by rivers, but the problem of how this could happen remained unanswered. Some canyons are 6000 m below sea level, and it is difficult to conceive how sea level changed enough for rivers to cut to that depth. It was suggested that the continents were uplifted thousands of meters, were dissected by streams, and then subsided, so that the canyons became submerged.

More recently, submarine canyons were thought to have been eroded by turbidity currents, but it is not known whether turbidity currents are capable of such erosion. It is clear, however, that much sediment has flowed through many of the canyons, be-

cause of the great volumes of sand and mud deposited at their lower ends as deep-sea fans.

The theory of plate tectonics can be applied to explain the origin of canyons as a product of stream erosion during the initial stage of continental rifting (*figure 21.9*). Convection currents beneath a continent arch up the overlying continental crust and begin to pull it apart. The initial response is the formation of rift valleys, like those in East Africa. The sides of the rift valleys are soon attacked by erosion, and many canyons and ravines are cut into them. As the continents separate completely and move off the uparched spreading center, the margins become submerged and the canyons are drowned. They then act as channels for turbidity flows and are further modified by them. The canyons, therefore, could be due to initial erosion by streams and later modification by turbidity currents.

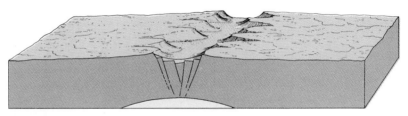

(A) In the initial stage of rifting, a continent is arched upward and pulled apart by convection currents in the asthenosphere. Erosion cuts canyons into the sides of the rift valley, where new continental margins will form as rifting continues.

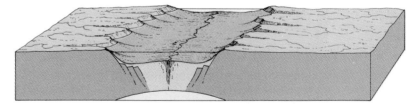

(B) The two fragments of the rifting continent move away from the uparched area, and the margins become partly submerged. The canyons cut by erosion on the sides of the rift valley become submarine canyons.

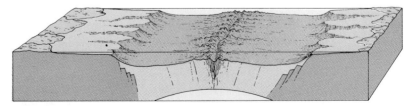

(C) As the continents drift further from the spreading center, the canyons are more deeply submerged and are modified by turbidity currents.

Figure 21.9 Submarine canyons may be formed by continental erosion along a young rift zone.

THE COMPOSITION AND STRUC-TURE OF THE OCEANIC CRUST

Statement

With the growth of knowledge about the ocean basins came considerable efforts to determine the composition and structure of the oceanic crust and how it differs from the continental crust. A variety of methods have been used, including seismic reflection, drilling, and studies of fragments of oceanic crust thrust up on the surface. Among the more significant facts we have learned about the oceanic crust are the following:

1. The oceanic crust is composed of four major layers: (*a*) a surface layer of marine sediment, (*b*) a layer of pillow basalt, (*c*) a zone of sheeted dikes, and (*d*) a layer of gabbro.
2. The oceanic crust and its topographic features are related in some way to igneous activity.
3. The rocks of the ocean floor have not been deformed by compression, so that their structure stands out in marked contrast to that of rocks in the folded mountains and shields of the continents.

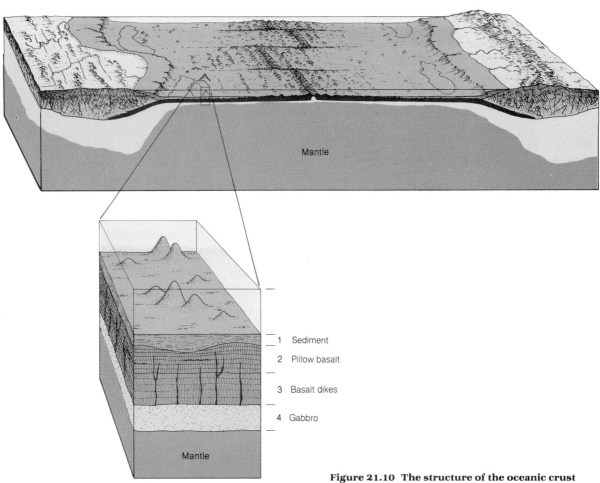

1 Sediment

2 Pillow basalt

3 Basalt dikes

4 Gabbro

Mantle

Mantle

Figure 21.10 The structure of the oceanic crust consists of four distinct layers. Layer 1 is a relatively thin sequence of oceanic sediments, composed of the shells of microscopic marine organisms mixed with red clay. Layer 2 is pillow basalt fed by numerous dikes. Layer 3 is almost entirely composed of basalt dikes in vertical sheets. Layer 4 is gabbro, which is believed to represent magma that was generated at a spreading center and cooled slowly at depth.

4. The rocks of the ocean floor are young in a geologic time frame. Most are less than 150 million years old, whereas the ancient rocks of the shields are more than 700 million years old.

Discussion

The diagrams in *figure 21.10* illustrate the main elements of the composition and structure of the oceanic crust, as it is presently understood. The typical rock sequence consists of four layers. At the top of the sequence (layer 1) is a relatively thin layer of sedimentary rock. This consists of calcareous and siliceous shells of microscopic marine organisms together with red clay, all of which typify deep-marine sediments. Layer 2 consists of pillow basalts that were originally fed by numerous dikes and intruded into vertical fractures. These now form a series of parallel sheets of coarse-grained basalt or gabbro. The pillow lava represents volcanic extrusions on the sea floor, and the dikes are feeder vents, which were produced as the plates split and moved apart. Below the pillow lava, the rock consists almost entirely of the sheeted dikes (layer 3), and many younger dikes are intruded into portions of earlier ones. Below the sheeted dike complex, the rocks are coarse-grained gabbro, which is believed to represent magma that was generated at a spreading center but cooled very slowly at some depth. Underlying the gabbro are peridotites, composed almost entirely of olivines and pyroxenes. This material is considered part of the mantle. The Moho represents the boundary between the gabbro and the peridotite.

Although seismic reflection studies and drilling on the ocean floor provide important data concerning the upper layers of the oceanic crust, our most direct and detailed information is obtained from areas where large fragments of the oceanic crust have been incorporated into a folded mountain belt at a convergent plate margin and are available for direct observations. The most notable exposures are in Cyprus, Greece, New Guinea, Newfoundland, and California. They show the exact sequence illustrated in *figure 21.10*.

A significant fact about the oceanic crust is that drilling samples brought up thus far are all much younger than the rocks that form the bulk of the continents. The oldest basalts retrieved from the ocean floor are only 50 million years old. There is little reason to suspect that any basalt on the ocean floor is older than 160 million years. Extensive drilling in recent years has produced no sediment older than 160 million years.

THE OCEAN BASINS

Statement

Although all of the ocean basins originated in the same manner (that is, by sea-floor spreading), they are quite different in size, shape, and topographic features. These differences are significant because they tell much about the ages, origins, and evolution of the ocean basins.

Discussion

The Atlantic Ocean

The regional topography of the Atlantic floor is basically simple and shows remarkable symmetry in the distribution of the major features (see the physiographic map on the endpapers of this book). The dominant feature is the mid-Atlantic ridge, which forms an S-shaped pattern down the center of the basin. It separates the ocean floor into two long, parallel subbasins, trending north and south, which are characterized by abyssal plains. Abyssal hills occur alongside the ridge, and the plains occur along the margins of the continental platforms. The symmetry of the Atlantic basin extends to the continental margins: the outlines of Africa and Europe fit those of South America and North America. This symmetrical distribution of the major features of the Atlantic has attracted scientific attention for many years and is one of the cornerstones of the theories of continental drift and sea-floor spreading. The structure and topographic features of the Atlantic are simple, reflecting the initial opening of this young ocean basin.

The Arctic Ocean

Although the Arctic Ocean has been considered by some to be an extension of the Atlantic, recent exploration shows that it is unique in several respects. (1) It is nearly landlocked. If the basin beyond the continental slope is considered, there is only one inlet, the Lena Trough (between Spitsbergen and Greenland). (2) The continental shelf north of Siberia is the widest shelf in the world (over 1100 km across). Most of it is deeper than 200 m, but very little exceeds 600 m in depth. (3) Two prominent submarine ridges trend roughly 140° east and divide the deep Arctic Ocean into three major basins.

The Indian Ocean

The Indian Ocean is the smallest of the three

great oceans. It connects with both the Atlantic and the Pacific through broad, open seas south of Africa and Australia. Like the Atlantic, its most conspicuous feature is the oceanic ridge. This continues from the Atlantic around southern Africa and splits near the center of the Indian Ocean to form a pattern similar to an inverted Y (see the physiographic map on the endpapers of this book). The northern segment of the ridge extends into the Gulf of Aden, apparently connecting with the African rift valleys and the Red Sea rift. Thus, the ridge divides the ocean basin into three major parts.

The topography of the Indian Ocean floor, unlike that of the other oceans, is dominated by scattered blocks and some remarkably linear plateaus called **microcontinents.** Most are oriented in a north-south direction. Prominent parallel fracture zones are numerous. The striking northward trend of the parallel fracture zones, together with the trend of the linear microcontinents, imparts a linear structural fabric to the Indian Ocean floor.

The Pacific Ocean

The Pacific differs somewhat from other oceans in that the oceanic ridge lies near its eastern margin. Its basin covers approximately half of the planet and is the largest single unit of oceanic crust. It is probably the oldest ocean basin, and it lacks the symmetry of the Atlantic and Indian basins.

The oceanic ridge continues in a broad sweep from the Indian Ocean, trends eastward between New Zealand and Antarctica, and then turns northward along the American side of the ocean. The crest of the ridge extends into the western United States at the head of the Gulf of California and is involved in the structure of the Basin and Range province in Utah and Nevada and the San Andreas Fault. It reappears off the coast of Oregon.

The floor of the western Pacific is studded with more seamounts, guyots, and atolls than all other oceans combined. As is apparent from the physiographic map (see the endpapers of this book), many of the seamounts form linear chains that extend for considerable distances. The margins of the Pacific are also different; they generally are marked by a line of deep, arcuate trenches. In the eastern Pacific, the trenches lie along the margins of Central America and South America, paralleling the great mountain systems of the Andes and the Rockies. Local relief from the top of the Andes to the bottom of the trench is 14,500 m (nearly nine times the depth of the Grand Canyon). In the western Pacific, a nearly continuous line of trenches extends from the margins of the Gulf of Alaska, along the

margins of the Aleutians, Japan, and the Philippines, and down to New Zealand.

Small Ocean Basins

A number of small deep-ocean basins are not directly connected to the major oceanic plates. Nevertheless, they are considered a direct consequence of plate tectonic processes. Most of them are nearly landlocked and connect to the main ocean by narrow gaps (*figure 21.11*). The Mediterranean Sea, the Red Sea, and the Sea of Japan are excellent examples. These small ocean basins originate in three ways: (1) by the growth of island arcs, (2) by the initial rifting of continental plates, and (3) by the convergence of continental plates. The small ocean basins are temporary features. Like the major oceanic plates, they can grow but are ultimately destroyed.

The growth of island arcs created the small ocean basins of the Pacific. They formed along subduction zones that are not directly adjacent to a continental plate, so that the associated volcanic arcs isolate segments of oceanic crust to form small independent basins, which are not directly attached to the major ocean basins. Several such basins extend from the Aleutian Islands to New Guinea. Most of them are shallower than the adjacent open ocean, partly because they trap the sediment washed in from the continents. Sediment in the Bering basin even overflows through gaps in the Aleutian arc and spills into the Aleutian Trench.

The Red Sea and the Gulf of California are examples of another type of small ocean basin, which develops in the initial rift zone, where the continents split and begin to spread apart. Basalt from the upwelling mantle commonly fills the new opening during the early stages of rifting, and sediment eroded from the adjacent continents can help keep the basin filled to near sea level. If spreading is rapid, a long, narrow ocean basin develops. An interesting and potentially important feature of these new basins is the circulation of hot water through the fresh magma. Rising through the sediments on the ocean floor, it forms pools of hot brine with concentrations of rare metals.

As the sea floor continues to spread, the intervening sea develops the characteristics of a major ocean basin: an oceanic ridge, abyssal plains, and continental slopes. The Arctic Ocean is apparently in this transitional stage from sea to ocean.

The Mediterranean and Black seas, in contrast to those previously described, are basins formed by the collision of continental plates, namely, Africa and Eurasia. They are remnants of the basin of the once vast Tethys Sea, which extended in an east-west direction and separated Pangaea into two large continental landmasses, **Laurasia** and **Gondwanaland,** 200 million years ago (*figure 21.12*). As Africa and India moved northward, the Tethys Sea gradually closed, although it continued to link the Atlantic and Indian oceans for many millions of years. Compression due to the northward movement of Africa and India produced the sinuous mountain belts of the Alps and the Himalayas, which are composed mostly of folded and uplifted deep-ocean sediments. If the present plate motion continues, the Mediterranean basin will eventually be completely closed, and Africa will be tucked under the Alps, much as India is thrust under the Himalayas.

Recent drilling in the Mediterranean basin indicates a fascinating history during the last 20 million years, the period when it became isolated from the major oceans. As is shown in *figure 21.11C*, the Mediterranean is almost landlocked, being connected with the Atlantic by a narrow gap through the Straits of Gibraltar. The deep basin, however, is completely isolated. This produces some very interesting hydrologic conditions. Evaporation removes more than 4000 km³ of water each year, but less than 500 km³ is replaced by rain and surface runoff from Europe and Africa. Each year approximately 3500 km³ of water flows in through the Straits of Gibraltar to maintain the Mediterranean at sea level. If the straits were closed, the

Figure 21.11 Small ocean basins originate in several ways. Those in the western Pacific and the western Atlantic developed when island arcs isolated part of the sea from the main ocean basins. The Mediterranean basin and the Black and Caspian seas represent remnants of the ancient Tethys Sea, which was closed by the convergence of India and Africa with Europe and Asia.

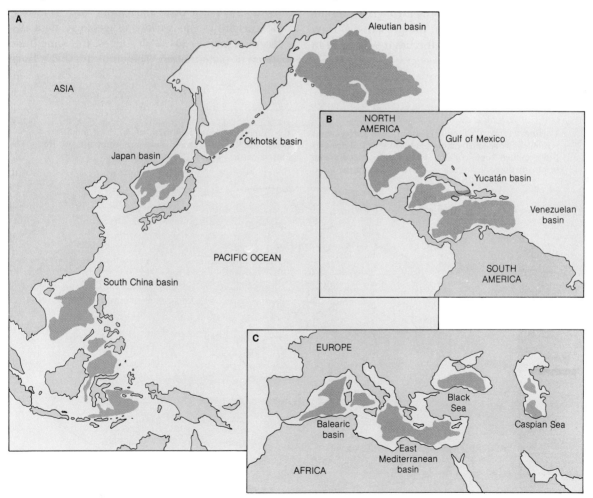

Mediterranean Sea would evaporate completely in about 1000 years.

Total evaporation of the Mediterranean actually happened between 5 million and 8 million years ago. Deep drilling by the *Glomar Challenger* has revealed that the Mediterranean basin was then a deep, desolate, dry ocean basin 3000 m below sea level. Core samples of windblown sand and salt deposits show that it was a desert region. Huge volcanoes (now islands) rose above the basin floor. The continental platforms of Africa and Europe surrounded the basin as huge plateaus. The area was undoubtedly devoid of life, since temperatures must have reached 65 °C.

The salt deposits and windblown sand that formed in the dry basin are covered throughout with deep-sea organic oozes, which imply that the basin was flooded almost instantaneously about 5.5 million years ago. The inundation probably resulted from erosion of the barrier at the Strait of Gibraltar, so that water from the Atlantic flowed into the Mediterranean basin over an enormous waterfall. Estimates based on fossils preserved in the sediments indicate that the flow through the straits was 1000 times greater than the present-day flow over Niagara Falls and that the basin was filled in about 100 years.

THE HISTORY OF PLATE MOVEMENT DURING THE LAST 200 MILLION YEARS

Statement

The tectonic system probably has operated during much of the earth's history, and it is believed to be responsible for the origin and evolution of continents as well as for the growth and destruction of ocean basins. Our understanding of the early history of plate movements comes mostly from evidence preserved in continental rocks. Ocean basins come and go, because the ancient oceanic crust is consumed at subduction zones and replaced by newer oceanic crust at spreading centers. Continents have drifted, with tectonic plates splitting and colliding a number of times, but details of the patterns of ancient plate movements are scanty.

There is a considerable amount of data on plate motion during the last 200 million years, however, and it is possible to reconstruct the position of continents and trace plate movement with considerable certainty. The variety of geologic data discussed on pages 330–32 all suggest the same basic pattern of plate motion. They indicate that a large

Figure 21.12 A map of the ancient Tethys Sea was made by reconstructing the positions of Africa and India before they collided with Eurasia. Sediments originally deposited in this sea have been deformed into the Himalaya-Alpine mountain chain. The Mediterranean, Black, and Caspian seas are remnants of the Tethys basin.

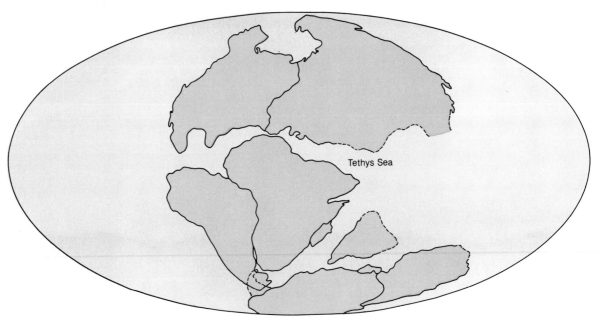

continental mass (Wegener's Pangaea) began to break up and drift apart about 200 million years ago. Dispersal and collision of the fragments have continued to the present time. The reconstruction of Pangaea in terms of absolute coordinates is now possible, and the directions and rates of plate movement have been determined.

Figure 21.13 The history of plate movement during the last 200 million years has been reconstructed from all available geologic and geophysical data. These maps show the general directions of movement from the time Pangaea began to break up until the continents moved to their present positions.

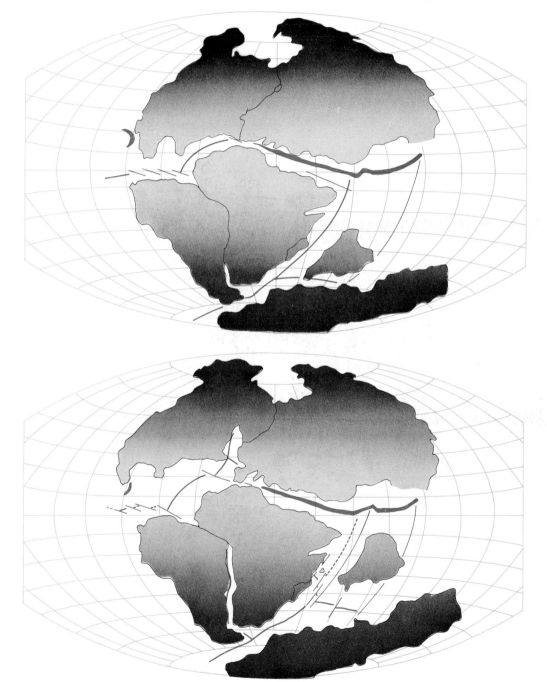

(A) Pangaea, 200 million years ago

(B) Plate movement, 150 million to 100 million ago

Discussion

The history of relative plate movement during the last 200 million years is shown in *figure 21.13*. These maps are adapted from those prepared by R. S. Dietz and J. C. Holden in 1970. Their maps were constructed with cartographic precision, using all available evidence, including the geologic fit of continents, paleomagnetic pole positions, and patterns of sea-floor spreading.

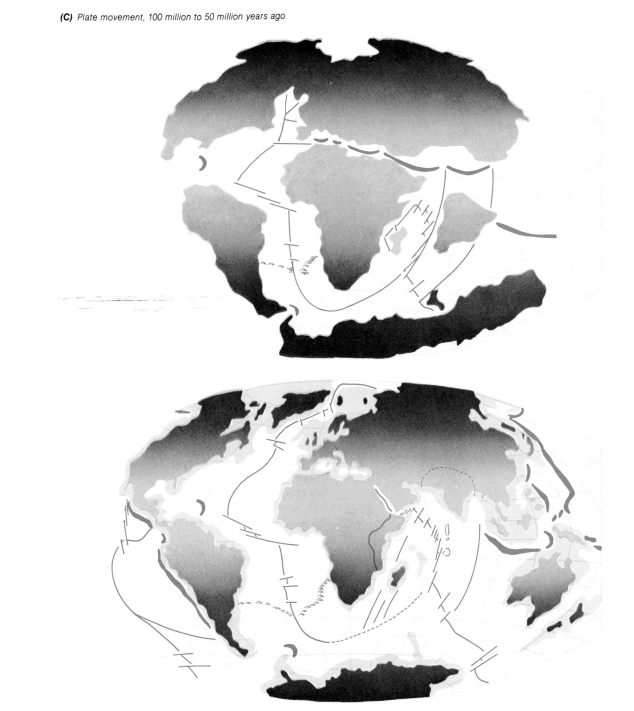

(C) *Plate movement, 100 million to 50 million years ago*

(D) *Plate movement, 50 million years ago to the present*

SUMMARY

The oceanic ridge is a broad, fractured swell, extending as a continuous feature for more than 65,000 km. It has a narrow, axial rift valley and is cut by numerous transverse fractures.

Abyssal hills are formed at or near the ridge and move away from it with the drifting plates. As they move, they become covered with sediment. They can eventually be buried to form abyssal plains.

Trenches form at subduction zones, where two plates converge and one is thrust down into the mantle. Typically, they are bordered by island arcs or young mountain ranges.

Volcanic islands or seamounts can originate near a spreading center and move with the spreading plates. They can also form over local hot spots in the mantle, so that linear chains of islands are produced as the plates move.

The continental shelf and slope are not parts of the oceanic crust but are submerged parts of the continental platform. The continental slope is dissected by many deep submarine canyons. These are similar in size and depth to the great canyons cut on the continents by rivers.

The oceanic crust is relatively thin (from 7 to 8 km) and is mostly basalt. It is composed of four major layers: (1) a surface layer of marine sediment, (2) a layer of pillow basalt, (3) a zone of sheeted dikes of basalt, and (4) a basal layer of gabbro. The oceanic crust is *not* deformed by tight folding or thrust faulting. It is geologically young: most of it was formed during the last 50 million years.

The three major ocean basins—the Atlantic, Indian, and Pacific—are believed to have originated by sea-floor spreading. They are different in size, shape, sea-floor topography, and age. The small ocean basins in the western Pacific, the Red Sea, the Caribbean, and the Mediterranean resulted from the growth of island arcs, from initial rifting, and from the convergence of continental plates.

We can outline the history of the ocean basins during the last 200 million years in some detail. The Atlantic, Indian, and Arctic oceans were formed by the rifting of a supercontinent (Pangaea). The major fragments of the supercontinent moved with the spreading plates, and as the ocean basins enlarged, the continents eventually arrived at their present positions.

Key Words

bathymetry	bathymetric chart
seismic reflection profile	oceanic ridge
rift valley	mantle plume
abyssal floor	continental margin
abyssal plains	continental shelf
abyssal hills	continental slope
pelagic sediment	trailing edge
deep-sea fan	leading edge
trench	submarine canyon
gravity anomaly	microcontinent
island	small ocean basin
seamount	Laurasia
guyot	Gondwanaland

Review Questions

1. Make a sketch map of the major types of landforms found on the ocean floor.
2. Study the seismic reflection profile in *figure 21.1* and label the following: (a) oceanic sediment, (b) bedrock, and (c) probable faults.
3. Explain the origin of the following: (a) the oceanic ridge, (b) fracture zones, (c) abyssal hills, (d) abyssal plains, (e) trenches, (f) seamounts, and (g) submarine canyons.
4. What is the major source of the sediment that covers the abyssal plains?
5. What is the significance of the fact that the thickness of pelagic sediments increases with distance from the crest of the oceanic ridge?
6. Many submarine volcanoes (guyots) have flat tops. Explain their origin.
7. Compare and contrast the Atlantic Ocean and the Pacific Ocean with respect to their shape, age, major structural features, and history.
8. Compare and contrast the Mediterranean Sea and the small ocean basins in the western Pacific with respect to such factors as structure, age, origin, and history.
9. Briefly outline the history of plate movement during the last 200 million years.

Additional Readings

Bullard, E. 1969. The origin of the oceans. *Scientific American* 221(3):66–75. Reprint no. 880. San Francisco: W. H. Freeman.

Heirtzler, J. R. 1968. Sea-floor spreading. *Scientific American* 219(6):60–70. Reprint no. 875. San Francisco: W. H. Freeman.

Menard, H. W. 1969a. The deep-ocean floor. *Scientific American* 221(3):126–42. Reprint no. 883. San Francisco: W. H. Freeman.

———. 1969b. Growth of drifting volcanoes. *Journal of Geophysical Research* 74(20):4827–37.

Moore, J. R., ed. 1971. *Oceanography: Readings from Scientific American.* San Francisco: W. H. Freeman.

Scientific American editors. 1969. *The ocean, A* Scientific American *book.* San Francisco: W. H. Freeman.

22

EVOLUTION OF THE CONTINENTS

Although the theory of plate tectonics was developed largely from new studies of the sea floor, a record of plate motion during most of geologic time is preserved only in continental rocks. The oceanic crust is a temporary feature, continually created along the oceanic ridge and destroyed along subduction zones. Plate movement can be studied and measured in the relatively young oceanic rocks, but all of the oceanic crust is less than 160 million years old.

By contrast, the rocks of the continents are as old as 3.8 billion years, and a record of plate movement during the early history of the earth is preserved only in the ancient metamorphic and igneous rocks of the continental shields. On the continental shields, geologists find evidence of a long and complex history of earth dynamics. This evidence suggests that the tectonic system has been operating during most of the time since the earth was formed.

In this chapter, we will consider the evolution of continental crust by a combination of deformation, metamorphism, igneous activity, erosion, sedimentation, and isostatic adjustment. The continents are products of the tectonic system. They grow by accretion but can also split and drift apart. All of these processes of continental evolution involve the fundamental process of segregation and differentiation of the materials of the earth.

MAJOR CONCEPTS

1. The continents are made up of three basic structural components: (a) shields, (b) stable platforms, and (c) young, folded mountain belts.
2. Orogenesis (mountain building) occurs at convergent plate margins.
3. The major features of orogenesis are (a) intensive crustal deformation, (b) metamorphism, and (c) igneous activity.
4. The characteristics of a mountain belt and the sequence of events that produced it depend on the rock types that are involved and the types of interactions at convergent plate boundaries.
5. Three convergent plate interactions are recognized: (a) convergence of two oceanic plates, (b) convergence of a continental and an oceanic plate, and (c) convergence of two continental plates.
6. Continents grow by accretion as new crustal material forms in an orogenic belt.

THE CONTINENTAL CRUST

Statement

The continental crust has been studied for more than 150 years, but only recently have we been able to synthesize the vast amount of data concerning continental geology and develop a reasonably clear concept of what continents really are. The more important geologic facts about continents can be summarized as follows:

1. Continents are composed of huge, flat slabs of granitic rock. This does not mean that all of the rocks of the continents are igneous, however. Much of the continental crust is metamorphic rock of roughly the same composition as granite. Seismic data suggest that there is a gradation from the granitic composition observed near the surface of a continent to a more basaltic composition near its base.

2. Continents range in thickness from 30 to 50 km. The thickest portions are beneath mountain ranges.

3. Continents cover roughly one-third of the planet and appear to have grown throughout most of geologic time.

4. The structure of continents is extremely complicated. It consists mostly of a highly deformed sequence of metamorphosed sediments and volcanic rocks, together with large volumes of granitic intrusions and granitic gneiss.

5. The continents contain the oldest rocks on the earth, which range in age to 3.8 billion years. By contrast, all of the oceanic crust is less than 160 million years old.

6. Preliminary studies indicate that the earth is

Figure 22.1 Structural trends in the metamorphic rocks of the Canadian Shield are emphasized by linear lakes. On a regional basis, the shield is a broad surface of low relief, eroded close to sea level.

the only terrestrial planet of the solar system with continental crust.

7. Continental crust results from planetary differentiation.

8. Although each continent may appear to be unique, they all have three basic components: (*a*) a large area of very old and highly deformed metamorphic and igneous rocks, known as the shield; (*b*) broad, flat stable platforms, where the shield is covered with a veneer of sedimentary rocks; and (*c*) young, folded mountain belts, located along the continental margins.

9. Geologic differences among continents are mostly in the size, shape, and proportions of the three basic components.

Discussion

Shields

The continental **shields** are the key to modern theories of the origin and evolution of continents. Take a moment to study *figures 22.1* and *22.2* and also *plate 3*.

The most striking characteristic seen in these photographs is the vast expanse of the low, rela-

Figure 22.2 Complex metamorphic rocks (dark tones) are intruded by granitic rocks (light tones) to form the complex structural characteristics of a shield. These rocks were formed in the roots of an ancient folded mountain belt during Precambrian time, approximately 1.8 billion years ago.

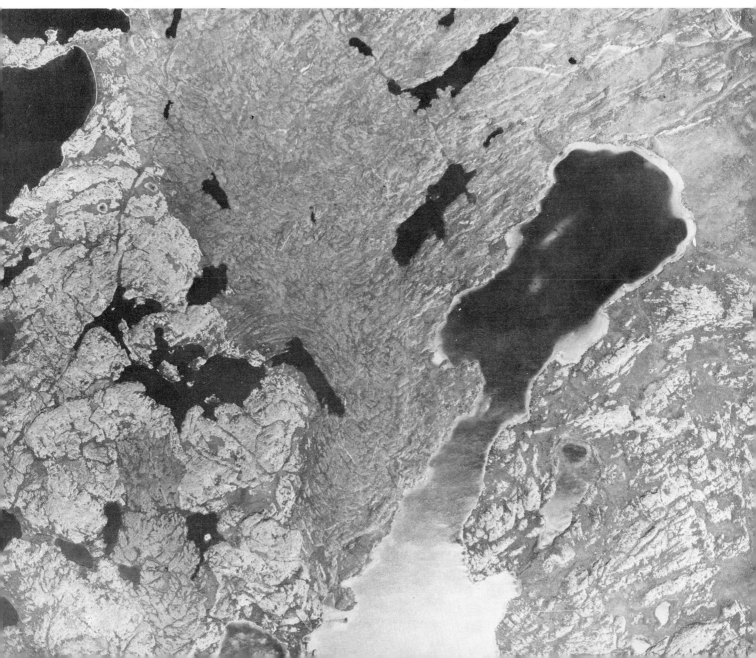

tively flat surface of the shield. Throughout an area of thousands of square kilometers, this surface lies within a few hundred meters of sea level. The only features that stand out in relief are resistant rock formations that rise a few tens of meters above the surrounding surface. On a regional basis, shields are flat and almost featureless. Their structural complexities are shown by patterns of erosion, the alignment of lakes, and differences in the tones of the photographs. Faults and joints are common. They are expressed at the surface by linear depressions, some of which can be traced for hundreds of kilometers.

A second fundamental characteristic of shields is their structure and composition. Shields are composed of a highly deformed sequence of metamorphic rocks and granitic intrusions, known as the **basement complex.** Most of the rocks are Precambrian, and most were formed under high temperature and highly compressive stresses several kilometers below the surface. In *figure 22.2,* metamorphic rocks appear in tones of dark gray. Granitic intrusions, which appear in lighter tones, have a more massive texture. *Figures 22.1* and *22.2* show evidence that the shields have been intensely deformed, so that sedimentary and volcanic rocks were converted into complex metamorphic rocks. They were also intruded by granitic magmas. Subsequently, erosion removed the upper cover of the sedimentary and metamorphic terrain, exposing what we now see at the surface.

From these facts, our present understanding of shields can be summarized graphically in *figure 22.3.* The basic structure of complex igneous and

Figure 22.3 The general characteristics of shields are depicted in these block diagrams. Shields are flat, low-lying surfaces eroded on complex igneous and metamorphic rocks.

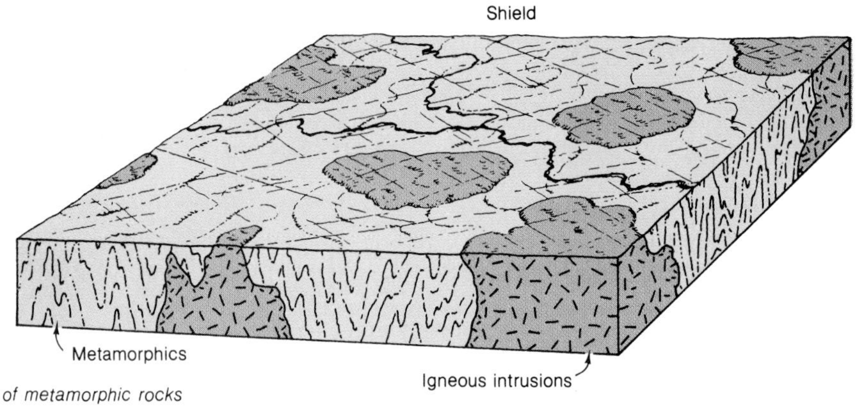

(A) The shield consists of a complex of metamorphic rocks and a variety of igneous intrusions. The upper surface is flat and is eroded down to near sea level.

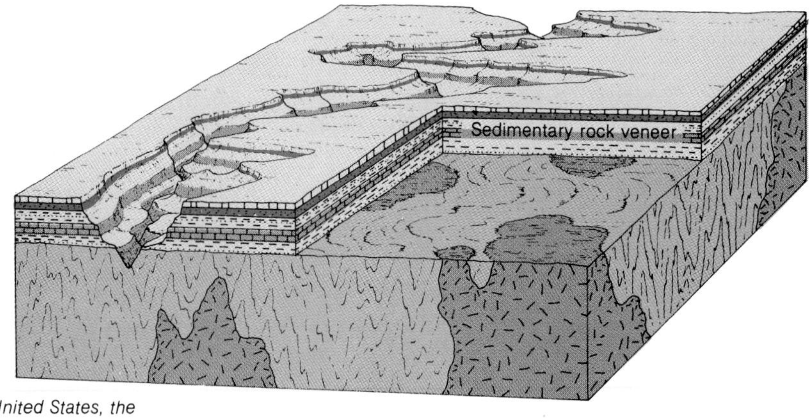

(B) Throughout most of the interior of the United States, the shield is covered with a veneer of horizontal sedimentary rock thousands of meters thick. In some places, such as the Grand Canyon, erosion has cut through the sedimentary veneer to expose the shield below.

Figure 22.4 Major structural trends and radiometric dates of rocks in the Canadian Shield show that the shield consists of several geologic provinces, each representing a mountain-building event. The numbers refer to the ages, in billions of years, of the major granites, and the lines represent the trends of the folds. The shield apparently grew by accretion as new mountain belts formed along its margins.

metamorphic rocks, eroded to a flat surface near sea level, forms the nucleus of every continent.

The belts of metamorphic rocks, together with igneous intrusions, indicate that the shields are composed of a series of zones that were once highly mobile and tectonically active. These facts have long been known. For many years, geologists have considered the shields to consist of a series of ancient mountain belts that have been eroded down to their roots and have since remained stable.

The Canadian Shield of North America is relatively well known. The map in *figure 22.4* summarizes many of its structural details, according to data compiled from years of fieldwork, geophysical studies, and radiometric dating. In a general way, North America is quite symmetrical. The oldest rocks, recording the first clearly recognizable geologic events, are in the Superior, Wyoming, and Slave provinces. They consist of volcanic flows and volcanic-derived sediments that are similar to those forming today in the island arcs of the Pacific. These sediments are metamorphosed and engulfed by granitic intrusions. The granite-forming event, as determined by radioactive dating, occurred between 2.5 billion and 3.0 billion years ago. Since the

earth has been estimated to be a little more than 4.5 billion years old, this segment of crust apparently formed from 1.5 billion to 2.0 billion years after the earth's origin. Generally speaking, the composition of this ancient continental crust (prior to granitic intrusion) was closer to basalt than to granite. Granitic intrusions now constitute three-fourths of the area. The metamorphosed sediments and volcanic rocks display structural trends in a northward direction, as is shown in *figure 22.4*.

This segment of crust is significant because it is thought to represent the first stable and resistant granitic crust in North America. Since its formation, the upper surface has been eroded down almost to sea level and has persisted as a stable unit. It has sometimes been partly submerged beneath a shallow sea and has sometimes rested slightly above sea level, but it was never again the site of mountain-building processes. This block of crust probably was once from 20% to 60% larger than it is now, because the trends of the metamorphic structures are terminated abruptly by younger provinces. We can assume that the original continental mass was split and fragmented by plate movement, with each segment subsequently acting as a center for future continental growth.

Surrounding the Superior province to the south, west, and north is the Central province, a vast area of gneiss and granite from 1.0 billion to 1.4 billion years old. The rocks in this younger province are also metamorphosed, but the original sediment was quite different from that which formed the rocks in the Superior province. These younger rocks contain less lava. They show a distinct increase in quartz-rich sandstone and limestone, which originally formed on a continental shelf. In addition, the structural trends are oriented in a different direction.

Beyond the Central province, the rocks are still younger, and the associated granitic intrusions range in age from 0.8 billion to 1.1 billion years. The rocks are mostly quartzites, marbles, and schists derived from well-sorted sandstone, shale, and limestone. They also include volcanic rocks that are richer in silica than the older volcanics of the Superior province.

The concentric pattern of the provinces in North America is considered strong evidence that the continent grew by the accretion of material around its margins during a series of orogenic events. Each province represents a mountain-building event during which sediments were deformed into a mountain range by converging plates. Subsequently, erosion removed the upper part of the deformed belt, exposing the deeper, highly deformed metamorphic rocks and the associated granitic intrusions.

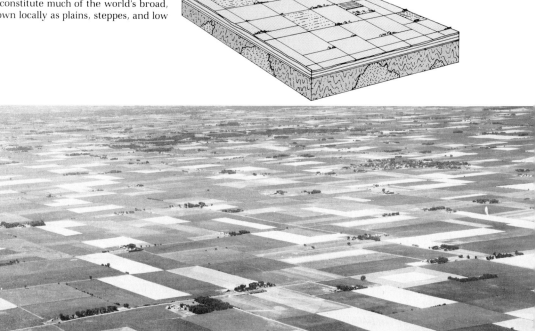

Figure 22.5 Stable platforms are areas where the continental shield is covered with a veneer of sedimentary rocks. They constitute much of the world's broad, flat lowlands, known locally as plains, steppes, and low plateaus.

Stable Platforms

Large areas of all of the shields are covered with a series of horizontal sedimentary rocks, as is shown in *figure 22.3*. These areas are removed from the tectonically active plate margins and have never been uplifted far above sea level or submerged far below it. They have remained stable and hence are called **stable platforms.**

The stable platforms form much of the broad, flat lowlands of the world and are known locally as plains, steppes, and low plateaus (*figure 22.5*; see also *plate 4*). The relationship between the sedimentary rocks of the stable platform and the underlying igneous and metamorphic complex of the shield (the basement complex) is known from thousands of wells that penetrate the sedimentary cover and from seismic studies, which reveal the rock structure beneath the surface. Although, locally, the rocks appear almost perfectly horizontal, on a regional basis they are warped into broad, shallow domes and basins (see *plate 4*).

The flat-lying sedimentary rocks that cover parts of the shields are predominantly sandstone, shale, and limestone, which were deposited in ancient shallow seas. These flat-lying marine sediments, preserved on all continents, show that large areas of the shields have periodically been flooded by the sea and have then reemerged as dry land. At present, more than 18% of the continental crust (namely, the continental shelf) is covered with water. At various periods in the past, however, shallow seas spread over a much greater percentage of the land surface.

There are several reasons for the periodic flooding of large areas of the shields during various periods of geologic time. The shields are broad, flat areas eroded down to within a few tens of meters

Figure 22.6 The structure of a folded mountain belt in southern Iran is clearly visible in this satellite photograph. The ridges are plunging anticlines, which are in the initial stage of erosion. Some are partly breached, and exposures of older rocks can be seen in their cores. Young mountain belts such as this occur along continental margins as a result of the motion of converging plates.

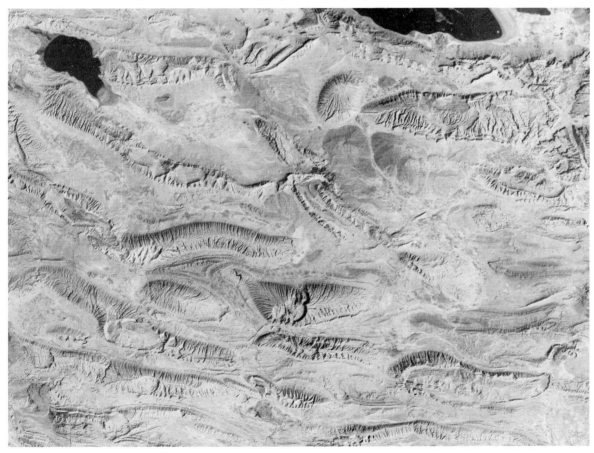

of sea level. Thus, uplifts or downwarps, which occur as the tectonic plates move across the globe, cause expansion or contraction of shallow seas over the continents. For example, rapid convection in the mantle would cause the oceanic ridge to arch up and thus displace water from the ocean basins to the continents. If the convection rate decreased, the oceanic ridge would subside and the seas would recede. Also, as a tectonic plate moves with the convecting asthenosphere, it can rise and fall slightly and thus cause expansion and contraction of shallow seas over the shields.

Mountain Belts

The great, linear **folded mountain belts,** which typify the continental margins, are one of the earth's most distinctive tectonic features. The rocks within these relatively narrow zones are compressed and folded and in many places are broken by thrust faults. As a result, the crust is shortened by as much as 30% (*figure 22.6;* see also *plate 5*).

The young, active mountain belts of today occur along convergent plate margins, coinciding with zones of intense seismicity and andesitic volcanism (*figure 22.7*). Two major mountain belts are still in the process of growing. They are (1) the Cordilleran belt, which includes the Rockies and the Andes, and (2) the Himalaya-Alpine belt, which extends across Asia and western Europe and into North Africa. Older mountain ranges can still be expressed by significant topographic relief, although deformation ceased long ago. Examples include the Appalachian Mountains of the eastern United States, the Great Dividing Range of eastern Australia, and the Ural Mountains of Russia.

The location of young mountains in long, narrow belts along continental margins is significant, because it implies that mountains could not result from a uniform worldwide force evenly distributed over the earth. They must be the result of forces concentrated along the margins of continents. Another important aspect of their location is that many mountain belts extend to the ocean and abruptly terminate at the continental margin. The northern Appalachians, the Atlas Mountains of Africa, and the mountains of Great Britain are excellent examples. The abrupt termination of the folded structures suggests that the mountain systems were once much more continuous and have been separated by continental rifting.

Figure 22.7 Active mountain belts of the earth occur along convergent plate margins and coincide with zones of intense seismicity and andesitic volcanism. The structural trends of the mountains extend into the oceans as island arcs.

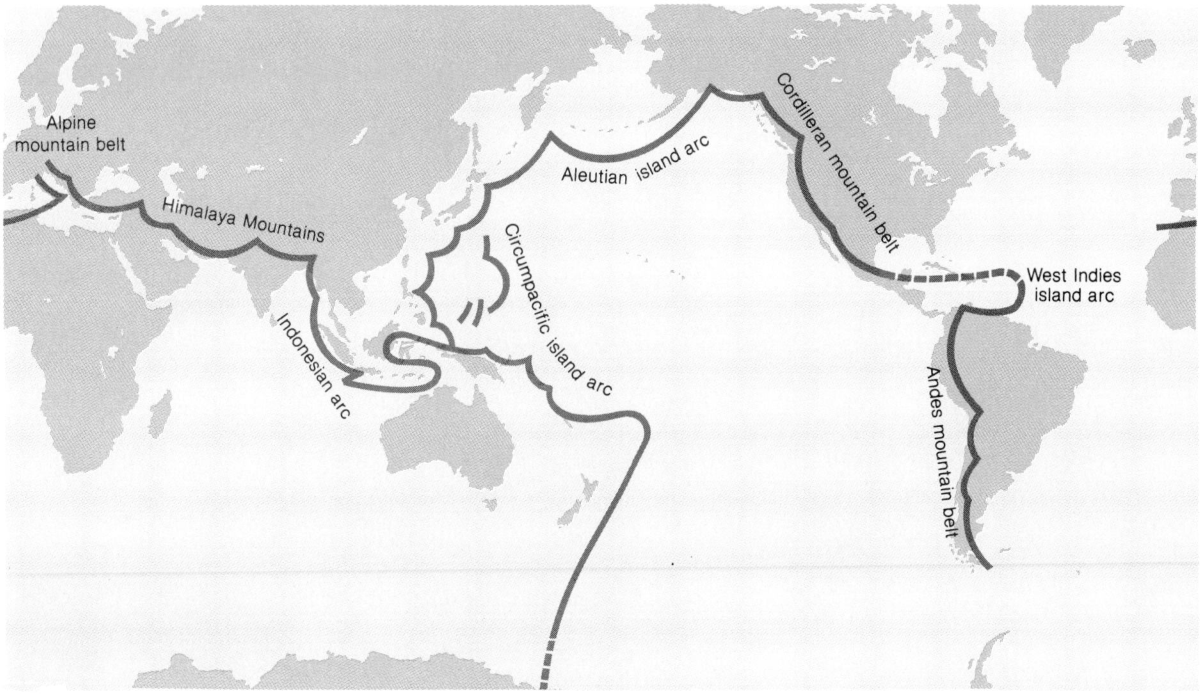

The geosynclinal theory of mountain building was further developed by James Dana (1813–1895), another American geologist, who also worked in the Appalachians. He proposed that mountain building involves a three-phase cycle consisting of (1) geosynclinal sedimentation and contemporaneous subsidence, (2) compression and deformation, and (3) uplift and erosion.

Studies of mountain belts in different parts of the world indicate that the Hall-Dana geosynclinal theory is too simple. Parallel belts of different types of geosynclinal sedimentation exist. One belt, called a **miogeosyncline,** is adjacent to the stable platform and is underlain by continental crust. It consists of clean, well-sorted, shallow-water sandstone, limestone, and shale, with no volcanic rocks. The other, called a **eugeosyncline,** consists of sediments deposited in deep-marine environments and typically includes volcanic rocks. Modern theories of geosynclinal sedimentation and mountain building utilize new information about sedimentation along continental margins and in the adjacent deep-marine basins and relate the formation of geosynclines to the plate tectonic theory.

The theory of plate tectonics proposes that **orogenesis** occurs along convergent plate margins. Unlike the geosynclinal theory, it does not require specific events to occur in the same order in all **orogenic belts.** Indeed, a variety of rock sequences are deposited along continental margins and on the ocean floor. When they are deformed at convergent plate margins, different sequences produce different styles of mountain belts. Modern oceanographic studies have determined that the classical miogeosynclinal sequence of sediments accumulates along the continental margins and that

the eugeosynclinal suites of rocks form in the deep water beyond. Probably the best known example is the present Atlantic continental margins of the eastern United States (*figure 22.9*). The sequence of thick sediments has recently been called a **geocline** by some geologists, because it is not a two-sided trough but is open toward the ocean.

Two distinctly different rock sequences are recognized. In **miogeoclines,** which lie in shallow water along the continental margins, the sequence consists of clean, well-sorted sandstone, shale, and limestone derived from erosion of the continents. In **eugeoclines,** in the deep water off the continental margins, the sequence consists of poorly sorted sandstone and shale deposited by turbidity currents, submarine slump blocks, and rock debris from submarine landslides. As is shown in *figure 22.9*, the turbidites of the eugeoclinal sequence grade seaward into deep-marine organic oozes.

Shallow-water miogeoclinal sediments also can form behind an island arc, and deep-marine eugeoclinal assemblages can form in trenches or on the seaward side of arcs.

A third rock assemblage, which is common in some mountain belts, is called an **ophiolite** sequence. It consists of peridotite, gabbro, pillow basalt, and the deep-marine sediments that form on the oceanic crust. These rocks are believed to have been scraped off a subducting plate and plastered against the upper plate as a chaotic melange, with complex folding and thrust faulting.

One feature of geoclinal sequences that always has been difficult to explain is the great thickness (as much as 10 km) of shallow-marine sediments. For shallow-water sediments to accumulate to such depths, the rate at which the crust subsides must

Figure 22.9 Two types of rock sequences accumulate along continental margins. Clean, well-sorted sandstone, shale, and limestone, deposited in shallow water, accumulate on the continental shelf. Poorly sorted, dirty sandstone and shale are deposited by turbidity currents in the deep water beyond the continental margins.

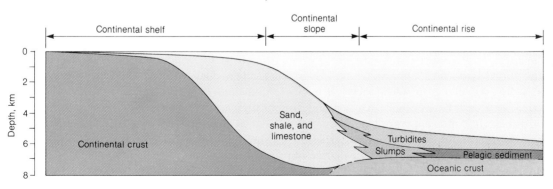

be essentially the same as the rate at which the sediment accumulates. What causes this gradual subsidence? The question has not been completely answered, but this subsidence may be related to vertical movements of the crust following extension and rifting of a continent. As the continent begins to split apart and a new ocean basin begins to form, the continental crust is uparched, extended, and split. The edges of the rift zone, which eventually become new continental margins and the site of geoclinal sedimentation, are soon leveled by erosion. As the continent moves off from the uparched spreading center, it begins to subside, and sediment derived from erosion of the continent accumulates along the margins. Gradual subsidence of the continental margins occurs for two reasons: (1) the continent moves off from the rising mantle that underlies the ridge, and (2) the weight of the deposited sediment causes the crust to be depressed.

In summary, the sedimentary rocks commonly involved in orogenesis are (1) shallow-marine miogeoclinal sediments, (2) deep-marine eugeoclinal sediments, and (3) ophiolite sequences of the oceanic crust. To study the history of mountain building and the plate movements that produce it, it is not enough to consider the direct results of orogenesis, such as deformation, metamorphism, and igneous activity. Geologists must also study the details of sedimentary rock assemblages that were deposited prior to plate collision. These assemblages provide insight into the nature of plate interactions.

Structural Deformation

The single most distinctive feature of a mountain belt is the structural deformation of the rocks. The nature of this deformation was discussed in chapter 20 (refer again to *figures 20.3, 20.4,* and *20.13* and to *plate 5*). Deformed structures result from compression, and the scale of deformation ranges from wrinkled grains or fossils in the rock to folds tens of kilometers wide.

The illustrations cited in the preceding paragraph show only small parts of mountain belts. From detailed field mapping, the structure of an entire mountain range can be determined and illustrated by geologic maps and cross sections. *Figure 22.10* includes three well-known examples.

In the Canadian Rockies, a major type of deformation is thrust faulting, in which large slices of rock have been thrust over others in a belt 60 km wide (*figure 22.10A*). It is apparent from the orientation of the faults and the direction of displaced beds that the rocks were thrust from the margin of the continent toward the interior.

A cross section of the Appalachian Mountains shows a different style and magnitude of deformation (*figure 22.10B*). The major structural feature is a series of tight folds. Deformation is most intense near the continental margins, dying out toward the continental interior. The compression shown by both the Rockies and the Appalachians is a result of plate collision.

The structure of the Alps is even more complicated (*figure 22.10C*). Great overturned folds called **nappes** (French, "tablecloths") show an enormous amounts of crustal shortening. The rocks are so intensely deformed that spherical pebbles were stretched out into rods as much as 30 times longer than the original diameters of the pebbles. Most of these structures can be explained only in terms of compressive forces.

Note that, in each mountain belt, the internal structures result from strong, horizontal compressive forces. Similar deformation exists in older rocks in places where erosion has reduced the topographic relief. This deformation provides evidence that regions such as the shields are the roots of ancient mountain systems.

Metamorphism

In the deeper parts of an orogenic belt, intensive plastic deformation and recrystallization at elevated temperature and pressure metamorphose the original sedimentary and volcanic rocks into schists and gneisses. The horizontal stress generated by converging plates causes the recrystallization of many minerals and develops foliation that is perpendicular to the direction of stress. Thus, slaty cleavage, schistosity, and gneissic layering in the deeper parts of a mountain range characteristically are vertical or dip at a high angle. Abnormally high heat in local areas within a mountain belt can produce a system of concentric metamorphic zones around a thermal center.

In the deeper parts of a mountain belt, metamorphism can become intense enough to produce granitic **migmatite** complexes. Migmatite is a complex mixture of thin layers of granitic material between sheets of schist or gneiss. It develops largely from the partial melting of preexisting rocks. Apparently, the magma generated by the compression and heat does not migrate far and is mixed with the unmelted material. In these zones, high temperatures and pressures soften the entire rock body, which behaves like a highly viscous liquid if it is subjected to stress. Consequently, metamorphic rocks in deeper parts of orogenic belts exhibit complex flow structures (see *figures 7.2, 7.3,* and *7.4*).

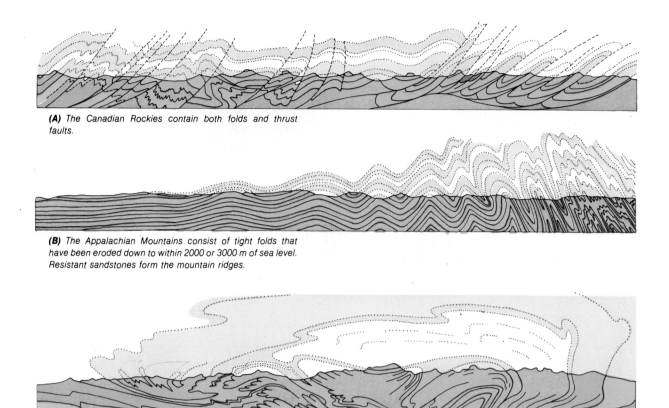

(A) *The Canadian Rockies contain both folds and thrust faults.*

(B) *The Appalachian Mountains consist of tight folds that have been eroded down to within 2000 or 3000 m of sea level. Resistant sandstones form the mountain ridges.*

(C) *The Alps are complex folds, many of which are overturned.*

Figure 22.10 The structure of folded mountain belts reflects intense compression at convergent plate boundaries, but each range can have its own structural style.

Igneous Activity

Igneous activity associated with mountain building is part of the fundamental process of differentiation, by which the earth's materials are separated and concentrated into layers according to density. In particular, this igneous activity is responsible for **magmatic differentiation.**

The concentration of the earth's lighter material in the continental crust occurs in two steps. The first phase begins at a spreading center, where partial melting of peridotite in the upper mantle generates a basaltic magma, which rises to form oceanic crust. Basalt is richer than peridotite in the lighter elements, particularly silicon and oxygen. The second phase occurs at a subduction zone. There, partial melting of the oceanic crust forms a silica-rich magma, which is then emplaced in the mountain belt as granitic intrusions or andesitic volcanic products. This process further separates the lighter elements, especially silicon and oxygen, and concentrates them in the continental crust. The granitic continental crust is less dense than the mantle and the oceanic crust are, and its buoyancy prevents it from being consumed at subduction zones. Once it is formed, the continental crust remains on the outer surface of the earth.

The generation of silica-rich magma at a subduction zone is shown in *figure 22.11*. The magma conceivably can result from partial melting in three different regions of the subduction zone: (1) in the subducting oceanic crust, (2) in the overlying mantle, as hot fluids percolate upward, and (3) near the base of the continental crust, as upward-migrating magma raises the temperature. Water in the pore spaces of the rock and in chemical combination in many minerals of the oceanic crust plays an important role in the type of igneous activity at con-

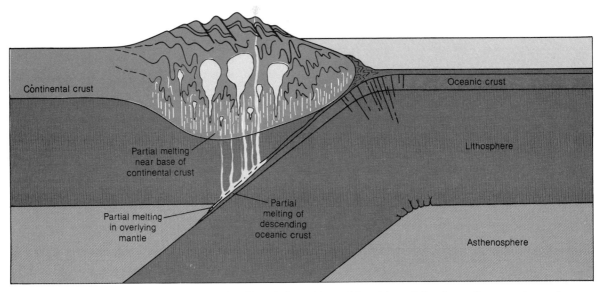

Figure 22.11 The generation of silica-rich magma can occur by partial melting in three different regions of a subduction zone. The process can involve (1) partial melting of the descending oceanic crust, (2) partial melting of the upper mantle, and (3) partial melting of the lower part of the continental crust.

vergent plate margins. As the crust is subducted and heated, this water is driven out. It percolates upward and enhances melting at higher levels in the crust.

The Evolution of a Mountain Belt

The general conception of how a mountain belt evolves into a segment of the shield is shown in *figure 22.12*. Folding and thrusting occur at relatively shallow depths, metamorphism occurs deeper, and partial melting occurs at still greater depths. Granitic magma initially forms deep within the crust, at points where the rock begins to melt. The magma is then injected into the foliation of the adjacent metamorphic rock, so that a migmatite is formed. Much of the magma migrates upward, because it is less dense than the solid rock. As it rises, it forms teardrop-shaped bodies, which collect into larger and larger masses. The boundaries of the body of magma are largely parallel with the broad zones of foliated metamorphic rock. The rising granitic magma cuts across the upper folded strata, which are not metamorphosed but are only deformed by folding and faulting. The magma can cool within a few kilometers of the surface, forming a batholith, or it can be extruded as andesitic volcanic material.

Erosion and Isostatic Adjustment

The history of a mountain belt does not end with deformation, metamorphism, and igneous activity. After the orogenic activity is terminated—presumably because of shifts of the convection cells in the mantle—deformation ceases, but erosion and isostasy combine to modify the orogenic belt. As is illustrated in *figure 22.12*, the entire crust is deformed by the orogeny, so that a mountain root, composed of the most intensely deformed rocks, extends down into the mantle. The high mountain range and its deep roots are in isostatic balance. As erosion wears away the high mountains, this balance is destroyed. Isostasy then causes the mountain belt to rise in a broad upwarp to reestablish the balance. In the early stages of erosion, the removal of 500 m of rock is accompanied by an isostatic uplift of approximately 400 m, with a net lowering of the surface of only 100 m. The rate of net lowering is believed to decrease as the mountain belt is eroded down and approaches a new state of isostatic equilibrium. Uplift continues as long as erosion removes material from the mountain range. Eventually, however, a balance is reached, when the mountainous topography is eroded to near sea level and the mountains' roots are removed.

It is important to note that the rocks exposed at the surface when isostatic balance is established are the metamorphic and igneous rocks formed at great depths in the orogenic belt. These rocks are tectonically stable and become part of the shield.

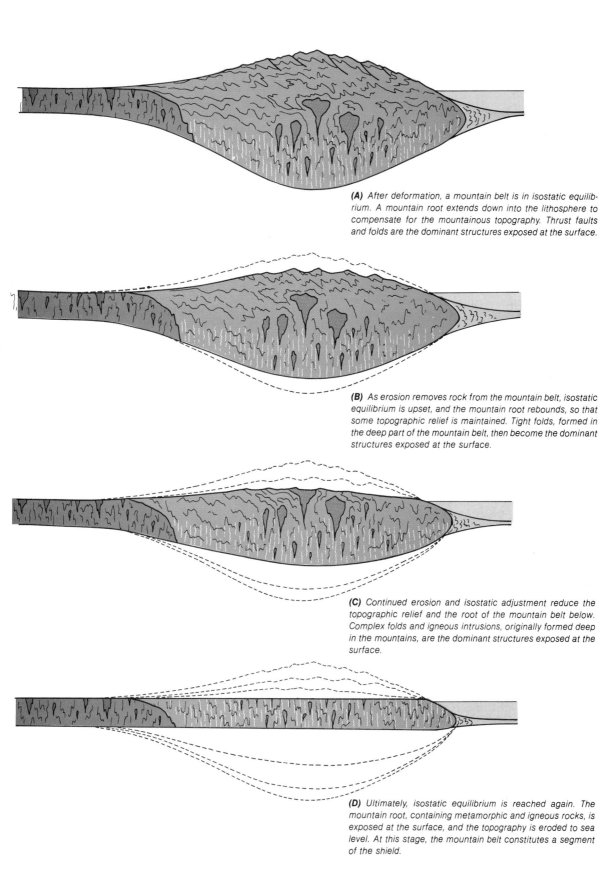

(A) After deformation, a mountain belt is in isostatic equilibrium. A mountain root extends down into the lithosphere to compensate for the mountainous topography. Thrust faults and folds are the dominant structures exposed at the surface.

(B) As erosion removes rock from the mountain belt, isostatic equilibrium is upset, and the mountain root rebounds, so that some topographic relief is maintained. Tight folds, formed in the deep part of the mountain belt, then become the dominant structures exposed at the surface.

(C) Continued erosion and isostatic adjustment reduce the topographic relief and the root of the mountain belt below. Complex folds and igneous intrusions, originally formed deep in the mountains, are the dominant structures exposed at the surface.

(D) Ultimately, isostatic equilibrium is reached again. The mountain root, containing metamorphic and igneous rocks, is exposed at the surface, and the topography is eroded to sea level. At this stage, the mountain belt constitutes a segment of the shield.

Figure 22.12 Erosion and isostatic adjustment combine to transform a newly deformed mountain belt into a segment of the stable shield.

TYPES OF OROGENIC ACTIVITY

Statement

In the preceding section, we saw that mountain building is the result of plate convergence and that it involves intensive deformation, metamorphism, and igneous activity. The characteristics of a mountain belt and the sequence of events in its development can vary. They depend on the types of interactions at convergent plate margins and the types of rock sequences that are involved in the deformation. There are three fundamentally different types of convergence:

1. Convergence of two oceanic plates
2. Convergence of continental and oceanic plates
3. Convergence of two continental plates

Each of these develops a mountain belt with distinctive characteristics.

Discussion

Convergence of Two Oceanic Plates

The main types of orogenic activity involving the convergence of two oceanic plates are shown in *figures 22.13* and *22.14*. The major result is an arc of volcanic islands above the subduction zone. In the first stages of convergence, the process can be relatively simple and restricted to volcanic activity on the overriding plate. It does not involve widespread metamorphism or granitic intrusion. The Tonga Islands are an example of a simple island arc in the present oceans (*figure 22.13*).

In more complex island arcs (such as Japan), crustal deformation, metamorphism, and igneous activity combine to produce distinctive rock associations and deformational patterns. Their major features, illustrated in *figure 22.14*, are the following:

1. Rock sequences consist of the oceanic sediments and pillow basalts of the oceanic crust, in addition to sediment derived from the erosion of the volcanic arc.
2. Partial melting of rocks in the subduction zone produces a silica-rich magma, which rises to form granitic intrusions and volcanic products in the island arc.
3. A zone of high-pressure, low-temperature metamorphism develops at the outer margins of the overriding plate.
4. An inner zone of higher-temperature metamorphism develops, with associated granitic intrusions.
5. Crustal deformation of the volcanic arc results from the collision of converging plates and the intrusion of granitic magma.

The zone of high-pressure, low-temperature metamorphism that develops on the outer margin of the overriding plate is easily understood in light of its dynamic setting. High pressure is produced because this zone is the focal point of the plates' convergence. Low temperature results because the cold oceanic plate descends at rates between 5 and

Figure 22.13 Simple ocean-to-ocean orogenesis is restricted largely to volcanic activity and does not involve widespread metamorphism or granitic intrusions. An example is the Tongan arc, in the South Pacific.

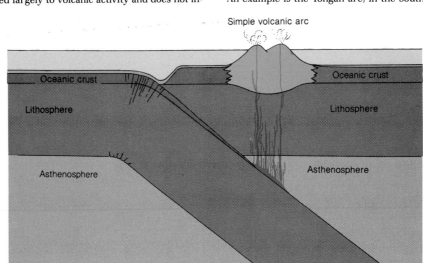

Simple volcanic arc

Oceanic crust

Lithosphere

Asthenosphere

Oceanic crust

Lithosphere

Asthenosphere

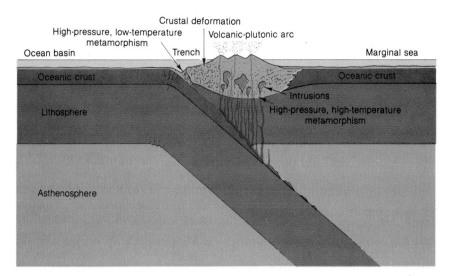

Figure 22.14 Complex ocean-to-ocean orogenesis involves crustal deformation, metamorphism, and granitic intrusions, as well as volcanic activity, producing a complex island arc. An example is the Japanese arc.

15 cm per year. These rates exceed the rate of heat flow from the mantle through the descending lithosphere. The rocks that are metamorphosed consist of sediment derived from erosion of the adjacent volcanic arc, oceanic sediments, and pillow basalts of the oceanic crust. This complex mixture is squeezed downward along the top of the descending plate at temperatures that generally are lower than 300 °C. The mixture of pillow basalts, oceanic sediments (ophiolite complex), and erosional debris from the island arc is metamorphosed into a distinctive, fine-grained schistose rock, characterized by high-pressure, low-temperature mineral assemblages. The rock typically develops a bluish color, because of the presence of the mineral glaucophane, and is referred to as a **blueschist.** There are no contemporaneous igneous intrusions in this zone, because magma is generated only where the descending plate reaches a greater depth.

Blueschists are found in Japan, California, and New Zealand and have been recognized in the Alps, but they represent only a small fraction of metamorphic rocks exposed on the continents. Their small proportion can be explained by the fact that these high-pressure, low-temperature assemblages lose their distinctive character if they are subsequently heated above 300 °C. Such heating occurs due to granitic intrusions that develop in continental orogenic belts.

In deeper parts of the subduction zone, magma generated by partial melting of the descending oceanic plate rises to form large granitic batholiths.

These intrusions make up the deeper parts of the island arc. They also produce a surrounding zone of high temperature, which develops a very different assemblage of metamorphic minerals. At shallow depths, high-pressure, low-temperature metamorphism occurs. At deeper levels, high-pressure, high-temperature metamorphism (high-grade metamorphism) occurs. The original rocks are the volcanic flows and the sediments that formed within the arc system proper. The resulting metamorphic rocks are quite distinct from blueschist, which is formed near the trench.

As a result of the collision of converging plates, the older rocks in the volcanic arc are compressed, folded, and broken by thrust faults. In addition, deformation results from the emplacement of granitic batholiths. The fold axes and thrust faults trend parallel to the long axis of the arc and the linear belts of granitic batholiths.

Orogenesis caused by the convergence of two oceanic plates is distinctive in (1) rock sequence (andesitic and volcanic sediments), (2) metamorphism (blueschist), and (3) zonation of ophiolite and volcanic rock types.

Orogenesis in island arcs is significant because the deformed andesitic volcanic rocks (and sediments derived from them) are differentiated from oceanic basalt. They represent the initial stage in the development of new continental crust in areas where only oceanic crust previously existed. This process provides important insight into the origin and evolution of continents.

Convergence of Continental and Oceanic Plates

The style of mountain building produced by the collision of a continental plate and an oceanic plate resembles that produced by the convergence of two oceanic plates, but it is distinctive in two main respects.

1. Thick sequences of geoclinal sediments, derived from the continent, commonly occur along the continental margins and are quite different from rock sequences deposited around a volcanic arc. The convergence of the plates produces a mountain belt characterized by folded sedimentary rocks with roots of granitic intrusions and metasediments.

2. Silica-rich magma generated by the partial melting of the descending oceanic plate is placed upon or within the continental crust, rather than upon the ocean floor. Consequently, a high topography is produced, with the resulting rapid erosion. Volcanic material is soon stripped off the mountain range, and granitic batholiths commonly are exposed.

A schematic cross section showing the major features produced by the collision of continental and oceanic plates is shown in *figure 22.15*. Before the continental crust arrives at the subduction zone, a considerable thickness of sandstone, shale, and limestone can accumulate in the miogeocline along the continental margins. During this same period, deep-marine sediments accumulate on the oceanic plate. When the plates collide, the buoyant granitic mass of the continental crust always overrides the adjacent oceanic plate. As the continental plate approaches the subduction zone, some of the deep-marine sediments on the oceanic plate can be crumpled and deformed. Slabs of oceanic crust shear off and are incorporated into the chaotic mass. This material, like that developed by the convergence of two oceanic plates, is subjected to high-pressure, low-temperature metamorphism.

The thick sequence of geoclinal sediments along the continental margin is then compressed and

Figure 22.15 Continent-to-ocean orogenesis involves the deformation of the thick sequence of sediment that accumulates along continental margins. This material is deformed into a folded mountain belt, intruded by granitic batholiths, and metamorphosed in the deeper zones of the orogenic belt. High topography is produced, and andesitic volcanism is common.

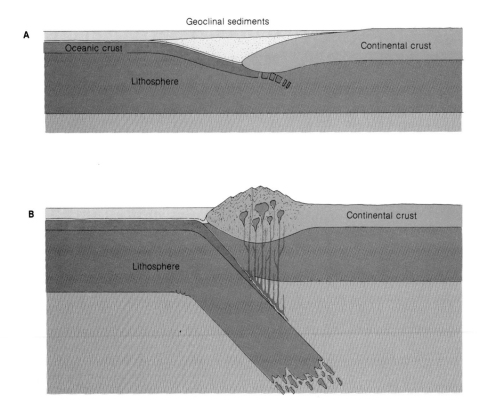

deformed. Thrust faults occur in the shallow zone of the mountain belt, where brittle rocks fail by rupture. At intermediate depths, plastic flow develops tight folds and nappes. Intense metamorphism occurs within the deeper zones, where temperatures and pressures are relatively high. The partial melting of the descending lithosphere generates silica-rich magma, which rises to form granitic intrusions in the deformed sediments in the upper plate or is extruded as volcanic material at the surface.

With continued compression, deformation of the geoclinal sediments and the resulting crustal thickening progress landward, and the edge of the deformed continental margin rises above sea level. Erosion of the mountain belt begins, and it continues contemporaneously with subsequent deformation. Sediments eroded from the growing mountain range can be shed in both directions, and thick sequences of coarse conglomerates and sandstone can be deposited as alluvial fans or deltas in front of the rising mountain belt.

As the orogenic belt evolves, resistance builds up. Convergence can stop, or the subduction zone can step seaward to form an ocean-to-ocean orogenic belt. After compression terminates, erosion continues to wear down the mountain range. Isostatic rebound occurs until, ultimately, the mountain roots are exposed at the surface, and the topography is eroded to near sea level.

There are many excellent examples of orogenic belts formed by the convergence of continental and oceanic plates. The Rocky Mountains of western North America were deformed during late Mesozoic and early Tertiary time. The Andes Mountains of South America have been deformed from the Tertiary to the present. The Appalachian Mountains of the eastern United States were deformed during the late Paleozoic. All represent continent-to-ocean convergence, although the structure and topography of each may be unique in some respects.

Convergence of Two Continental Plates

In the tectonic system, virtually all oceanic crust is destined to descend into the mantle by the subduction processes at convergent plate margins. As continents, carried by the plates, move toward a subduction zone, the oceanic basin is continually reduced in size. It is eliminated altogether if two continents collide. The collision of two continental plates generates an orogenic belt with several characteristics that are quite different from those of the other types.

If both converging plates contain continental crust, significant subduction cannot occur. The oceanic basin between the continents closes, and the bounding prisms of sediments are deformed. The buoyant, low-density continental masses cannot descend into the mantle, although one can override the other for a distance, so that the plates form a double thickness of continental crust, with exceptionally high topographic relief. Subduction in such a case is deactivated, and the continents are welded together.

Although the generation of mountain belts by continental collision is exceedingly complex in detail, the major events are similar to those outlined in *figure 22.16*.

1. Before the actual collision, the wedge of sediments along the margin of the continent above the subduction zone is deformed and plastered against the leading edge of the continental crust. The oceanic lithosphere is consumed at the subduction zone, and the ocean basin decreases in size (*figure 22.16A*).
2. As the continents approach collision, segments of the remaining oceanic crust are deformed by overthrusting and eventually are squeezed between the converging plates (*figure 22.16B*).
3. As the continental crust moves into the subduction zone, its buoyancy prevents it from descending into the mantle more than perhaps 40 km below its normal depth. It can be thrust under the overriding plate, however, so that a double layer of low-density continental crust is produced. This layer rises buoyantly to create a wide belt of deformed rock with an adjacent high pleateau (*figure 22.16C*).
4. Alternatively, the continental masses can become welded together, and fragments of ophiolite assemblage (oceanic crust) can be caught between them and squeezed upward.
5. The oceanic slab of lithosphere, descending down into the mantle, ultimately becomes detached and sinks independently. When the slab has been consumed, the volcanic activity and earthquakes it generated cease.
6. Eventually, convergence stops as resisting forces build up, and the mountain belt is eroded and adjusts isostatically.
7. The welding together of two continents produces a single large continental mass with an internal mountain range.

The Himalaya Mountains are an example of orogenesis due to continental collision. They were formed during the last 100 million years as India moved northward and destroyed the oceanic litho-

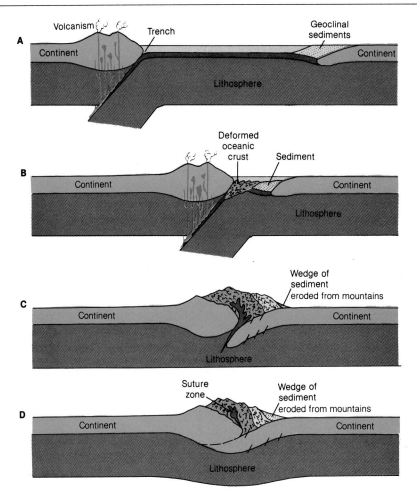

Figure 22.16 Continent-to-continent orogenesis involves the deformation of oceanic and geoclinal sediments and commonly produces a complex, high mountain range. As two continents converge, the oceanic crust between them is caught and deformed. A double layer of continental crust can be produced, resulting in abnormally high topography. The continents are welded together. The descending oceanic plate becomes detached in the subduction zone and sinks independently. When this slab is consumed, volcanic activity and deep earthquakes end.

sphere that formerly separated it from Asia. As the two continents collided, India was thrust under the Asian plate, and the Himalaya Mountains and the extensive highlands of the Tibetan Plateau were formed. Earthquakes are frequent in the region, but they are shallow and occur in a broad, diffuse zone because there is no descending oceanic plate.

The Alps and the Ural Mountains are other examples. The European Alps result from the convergence of the African and Eurasian plates. In many ways, Africa, moving northward against Eurasia, is like India, but it has not evolved to the point at which the oceanic crust (the Mediterranean Sea) is completely consumed. The Urals were formed much earlier, during late Paleozoic time, when the Siberian continental mass collided with Europe.

They are in a late stage of development, in which erosion and isostatic adjustments are the main processes.

In summary, mountain building is a fundamental process in the differentiation of the earth. Orogenesis occurs along convergent plate boundaries. Processes of compressional deformation, metamorphism, and igneous activity are involved in mountain building. The style of the orogenesis and the specific events can vary significantly, depending on the interactions that occur along the convergent plate boundaries. The three fundamental types of convergence include ocean-to-ocean, ocean-to-continent, and continent-to-continent collisions. Each produces its own distinctive style of mountain building.

THE ORIGIN AND EVOLUTION OF CONTINENTS

Statement

Continents are believed to originate and evolve through plate tectonic processes operating along convergent plate boundaries. A continent begins as an island arc, composed of silica-rich volcanic material produced by the partial melting of a descending plate. This low-density material resists subduction because of its buoyancy. It remains on the surface of the lithosphere, continually drifting with the moving plates. The small, embryonic continent grows by accretion, accumulating new granitic material with each subsequent mountain-building event. Erosion and isostatic adjustment reduce the mountainous topography to a surface of low relief and develop a stable shield. Subsequent rifting can split and separate a continent, and each segment then acts as a separate center for future continental growth.

Discussion

The origin and evolution of continents are shown graphically in the series of schematic diagrams in *figure 22.17*. The initial stage of development is the formation of a subduction zone (*figure 22.17A*). Magmatic differentiation occurs there by the partial melting of the descending plate. The silica-rich minerals melt first, producing a silicic magma, which erupts as andesitic volcanoes or forms granitic intrusions. Sediment derived from erosion of the island arc further separates the low-density material and concentrates it by sedimentary differentiation. The new igneous and sedimentary rocks produced in the island arc are less dense than the basaltic oceanic crust, and their buoyancy prevents any significant consumption in a subduction zone.

The volcanic material and sediment in the island arc are deformed into an orogenic belt by subsequent collision with another plate. Metamorphism occurs in the mountain roots, and granitic intrusions result from partial melting in the subduction zone (*figure 22.17B*). Erosion and isostatic adjustment produce a small, stable shield, in which igneous and metamorphic rocks formed in the mountain roots are exposed at the surface (*figure 22.17C*). Sediment derived from the erosion of the mountains is deposited along the margins of the new continent. Repetition of this process deforms the geoclinal sediments along the continental mar-

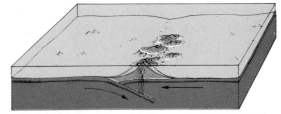

(A) Partial melting in the subduction zone produces an island arc composed of andesitic volcanic material, which is too light to descend into the mantle.

(B) The sediment and andesitic volcanics are deformed by compression at converging plates. Mountain belts and metamorphic rocks form an embryonic continent.

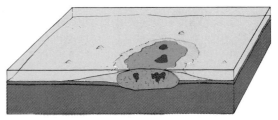

(C) The erosion of mountains concentrates the light minerals (quartz, clay, and calcite) as geoclinal sediments along the margins of the small continent.

(D) The geoclinal and orogenic cycle is repeated. New material is added to the continental mass in the form of granitic batholiths and andesitic flows.

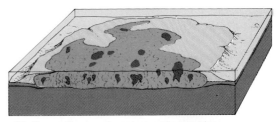

(E) The continent continues to grow by accretion.

Figure 22.17 Continents evolve by accretion of low-density rock material during a series of orogenic events. They begin as island arcs and grow by the addition of new material, which is metamorphosed into a segment of shield with each orogenic event.

gins into a new orogenic belt. New igneous material is added from the partial melting of the oceanic crust in the subduction zone (*figure 22.17D*). The continent continues to grow by **continental accretion** of material along its margins during each subsequent orogenic event (*figure 22.17E*).

The major processes of continental growth are: (1) mountain building along convergent plate margins, involving deformation and metamorphism of geoclinal sediments; (2) igneous activity and emplacement of new material in the orogenic belt; and (3) erosion and isostatic adjustment, which produce a shield. Broad upwarps or slight vertical movements of the shield permit the ocean to expand and contract over the continent and deposit a sediment cover, which forms the stable platform.

At some stage, rifts can form within the stable continental block as a new spreading center develops, presumably as a result of shifting convection currents in the mantle. The continent then splits, and its fragments drift apart with the spreading plates (*figure 22.18*). Each fragment acts as a separate center for future continental growth.

Since the earth is a sphere, the moving plates carrying continental blocks sometimes converge, and continents can collide and become welded together.

SUMMARY

Extensive geologic mapping shows that continents consist of three major components: (1) shields, composed of complexly deformed and recrystallized metamorphic and granitic rocks, generally eroded to a flat surface near sea level; (2) stable platforms, or areas of the shield that are covered with a veneer of essentially horizontal sedimentary rocks; and (3) young, folded mountain belts. Continents can also be split by rift systems.

The characteristics of a mountain belt and the sequence of events that produced it depend largely on the type of crust carried by converging plates and the type of sediments involved in the deformation. Three distinctive types of convergence are recognized: (1) convergence of two oceanic plates, (2) convergence of a continental plate and an oceanic plate, and (3) convergence of two continental plates. Convergence of two oceanic plates produces island arcs with andesitic volcanism, metamorphism of ophiolite rock sequences, and emplacement of granitic batholiths in the deeper parts of the orogenic belt. Convergence of a continental plate and an oceanic plate involves deformation of a thick geoclinal sequence of sediment, andesitic

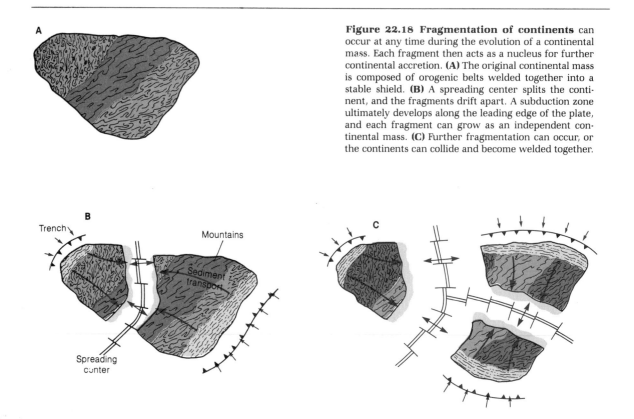

Figure 22.18 Fragmentation of continents can occur at any time during the evolution of a continental mass. Each fragment then acts as a nucleus for further continental accretion. **(A)** The original continental mass is composed of orogenic belts welded together into a stable shield. **(B)** A spreading center splits the continent, and the fragments drift apart. A subduction zone ultimately develops along the leading edge of the plate, and each fragment can grow as an independent continental mass. **(C)** Further fragmentation can occur, or the continents can collide and become welded together.

volcanism, and emplacement of granitic batholiths in the deformed mountain belt. Convergence of two continental plates produces a continental suture or a double layer of continental crust with high mountainous topography and limited igneous activity.

After orogenesis, erosion and associated isostatic adjustment reduce the mountainous topography to a flat surface near sea level. These processes expose the metamorphic rock of the deep mountain roots as a new segment of the shield.

The origin of continents is intimately involved with mountain building at convergent plate boundaries. In this process, silica-rich, low-density material is concentrated on the outer part of the crust through the partial melting of the oceanic plate in the subduction zone.

Thus, continents grow by accretion, as new silica-rich material is added to an orogenic belt through igneous activity at convergent plate margins. This low-density crust resists any further subduction and remains on the outer lithosphere. Subsequent rifting can split and separate continental blocks, but each block then acts as a separate center for future continental growth.

Continental growth is simply part of the process of planetary differentiation, in which the lighter elements are separated and concentrated in the outer layers of the earth. The process of differentiation has produced the ocean basins, the continental platforms, and the surface fluids (air and water). Thus, differentiation is directly or indirectly responsible for the origin of all surface features and the internal structure of the planet. In the next chapter, we will consider the environmental effects of processes operating in and on the earth and the production of natural resources by these processes.

Key Words

shield
stable platform
folded mountain belt
basement complex
orogenesis
orogenic belt
geosyncline
miogeosyncline
eugeosyncline
geocline
miogeocline
eugeocline
ophiolite

nappe
migmatite
magmatic differentiation
blueschist
isostasy
continental accretion

Review Questions

1. What is the composition and structure of the continental crust?
2. Describe the Canadian Shield. How does it differ from the Ethiopian Shield (of Africa) or the Australian Shield?
3. How does a stable platform differ from a shield?
4. How are young mountain belts related to island arcs?
5. Study the tectonic map of the world (*figure 22.8*) and label (a) shields, (b) stable platforms, (c) Mesozoic and Cenozoic mountain belts, (d) the oceanic ridges, and (e) trenches.
6. What is orogenesis? What factors are most important in this process?
7. Describe and diagram the two types of rock sequences that accumulate along continental margins.
8. Describe the styles of structural deformation that occur during orogenesis.
9. Explain the processes of metamorphism and igneous activity that accompany orogenesis.
10. What role does isostatic adjustment play in the evolution of a mountain belt and the development of a new segment of the shield?
11. Sketch a series of diagrams to show the type of orogenesis that involves the convergence of two oceanic plates.
12. Compare and contrast orogenesis involving two oceanic plates and orogenesis involving a continental and an oceanic plate.
13. Describe orogenesis involving two continental plates. How can this process be recognized in the rock record?
14. Outline the steps in the evolution of a continent.

Additional Readings

Condie, K. C. 1976. *Plate tectonics and crustal evolution.* Elmsford, N.Y.: Pergamon Press.

Engel, A. E. J., and C. G. Engel. 1964. Continental accretion and the evolution of North America. In *Advancing frontiers in geology and geophysics: A volume in honour of M. S. Krishman,* ed. A. P. Subramanian and S. Balakrishna, pp. 17–37e. Hyderabad: Indian Geophysical Union. Reprinted in *Adventures in earth history,* ed. P. E. Cloud, pp. 293–312. San Francisco: W. H. Freeman, 1970.

Hallam, A. 1973. *A revolution in the earth sciences.* Oxford: Clarendon Press.

Wyllie, P. J. 1976. *The way the earth works.* New York: John Wiley and Sons.

ENVIRONMENT, RESOURCES, AND ENERGY

In the preceding chapters, we have studied the earth as a dynamic system that produces a constantly changing environment. Some rapid changes, such as earthquakes, volcanoes, and floods, have catastrophic effects. Most changes, however, are so slow by human standards that they are barely noticeable, even during a human lifetime. Perhaps there is comfort in realizing that most geologic processes are extremely slow and are of little concern to us as individuals. Still, the last 10,000 to 20,000 years have brought many changes that significantly affect human history. For example, the present sites of many northern cities, including Chicago, Detroit, Montreal, and Toronto, were buried beneath thousands of meters of glacial ice as recently as 15,000 to 20,000 years ago. At the same time, the sites of most major port cities, such as New York, were located many kilometers inland. San Francisco Bay did not exist. The Missouri and Ohio rivers had not been completely established, and the valley where Salt Lake City is located contained a large freshwater lake 300 m deep. These changes were largely the results of worldwide climatic changes associated with the ice age.

In contrast, man can simulate drastic geologic changes within a few years and can radically alter the environment in which we live. If this happens, the balance of the natural systems, established over thousands of years, is instantaneously upset, and adjustments to the changes cannot always be anticipated.

Not only is the environment of great concern, but people are beginning to realize that resources and energy are finite and are being depleted. During the next 10 years, we will use more oil, gas, iron, aluminum, and other mineral resources than were consumed throughout all of history.

In this chapter, we will discuss the environment, resources, and energy from a geological point of view.

MAJOR CONCEPTS

1. Our environment consists of natural systems that have operated in a delicate balance for a long period of time.
2. We can manipulate many natural systems, such as rivers and shorelines.
3. Natural systems adjust to artificial changes in ways that cannot always be anticipated.
4. Waste disposal involves natural systems. Waste products cannot be "thrown away." They remain in some form in the earth's systems.
5. Mineral resources are concentrated by geologic processes operating in the hydrologic and tectonic systems. These resources are finite and nonrenewable.
6. There are limits to growth in the earth's natural systems.

ENVIRONMENTAL GEOLOGY

Statement

We are a part of nature, one of many species of animals adapted to the present natural environment. We are so well adapted that we dominate all other species, and our numbers are so great that we have become an effective agent of physical, chemical, and biological change. Presently, we are capable of significantly modifying the earth's natural systems. Many of our modifications, unfortunately, conflict with the normal evolution of the earth's environment. For example, since the development of an agrarian culture, we have drastically changed large areas of the earth's surface from natural wilderness to controlled, cultivated land. This has greatly strained the food supply of many animal populations, which previously had established a degree of balance. The advancements of science and technology have telescoped thousands, if not millions, of years of normal evolutionary changes into days. Many species simply cannot adapt over such a short period of time.

We have also (1) diverted and manipulated drainage systems; (2) altered the quality and flow of water, both on the surface and the subsurface; (3) changed the atmosphere; and (4) modified coastal and oceanic waters. The changes we institute, almost without exception, are intended to produce improvements and advantages for society. Frequently, the results are quite the opposite. At best, some are detrimental on a short-term basis. Others are catastrophic and irreversible in the long run. Therefore, we are forced to adapt to a rapidly changing environment, one that we have in part created. Changes are necessary. At times, we must tolerate undesirable side effects (anticipated or unanticipated) as the "necessary price" of change.

Clearly, a decision to alter our environment should be made carefully. First, we should have a sufficient geological and biological understanding of how natural systems operate and how they will be affected by artificial modifications. If we decide to dam a river, we should understand the side effects of altering part of a natural river system. We should ask how the damming will affect erosion and sedimentation downstream. How will the ground-water system respond? How will lakes be altered? How will the beaches along shores far removed from the dam be affected? How will marine and terrestrial life be changed? If we develop a great volume of waste, we should know how it will be assimilated into the earth's system.

How will it affect the quality of surface and ground water, the atmosphere, and the oceans? If we build cities, we should fully understand how this development will affect the terrain, the atmosphere, and the hydrosphere. In too many cases, when we spend money to correct or alter a natural condition to suit our needs, we upset a natural balance. Then we have to spend more money to reestablish the balance.

The human race exists by geologic consent. The basis of our entire environment is the geologic system. Therefore, it is incumbent on us to understand the system and to live within it.

Discussion

In many of the preceding chapters, we have seen examples of artificial modifications of natural systems. We will review these from the perspective of environmental geology.

The Modification of River Systems. Water is perhaps our most important natural resource. As the population has grown, we have increasingly modified river systems to better utilize this finite resource. Most of our attempts to modify river systems have involved the construction of dams and canals, but urbanization also affects river systems. The important point to be considered in all of these manipulations is that a river is a system that has approached equilibrium among a large number of variables over a long period of time. If this equilibrium is upset, a number of rapid adjustments occur. Recall the example of the Aswân High Dam, in Egypt (see page 165), and how the Nile adjusted to it. Whenever a dam is built, several major geologic adjustments occur:

1. Sediment is trapped in the reservoir, and the clear water discharged downstream is capable of accelerated erosion.
2. Deltas are deprived of their source of sediment, and coastal erosion accelerates.
3. Ground-water systems are modified by artificial lakes created behind the dam.

A more universal problem is the subtle modification of a river system by urbanization (see page 166). Where cities are constructed, the surface runoff is modified by streets, sidewalks, parking lots, and roofs of buildings, which make large areas of the surface impermeable. This change produces two major effects. First, the volume of surface runoff is greatly increased and is channeled into gutters and storm drains. As a result, flooding increases in intensity and frequency. Second, ground-water systems are altered by reduced infiltration.

The Modification of Slope Systems. Slopes are also dynamic systems that have some degree of balance (see page 182). Construction on hills and slopes modifies the system, causing an increase in the magnitude and frequency of mass movement. A landslide in northern Italy that resulted from slope modification associated with the Vaiont Reservoir was cited earlier (page 182). The problem exists wherever uncontrolled hillside construction occurs, especially where the slope is underlain by unconsolidated rock material.

The Modification of Ground-Water Systems. The use of ground-water resources is constantly increasing. As we drill wells and pump ground water, we create a new and unnatural ground-water discharge. Subsidence is one of the more important results. Mexico City is an excellent example: many of its buildings have subsided more than 3 m (see page 238). Also, the encroachment of salt water into the lens-shaped body of fresh water on an island or a peninsula commonly results from excessive pumping (see page 236).

Alteration of the water table can result from the modification of surface infiltration. An example is the canals in southern Florida, whose construction caused a variety of problems in the Everglades (see pages 236–37). In contrast, the irrigation of arid lands causes the water table to rise. This condition can produce unstable slopes and accelerated mass movement, such as occurred in the Pasco basin of Washington (see page 237).

The Modification of Shorelines. Perhaps even more dramatic are the results of modifying shorelines. Santa Barbara, California, is an example in which the process of longshore drift was manipulated by constructing a breakwater to form a harbor (see pages 284–85). If longshore drift is disrupted, the supply of sand is eliminated or reduced, and beach erosion results.

These and other examples illustrate a great central theme: geologic systems are complex dynamic systems in which balance has been established among a number of variables. Manipulating these systems destroys the balance, often resulting in many serious side effects—some detrimental over the short term, others catastrophic and irreversible in the long run.

Geologic Problems with Waste Disposal

Our industrialized society produces an ever-increasing variety and quantity of toxic waste. Traditionally, people have used fresh water to remove and dilute solid and liquid wastes and have used the atmosphere to dilute the gaseous waste products of combustion. But, until recently, people generally have been unaware that a local natural system of waste disposal can become saturated, so that an unhealthy environment is created. People cannot take their waste "away," as has been suggested by some politicians. The waste products are in the earth's natural systems and will remain in them. The problem is particularly acute because of such business practices as planned obsolescence, the use of throwaway containers, and the hard sell of new models of old products. In addition, high labor costs often make it uneconomical to repair, reclaim, and recycle used items, so that the volume of waste grows unnecessarily, at a staggering rate. The replacement of products, of course, greatly reduces natural resources. Unfortunately, waste is not just a by-product; eventually, it is a product itself.

Waste disposal has many geologic ramifications. If waste is buried, the quality of ground water is threatened. If it is dumped into streams and rivers, not all of it is carried away and lost; it accumulates on beaches and in estuaries, altering the environment of the oceans. Previous methods of elimination have not been "waste disposal"; they have been "waste dispersal." Any significant solution to the problem of elimination must consider what kinds of waste disposal and dispersal a given geologic environment can accommodate without critical alterations in geologic and biological conditions.

Most people are aware of the great variety of waste products of modern culture. Commonly, they are labeled as either municipal, agricultural, or industrial. It is probably more important to consider waste according to its composition rather than its source. The materials listed in *table 23.1* are society's contribution to the hydrologic system.

Solid Wastes. Solid wastes are disposed of in many ways, including landfill, incineration, composting, open dumping, animal feeding, fertilizing, and disposal in oceans. The geologic consequences include changes in the surface of the land where the waste is deposited and changes in the environment (rivers, lakes, oceans, and ground water) where the mass of waste is concentrated. The major problems with solid waste disposal involve the hydrologic characteristics of the site. These include the porosity and permeability of the rock in which the fill is located and whether or not the waste deposit intersects the water table. The altered topography associated with dumps and landfills is also critical, because it can change the drainage and ground-water conditions. Perhaps the most critical contamination problem is created by leachates,

Table 23.1 Wastes That Enter the Hydrologic Cycle

Solid Wastes	Solid Wastes (cont.)
Garbage	Sewage and sewage
Waste from prepara-	treatment residues
tion of food	Sludge
Market refuse—	Mining wastes
handling, storage	Concentrations of
of produce and	various minerals
meat	Strip-mine tailings
Rubbish (household)	Placer-mine tailings
Paper—boxes,	Quarries, rock piles
cartons	**Liquid Wastes**
Plastics	Domestic sewage
Rags	Raw sewage
Grass, leaves	Treated residues
Metal cans	Industrial waste
Dirt	Liquid chemical
Stones, brick,	residues
ceramics	Saline brines
Glass	Paint sludge
Ash	**Gaseous Wastes**
Fly ash	Carbon monoxide
Residue from com-	Emissions from
bustion of coal	internal
Bulky waste	combustion of
Large auto parts	automobiles
Appliances	Sulfur dioxide and
Furniture	sulfuric acid
Trees, stumps,	Sulfur-bearing gases
branches	Emissions from
Dead animals	power plants,
Construction and	smelters
demolition	Hydrogen fluoride
Lumber and	Production of
sheeting scraps	fluorine ores
Broken concrete,	Particulate matter
plaster, etc.	Dust, soot, ash
Pipes, wire	Metals
Asphalt paving	Insecticides,
fragments	herbicides
Industrial waste	Lead
Food processing	Jet exhaust
(slaughterhouse,	**Radioactive Wastes**
etc.)	Explosions of nuclear
Wood, plastic scraps	devices
Slag	Radioactive fallout
Agricultural waste	Atomic energy plants
Manures	Radioactive waste
Crop residues	
Pesticide residues	

which form as circulating water passes through a landfill. Organic and inorganic compounds dissolved in a leachate enter ground-water reservoirs.

Liquid Wastes. Traditionally, liquid wastes have been discharged into surface drainage systems and diluted. They accumulate ultimately in lakes and oceans, where they are stored. As the volume of liquid waste increases, the capacity of the natural water system to dilute it is overwhelmed, and the drainage system becomes a system of moving waste.

One very subtle type of liquid pollutant is the hot water created by cooling systems in power plants and factories. Although the water itself is not contaminated, the temperature alone is enough to alter the biological conditions in streams and lakes into which it flows. Such pollution is called thermal pollution.

Gaseous Wastes (Air Pollution). The population explosion, with the consequent industrial expansion, has produced a variety of gaseous wastes and pollutants in the form of minute liquid and solid particles that are suspended in the atmosphere. In the past, pollutants were expelled into the air with the reasonable assurance that normal atmospheric processes would disperse and dilute them to a harmless, unnoticeable level. In many heavily industrialized areas, however, the atmosphere's capacity for absorption and dispersal has been exceeded, and the composition of the air has been severely altered. The problem is so severe in some areas that acid rain develops. If the troposphere (the lower part of the atmosphere, which is involved in most human activities) extended indefinitely into space, there would be little problem from air pollution. The troposphere, however, extends only to an altitude of 10 or 15 km, and few pollutants move out of it into the overlying stratosphere for any great length of time. Thus, a steadily increasing volume of pollutants is concentrated mostly in the lower part of the troposphere.

Radioactive Wastes. All industries face waste-disposal problems, but none are greater than those of the nuclear energy industry. The generation of nuclear energy creates numerous radioactive isotopes—some with short half-lives, others with very long ones. Nuclear waste is extremely hazardous itself, but another waste product is large amounts of heat. Thus, any disposal system must be capable of removing the waste while completely isolating it from the biological environment. In addition, containment must be maintained for exceptionally long periods. Compared to the waste produced by many other industries, the volume of radioactive waste is not large, but the hazards and the heat that is generated are considerable.

In many respects, the problems with radioactive waste disposal are similar to those with other pollutants produced by American society. What long-range effect will it have on the environment? Unlike other pollutants, radioactive waste can induce mutations in living species. Since we may be forced to look more and more to nuclear energy to solve many of our resource shortages, the long-term effect of radioactive waste could be catastrophic.

One of the more promising methods of radioactive waste disposal involves storage in thick salt formations. Salt deposits are desirable because they are essentially impermeable and are isolated from circulating ground water. In addition, salt yields to plastic flow. It is thus unlikely to fracture and make contact with leaching solutions over extended periods of time. Salt also has a high thermal conductivity and thus can absorb heat from the waste, and it has approximately the same shielding properties as concrete.

In theory, radioactive wastes would be solidified and sealed in containers from 15 to 60 cm in diameter and as much as 3 m in length. The containers would then be shipped to salt mines in the stable interior of the continent, where seismic activity is minimal. There, they would be placed in holes drilled into a salt formation deep in a mine. When filled with waste, the hole would be packed with crushed salt and closed.

Mining Wastes. The waste products from mining operations include (1) tailings and dumps, (2) altered terrain (due to openpit mining and **strip mining**), (3) changes in the composition of the surface, and (4) solid, liquid, and gaseous wastes produced by refining.

In the United States, approximately 3 billion metric tons of rock are mined each year. About 85% comes from openpit and strip mines, which require the removal of an additional 6 billion metric tons of rock as overburden. Such surface mining operations have affected about 12,000 km² of land in the United States. The principal geologic problem arises from the alteration of the terrain by the creation of open pits and artificial mounds and hills of tailings.

Most mine dumps are unstable. They are highly susceptible to mass movement unless they accumulate under proper engineering supervision. In 1966, a mudflow from coal-mine dumps in Aberfan, Wales, completely destroyed a school, killing and injuring many children.

An additional problem arises if mine tailings enter the drainage system. They can choke a stream channel, increasing the flood hazards. Alteration of a stream system also can be produced from placer mining, in which the movement of large quantities of sediment upsets the balance of the stream.

RESOURCES

Statement

The important minerals on which modern civilization depends constitute an infinitesimally small part of the earth's crust. Whereas the rock-forming minerals (such as feldspar, quartz, calcite, and clay) are abundant and widely distributed, copper, tin, gold, and other metallic minerals occur in quantities measured in parts per million (and, in most cases, a very few parts per million). The important question then is, How are these very small quantities of important minerals concentrated into deposits large enough to be used? The answer is simple: they are deposited by the various geologic processes operating in the earth's system.

It may surprise you to learn that essentially every geologic process—including igneous activity, metamorphism, sedimentation, weathering, and deformation of the crust—plays a part in the genesis of some valuable mineral deposits. The occurrence or absence of most mineral deposits, therefore, is controlled by the specific geologic conditions of a region.

It is critical to understand that the processes that form these minerals operate so slowly (by human standards) that the rates of replenishment are infinitesimally small in comparison to rates of consumption. We must clearly understand that mineral deposits are finite and, therefore, are exhaustible and nonrenewable. If the approximate extent of a deposit and its rate of consumption are known, we can predict how long it will last. Our resources are like a checking account that will never receive another deposit. The faster we withdraw or the larger the check we write, the sooner the account will be depleted. Moreover, today relatively few areas of potential mineral deposits are still unexplored. Most of the continents have been mapped and studied extensively, so that the inventory of natural resources is nearly complete. We basically know the extent of our mineral resources and the rates of consumption. It is not difficult to project how long they will last.

Discussion

Most natural resources are finite and nonrenewable. To best appreciate that fact, we will consider some of the principles that govern the concentration of rare minerals into ore deposits and the origin of some nonmetallic resources. These principles are complex and diversified, but the origin of most mineral deposits is somehow related to the various tectonic and hydrologic processes. Therefore, we can recognize major groups of mineral deposits formed by (1) igneous processes, (2) metamorphic processes, (3) sedimentary processes, and (4) weathering processes.

Igneous Processes

Many metallic ores are concentrated by magmatic processes, much as silicate minerals are. That is, concentration results from differences in crystal structure, order of crystallization, density, and other factors. Magmatic intrusions themselves are concentrations, on a regional scale, of silicate minerals and metallic ores. They characteristically are generated at plate boundaries. On a local scale, metallic minerals are concentrated in specific areas of intrusion in a variety of ways. One process is direct **magmatic segregation,** in which heavy mineral grains that crystallize early sink down through the fluid magma and accumulate in layers near the base of the igneous body. Deposits of chromite, nickel, and magnetite are good examples of this type of concentration (*figure 23.1*).

Late crystallization is another process by which rare minerals are concentrated. Many elements that occur in amounts of only a few parts per million in the original magma do not fit readily into the crystalline structure of the silicate rock-forming minerals. These are concentrated in the residual liquid as the feldspar, amphibole, mica, and other rock-forming minerals crystallize. Through this process, much of the magma's silicon, oxygen, and aluminum is incorporated into the crystal structure of rock-forming minerals. Thus, as a magma cools, rare elements, such as gold, silver, copper, lead, and zinc, become concentrated in the last remaining

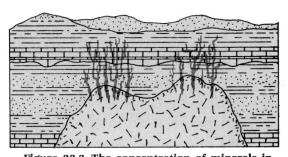

Figure 23.2 The concentration of minerals in hydrothermal vein deposits occurs during the late stage of crystallization. Many rare minerals, such as gold, silver, copper, lead, and zinc, do not fit into the crystalline structure of silicate minerals, so that they are concentrated in residual fluids of the magma. This material can then be squeezed into fractures in the surrounding rock, and the rare metallic elements are precipitated as vein deposits.

fluid. These late-stage, metal-rich solutions can then be squeezed into the fractures of the surrounding rock. As they cool, the rare elements are precipitated as veins of metallic minerals, which differ little from ordinary dikes and sills except in their chemistry and mineralogy. The residual hot solutions of the magma contain much water (because of the low temperature at which water crystallizes), so that the deposits are known as **hydrothermal deposits** (*figure 23.2*).

Hydrothermal solutions are not necessarily injected into the surrounding rock to form veins of mineral deposits. They can also permeate the early-formed crystals in the granitic rock and disseminate metallic minerals throughout the body. This process produces large deposits of low-grade ore, because practically the entire igneous body is mineralized. If the intrusion is exposed at the surface, these low-grade deposits can be mined profitably by special equipment and openpit mining techniques. An important example of this type of concentration is the disseminated copper in porphyritic intrusions. These deposits, called **porphyry coppers,** presently account for more than 50% of the world's copper production.

Porphyry copper and many other metallic minerals are associated with granitic intrusions, which in turn are associated with convergent plate boundaries. The metals are believed to derive from partial melting of the oceanic plate in the subduction zone, and their final concentration and emplacement result from the differentiation of magma as it cools. Mineral deposits, therefore, are products of the tectonic system. The plate tectonic theory promises to be an important key to a better

Figure 23.1 The concentration of minerals by magmatic segregation occurs as heavy, early-formed crystals sink through the fluid magma and accumulate in layers near the base of the magma chamber. The chromite deposits of South Africa are an example.

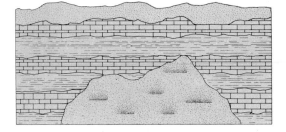

understanding of the genesis of minerals and to future exploration for minerals.

Metamorphic Processes

Ore deposits are also formed by **contact metamorphism,** or metamorphism along the contact between an igneous intrusion and the surrounding rocks. In this process, variations in heat and pressure and the presence of chemically active fluids in the cooling magma alter the adjacent rock by adding or replacing certain components. Limestone that surrounds granitic intrusions is particularly susceptible to alteration and replacement by chemicals in the hot, acid magma solutions. For example, large volumes of calcium can be replaced by iron to form a valuable ore deposit (*figure 23.3*).

Regional metamorphism changes the texture and mineralogy of rocks and in the process forms deposits of nonmetallic minerals, such as asbestos, talc, and graphite.

Mineral deposits formed by metamorphism again show the close association of the generation of ores with the dynamics of convergent plate margins.

Sedimentary Processes

A significant result of the erosion, transportation, and deposition of sediments is the segregation and concentration of material according to size and density. As was emphasized in chapter 6, soluble minerals are transported in solution, silt and clay-size particles are transported in suspension, and sand and gravel are moved mostly as bed load by strong currents. As a result, sand and gravel are concentrated in river bars and beaches. These deposits, both modern and ancient, are a valuable resource of the construction industry. In the United States alone, over $1 billion worth of sand and gravel is mined each year, making this operation the largest industry in the country not associated with fuel production. In areas where sand and gravel have not been concentrated by natural processes, they must be made by crushing and screening, an operation requiring considerable energy and expense.

Sedimentary processes also concentrate other valuable materials, such as gold, diamonds, and tin. Originally formed in veins, volcanic pipes, and intrusions, these minerals are eroded and transported by streams. Because they are heavy, they are deposited and concentrated where current action is weak, such as on the inside of meander bends or on protected beaches and bars (*figure 23.4*). Such layers and lenses of valuable minerals are known as **placer** deposits, and large operations mine them along both modern and ancient rivers and beaches.

Sedimentary processes concentrated the great bulk of iron ore mined today. Banded iron formations of Precambrian age are especially significant because of their abundance. These deposits consist of alternating layers of iron oxide and chert formed during a unique period of earth's history, from 1.8 billion to 2.2 billion years ago, when oxidation conditions on the earth were much different. Iron was transported by streams in a reduced state and accumulated in lakes and seas as a sedimentary deposit. The combination of conditions that are necessary to produce such deposits has never been repeated.

Another way in which sedimentary processes concentrate valuable minerals is the evaporation of saline waters in large lakes and restricted embayments of the ocean. As evaporation proceeds, dissolved minerals are concentrated and eventually precipitated. These **evaporite** deposits include several minerals of important commercial value, including potassium, sodium, and magnesium salts; gypsum; sulfates; borate; and nitrates. Many marine evaporites, occurring in extensive layers interbedded with shale and limestone, have been mined for thousands of years. Thick salt deposits are especially significant. Where they are buried at sufficient depth, the pressure from the overlying beds commonly is great enough to cause the salt to flow, and it rises like an intrusive body of magma into the overlying strata. These intrusions, known as **salt domes,** are common in the Gulf Coast of the United States, Germany, and Iran. Salt domes are economically important, not only because they constitute a source of nearly pure salt, but also because they are important traps for petroleum.

Figure 23.3 The concentration of ore deposits by contact metamorphism occurs as fluids from the cooling magma replace parts of the surrounding rock. Iron deposits commonly are formed by this process.

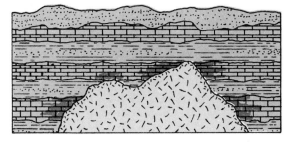

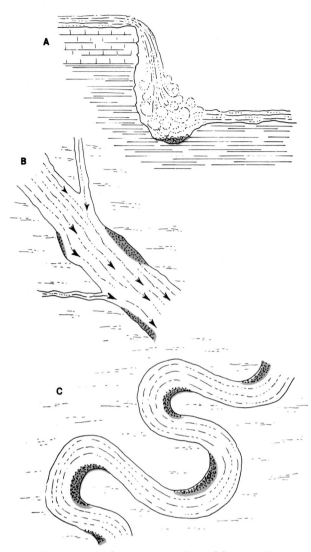

Figure 23.4 The concentration of heavy minerals in placer deposits occurs by stream action. Gold, diamonds, and tin, for example, can be eroded from their original deposits and accumulate in areas where stream currents are weak, such as the base of a waterfall or the inside of a meander bend. Early prospectors who panned for gold exploited this type of deposit.

Weathering Processes

The simple process of weathering also concentrates minerals. It removes soluble material from rocks and leaves the insoluble material as a residue. Weathering, therefore, can enrich ore deposits that were originally formed by other processes (*figure 23.5*), or it can concentrate in the regolith material that was originally dispersed throughout a rock body. For example, extensive weathering of granite in tropical and semitropical zones commonly concentrates relatively insoluble metallic oxides in the thick regolith as it removes the more soluble material. Deposits of aluminum, nickel, iron, and cobalt are formed in this way.

The Nonrenewability of Mineral Deposits

The few examples cited here show that the world's valuable mineral deposits were formed in a systematic way by the major geologic processes during various periods in the geologic past. Their formation required long intervals of time and occurred under specific geologic conditions. Some deposits were formed in such restricted geologic settings that they approach uniqueness. For example, 40% of the world's reserves of molybdenum is in one igneous intrusion in Colorado, 77% of the tungsten reserves is in China, more than 50% of the tin reserves is in Southeast Asia, and 75% of the chromium reserves is in South Africa. If resources are depleted, we cannot "just go out and find some more." There simply are no more. Fortunately, most metals, unlike fossil fuels, can be recycled. We obviously must emphasize recycling to conserve metals.

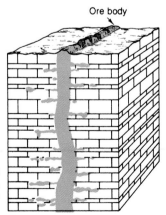
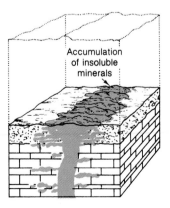

Figure 23.5 The concentration of ore deposits by weathering processes occurs as soluble rock, like limestone, is removed in solution, so that insoluble minerals are concentrated as a residue.

ENERGY

Statement

The technological progress and standard of living of modern society are intimately related to energy consumption. Until recently, energy resources and the capacity for growth seemed unlimited. Now, for the first time, people all over the world are experiencing a serious energy shortage, one that is sure to become more acute in the future. Understanding the sources of energy and how they can be used most effectively is one of the more pressing problems of the twentieth century.

If we consider the earth's entire dynamic system, we see that our sources of energy are found in both renewable and nonrenewable forms. A renewable energy source is either one that is available in unlimited amounts, for all practical purposes, or one that will not be appreciably diminished in the foreseeable future. Solar energy, tidal energy, and geothermal energy are the most important examples. In contrast, nonrenewable energy sources, like mineral resources, are finite and exhaustible. They cannot be replaced once they are consumed. The fossil fuels, coal and petroleum, on which modern culture relies so much, are nonrenewable. These energy sources have been concentrated and preserved by geologic processes that operated during past periods of geologic time. Although the same processes may function today, they operate too slowly to replenish our supplies of these fuels.

An important aspect of today's energy picture is that over 90% of the energy we use is produced from nonrenewable fossil fuels. The exponential growth in consumption of the world's fossil fuels has brought on the present energy crisis. Few technical analysts, whether they be economists, geologists, or engineers, doubt that the problem exists. They see the trends and events that indicate, in the relatively near future, difficulties ranging from an awkward situation to disaster and economic peril. By contrast, few of the general public—if we are to accept the results of public opinion polls—believe that a problem exists. They see only gas pumps and energy bills, not the statistics of the analysts.

The energy crisis, however, was accurately predicted. In 1956, M. King Hubbert, an eminent research geologist for Shell Oil Company, analyzed the reserves, production, and rate of consumption of petroleum. From these data, he predicted a continuous decline in production beginning in 1970 (*figure 23.6*). Look at these curves carefully and note that they show a peak in production at 150 billion barrels in about 1965. If our actual reserves exceeded this estimate by a third, the whopping incease would only postpone the "day of reckoning" by 5 years. Hubbert's projection caused some alarm in the petroleum industry 25 years ago, but it was mostly ignored by the government. It is estimated that petroleum production will decline to near exhaustion by the year 2070. Coal is expected to replace petroleum as the main hydrocarbon resource, with peak production by the year 2100 (plus or minus 100 years). It is quite clear, there-

Figure 23.6 Projected rates of petroleum consumption were calculated in 1956 by M. King Hubbert. His analysis of reserves, production, and consumption rates predicted a decline in production beginning in 1970, with supplies declining to near exhaustion by the year 2070.

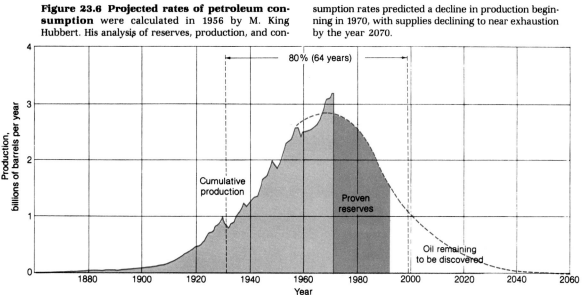

fore, that in the very near future our reliance on fossil fuels must be greatly curtailed. Fossil fuels will have to be replaced by other energy sources.

Discussion

Renewable Energy Sources

Solar Energy. Solar radiation is the most important renewable, or sustained-yield, energy source. It has the added benefits of being clean, constant, and reliable. The major problem, of course, is that solar energy is distributed over a broad area, so that it must be concentrated into a small control center where it can be converted to electricity and distributed. Panels for collecting solar radiation have been mounted on buildings to provide heating or cooling, but collecting systems for large-scale use of solar energy, although technically feasible, are not at present economical. For example, to satisfy the present need for electrical energy in the United States, a collecting system covering 25,000 km² (one-tenth the size of the state of Nevada) would be required. Large-scale use of solar energy, therefore, is a long way in the future, although local use in homes and buildings would help alleviate the need for other forms of energy.

Hydropower. Hydropower is another sustained source of energy. It has been developed in the United States to approximately 25% of its maximum capacity. With full development by the year 2000, hydropower would still provide only 10% or 15% of the energy needed in the United States. A problem with hydropower is that, when a river system is modified by a dam, unforeseen side effects can occur. Moreover, the reservoir behind the dam is a temporary feature, destined to become filled with sediment. The useful life expectancy for most large reservoirs is only 100 or 200 years, so that this source of energy is, in reality, limited.

Tidal Power. Another sustained energy source is the ocean's tides. Tidal power can be harnessed by a dam built across the mouth of a bay. At the narrowed entrance of the bay, the rise and fall of the tides produces a strong tidal current flow, which can be used to turn turbines to generate electricity. Even with maximum development, however, tidal power could supply only 1% of the energy needed in the United States.

Geothermal Energy. The earth has its own internal source of heat, which is expressed on the surface by hot springs, geysers, and active volcanoes. In general, temperature increases systematically with depth, at a rate of approximately 3°C per 100 m. Temperatures at the base of the continental crust can range from 200°C to 1000°C, and at the center of the earth they are perhaps as high as 4500°C. Unfortunately, most of the earth's heat is far too deep ever to be artificially tapped, and what could be reached by drilling typically is too diffuse to be of economic value. Like ore deposits, however, geothermal energy can be concentrated locally and has been used for years in Iceland and in areas of Italy, New Zealand, and the United States. Geothermal energy is concentrated along active plate margins or in mantle plumes, and concentrations are most likely to occur in regions where magma is relatively close to the surface.

It is estimated that, at its maximum worldwide development, geothermal energy would yield 10 times as much as it did in 1969, but this would still amount to only a small fraction of the world's total energy requirements. Locally, however, geothermal energy could be significant.

Fossil Fuels

Coal, petroleum, and natural gas commonly are referred to as **fossil fuels,** because they contain solar energy preserved from past geologic ages. The idea that we presently use energy released by the sun more than 200 million years ago may seem remarkable, but the basic process of storing solar energy is quite simple. Heat from the sun is converted by biological processes into combustible, carbon-rich substances (plant and animal tissues). These are subsequently buried with sediment and preserved.

Coal. Coal originates from plant material that flourished in ancient swamps typically found in low-lying coastal plains. A modern example of such an area is the present Great Dismal Swamp, along the coast of Virginia and North Carolina (*plate 12*). In this area, the lush growth of vegetation has produced a layer of **peat** more than 2 m thick, covering an area of over 5000 km². (Peat is an accumulation of partly decomposed plant material containing approximately 60% carbon and 30% oxygen.) In such an environment, the layer of peat can be covered with sand and mud from the adjacent lagoon and beach, as sea level slowly rises (*figure 23.7*). Under pressure from the overlying sediment, water and organic gases (volatiles) are squeezed out, and the percentage of carbon increases. By this process, peat is compressed and is eventually transformed into coal.

If sea level rises and falls repeatedly, a series of coal beds can develop, interbedded with beach sand and nearshore mud. The essential require-

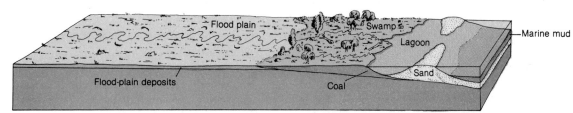

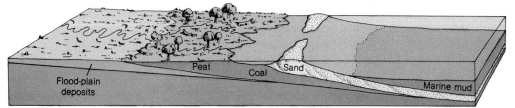

(A) *The sequence of sedimentary environments along a coast grades seaward, from flood plain to swamp and lagoon, to beach, to offshore mud.*

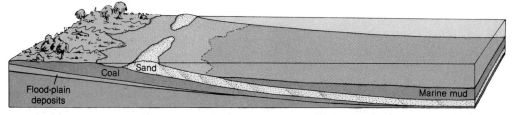

(B) *As the sea expands inland, swamp vegetation is deposited over the flood plain, beach sand is deposited over the previous swamp (peat) muck, and marine mud is deposited over the beach. The heat and pressure of the overlying sediment changes the peat to coal.*

(C) *Continued expansion of the sea superposes coal over flood-plain sediments, beach sand over coal, and mud over beach sand.*

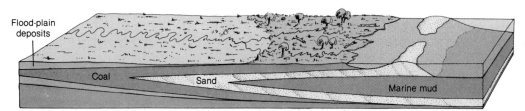

Figure 23.7 Coal deposits are formed from plant material that once flourished in ancient swamps. This series of diagrams shows how expansion and contraction of the adjacent sea buries the swamp vegetation and transforms it into coal.

(D) *As the sea recedes, the sequence is reversed. That is, beach sand is deposited over offshore mud, coal is deposited over beach sand, and flood-plain sediments are deposited over coal. Thus, by expansion and contraction of the sea, layers (lenses) of sediments and coal are deposited in an orderly sequence.*

ments for the development of coal deposits, therefore, are a luxuriant growth of vegetation and relatively rapid burial by sediment to prevent decomposition. Coal deposits are thus restricted to the latter part of the geologic record, when plant life became plentiful. The most important coal-forming periods in the earth's history were the Pennsylvanian and Permian periods. Great swamps and forests then covered large parts of most of the continents. (The Western industrial nations, such as Great Britain, Germany, and the United States, developed with energy from these coals.) Other important periods of coal formation were the Cretaceous and Tertiary periods.

With the completion of at least a reconnaissance geologic mapping of most of the continents, all of the major coalfields are believed to have been discovered, and a reasonably accurate inventory of the world's coal reserves can therefore be made. Most of the reserves are found in the United States and Russia, so that developing nations are not likely to grow by using vast amounts of coal.

Coal is important because there are reasonably large reserves of it. With the depletion of oil and gas, we are already witnessing an important shift to reliance on coal. The real pinch on oil and gas is now upon us and will put tremendous pressure on the coal industry, which undoubtedly will bring serious environmental problems.

Petroleum and Natural Gas. Petroleum and natural gas are hydrocarbons (molecules composed only of hydrogen and carbon, in various combinations). In contrast to coal, the hydrocarbons forming oil and gas deposits originate largely from microscopic organisms that once lived in the oceans or in large lakes. The remains of these organisms accumulated with mud on the sea floor. Because of their rapid burial, they escaped complete decomposition.

Deposits of oil and gas form if four basic conditions are met.

1. There must be sufficient organic material in the fine-grained sediments (source beds).
2. The source beds must be buried deep enough (usually at least 500 m) for heat and pressure to compress the rock and cause the chemical transformations that break down organic debris into hydrocarbons.
3. Once they are formed, the hydrocarbons must migrate upward from the source beds into more porous and permeable rock (usually sandstone or porous limestone or dolomite) called reservoir beds.
4. As the oil and gas migrate through the reservoir beds, they must encounter a trap, or a barrier that causes them to accumulate.

If the reservoir beds provide an unobstructed path to the surface, the oil and gas seep out and are lost. This is one of the reasons that most oil and gas deposits are found in relatively young rocks. In older rocks, there has been more time during which erosion and earth movements have provided means for oil and gas to escape. Barriers, or traps, can result from a variety of geologic conditions, such as those shown in *figure 23.8.* Exploration for oil and gas, therefore, is based on finding sequences of sedimentary rocks that provide good source and reservoir beds and then locating an effective trap.

Figure 23.8 The accumulation of oil and gas requires (1) a permeable formation, such as a porous sandstone, into which the petroleum can migrate, and (2) impermeable cap rock, to trap the fluids. Some of the geologic structures that trap oil and gas are shown here.

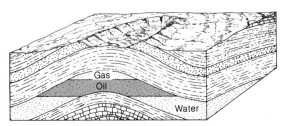

(A) Anticline. *Oil, being lighter than water, migrates up the dip of permeable beds and can be trapped beneath an impermeable shale bed in the crest of an anticline.*

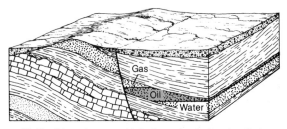

(B) Fault trap. *Impermeable beds can be displaced against a permeable stratum and trap the oil as it migrates updip.*

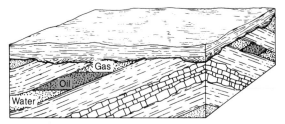

(C) Unconformity. *An impermeable layer can cap inclined strata to form a seal that traps the upward-migrating oil.*

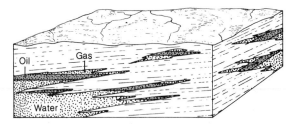

(D) Impermeable barrier. *Shale surrounding a sandstone lens can form a barrier and prevent the oil from escaping.*

In some instances, hydrocarbons can remain as solids in the shale in which the organic debris originally accumulated. These deposits are known as **oil shales.** They are reservoirs of oil that may become important in the future. The problem with oil shale, however, is that it must be mined and heated to extract the oil. This process requires considerable energy and is not economically feasible at present.

Oil and gas are convenient forms of energy because they are easy to handle and transport. Unfortunately, at the present rate of consumption, the known reserves will soon be depleted. If the present trends continue, we will soon be forced to begin large-scale gasification and liquification of coal and oil shale deposits and to rely more on nuclear and solar energy. Clearly, we can expect to pay a great deal more for petroleum in the future and should use alternate sources of energy wherever possible.

Nuclear Energy

The ever-increasing demands for energy and the decreasing supply of fossil fuels naturally put the spotlight on nuclear power as the answer to our energy requirements. The technology of nuclear energy production is well developed. Modern society, however, has hesitated to move toward large-scale production of nuclear energy because of the possibility of serious long-range environmental problems. Radiation hazards, problems of waste disposal, and thermal pollution of fresh and marine waters, as well as potential terrorist activities, are among the greatest concerns.

The key element in the development of nuclear energy is uranium. The average uranium content in the rocks of the earth's crust is only 2 parts per million. It is concentrated into deposits, however, by late magmatic segregation, and it occurs in veins associated with intrusive igneous rocks (see *figure 23.2*). As the veins oxidize, the water-soluble uranium is leached out and transported by surface and ground water. It can later be deposited in permeable sedimentary rocks, as it is absorbed by clay minerals and organic matter. The rich uranium deposits in the Colorado Plateau are concentrated in ancient stream channels, especially where fossil wood and bones are found.

Important uranium deposits occur in Canada, the United States, South Africa, and Russia. If the hazards and environmental problems can be solved, nuclear energy may indeed become an important source of energy.

LIMITS TO GROWTH

Statement

The consumption of natural resources is proceeding at a phenomenal rate. A moment's reflection shows that the rapid population and industrial growth that has prevailed during the last few hundred years is not normal. Indeed, this period is one of the most abnormal phases of human history. Current rates of growth and the associated consumption of natural resources present one of our most serious problems. The problem is basically one of changing from a period of growth to a period of nongrowth. This change will require a fundamental revision of current economic and social thinking based on the assumption that growth can be permanent.

Discussion

The preceding section emphasized how mineral and energy resources are systematically formed by geologic processes. They do not occur haphazardly, nor are they distributed evenly throughout the continents. Thus, some continents and nations have resources and some do not. Iceland and Hawaii, for example, being built exclusively of basaltic lava, cannot be considered good potential sources of petroleum and natural gas, no matter how long those areas are explored. In addition, mineral deposits are nonrenewable resources. Once a deposit is mined out and used, it is gone forever.

Modern industrial civilization has developed principally during the last century. It differs from all previous civilizations in the amounts of energy and resources it uses and in its rate of growth. It is important, therefore, to consider how resources are used and the rates at which they are being depleted. Some of the more important mineral and energy resources used in today's major industries are listed in *table 23.2*. Column 3 shows the number of years known reserves will last at the current rate of consumption. These figures assume no growth and constant rates of usage. Such figures are called **static indices** and normally are used to express the future availability of resources.

The rate of resource consumption is not static, however. It increases exponentially. The exponential rate results both from population growth and from the growth of the average annual consumption per person, which rises each year. In other words, growth is exponential because of an increasing population and a rising standard of living. A more accurate estimate of the lifetime of a given mineral resource is the **exponential index,** or the number of years known reserves will last at an exponential rate of consumption. For many minerals, the rate of usage is growing even faster. For the resources listed in *table 23.2*, the exponential index is shown in column 4. Exponential rates of consumption drastically reduce the estimated lifetime of a reserve. Coal will be depleted in 111 years rather than 2300, chromium will be gone in 95 years rather than 420, and so on.

We can assume that known reserves can be expanded by exploration and new discoveries to five times their presently known amounts. This increase would not extend the lifetime of a deposit by a factor of five, however, because the effect of ex-

Table 23.2 Nonrenewable Natural Resources

1	*2*	*3*	*4*	*5*
				Exponential Index Calculated Using Five Times
Resource	*Known Global Reserves*	*Static Index (years)*	*Exponential Index (years)*	*Known Reserves (years)*
Aluminum	1.17×10^9 tons	100	31	55
Chromium	7.75×10^8 tons	420	95	154
Coal	5×10^{12} tons	2300	111	150
Cobalt	4.8×10^9 lb.	110	60	148
Copper	3.08×10^8 tons	36	21	48
Gold	3.53×10^{10} troy oz.	11	9	29
Iron	1×10^{11} tons	240	93	173
Lead	9.1×10^9 tons	26	21	64
Manganese	8×10^8 tons	97	46	94
Mercury	3.34×10^6 flasks	13	13	41
Molybdenum	1.08×10^{10} lb.	79	34	65
Natural gas	1.14×10^{15} ft.^{3}	38	22	49
Nickel	1.47×10^{11} lb.	150	53	96
Petroleum	4.55×10^{11} bbl.	31	20	50
Platinum group	4.29×10^8 troy oz.	130	47	85
Silver	5.5×10^9 troy oz.	16	13	42
Tin	4.3×10^8 long tons	17	15	61
Tungsten	2.9×10^9 lb.	40	28	72
Zinc	1.23×10^8 tons	23	18	50

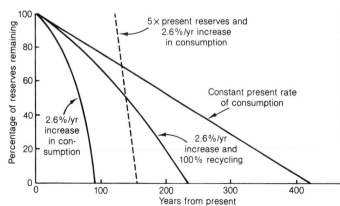

Figure 23.9 The lifetime of chromium reserves depends on future rates of use. If usage remains constant, reserves will be depleted linearly and will last 420 years. If usage increases exponentially at its present rate of 2.6% per year, reserves will be depleted in just 95 years. If actual reserves are five times greater than the present known reserves, chromium ore will be available for 154 years, with an exponential growth in usage. Even if all chromium were perfectly recycled, from 1970 onward, exponentially growing demand would exceed the supply after 235 years.

ponential growth is to consume resources at ever faster rates. Column 5 shows the number of years that five times the known global reserves will last with consumption growing exponentially at the average annual rate of growth.

Figure 23.9 illustrates how an exponential increase in consumption affects a nonrenewable resource. Chromium is taken as an example, because it has one of the largest static indices of all the resources listed in *table 23.2*. At the current rate of usage, the known reserves will last about 420 years. Reserves of chromium, however, are actually being consumed at a rate that increases by 2.6% annually. Known reserves will be depleted in

Figure 23.10 A computer model of resource consumption and its influence on other variables assumes no major changes in the physical, economic, and social relationships that historically have governed the development of the world system. All variables plotted here follow historical values from 1900 to 1970.

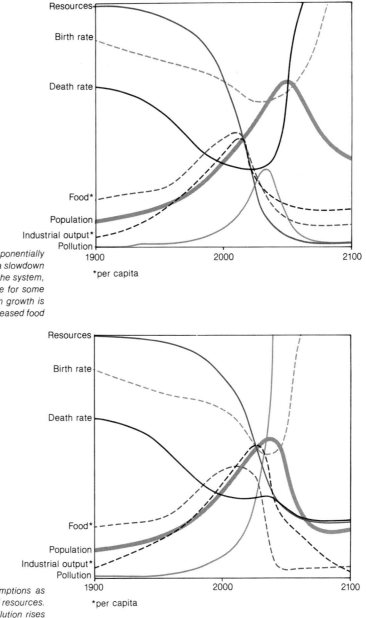

*per capita

(A) *Food, industrial output, and population grow exponentially until the rapidly diminishing resource base forces a slowdown in industrial growth. Because of natural delays in the system, both population and pollution continue to increase for some time after the peak of industrialization. Population growth is finally halted by a rise in the death rate due to decreased food and medical services.*

*per capita

(B) *These curves are based on the same assumptions as those in diagram **A**, but with double the reserves of resources. Industrialization can reach a higher level, but pollution rises much more rapidly, causing an immediate increase in the death rate and a decline in food production.*

95 years at this rate. If we assume that deposits that are yet undiscovered could increase the known reserves by a factor of five, this increase would only extend the lifetime of the reserves from 95 to 154 years. Even with 100% recycling of chromium, so that none of the initial reserves were lost, the demand would exceed the supply in 235 years. People would simply want more than there is.

A similar graph for each resource could be drawn from the data in *table 23.2.* The time scale would vary, but the general shape of the curves would be the same. The exponential index, shown in column 4 of the table, is the critical figure. At the present rate of growth in consumption, copper will be depleted in 21 years or, if reserves are multiplied by five, in only 48 years. Gold has a projected lifetime of 9 years; natural gas, 22; tin, 15. These figures are projections based on the best available data on consumption rates, reserves, and growth. They are not predictions. They are extrapolations from known facts.

Obviously, with the present rates of consumption and the projected increase in these rates, the great majority of the currently important nonrenewable resources will be extremely costly 100 years from now, regardless of the most optimistic assumptions about undiscovered reserves. The limits of growth in the world system probably will not be imposed by pollution. The limits will be set by the depletion of natural resources. The projected interaction of some of the major variables as the world system grows to its ultimate limits is shown in *figure 23.10.* On the assumption that no major changes will occur in the physical, economic, and social relationships that have historically governed the world system, depletion of natural resources will be the main factor in limiting the growth of industrialization and population. According to *figure 23.10,* food, industrial output, and population will continue to grow exponentially until the rapid depletion of resources forces a sharp decline in industrial growth.

In our finite earth's system, unlimited exponential growth is impossible. In fact, the transition into a stable or declining phase has already begun. This does not pose insurmountable technological, biological, or social problems. It does require some fundamental adjustments in our present growth culture. If we can achieve appropriate cultural adjustments, the steady-state system could foster a social environment that would be conducive to the flowering of one of humanity's greatest intellectual advances. The alternative could be catastrophic.

SUMMARY

The earth's natural systems are intricately related, and they maintain a delicate balance with one another. We are capable of modifying and manipulating most natural systems, including rivers, slopes, ground water, and shorelines. Such systems sometimes adjust to artificial modifications with many unanticipated side effects.

Waste disposal is a major environmental problem. Matter cannot be created or destroyed. It can only change from one state to another or from one chemical combination to another. Problems with waste disposal involve solid, liquid, gaseous, and radioactive wastes.

Mineral resources have been concentrated by a variety of geologic processes. These processes are related to the plate tectonic and hydrologic systems. Most resources are therefore finite and nonrenewable.

Igneous processes concentrate valuable minerals by (1) magmatic segregation of early-formed heavy minerals, which settle to the base of the magma chamber, and (2) hydrothermal activity during the late stages of cooling.

Metamorphic processes concentrate minerals by (1) contact metamorphism around igneous intrusions (replacement) and (2) regional metamorphism, which changes the texture and mineral composition of rock bodies.

Sedimentary processes concentrate minerals by (1) sedimentary differentiation (for example, of sand and gravel), (2) concentration of placers (such as gold and diamonds), and (3) evaporation. Weathering concentrates minerals by leaching.

Energy resources are classified as renewable and nonrenewable forms. Renewable energy sources include (1) solar energy, (2) hydropower, (3) tidal energy, and (4) geothermal energy. None of these can be expected to supply a large percentage of the energy that will be required worldwide in the near future.

Nonrenewable energy sources include fossil fuels and nuclear energy. Most industry is based on fossil fuels, particularly oil and gas. Fossil fuels are essentially solar energy from past geologic ages stored in the form of organic matter. Coal originates principally from vegetation in coastal swamps that became buried by other sediments. Oil and gas originate from microscopic organisms that accumulate with sediment in the oceans or in large lakes. In order for oil and gas deposits to

PLANETARY GEOLOGY

The geologic exploration of the planets of the solar system is one of the most exciting adventures ever experienced by geologists, because for the first time whole new worlds can be compared geologically with the earth. The study of planetary geology, however, does much more than merely satisfy scientific curiosity. By comparing in detail the geologic nature and evolution of different planets, we can better recognize which principles and processes are fundamental to the geology of the earth and which are of secondary importance.

We wonder why our moon has no atmosphere, no folded mountain belts, and no recent volcanism. Why does Mars have such large shield volcanoes and an immense canyon? How did stream channels develop and how did catastrophic flooding occur on Mars, if water cannot exist as a liquid on the planet today? Why did the moon, Mercury, Venus, and Mars all have major thermal events and outpourings of basalt early in their history? Why are the moons of Jupiter so different, and what are the origin and history of the small, icy moons of Saturn? To gain an insight into these and other questions about the geology and evolution of the planets, we will venture into a brief study of planetology.

MAJOR CONCEPTS

1. Cratering was the dominant geologic process in the early history of all planetary bodies.
2. The frequency of craters on a planet's surface and the superposition of crater ejecta provide a basis for establishing sequences of events that occurred early in the history of the planets.
3. The moon, Mercury, Venus, and Mars form a family of related planets, known as the terrestrial planets, which probably experienced similar sequences of events in their early history. Each of these planets records (a) an early period of intense bombardment, (b) a thermal event in which the extrusion of fluid lava flows formed extensive plains, and (c) a subsequent period of light bombardment.
4. Cratering on the moons of Jupiter and Saturn suggests that the period of intense bombardment affected the entire solar system.
5. Both the moon and Mercury are primitive bodies, and their surfaces have not been modified by hydrologic or tectonic systems. Mars, in contrast, has evolved beyond an impact-dominated planet. It has a varied and complex history of changes resulting from its own hydrologic and tectonic systems.
6. The surface of Venus has been modified by all three major geologic processes: impact, tectonism, and surface erosion. The crust of Venus does not appear to be broken into tectonic plates, however, and tectonic motion does not appear to have played a dominant, ongoing role in altering the surface.

THE GEOLOGY OF THE MOON

Statement

Before the exploration of space, many scientists thought of the moon as a geologic Rosetta stone: an airless, waterless body untouched by erosion. As a result, they expected that the moon would contain clues to events that occurred in the early history of the solar system, which would help reveal details of its origin and history and provide new insight into the evolution of the earth. Now, with the Apollo missions completed, many of their fondest hopes have been realized. Thousands of orbital photographs of the moon have permitted us to map its surface with greater accuracy than we could map the surface of the earth a few decades ago. We now have over 380 kg of rocks from eight locations on the moon, plus data from a variety of experimental packages. As it turns out, the moon is truly a whole new world. The rocks (analyzed by hundreds of scientists from many countries) and surface features record events that occurred during the first billion years of the solar system—events that have been completely erased from the earth.

We have found that the moon is a planet* without hydrologic and tectonic systems. Cratering has been the most important surface process and is responsible for the formation and degradation of most of the lunar landscape. A major thermal event did occur on the moon, however, when basaltic lava was extruded in a series of eruptions to form the lunar maria. Only minor structural features, such as grabens and wrinkle ridges, have been found. The lack of large-scale compressional features suggests that the crust has not been deformed by a tectonic system.

A geologic time scale has been constructed for the moon on the basis of the principle of superposition, which was devised in the early nineteenth century by geologists studying the earth. Radiometric dates of lunar rock samples provide benchmarks of absolute time, and the major events in lunar history have been outlined.

Perhaps the most important aspect of the moon's geologic evolution is that the most dynamic events occurred during the early history of the solar system, before the oldest rock on the earth was formed. Thus, the moon provides important insights into planetary evolution that are unobtainable from studies of the earth.

Discussion

Major Physiographic Divisions

When Galileo first observed the moon through a telescope in 1609, he discovered that the dark areas are fairly smooth and the bright areas are rugged and densely pockmarked with craters. He called the dark areas **maria** (Latin, "seas") and the bright areas **terrae** ("lands"). These terms are still used today, although we know that the maria are not seas of water and the terrae are not geologically similar to the earth's continents.

Maria and terrae are easily visible from the earth through a small telescope or binoculars. As is shown in *figure 24.1*, the maria on the near side of the moon appear to be flat, with only a few large craters. Some occur within the walls of large, circular basins. Others occupy much larger, irregular depressions. We know from rock specimens and surface features that the maria are vast floods of lava, which flowed into low depressions on the lunar surface.

The terrae, or highlands, constitute about two-thirds of the visible surface of the moon. They exhibit a wide range of relief forms and the highest and most rugged topography on the moon, with local relief in many areas being as much as 5000 m. The single most important characteristic of the lunar highlands is that they are completely saturated with **craters,** many of which range from 50 to 100 km in diameter. The far side of the moon is composed almost entirely of densely cratered highlands (see *plate 1*).

The maria and highlands not only represent different types of terrain, but they broadly represent two different periods in the history of the moon. For the most part, the highlands are old surfaces,

---

*The term *planet* is used by geologists to refer to any large object in the solar system. Satellites can be considered planets because their origin and evolution are similar to the origin and evolution of the earth and the other planets.

Figure 24.1 The major physiographic divisions of the moon are (1) the densely cratered highlands and (2) the dark, relatively smooth maria. The highlands occur in the Southern Hemisphere and on the far side of the moon, occupying about 80% of the lunar surface. They are, for the most part, an old surface, formed during a period of intense meteorite bombardment, which occurred early in the history of the solar system. The dark areas are composed of vast floods of basalt, which filled large multiringed basins and lapped up on the older highlands. Young craters on the mare surface have bright rays.

which became extensively pockmarked with craters early in the moon's history. After the highlands were formed, the mare basins were produced as large impact structures and subsequently were filled with lava. In places, the lava overflowed the basins and spread over parts of the lunar highlands. Thus, the maria are relatively young features of the lunar surface, although they were formed from 3 billion to 4 billion years ago.

Craters

Although cratering is a rare event on the earth today, it is a fundamental and universal process in planetary development. The moon is pockmarked with billions of craters, which range in size from microscopic pits on the surface of rock specimens to huge, circular **basins** hundreds of kilometers in diameter. Craters are also abundant on Mercury, Venus (in spite of its dense atmosphere), Mars, the **asteroids,** and many of the satellites of the outer planets. Indeed, cratering was undoubtedly the dominant geologic process on the earth during the early stages of its evolution, and probably the surface of the earth once looked much like that of the moon today.

The Mechanism of Crater Formation. Impact processes are nearly instantaneous, but they can be studied in the laboratory with high-speed motion pictures. Conceptually, the process is simple, as is illustrated in *figure 24.2*. As a meteorite strikes the surface, its kinetic energy is almost instantaneously transferred to the ground as a shock wave, moving downward and outward from the point of impact. This initial compression wave is followed by a rarefaction wave, rebounding in the opposite direction, which causes material to be ejected from the surface and thrown out along ballistic trajectories. This fragmented material accumulates around the crater, forming an **ejecta blanket** and a system of splashlike rays. Such impact structures are known as **ray craters.** A central peak on the crater floor can result from the rebound, and the crater rims can be overturned. Many large basins (more than 300 km is diameter) contain a series of concentric ridges and depressions and hence are called **multiringed basins.**

Meteorite impacts, like other rock-forming processes that operate at the surface of a planet, produce new landforms (ray craters) and new rock bodies (ejecta blankets). The rock-forming processes associated with impact include (1) fragmentation, transportation, and deposition of rock particles and (2) shock metamorphism, in which new high-pressure minerals are created and partial

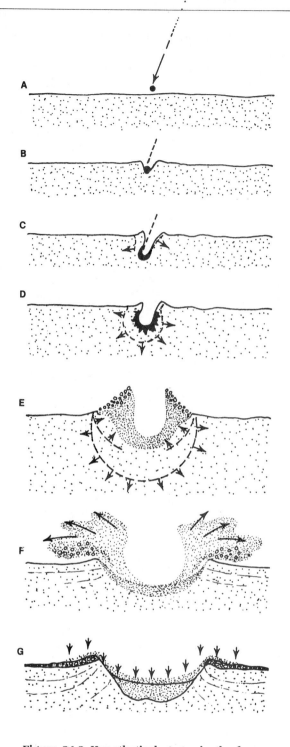

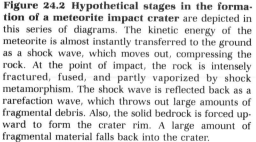

Figure 24.2 Hypothetical stages in the formation of a meteorite impact crater are depicted in this series of diagrams. The kinetic energy of the meteorite is almost instantly transferred to the ground as a shock wave, which moves out, compressing the rock. At the point of impact, the rock is intensely fractured, fused, and partly vaporized by shock metamorphism. The shock wave is reflected back as a rarefaction wave, which throws out large amounts of fragmental debris. Also, the solid bedrock is forced upward to form the crater rim. A large amount of fragmental material falls back into the crater.

melting and vaporization occur as results of the force of impact.

Types of Craters. The surface of the moon has been described as a forest of craters, and at first glance all craters look alike. After a few moments of thoughtful study, however, one can identify various kinds of "trees" in the forest. We recognize various types of craters by certain characteristics of their size and form. Detailed study of lunar craters shows that many morphologic differences are associated with size. Craters of similar size and shape probably originated from the impact of meteorites of similar size. It is therefore useful to classify craters according to their size and shape (*figure 24.3*).

Crater Degradation and Modification. The

Figure 24.3 Four types of craters are recognized on the basis of size and morphology.

(A) Craters smaller than 20 km in diameter are almost perfectly circular and typically are bowl-shaped. In this example, the raised crater rim is well defined and is surrounded by a prominent ejecta blanket.

(B) Craters from 20 to 200 km in diameter are the most common type of crater on the moon. They typically have a central peak, a relatively flat floor, and a series of terraces on their inner walls.

(C) Craters from 200 to 300 km in diameter are transitional in morphology to large multiringed basins. As crater size increases, the configuration of the central peaks changes from a single promontory to a cluster of peaks.

(D) Craters larger than 300 km in diameter, called multiringed basins, contain a series of concentric ridges and depressions resembling a giant bull's-eye. Beyond the outermost ring are ridges oriented radially to the basin. Multiringed basins are the largest structures on the moon. They are believed to have been formed by the impact of asteroid-size bodies, which generated large shock waves that permanently deformed the crust.

crater morphology illustrated in *figure 24.3* refers to the original size and shape produced by impact. After a crater is formed, however, it is subject to certain types of modification. The steep crater walls soon begin to slump and move downslope by mass movement. This process results in partial filling of the depression and the formation of concentric terraces (slump blocks) inside the crater rim. Isostatic adjustment can arch up the crater floor to compensate for the removal of material in the excavation of the crater. Further modification results from three main processes: (1) subsequent impact can partly destroy or obliterate the crater and ejecta blanket; (2) the crater can be covered with ejecta from the formation of younger craters; and (3) the crater and ejecta can be buried by lava flows. Examples of crater degradation and modification are shown in *figure 24.4*.

Volcanic Features

The nature and distribution of volcanic activity are especially important in studies of planetary dynamics, because volcanic products constitute a window into a planet's interior, providing valuable insight into how the planet operates. Although craters dominate the lunar landscape, a variety of volcanic features have been discovered, showing that the moon has had a thermal history.

The most important volcanic features on the moon are the maria. These vast floods of lava fill the large multiringed basins on the near side of the moon and commonly overflow, spilling into the surrounding areas. From distant views, the maria appear to represent one huge flood that filled all the basins to the same level, just as water fills the oceans on the earth. Upon closer examination, however, we find that the maria surfaces are not at the same level and that the basins have been filled with innumerable lava flows. Absolute dating of lunar samples implies that the lava was extruded over a period of nearly 1 billion years. The maria were formed during a specific interval of time, however, extending from 3.9 billion to 3.1 billion years ago, and record a major thermal event in

Figure 24.4 Crater degradation and modification can occur by means of slumping, isostatic adjustment, subsequent impact, and burial by ejecta or lava. Examples of the modification of craters ranging from 20 to 45 km in diameter are shown in this series of photographs. Studies of crater degradation and modification are important in reconstructing the sequence of events in lunar history.

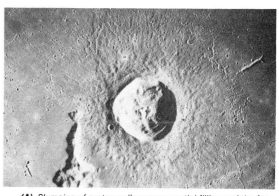

(A) Slumping of crater walls causes partial filling and the formation of terraces, and isostatic adjustment results in uplift of the crater floor.

(B) Crater rims are partly obliterated by subsequent impact.

(C) The large crater in the center is partly covered by ejecta from younger craters.

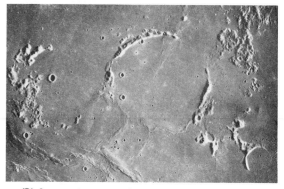

(D) Some craters are partly covered by lava flows.

lunar history. Since then, the moon has not experienced significant volcanic activity.

Tectonic Activity

Almost no global tectonic activity has occurred on the moon during the last 3 billion years, and very little has occurred during its entire history. We know this because the billions of craters that cover the lunar surface provide an excellent reference system for even the most subtle structural deformations. Like lines on graph paper, the circular craters would show the effects of deformation by compression, tension, or shear stresses and would record even the slightest disturbances. The network of craters, however, is essentially undeformed, so that the lunar crust appears to have been fixed throughout time. There is no evidence of intense folding or thrust faulting and no indication of major rifts. The major features that can be attributed to structural deformation are linear rilles, wrinkle ridges, and other **lineaments.**

Many **rilles** are sharp, linear depressions that generally take the form of flat-floored, steep-walled troughs, as much as several kilometers in width and hundreds of kilometers in length (*figure 24.5*). The valley walls are straight or arcuate, and they stand at the same elevation, as though they were pulled apart and the floor subsided as a graben. Straight rilles are parallel or arranged in an echelon pattern. Some intersect, and others form a zigzag pattern similar to that of normal faults on the earth.

Other linear features are common on the moon's surface, although their origins are a matter of conjecture. Some radiate to form multiringed basins and are undoubtedly faults related to impact. Others are believed to be surface expressions of fracture systems in the old rocks beneath the surface cover.

The only possible expressions of compression in the lunar crust are the **wrinkle ridges** of the maria (*figure 24.6*). These long, narrow anticlines have a sinuous outline and extend for considerable distances parallel to the margins of the mare basins. Some geologists consider them compressional features that were produced as the maria subsided. Others believe that they represent differential compaction.

Compared to the extensive compressive structures found in the rocks of the earth, the moon's structural features are minor indeed. The moon has no folded mountain belts and no great rift valleys. Its crust has been rigid and essentially undeformed by tectonic activity throughout all of its recorded history.

Figure 24.5 Linear fault blocks show that the lunar crust has been subjected to tension in some local areas. There is no evidence of thrust faults or strike-slip faults, however, which would indicate regional compressional forces.

Figure 24.6 Wrinkle ridges in the lunar maria are long and straight or sinuous anticlines. They are probably the only compressional features in the lunar crust.

Lunar Rocks

To the public, the rocks returned from the moon may seem to be a great disappointment. Moon rocks are not exotic and mysterious but are much like basalts and other common rock types found on the earth. Yet, the lunar rocks are a key to the history and chronology of planetary development and a record of the sun's activity. They were sampled from a variety of structural settings and have been studied by about 1000 scientists from 19 countries. Lunar rocks can be classified into the following types:

1. Basalt (*figure 24.7*)
2. **Anorthosite** and associated coarse-grained crystalline igneous rocks (*figure 24.8*)
3. **Breccia,** composed of angular particles of surface material formed by meteorite impact and subsequently compacted into a coherent rock by shock compression during cratering (*figure 24.9*)
4. Regolith, or soil, a mixture of crystalline fragments and **glass** particles formed by repeated fragmentation of the surface material by meteorite impact (*figures 24.10* and *24.11*)

Rock samples returned from the moon tell exactly what one would expect from studies of the surface features. They record the details of the major lunar

Figure 24.8 Lunar anorthosite, seen through a microscope, consists of a meshwork of plagioclase crystals. These are characteristically lath-shaped, with some pyroxene occupying the interstitial spaces between the plagioclase crystals. Olivine occurs in amounts as great as 1%, with small traces of opaque minerals and glass. This is an older lunar rock type, which is abundant in soil and breccia from the highlands. It is important because it records a major thermal event early in the moon's history, before the period of intense bombardment. The width of the field of view is approximately 4 mm.

Figure 24.7 Lunar basalt, seen through a microscope, is similar to terrestrial basalt but contains greater amounts of heat-resistant elements (titanium, zirconium, and chromium). Basalt is the most common rock collected by the Apollo missions. The width of the field of view is approximately 4 mm.

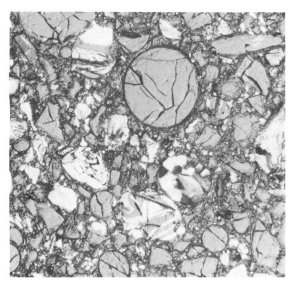

Figure 24.9 Lunar breccia, seen through a microscope, consists of angular fragments of broken rocks from a variety of sources. Typically, the fragments are angular and show essentially no evidence of modification by abrasion. Some lunar breccias contain large amounts of glass, with particles that are remarkably spherical. The dark gray and black materials are glass particles. Breccia results from fragmentation and subsequent compression due to repeated impact. It is the main component of the lunar regolith.

Figure 24.10 Glass particles are common in the lunar regolith. They are formed by the shock melting of rock fragments during impact.

Figure 24.11 Micrometeorite craters on the surface of crystal fragments in lunar breccia show that the products of impact on the moon range from the large multiringed basins to microscopic pits. Practically every rock fragment on the moon appears to have been involved in one or more impact events.

Since the density of basalt is about 3.0 g/cm³, the surface material is only slightly less dense than the moon as a whole. There is thus little possibility for a significant increase in density with depth. (The earth, in contrast, has a mean density of 5.5 g/cm³, and the density of its surface rocks is only 2.7 g/cm³, so that the interior clearly is much denser than the crust.) Nevertheless, from studies of lunar rocks, surface features, and lunar seismicity, it is also clear that the moon is layered and that the composition of the interior is different from that of the surface material. The favored model of the moon's interior is shown in *figure 24.12*. The major units are (1) a crust, (2) a rigid mantle (lithosphere), and (3) a central core (asthenosphere). Our present understanding of the structure of the lunar crust, based on measurements of density and variations in seismic wave velocities with depth, is illustrated in *figure 24.13*.

Mascons. Studies of the orbits of Apollo and lunar orbiter spacecraft close to the moon have revealed significant irregularities in the lunar gravitational field. Relatively strong gravitational attraction, attributed to high concentrations of mass, have been discovered in areas called mascons. Most

Figure 24.12 Our present understanding of the internal structure of the moon is based on measurements of density and seismic studies. The thickness of the crust ranges from 60 to 100 km. The rigid mantle (lithosphere) extends to a depth of about 1000 km. Most moonquakes originate in the region near the base of the lithosphere. A central core (asthenosphere) may lie beneath the moonquake zone.

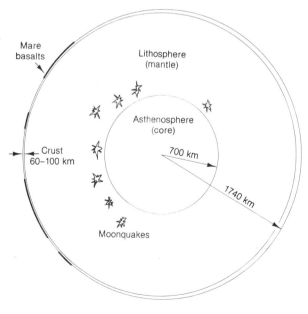

processes of impact and volcanism. Their greatest value, however, is the information they provide about the absolute dates of major lunar events, which can be used to establish a radiometric time scale for the moon and other planets.

The Structure of the Moon

Our present understanding of the moon's internal structure is based on observations of a variety of physical characteristics, including density, magnetism, and seismicity. Much remains uncertain, but several facts place significant constraints on what the internal structure can or cannot be. First, the bulk density of the moon is 3.34 g/cm³.

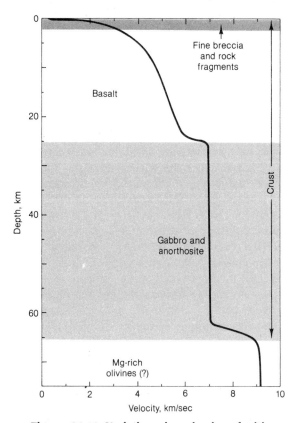

Figure 24.13 Variations in seismic velocities provide a basis for interpreting of the structure of the lunar crust. Three major discontinuities in seismic velocities indicate that the crust is layered. Seismic velocities increase very rapidly with depth to about 1 or 2 km beneath the surface. A very sharp increase occurs at a depth of about 25 km. Between 25 and 60 km below the surface, the velocities are nearly constant, but a significant increase occurs at a depth of 60 km. This discontinuity is interpreted as representing the base of the crust. From comparisons of these velocities with the velocities of seismic waves in major rock types, it appears that, from near the surface to a depth of about 1 or 2 km, the material is fine breccia and broken rock fragments. Below this, to a depth of about 25 km, is a layer composed largely of basalt. A second layer of lunar crust, between 25 and 60 km below the surface, apparently is composed of gabbro and anorthosite. Below 60 km is the mantle of the moon, which is believed (tentatively) to be composed of magnesium-rich olivines.

mascons are found in the circular mare basins; they do not occur prominently in the highland regions. The origin of mascons is not fully understood. The excess mass, which is responsible for the high gravitational values, is generally believed to result from a combination of lava flooding and isostatic adjustment of dense mantle material following the great impacts that excavated the mare basins.

Developing a Lunar Geologic Time Scale

Ejecta deposits from craters, together with lava flows and other volcanic deposits, form a complex sequence of overlapping strata covering most of the lunar surface. The individual deposits can be recognized by their distinctive topographic characteristics and by their physical properties (color, tone, and thermal and electrical properties), determined from measurements made with optical and radio telescopes.

In 1962, the United States Geological Survey developed a geologic time scale for the moon, so that the major geologic events could be arranged in their proper chronologic order. The basic principles used by the survey to interpret lunar history are essentially the same as those used to study the history of terrestrial events. The most important of these principles are the law of superposition and the law of crosscutting relations. These principles for determining relative ages are, of course, as valid on any planet as they are on the earth. In addition, a second method of determining the relative ages of lunar features is based on the abundance of craters, or crater frequency. The development of a lunar geologic time scale was a major advancement in the study of geology because, for the first time, the sequence of events in the history of another planet was firmly established.

The first lunar chronology was developed in 1962 by Eugene M. Shoemaker and R. J. Hackman, who interpreted the sequence of events in the vicinity of the crater Copernicus. In many ways, their pioneering work is comparable to what Smith, Lyell, and their contemporaries did in the early 1800s to establish the geologic time scale for the earth. The planet is different and the nomenclature is different, but the logic remains the same. We will carefully study ejecta from the major craters examined by Shoemaker and Hackman and will see how this material represents a specific sequence of events. As you read the following discussion, study *figures 24.1* and *24.14.* Only by recognizing the physical relationships among the moon's features can you appreciate the relative time involved in their formation.

Copernicus. One of the more outstanding features on the near side of the moon is the crater Copernicus, whose spectacular system of bright rays extends outward in all directions and under proper lighting can be traced for hundreds of kilometers. The rays and ejecta blanket surrounding Copernicus are superposed on essentially every feature in their path *(figure 24.14). From this superposition,*

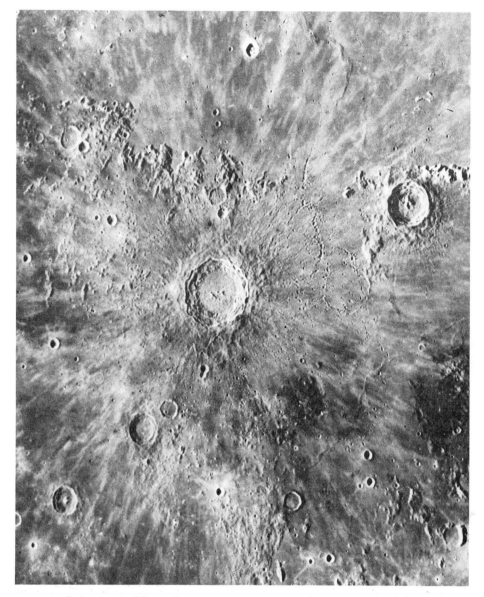

Figure 24.14 Relative ages of lunar features in the vicinity of the crater Copernicus (in the center of the photograph) are indicated by superposition of ejecta and mare basalts. Ejecta from Copernicus are superposed on all other features and are therefore the youngest materials. Ejecta from Eratosthenes (northeast of Copernicus) rest on the mare basalts and are younger than the mare material but older than Copernicus. Ejecta from the Imbrium basin contain mare lava and rest on the densely cratered highlands, and thus the basin and the lava flows are younger than the complex ejecta of the highlands.

it is clear that Copernicus and its associated system of ejecta and ray material are younger than the mare basalts and younger than rayless craters, such as Eratosthenes. The ejecta from rayed craters, the uppermost and youngest system of strata on the moon, are called the **Copernican System**, and the time during which rayed craters and their associated rim deposits were formed is called the **Copernican Period**.

Eratosthenes. About half of the craters larger than 10 km in diameter that occur on the maria are rayed craters belonging to the Copernican System. Most other craters of this size on the maria are similar, but their ejecta blankets are dark, and they lack rays. An example is the crater Eratosthenes, just northeast of Copernicus (*figure 24.14*). It has terraced walls, a roughly circular floor with a central peak, a hummocky rim, and a distinctive pattern of secondary craters, just as Copernicus does. Unlike Copernicus, it does not have a visible ray system. *Eratosthenes and similar craters, together with their ejecta, are superposed on the maria and are therefore younger than the mare lava flows on which they are formed. They are older than the rayed craters of the Copernican System, however.* The deposits of the dark-rimmed craters are called the **Eratosthenian System**, and the time during which they were formed is referred to as the **Eratosthenian Period**.

The Imbrium Basin. In the northwestern part of the near side of the moon is an enormous multi-ringed crater, mostly filled with lava flows, called Mare Imbrium, or the Imbrium basin (see *figure 24.1*). It is surrounded by ejecta deposits similar to those formed by smaller craters, the best exposures being in the Apennine Mountains, which extend outward from the basin's southeastern rim. *The ejecta from the Imbrium basin are partly covered with lava, as is most of the interior of the basin, and therefore the basin is older than the mare lava flows.* The lava and ejecta deposits constitute the **Imbrian System** of strata, and the time during which they were formed is called the **Imbrian Period**.

The Ancient Terrae. *The ejecta from the Im-*

Figure 24.15 Crater frequency can be used to determine relative ages of lunar surfaces, because an older surface has more craters than a younger one does. In the photographs, **A** has the fewest craters and is therefore the youngest surface. **B** and **C** are progressively older, and **D** is the oldest.

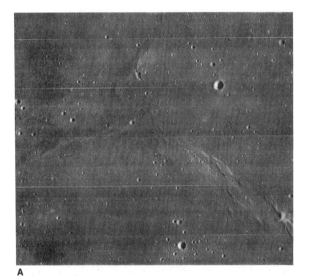

A

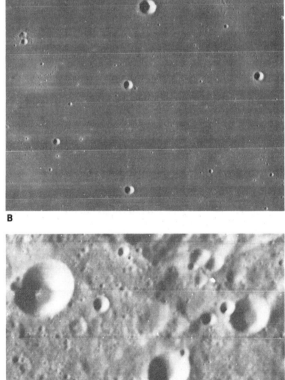

B

C

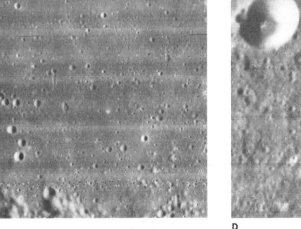

D

brium basin partly overlie a complex sequence of ejecta formed in the lunar highlands, which constitute the oldest material on the lunar surface. These strata are collectively called the **Pre-Imbrian System,** and the time during which they were formed is called the **Pre-Imbrian Period.** The stratigraphic and structural relationships of Pre-Imbrian strata are very complex. Large craters belonging to this system are closely spaced and have been modified by impact. Apparently, the surfaces of the terrae have been churned by the repeated formation of large craters early in lunar history.

Crater Frequency as a Measure of Relative Age. Cratering has been the most universal process in the solar system and provides a wealth of information concerning the history of a planet and the relative age of its surface features. Impact cratering can be used to determine the relative age of a surface because of the simple fact that more craters are present on an older surface than on a younger one. This holds true regardless of whether the rate of cratering is constant, steadily decreasing, or erratic.

A simple example illustrates the significance of crater frequency. Assume that it has been snowing for several days and that the snow is 1 m deep on undisturbed lawns throughout the neighborhood. If the snow is 1 m deep on the sidewalk in front of a house, you can conclude that nobody has shoveled the walk since the storm began. If the snow is about half a meter deep in front of another house, it is obvious that the walk has been shoveled, possibly midway through the storm. If another walk has only a few centimeters of snow, it obviously has been shoveled only an hour or two before. A walk with only a few scattered snowflakes has evidently just been shoveled.

Crater frequency is analogous to the depth of the snow in this example. To understand how cratering can be used to determine relative ages, one needs only to substitute a planet's surface for the sidewalk and meteorites for snowflakes. An example of how crater frequency has been used in studies of the moon is shown in *figure 24.15.* Crater counts on the rims of the Orientale, Imbrium, and Humorum basins show that Humorum has the most craters and Orientale the least. They are, respectively, the oldest and youngest basins.

Radiometric Dates of Lunar Events. Some of the most critical information about lunar geology has been obtained from radiometric dating of lunar rocks. Many geologists expected that the lunar surface was old, but the fresh lava of the maria and bright-rayed craters appeared as though they could

have formed as recently as the earth's ice age. The first samples to be dated were basalts from one of the maria. Their isotopic age was determined to be 3.65 billion years, nearly as old as any rock found on the earth. Radiometric dates of other mare basalts indicate that the extrusion of the lava forming the maria began about 3.9 billion years ago and continued for about 800 million years. The lunar highlands, of course, are older. Samples collected by *Apollo 17* show that the anorthosite crystallized 4.6 billion years ago (see Taylor, 1975, page 271). This is the age of the oldest meteorites. The crystallization of anorthosite is believed to signify the development of the lunar crust, soon after the formation of the moon.

The age of the crater Copernicus was determined from ray material collected at the *Apollo 12* landing site. The formation of this crater, one of the more recent major events in lunar history, occurred 0.8 billion or 0.9 billion years ago.

These and other radiometric dates have been integrated into the relative geologic time scale of the

---

Figure 24.16 Variations in the number of craters formed on the moon's surface during different periods show that the rate of cratering on the moon has decreased exponentially from Pre-Imbrian time to the present.

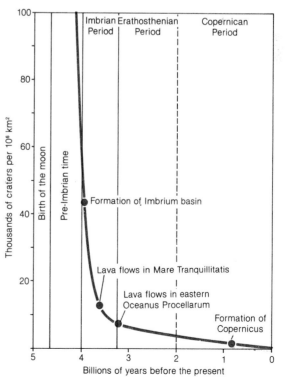

moon, determined from superposition and crater frequency. It is therefore possible to construct an absolute time scale for the moon and to graph rates of cratering (figure 24.16). The graph shows that, early in lunar history, the rate of cratering was hundreds or even thousands of times greater than it is today. The rate of impact declined very rapidly until about 3 billion years ago, and since then it has been relatively constant.

It is believed that other planets and satellites experienced similar variations in the rate of cratering. If this is the case, crater frequency may be a means of correlating interplanetary events.

The Geologic History of the Moon

Data obtained from studies of the relative ages of lunar landforms, the absolute ages of lunar rock samples, and the chemistry of lunar rocks allow us to reconstruct a sequence of events outlining the geologic evolution of the moon. Four major stages are recognized, each containing several distinct but overlapping events or processes (figure 24.17): (1) accretion, **planetary differentiation,** and intense meteorite bombardment; (2) the formation of multiringed basins; (3) volcanism and extrusion of mare lava; and (4) light meteorite bombardment.

Figure 24.17 The major events in lunar history include intense meteorite bombardment during an early period, the formation of multiringed basins, extrusion of mare basalts, and, subsequently, light meteorite bombardment.

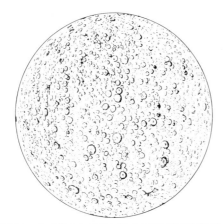

(A) Stage 1. Intense meteorite bombardment (from 4.6 billion to 3.9 billion years ago) formed densely cratered terrain over the entire surface of the moon. Remnants of this cratered surface are preserved in the lunar highlands.

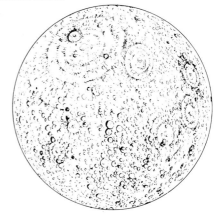

(B) Stage 2. The formation of mutilringed basins (3.9 billion years ago) is attributed to the impact of asteroid-size bodies.

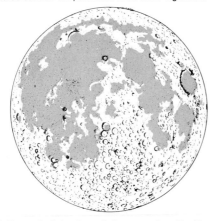

(C) Stage 3. Extrusion of the mare basalts (from 3.9 billion to 3.1 billion years ago) was a major thermal event in lunar history. Lava flows filled most of the multiringed basins and overflowed in some areas to cover parts of the highlands.

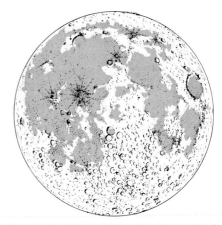

(D) Stage 4. Relatively light meteorite bombardment (from 3.1 billion years ago to the present) formed some ray craters on the maria, but the rate of cratering has been greatly reduced. The lunar landscape has changed little during the last 3 billion years.

THE GEOLOGY OF MERCURY

Statement

The surface features of the planet Mercury, as observed by the *Mariner 10* space probe, are strikingly similar to those of the moon. The early geologic histories of the two planets appear to be remarkably similar, and the differences that do exist can be explained by differences in mass and internal structure.

Discussion

Surface Features

The similarities between Mercury and the moon are apparent from the photomosaic of Mercury reproduced in *figure 24.18*. The photographs (taken from a distance of approximately 230,000 km) show that the entire surface is heavily crateral,

much like the far side of the moon. A number of bright craters, comparable to Copernicus, are evidently the youngest features on Mercury, because their bright halos of ejecta and long, splashlike rays appear to be superposed on everything they come in contact with. Craters range in size from small pits (whose diameters equal the limits of resolution of the photographs) to large multiringed basins. As is the case on the moon, many differences in crater morphology are associated with size. Two important differences can be noticed, however: (1) the progression of changes in crater morphology occurs at smaller crater diameters on Mercury than it does on the moon, and (2) for a given crater size, the extent of the ejecta blanket and secondary craters is systematically smaller on Mercury than on the moon. The differences result from the greater gravitational pull of Mercury, which is more than twice that of the moon. Although the style of cratering is slightly different, the process is fundamentally the same.

Figure 24.18 A photomosaic of Mercury, made from photographs taken at a distance of 234,000 km, shows that Mercury and the moon are strikingly similar. Each has a densely cratered terrain, multiringed basins, a younger area of dark plains (maria), and young rayed craters.

The largest structure on Mercury photographed by *Mariner 10* is the Caloris basin (*figure 24.19*), a large multiringed basin similar in size and form to the Imbrium basin on the moon. Its ejecta radiate from the main ring, forming elongate hills and valleys that extend outward over a distance roughly equal to the diameter of the basin. These are best expressed beyond the northeast part of the basin, which may be the most rugged topography on Mercury. Other multiringed basins are smaller (*figure 24.20*).

Smooth plains material covers the floor of the Caloris basin inside the main scarp, as well as the lowlands beyond. It cannot be said unequivocally that the plains material is volcanic. Its similarity to the lunar maria, however, suggests that it also formed by the extrusion of fluid basaltic lava. Evidence supporting this conclusion comes from

Figure 24.20 Small multiringed basins on Mercury represent a transition in morphology from large craters, with clusters of central peaks, to large multiringed basins, which are found on Mercury, Mars, and the moon.

Figure 24.19 The Caloris basin is a large multiringed basin on Mercury, which closely resembles basins on the moon and Mars. Radiating valleys and ridges are formed in the ejecta beyond the outer rim. The interior of the basin is completely flooded with plains material, which spreads out over the surrounding lowlands.

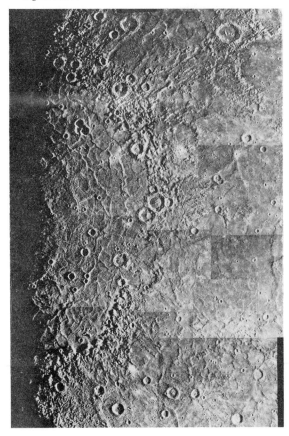

the great volume of material that forms a level surface and fills older depressions. The volume is far too great to have been produced by mass movement, and there are no surface processes to transport that much sediment. In addition, the plains material differs from the surrounding areas in tone and texture, and this difference suggests that it is composed of different material. The plains material is highly ridged and fractured. This characteristic appears to be unique, since similar features have not been found on the moon or Mars.

Careful mapping of the surface of Mercury shows that there are two major generations of plains material. The older type, referred to as the intercrater plains, is the most extensive terrain type observed on the planet. It forms rolling surfaces between and around areas of densely cratered terrain (*figure 24.21*). The distinguishing characteristic of the intercrater plains is the high density of small craters, ranging from 5 to 10 km in diameter. These small craters commonly are elliptical and are partly superposed on one another to form irregular ridges. The intercrater terrain predates, at least in part, the period of intense bombardment. This older sequence of plains material is tentatively interpreted to be volcanic, a product of the early differentiation of the planet. To establish its definite origin will probably require more data than are available from *Mariner 10*.

Younger plains material forms smooth, nearly level surfaces covering approximately 15% of the area viewed by *Mariner 10*. This surface is only sparsely cratered and closely resembles the lunar maria. The density of the relatively small, superposed craters is approximately the same wherever the younger plains material is observed, so that most of it is assumed to be approximately the same age. A notable feature of the plains material in many large craters is the large scarps that terminate at the enclosing crater wall (*figure 24.22*). These may be flow fronts. If they are, their height is enormous in comparison to flows on the earth and suggests the extrusion of high-viscosity magmas. Volcanic cones and domes have not been found on Mercury. The only likely volcanic features are the two generations of plains material.

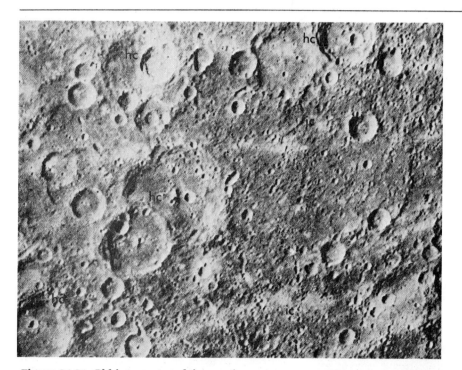

Figure 24.21 Old intercrater plains are the most extensive terrain type observed on Mercury. This terrain is distinguished by the high density of small craters, many of which have elliptical forms. A prominent scarp formed by a thrust fault cuts both older and younger craters.

Figure 24.22 High flow fronts on the floors of some craters on Mercury suggest that thick, viscous lava flows were extruded within the confines of the crater walls.

Tectonic Features

Although most of its surface features are quite similar to those of the moon, Mercury appears to have a tectonic framework that is not found on the other terrestrial planets. In contrast to the earth, Mars, and (to a lesser extent) the moon, Mercury shows evidence of tensional stress only in two localized areas, both apparently related to the Caloris basin. Mercury appears to be unique in that its dominant tectonic feature is a series of lobate scarps, many of which are more than 500 km long and rise more than 3000 m above the surrounding area (*figure 24.23*). The scarps are lobate in outline and have rounded crests, in contrast to the sharp crests and straight ridges formed by normal faults and grabens on the moon and Mars. These characteristics seem to indicate reverse and thrust faults resulting from compression in the Mercurial crust. The scarps transect both the old intercrater plains and some craters of the densely cratered terrain, although some old craters and smooth plains interrupt the scarps. This configuration indicates that the scarps formed around the time of the final stage of heavy bombardment and continued to form for some time after the smooth plains were produced.

An important characteristic of the scarps is their apparent distribution over the whole planet. Preliminary maps show that they extend from pole to pole over most of the visible surface. Therefore, the entire planet seems to have been subjected to the compressive forces that caused crustal shortening. The relatively uniform global distribution of the scarps suggests that minor crustal shortening occurred as the planet contracted during the process of planetary differentiation.

The Geologic History of Mercury

Only part of the surface of Mercury has been photographed, so that our knowledge of it is limited. With the same methods and techniques used to study the moon, however, geologists have been able to establish a preliminary geologic time scale for Mercury and have developed a working hypothesis for its geologic evolution. Mercury's surface features record a sequence of four major events, broadly similar to the sequence of events recorded on the moon: (1) accretion, planetary differentiation, and intense meteorite bombardment; (2) the formation of the Caloris basin; (3) flooding of basins, probably by extrusion of basaltic lava, and the formation of the smooth, young plains; and (4) light meteorite bombardment.

Figure 24.23 Fault scarps across the cratered surface of Mercury deform or displace crater rims. The lobate form of the scarps suggests thrusting.

THE GEOLOGY OF MARS

Statement

In November 1971, *Mariner 9* attained orbit around Mars, becoming the first artificial satellite of a planet beyond the moon. During its period of operation, which lasted nearly a year, it provided some 7300 spectacular photographs and a wealth of other scientific information. The mission was one of the most significant scientific advancements of the space program. The Viking missions were equally successful, with *Viking 1* landing on Mars in July 1976 and *Viking 2* the following September. Not only did these spacecraft send back beautifully detailed photographs taken from the surface of Mars, but the Viking orbiting satellites provided additional detailed photographs with much greater resolution than those taken by *Mariner 9*. With these data, photomosaics of the entire planet have been made, shaded relief maps of the surface have been published, and geologic maps have been compiled in a variety of scales. This newly acquired information has drastically changed our understanding of the planet that generations of astronomers have thought to be the most like the earth.

Mars is a planet of enormous geologic interest. Almost every geologic feature there is gigantic—immense volcanoes, enormous canyons, and giant landslides. There is evidence not only of stream action but of catastrophic flooding, which modified large areas of the planet. Wind action also appears to be an important process on Mars, and it has modified most of the planet's surface features to some extent. The polar regions are covered with alternating layers of ice and windblown sediments, and details of the advancing and retreating polar caps have been recorded by a type of time-lapse photography. It is also apparent that the geologic agents operating on Mars have varied both from place to place and from time to time throughout the planet's history.

Discussion

Physiographic Provinces

The whole surface of Mars was photographed during the epic flight of *Mariner 9*. Accurate terrain maps of the entire planet have been made, some to the scale of 1:5,000,000. From these maps and photographs, we see that Mars is divided into two major regions, or physiographic provinces: (1) the old, densely cratered highland to the south and (2) the smooth, sparsely cratered plains in the north (*figure 24.24*). The boundary between these provinces forms a great circle inclined at approximately 35° to the equator.

The densely cratered terrain in the Southern Hemisphere is a highland somewhat like the cratered highlands of the moon. It rises abruptly above the low volcanic plains of the Northern Hemisphere and is separated from them over much of the planet by a prominent escarpment from 2000 to 3000 m high. The scarp is well defined in the east, but it is masked in the west by the great volcanic field in the Tharsis region. It is dissected and modified by a unique type of stream erosion and mass movement, which produced various types of terrain, referred to as chaotic terrain, hummocky terrain, and fretted terrain. The major channels are believed to have been produced by running water, draining from the southern highland and emptying into the low northern plains.

The northern plains are believed to be the products of lava flows, because in many respects they resemble the maria of the moon and the plains regions of Mercury.

Two large domal upwarps, capped with huge volcanoes, are located in the Northern Hemisphere near the escarpment. The larger is in the Tharsis region, to the west, where the volcanic field appears to sit astride the global escarpment. The smaller—the Elysium field—is in the northern plains to the east.

The large structural upwarp in the Tharsis region produced a system of radial fractures extending nearly halfway around the planet. Erosion and enlargement of a segment of the fracture system created Valles Marineris, the Grand Canyon of Mars. This huge canyon system extends eastward from the fractured bulge in the Tharsis area across the cratered highland, breaking up into a number of distributaries near the global escarpment.

Several large basins are located in the densely cratered plateau. These are believed to be multiringed basins, similar to those on the moon and Mercury.

Wind action, a major surface process, has played a role in modifying all of the surface features of Mars.

The polar regions have several unique features, resulting from processes that are restricted to those areas. These include sections of pitted and etched terrain, regions of layered sedimentary deposits, and regions of permanent ice.

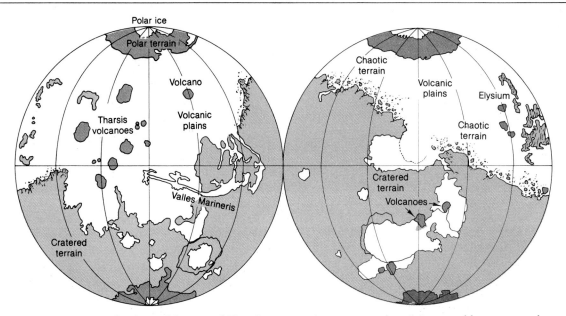

Figure 24.24 A physiographic map of Mars has been compiled from photographs taken during various space probes. Two major regions are recognized: (1) the Southern Hemisphere is an old, densely cratered highland, and (2) the Northern Hemisphere is a younger, relatively smooth plain, apparently formed by floods of basalt. A well-defined escarpment separates the two provinces, except where it is covered by younger volcanics in the Tharsis region. Several unique terrain types are formed by erosion and slope retreat along this global escarpment. The major river channels are associated with drainage from the southern highlands across the escarpment to the northern lowlands.

Craters

Although the photographs from *Mariner 9* reveal a variety of landforms on the surface of Mars, cratering still appears to have been a dominant process, especially during the planet's early history. The craters on Mars record impacts from a population of meteorites similar to those that bombarded the surfaces of the moon and Mercury. The Martian craters, however, have many distinctive features, which reflect the planet's particular gravitational attraction, surface processes, and erosional history. Most of the craters are shallow, flat-floored depressions. They show evidence of much more erosion and modification by sedimentation than craters on the moon and Mercury do. Rayed craters are rare, because the Martian winds can easily erode the loose ray material.

Photographs from the Viking missions, with much higher resolution than those from *Mariner 9*, show that fresh craters do exist on Mars. To the surprise of most scientists, however, the ejecta blankets appear to have flowed over the surface as avalanches and mudflows do on the earth. This phenomenon is unique to Mars. A spectacular example is the crater Yuty, shown in *figure 24.25.*

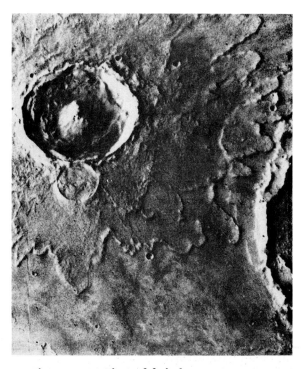

Figure 24.25 Ejecta debris from many craters on Mars has flowed like a mudflow. This suggests that the ejecta were fluidized at the time of impact, probably by the melting of near-surface ground ice.

Tongues of ejecta debris, like huge splashes of mud, have flowed outward from the crater rim. Near the bottom of *figure 24.25*, the ejecta flow can be seen overlapping the older eroded debris from the large adjacent crater, and it actually flows up and over the cliff formed on the older, eroded ejecta blanket.

The flow of ejecta material is believed to result from a global distribution of ground ice, similar to permafrost in the polar regions of the earth. Upon impact, the ground ice melts, so that material ejected from craters moves like great globs of mud.

An important finding is that the cratering history of Mars parallels that of the moon and Mercury. Each planet experienced an early period of intense bombardment and a subsequent period of impact by asteroid-size bodies, which produced large multiringed basins. Since then, the rate of impact has rapidly decreased to a minimum.

Volcanic Features

One of the most spectacular features observed on the surface of Mars is a series of giant volcanoes in the Northern Hemisphere. They are much larger than anything seen on the earth, and their freshness suggests that Mars has just been "turned on" volcanically. Studies of the available photographs however, reveal a number of older volcanoes, which are greatly eroded. Therefore, it appears that Mars has had a long and interesting volcanic history.

The enormous shield volcanoes on Mars, most of which occur in the Tharsis and Elysium regions (*figures 24.24* and *24.26*), are at least twice the size of the largest volcano on the earth. Olympus Mons, the largest, is from 500 to 600 km in diameter and rises 23 km above the surrounding plains. This is

Figure 24.26 Volcanoes in the Tharsis region include huge structures, much larger than any found on the earth. Olympus Mons, the largest volcano on Mars, is in the upper left corner of the map. It is 500 km across at the base and 23 km high. The complex caldera at the summit is 65 km in diameter.

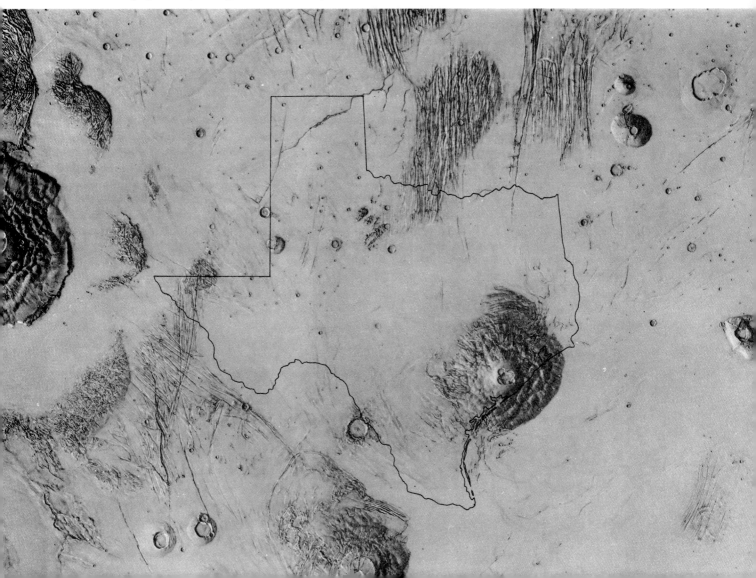

nearly half again the distance from the deepest spot on the earth—the Mariana Trench—to the top of Mount Everest. At the crest of Olympus Mons is a huge, complex, collapsed caldera, similar to those on the volcanoes of Hawaii. Other Martian volcanoes resemble Olympus Mons but differ in size and detail.

The huge, young shield volcanoes on Mars are not scattered at random. They are located high on the summit of large domes, or crustal upwarps, in the Tharsis and Elysium regions. Both upwarps are broken with major fracture systems that radiate from the crest and extend across the surface for great distances. Thus, the origin of the volcanoes is very likely associated with major upwarps of the crust.

Much older shield volcanoes have been discovered in the densely cratered terrain of the Southern Hemisphere. These, however, have been extensively modified by impact, wind erosion, and partial burial. Thus, volcanic activity apparently has occurred over a large portion of the planet's history, as far back as the period of intense bombardment.

In addition to the giant volcanoes, Mars has a number of smaller volcanic domes and extensive lava plains, which cover the Northern Hemisphere. These plains are similar to those on Mercury and the moon.

The Martian volcanoes are important because they indicate that the planet has been thermally active during much of its history. Whereas volcanic activity on the moon and Mercury terminated after the period of extensive extrusion of basalt, it has continued on the larger planets Mars and the earth.

Tectonic Features

The presence of undeformed craters across the surface of Mars clearly indicates that the crust has not been subjected to extensive horizontal compression. There are no folded mountains and no system of moving plates. Yet, some important Martian tectonic features suggest tensional stress and extension of the rigid crust. These, together with the volcanic features, provide convincing evidence that significant tectonic activity has occurred and that, from the standpoint of tectonics, Mars has a style of its own.

Two types of structures are the dominant tectonic features of Mars: (1) large domal upwarps and (2) grabens. Both are concentrated in the Northern Hemisphere, and they appear to be genetically related.

The large domal upwarps of the Tharsis and Elysium regions have been briefly mentioned in connection with the giant volcanic shields. The Tharsis dome is a broad bulge 5000 km in diameter and 7 km high. It is asymmetrical, being much steeper on its northwestern flank. The Elysium dome is smaller, only 2000 km across and from 2 to 3 km high. Both, of course, are much larger than most domes on the stable platforms of the earth.

The most significant tectonic feature associated with the Tharsis dome is a vast radial pattern of fractures, which extends over about a third of the planet. The most extensive fracture system is northeast of Tharsis, where a fanlike array of grabens and normal faults converges toward the line of shield volcanoes. The fractures form magnificent sets of grabens, which typically are from 1 to 5 km wide and may be several thousand kilometers long (*figure 24.26*).

The western part of the fracture system deserves special attention because of its controlling influence on the development of Valles Marineris. The dimensions of this huge canyon system—about 5000 km long and at least 6 km deep—are vividly shown in *figure 24.27*, on which an outline map of the United States is superposed for scale. A minor side canyon is similar in length and depth to the Grand Canyon, in Arizona. Valles Marineris consists of a series of parallel canyons with steep walls and a sharp brink at the lip of each. The troughs are highly irregular in detail, with sharp indentations, large embayments, and dendritic tributaries, similar to the features of canyons cut by running water on the earth. The configuration of the canyon walls indicates that erosion has considerably widened and modified the original rift valleys. The dominant processes were probably landslides, debris flows, and wind, and possibly running water.

The features on the earth that are most comparable in size to Valles Marineris are the Red Sea and the rift valleys of East Africa. Like the rift system on the earth, Valles Marineris was initiated where the crust was pulled apart and the interior block subsided. Subsequently, the rim of the rift valley was sculptured by erosion. The head of Valles Marineris is near the crest of the upwarp, nearly 9000 m above the plains. This high relief probably accelerated erosion and the widening of the canyon. Other fault systems might have developed similar canyons had they been favorably located on upwarps.

The development of the domal upwarps and the associated radial fracture systems represents a major event in the history of Mars. That event clearly

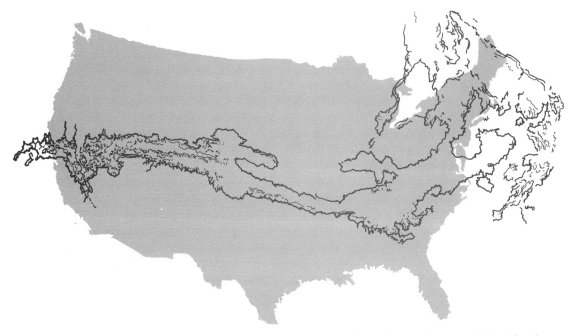

Figure 24.27 Valles Marineris, the Grand Canyon of Mars, extends across an area longer than the United States. This huge chasm is about 5000 km long and at least 6 km deep. The Grand Canyon of the Colorado River is about the size of a minor tributary of the canyon on Mars.

had some relationship with the focus of volcanic activity on the planet.

Fluvial Features

Perhaps the most startling result of the *Mariner 9* project was the discovery of numerous channels on Mars, which closely resemble dry river beds on the earth. These channels (which are unrelated to the well-known Martian canals) originate in the southern highlands, near the erosional escarpment, and empty into the low volcanic plains to the north. If they were located on the earth, no one would hesitate to call them dry river beds. Their presence on Mars, however, presents some of the most intriguing and perplexing questions about the planet.

The existence of water in the liquid state depends on the temperature and pressure regime at the planet's surface. The present Martian atmosphere is characterized by low pressures and low temperatures. As a result, water in the liquid state cannot exist in large amounts at the surface. It would either freeze or evaporate. If water is present as a liquid, it occurs in very small quantities for a short time when frost melts, before reentering the atmosphere as vapor. Most water in the Martian system is locked up either as ice in the polar regions or as ground ice, similar to permafrost on the earth.

The widespread river channels on Mars, however, imply that not only was running water present but, in places, large-scale flooding occurred. Therefore, either some unknown and presently unimaginable processes formed the channels or, at some time in the past, geologic conditions on Mars permitted running water. The question is, When did such water exist, where did it come from, and where has it gone? These questions relate directly to the problem of whether there is or has been life on Mars. The discussion will continue for some time.

The best evidence for the previous existence of rivers on Mars is the channel system along the eroded escarpment that separates the cratered highlands (the source region of the channels) from the low volcanic plains. The map in *figure 24.24* shows the size and distribution of the major channel systems. Many channels are wide and deep and display evidence of catastrophic flooding, like the Channeled Scablands of Washington (*figures 24.28 and 24.29;* compare *figures 14.30* and *14.31*).

The Martian channels, however, are not entirely like the river systems of the earth, which have delicate networks of tributaries, main trunk streams, and large deltas where they empty into the oceans. The tributaries of the Martian channels are short and stubby. Many channels have their

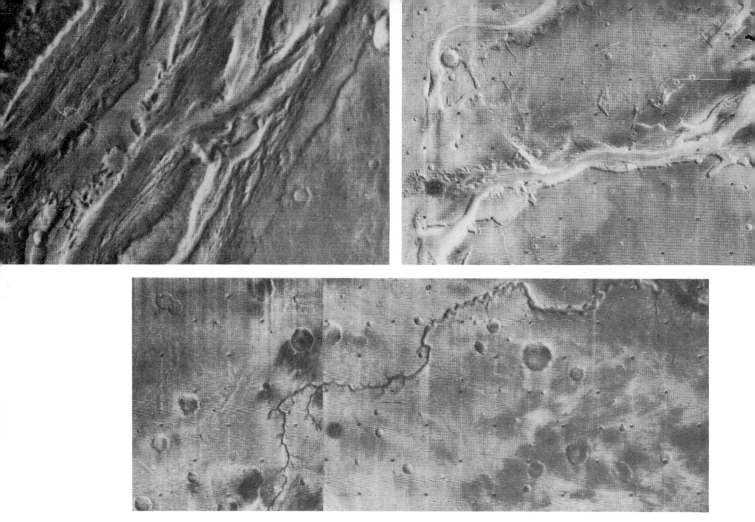

Figure 24.28 Stream channels on Mars are similar in many respects to dry river beds in arid regions of the earth. Some have a typical braided pattern, others meander. They generally lack well-integrated systems of branching tributaries, however, which characterize rivers on the earth.

headwaters in a highly fractured region or in a jumbled mass of angular blocks called chaotic terrain. Most of the chaotic terrain is located in the southern highlands, near the boundary with the northern low plains.

In summary, it is apparent that the surface of Mars has been modified by flowing liquid (presumably water) and that, at some earlier stage, fluvial processes operated on the surface. The change from ancient fluvial periods to the present dry period may be similar to the change from interglacial periods to glacial periods on the earth.

Mass Movement

High-resolution photography from the Viking orbiters provides dramatic evidence that mass movement is an important process in the evolution of the Martian landscape. It is especially important in the enlargement of canyons.

The importance of mass movement can be seen in the oblique view in *figure 24.30.* The canyon shown in this photograph is approximately 2 km deep. Multiple massive landslides and debris flows are visible on both walls. Several resistant rock layers form a steep cliff at the canyon rim and appear to break up in a series of large slump blocks. The lower units, however, are nonresistant and form a slope, which appears to have moved as a debris flow. Some flows have moved more than halfway across the canyon floor, at least 30 km from their point of origin. They have a lobate pattern, similar to that of debris flows on the earth, and flow lines show the direction of movement. These features show significant mobility and suggest the presence of a fluid (probably water) mixed with the fine debris. The low hills and knobs on the flat canyon floor may be remnants of the resistant rimrock that slumped into the canyon. The bright

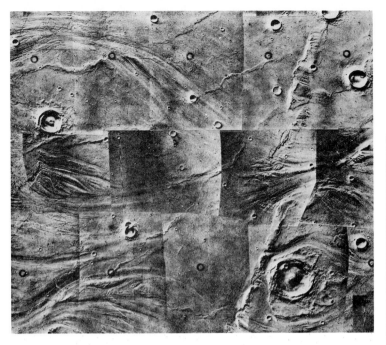

Figure 24.29 Details of stream channels on Mars show evidence of large-scale flooding. On the earth, similar features are found in the Channeled Scablands, where catastrophic floods occurred during the ice age.

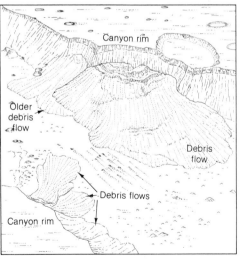

Figure 24.30 Mass movement along the walls of Valles Marineris is shown in this mosaic of oblique photographs taken by a Viking orbiter. Multiple massive landslides and debris flows can be seen on both walls. The layers visible in the canyon walls probably represent alternating strata of basalt, ash, and windblown deposits.

streaks behind the low hills are the result of wind shadows, on the lee side of obstructions. These streaks indicate that sediment is actively transported by wind after it slumps from the valley walls. The features shown in this photograph imply that the canyon is enlarged by mass movement and that some material is subsequently removed by wind. Perhaps this is why we see so little evidence of stream deposits such as deltas and alluvial fans on Mars.

Eolian Features

Wind is known to be a dominant geologic agent on the surface of Mars, and its scale of activity is much greater there than in desert regions on the earth. When *Mariner 9* entered orbit around Mars in 1971, a raging dust storm completely obscured the entire planet. The storm could be seen through telescopes from the earth. It had begun two months before and continued for several months after the spacecraft was in orbit. The magnitude of a global storm on Mars is hard to imagine. Winds comparable to a strong hurricane on the earth raged for several months with gusts to 500 km per hour. The storm blew dust high into the atmosphere, so that only the tops of the highest volcanoes could be seen, appearing as dark spots above the global cloud of dust.

When the storm cleared, a variety of features resulting from wind activity were photographed. Behind craters that created wind shadows, dark streaks covered broad areas (*figure 24.31*). Large dune fields, remarkably similar to those of the great deserts of the earth, were found on crater floors and in the North Polar regions (*figure 24.32*). In addition, surface photographs taken by the Viking landers show many wind-generated features, such as small dunes, ventifacts, lag deposits, and sand streaks behind boulders (*figure 24.33*).

Figure 24.31 Eolian features associated with craters in the Hesperia region of Mars (23° south latitude, 242° west longitude) consist of light and dark streaks extending downwind from craters. The streaks are believed to be fine, bright dust, transported into the craters in the waning stages of a dust storm and then blown out by high-velocity winds having a constant direction.

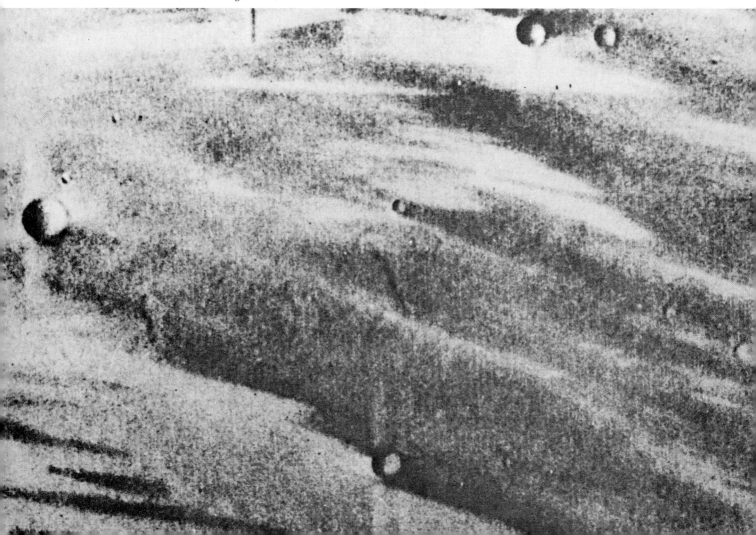

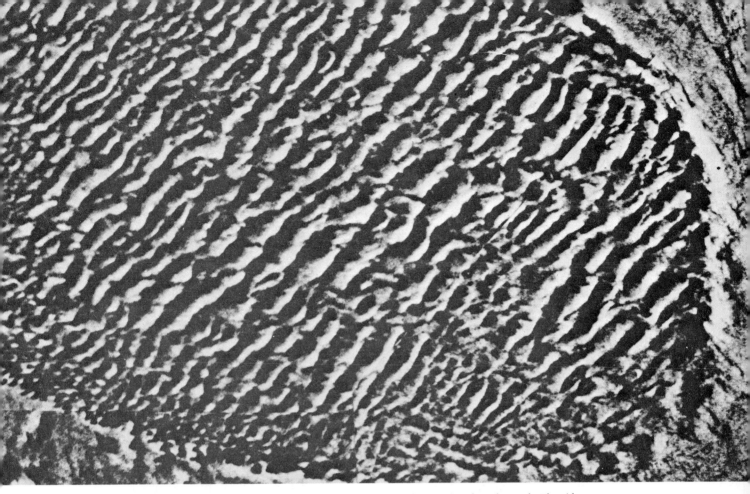

Figure 24.32 A dune field in the Hellespontus region of Mars (near 75° south latitude, 331° west longitude) was discovered by *Mariner 9*'s narrow-angle camera. The wavelike structure is highly suggestive of transverse sand dunes found on the earth. The ridges are 1.5 km apart. Note the tendency of the dunes to decrease in size toward the margins of the field.

Figure 24.33 This view of the Martian surface was photographed by a Viking lander. The large boulder in the left foreground is approximately 2 m across. Many of the features shown here indicate the importance of wind activity. The sand dunes have been eroded so that their internal cross-stratification is exposed. The gravel surface is probably lag deposits, resulting from deflation.

Polar Regions

The polar areas of Mars are of special interest because they are the centers of contemporary geologic activity. The ice caps are believed to be very thin, probably only about 30 m thick. Contrary to previous hypotheses, it now appears that the Martian ice caps are composed of frozen water, not frozen carbon dioxide.

Of particular interest are unique layered deposits that underlie the ice caps and are well exposed along their margins. These deposits, called laminated terrain, consist of alternating light and dark layers from 20 to 30 m thick (*figure 24.34*). As many as 50 beds have been observed in a single exposure. These deposits are believed to be alternating layers of ice and windblown dust and sand, which formed during glacial and interglacial periods, respectively. At present, a great dune field of dark sand surrounds the North Polar cap. The dunes will be covered by the next expansion of the ice.

The Moons of Mars

Photographs of Phobos and Deimos, the two little moons of Mars, have provided the first information concerning their size, shape, and surface morphology. Both satellites appear to be irregular fragments of once larger bodies, and both have been severely battered by repeated impact. On the basis of crater population (both appear to be nearly saturated), Phobos and Deimos are at least 1.5 bil-

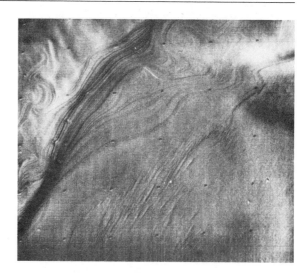

Figure 24.34 The laminated terrain near the South Pole of Mars consists of a series of layers of light and dark sediments. These may represent layers containing dust or volcanic ash alternating (possibly) with carbon dioxide frost and water ice. The photograph shows an area of about 57 km by 60 km.

lion years old, and they may date back to the birth of the solar system, some 4.5 billion years ago.

Typical close-up photographs of the Martian moons are reproduced in *figure 24.35*. Both are potato-shaped and show a profusion of rimmed craters, ranging in size from 5.3 km in diameter down to the limits of resolution.

The origin of the moons of Mars may be ex-

Figure 24.35 The moons of Mars are small, potato-shaped objects marked by impact structures.

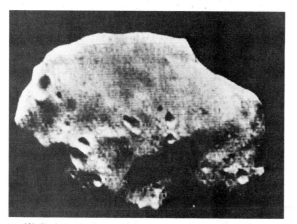

(A) Phobos, the larger of the two moons, is approximately 25 km long and 21 km wide.

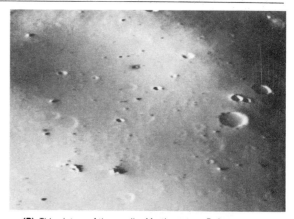

(B) This picture of the smaller Martian moon, Deimos, covers an area of 1.2 km by 1.5 km and shows features as small as 3 m. The surface of this moon is saturated with craters, but a layer of dust appears to cover craters that are less than 50 m in diameter, making Deimos look smoother than Phobos. Boulders as large as houses—from 10 to 30 m in diameter—are strewn across the face of Deimos. These are probably blocks ejected from nearby craters.

plained by one of two hypotheses: (1) they represent material left over from the period of intense bombardment, or (2) they were captured, probably from the asteroid belt, by the gravitational pull of Mars. Whatever the origin of the moons, the close-up photographs of Phobos and Deimos provided by *Mariner 9* and the Viking missions give us our best views of the type of bodies that populate the asteroid belt in uncountable numbers.

The Geologic History of Mars

The relative ages of the major rock bodies and terrain types on Mars can be determined by the principle of superposition, by crater density, and by the degree of erosion, much as these have been used in the study of the moon and Mercury. Many uncertainties remain about details, but the major events in the geologic history of Mars can be outlined in five stages (*figure 24.36*): (1) accretion, planetary differentiation, and intense meteorite bombardment; (2) uplift, faulting, and erosion; (3) the formation of volcanic plains; (4) renewed uplift and the formation of Valles Marineris; and (5) volcanism and eolian activity.

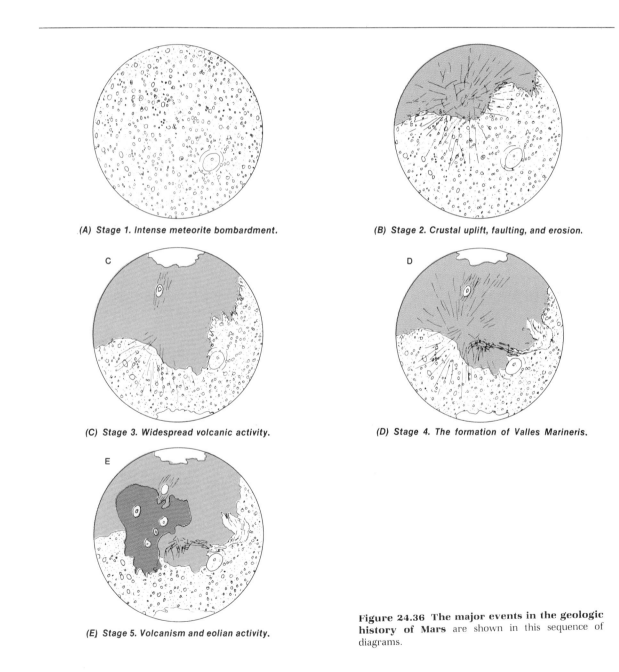

(A) Stage 1. Intense meteorite bombardment.

(B) Stage 2. Crustal uplift, faulting, and erosion.

(C) Stage 3. Widespread volcanic activity.

(D) Stage 4. The formation of Valles Marineris.

(E) Stage 5. Volcanism and eolian activity.

Figure 24.36 The major events in the geologic history of Mars are shown in this sequence of diagrams.

THE GEOLOGY OF VENUS

Statement

Of all the planets, Venus is most like the earth in size, density, and distance from the sun. It probably formed at the same time as the earth and out of the same mix of materials, but its atmosphere consists of carbon dioxide and is almost devoid of water vapor. This dense atmosphere has kept the surface of Venus completely hidden from view, but radar observations from the earth and from an orbiting spacecraft have finally parted the veil of clouds and have given us our first comprehensive glimpse of the planet's surface. The resolution achieved so far in radar images is too coarse to answer many questions, but the major landforms have been mapped over 93% of the surface. Briefly, this is what has been found.

1. Venus looks somewhat like the earth drained of its seas, but it does not have an active system of tectonic plates.
2. About half of the surface of Venus consists of relatively flat, rolling plains, with local relief of less than 1000 m.
3. Two highlands, rising above the great plains, may resemble the earth's continents, which are composed of light rocks.
4. High, conical mountains, probably volcanoes, occur on both "continental" highlands and on the great plains.
5. Circular landforms resembling impact craters are common, ranging from 100 km to 1500 km in diameter.
6. Moderately straight valleys with sharp vertical relief suggest that block faulting is the major tectonic feature.

Discussion

A radar map of the surface of Venus (*figure 24.37*) has been made on the basis of extensive data sent back by the Pioneer Venus spacecraft, which was placed in orbit around the planet in December 1978. The surface of Venus consists mostly of gently rolling plains with several large highlands, which rise like continents above the surrounding surface. The terrain is therefore similar to that of the earth, and yet Venus is also different in some ways from any of the known planets.

Figure 24.37 A shaded relief map of Venus was generated by computer from radar altimeter data collected by the Pioneer Venus orbiter.

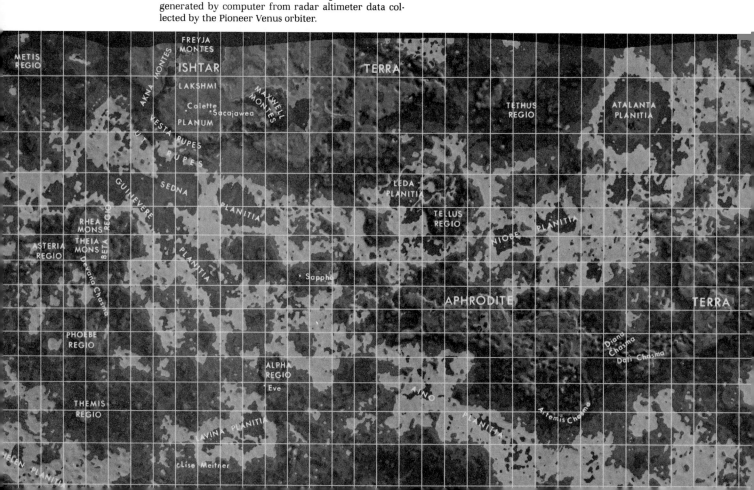

The Great Plains

The radar map (*figure 24.37*) shows that the plains encompass the entire planet and appear to be remarkably flat. The average elevation of the plains is 6050 km from the center of the planet, and most of their surface lies within 1000 m of this elevation. Small, isolated hills rise above the surface of the plains, and many of them are situated along distinct linear trends. Circular depressions, which may be impact structures, also occur on the plains, together with several large, deeper basins, which are roughly circular and have smooth, flat floors. The largest basin is centered at 65° north latitude and 165° west longitude. In their general appearance, the great plains of Venus resemble the mare basins of the moon and the northern plains of Mars. The terrain is therefore likely to be relatively young and may be covered with basaltic flows.

The "Continental" Highlands

Perhaps the most intriguing features of Venus are the highlands, which rise 2000 m or more above the great plains. They are surrounded by steep escarpments, resembling the continental slopes on the earth. Moreover, gamma-ray spectrometry conducted by the Soviet Union's *Venera 8* lander found a "granitic" composition in rocks from the "continents," so that the highlands may float on denser basaltic rock of the low plains. Direct comparison between the highlands of Venus and the continents of the earth is highly premature. What we do know from radar studies is something about the size and shape of the highlands and how they are related to the surrounding terrain.

The largest highland on Venus is named Aphrodite Terra, after the Greek goddess of love and beauty. It is located about 10° south of the equator and extends from 60° to 140° west longitude, a distance of more than 9500 km.

The highest and most spectacular "continent" is the northern highland, Ishtar Terra, named after the Assyrian goddess of love and war (*figure 24.38*).

Volcanoes

Several huge, isolated mountain masses have been identified on Venus and are interpreted to be volcanoes. The largest, Maxwell Montes (named af-

Figure 24.38 An artist's conception of Ishtar Terra, based on topographic measurements taken by the Pioneer Venus spacecraft, shows a high plateau, resembling a continent the size of the United States. Its western and central areas are smooth and may be covered with relatively young lava. The large mountain to the east is Maxwell Montes, the highest point on Venus, about 11,800 m above the mean surface ("sea level") of the planet. This area contains the roughest terrain on Venus.

ter the physicist James Clerk Maxwell), occupies most of the eastern part of Ishtar Terra (*figure 24.38*). Its highest point is 9000 m above the adjacent Lakshmi Plain and 11,800 m above the mean surface of the planet.

Another probable volcanic complex is centered at 30° north latitude and 280° west longitude. This region, called Beta Regio, consists of two huge shield-shaped mountains, which could be volcanic peaks. These mountain masses are larger than the Hawaiian chain, and two major peaks rise 4000 m above the surrounding great plains.

The belief that these mountains are volcanic is supported by information obtained by the Soviet spacecraft *Venera 9* and *Venera 10*, which landed just east of the Beta region. Data sent back from the landers indicate that the concentrations of radioactive elements in the rocks in this area are similar to those in basalt. Thus, a volcanic origin is suggested by both the nature of the landform and information about the composition of the surface material.

Craters

Close examination of radar images shows many dark, circular features with bright central spots and bright rims (*figure 24.39*). These are believed to be impact craters. Craters whose diameter is less than 20 km are too small to be detected by radar, so that the full range of crater sizes is not known. Most craters recognized with available radar images range in diameter from 200 to 600 km. Craters with diameters less than 80 km are rare. This is to be expected, because small meteorites probably burn up in the dense atmosphere, and only large ones are likely to strike the surface.

Note that the radar map of Venus (*figure 24.37*) shows a number of large, circular features, which may be multiringed basins. The most spectacular one is just southwest of Aphrodite, centered at 35° south latitude and 125° west latitude. Others may form some of the deeper basins in the great plains area.

A better understanding of the craters on Venus may come with better radar maps and better theoretical models of the mechanical properties of planets with hot surfaces.

Tectonic Features

Suggestions of tectonic features resulting from tension appear in a number of areas of Venus, but no global tectonic pattern can be detected from the present data, and there is no evidence of trenches, island arcs, or folded mountain belts, which would indicate compression and subduction. The most

common tectonic features on Venus are what appear to be normal faults and rift valleys. These are best developed in the area east of Aphrodite. There, the terrain is particularly rough and complex, cut by a number of relatively staight valleys and adjacent high ridges with sharp vertical walls nearly 4000 m high. These structures are similar to the rift valleys in Africa and the western United States. In this area lies the lowest part of the planet, a trench about 2900 m below the mean surface, which is slightly lower than the basin in the Northern Hemisphere. This trench is deeper than the Red Sea rift but only one-fourth the depth of the Mariana Trench, in the western Pacific. It is roughly half as deep as Valles Marineris, the great canyon on Mars.

Other distinct linear features on Venus suggest a

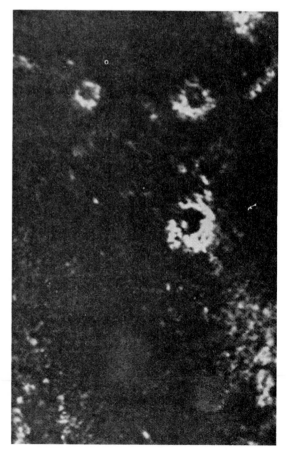

Figure 24.39 Ring-shaped features on Venus have been mapped by radar and are tentatively interpreted to be craters. The largest circular structure shown here is about 100 km in diameter. The radial extent of the rim roughness and the relief profile are unlike those of craters on the moon and other planetary bodies.

tectonic control, probably normal faulting. For example, note on the relief map (*figure 24.37*) the alignment of hills along the meridian at 220° west longitude.

The Geologic Interpretation of Venus

From the data now available, it is possible to make a tentative summary of the geology of Venus and compare it geologically to other planets of the solar system.

Venus is nearly the size of the earth, and its surface has been modified by all three major geologic processes: impact, tectonic deformation, and surface degradation. If the ring-shaped structures prove to be impact craters, we can infer that much of the surface of Venus is ancient terrain, about 4 billion years old. Significant planetary differentiation has occurred to produce the continental highlands, the great plains, and volcanic features. The large and spectacular topographic structures on Venus do not include trenches and island arcs, however, and the distribution and amount of light "continental" material are far different on Venus from what they are on the earth. The crust of Venus does not appear to be broken into plates, and tectonic motion apparently has not played a dominant, ongoing role in altering the surface. Tentatively, we can conclude that the crust of Venus is not as thick as those of the moon, Mercury, and Mars but is thicker than that of the earth. The topography of Venus may be influenced by lithospheric strength rather than by convection.

Venus has not preserved as much of its early history as the moon and Mercury have. Its state of crustal evolution probably more closely resembles that of Mars than that of any other planetary body.

Many questions remain. What are the relative roles of the major processes—meteorite impact, volcanism, tectonic activity, and the activity of atmospheric gases and fluids—that appear to have modified the surface of Venus? Does convection occur in the planet's mantle? Is Venus still volcanically active? Does it have a core? What small-scale landforms exist, and what could they tell us about the planet's history?

Later in the 1980s, a spacecraft carrying advanced radar equipment is to be launched into a low, circular orbit around Venus. This space probe, called VOIR (Venus Orbiting Imaging Radar), will map the entire surface of the planet at a resolution of about 500 m, providing data comparable to those obtained from Mars by *Mariner 9*. With this information, we could look forward to answering many questions about the geology of Venus.

THE GEOLOGY OF JUPITER AND ITS MOONS

Statement

Jupiter and its moons are a planetary system of incredible beauty and mystery. The giant planet has a volume 1300 times greater than the earth's, but it is composed mostly of gas and liquid swirling in complex patterns. Jupiter has 14 satellites. In a sense they resemble a miniature solar system.

The exotic photographs obtained in March and July 1979 by *Voyager 1* and *Voyager 2* revealed geologic wonders of the Jupiter system far beyond anyone's expectations. Colored time-lapse motion pictures were made, showing the dynamics of Jupiter's atmosphere, and a ring like Saturn's was discovered. But from a geological standpoint, the exploration of Jupiter's moons is the most exciting aspect of the Voyager missions. The largest—Io, Europa, Ganymede, and Callisto—were discovered by Galileo in 1610 and are referred to as the Galilean satellites. All are named after lovers of Jupiter in classical mythology. These bodies were previously known only as blurry specks of light in a telescopic field of view. Now we see them as new worlds available for geologic study, each unique and totally different from the planets of the inner solar system.

The importance of the moons of Jupiter is that we are now able to compare them to the earth to better understand how planetary evolution proceeds.

Discussion

Jupiter

Even before the Voyager missions, Jupiter was known to be a very strange object. Unlike the inner planets of the solar system, it is not composed mostly of solid rock but is a huge, rapidly spinning ball of liquid and gas, 142,800 km in diameter—more than 10 times the diameter of the earth. The mass of Jupiter is greater than the combined masses of all of the other planets.

Jupiter is composed mostly of hydrogen and helium, but it does have a solid rocky core about the size of the earth. If the core were exposed, it would probably look something like the planets with which we are familiar. Surrounding the rocky core is a sea (or mantle) of molten hydrogen 70,000 km thick, which makes up the bulk of the planet. The inner layer of this mantle consists of hydrogen in

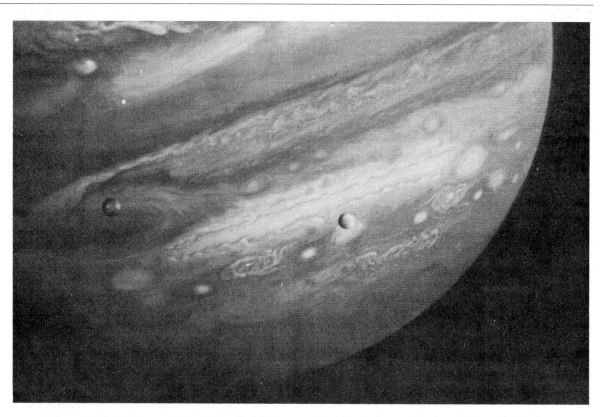

Figure 24.40 Jupiter and two of its planet-size satellites, Io (to the left) and Europa (to the right), are shown in this photograph taken by *Voyager 1.* Behind Io is the Great Red Spot, a tremendous cyclonic atmospheric storm on Jupiter. The diameter of this storm is about twice that of the earth. Dominant large-scale atmospheric patterns move from west to east, and small-scale eddies develop between the major bands.

the liquid metallic state, in which hydrogen molecules are dissociated into single atoms, because of the high pressure.

The patterns of atmospheric circulation on Jupiter are shown in *figure 24.40.* The bright, white bands are called zones. The darker, brown ones are called belts. These are semipermanent features, which are similar to cyclones and anticyclones (masses of rising and descending air). Other patterns include loops, streaks, plumes, and oval patterns, which grow and develop within the zones and belts and then dissipate.

Perhaps the most extraordinary feature seen in Jupiter's atmosphere is the Great Red Spot, an orange red cyclonic storm whose diameter is roughly twice that of the earth. This mammoth hurricane has persisted for at least 300 years and appears to be anchored in one place.

Another important discovery is that Jupiter is surrounded by a thin ring, 260,000 km in diameter, composed of rock fragments. It is the third planet (after Saturn and Uranus) known to have such a feature. The rings are believed to originate from satellites that were broken into innumerable small fragments; if a satellite's orbit brings it close enough to the planet, the satellite's internal force of gravity (which causes it to hold together as a moon) is overcome by the planet's gravitational force, so that the satellite breaks apart.

The Satellites of Jupiter

Amalthea. Amalthea, the second innermost moon of Jupiter, whirls around in a complete orbit every 12 hours. The Voyager photographs reveal that Amalthea is a potato-shaped chunk of rock 270 km long and 155 km wide. It is so small that its internal force of gravity is too weak to form it into a sphere. In a general way, it resembles the tiny moons of Mars, but it is 10 times larger. Amalthea is therefore the first intermediate-size planetary object to be observed in detail.

Io. Beyond Amalthea is the satellite Io. It is only slightly larger than the earth's moon, but the exploration of Io has proved to be one of the most

Figure 24.41 The surface of Io is dominated by volcanic features. An active volcano, identified by the surrounding spray of light-toned sulfur and ash, is visible near the center of the photograph. Dark lava flows are evident near the left margin. The bright, irregular patches are believed to be plains formed from ash and sulfur. The lack of impact craters indicates that most of the surface of Io was formed after the period of intense bombardment.

significant results of the Voyager missions. Io is volcanically active, and much of its surface is influenced in some way by molten sulfur spewed from huge volcanoes scattered across its surface (*figure 24.41*).

Your first impression from studying photographs of Io is that its surface is a mass of confused splotches. Closer examination reveals some degree of order. More than 100 depressions resembling calderas, as much as 200 km across, have been identified throughout the region that was photographed. All are believed to be centers of volcanic eruptions. In addition, a variety of shield volcanoes, low domes, numerous lava flows, and vast plains of sulfur and ash can be recognized. Eight volcanoes were seen actually erupting by *Voyager 1*. Four months later, *Voyager 2* revealed that six were still active.

The volcanic eruptions on Io are like huge fountains, with sprays in the form of an umbrella or in jetlike streaks. The form of the spray depends on the size and shape of the vent, as is typical of fountains. Solid fragmental material from the eruptions reaches heights of 300 km with ejection velocities as great as 1 km per second—an enormous rate of speed in comparison to ejection velocities on the earth.

It appears that Io is the most volcanically active body in the solar system and has been throughout much of geologic time. The big question is, Why? The moon and Mercury, which are about the same size as Io, have been volcanically dead for at least 3 billion years. The answer may be found in the internal friction generated in Io as it circles Jupiter. Europa and Ganymede, as well as Jupiter, exert a gravitational pull on Io, forcing it into an eccentric orbit. The eccentricity causes variations in the amplitude of the tidal force due to Jupiter's gravitational field. Io is therefore constantly wrenched about by variable gravitational forces. These conditions cause the interior of the satellite to yield somewhat, and thus friction and heat are produced. Indeed, Io may be molten at a depth of only 20 km.

Europa. Europa has a density of 3.03 g/cm³ and must therefore be composed mostly of rock. Spectroscopic measurements, however, indicate that Europa is surrounded by a crust of ice, which makes up perhaps 5% or 10% of its bulk. The surface shows great systems of long linear fractures, over 1000 km long and as much as 200 km wide (*figure 24.42*). On a regional scale, the surface resembles that of a fractured ice pack in the Arctic Ocean. The maximum relief on Europa is on the order of 1000 m, in contrast to features that rise from 4000 to 8000 m on Io. This low relief suggests that the subsurface material is partly liquid and slushy, so that isostatic adjustment of the ice prohibits the development of high landforms.

Ganymede. Jupiter's largest satellite is Ganymede, whose diameter is approximately 1.5 times that of the earth's moon. It has a bulk density of only 1.93 g/cm³, which is much less than that of any rock. Spectroscopic measurements indicate that the surface consists of large amounts of water in the form of ice, a conclusion supported by the planet's high reflectivity. This evidence suggests that Ganymede is composed of approximately 50% water. A preliminary model of Ganymede's internal structure suggests a silicate rock core surrounded by a mantle of liquid water or slushy ice and a rigid crust of ice.

The view shown in *figure 24.43* reveals that, on a global scale, Ganymede has a cratered surface. Two terrain types are recognized. The older of the two

Figure 24.42 The surface of Europa is dominated by a complex array of fractures. Europa is about the size of the earth's moon but has a crust of ice, perhaps 100 km thick. The absence of visible relief, such as mountains or craters, suggests that the crust is young and is probably warm a few kilometers below the surface. Tidal heating, which has been suggested to explain the intense volcanic activity on Io, may also be heating Europa's interior, but at a lower rate.

Figure 24.43 The surface of Ganymede shows a variety of impact structures, the most striking of which are bright-rayed craters. Ganymede has an ice crust, which has darkened with age. Recent meteorite impacts have splashed out clean ice from beneath the surface to form bright ejecta blankets and rays. Older craters have only faint remnants of the bright ejecta patterns. Two terrain types can be seen in this photograph: (1) an old terrain of dark, dirty ice, which is fractured and broken into large blocks, and (2) a young terrain characterized by a peculiar system of streaks and grooves.

Figure 24.44 The grooved terrain on Ganymede appears in this photograph as bright bands, which traverse the surface in various directions. These bands contain an intricate system of alternating bright and dark lines, which may represent deformation of the younger ice crust. These lineaments are particularly evident near the top of the picture. A bright band trending from north to south in the lower left of the picture is offset by a strike-slip fault. Numerous impact craters occur throughout the area. Those with bright rays are younger; those that lack a ray system are older.

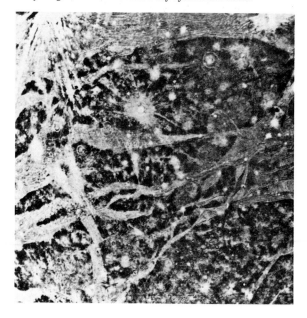

is nearly saturated with craters. It is dark and believed to be composed of "dirty" ice, containing fragments of dust and particles from outer space. This older crust appears to have been fractured and split apart, and many of the fragments have shifted about.

The younger terrain is the most peculiar structure on Ganymede (*figure 24.44*). It consists of a series of grooves and complex strips, or bands, which are closely spaced and roughly parallel. Crater density on the grooved terrain is much lower than that of the darker, cratered terrain, and therefore the grooved terrain is believed to be much younger. This younger surface presents some perplexing problems, because nothing like it has ever been observed on the earth or the other terrestrial planets. Its features result from tectonic processes on an ice crust, a subject about which we know very little. First impressions suggest that the old, cratered terrain was fractured and split apart sometime late in the period of intense bombardment, and cleaner ice from below was extruded into the fractures to form the grooved terrain.

Callisto. Callisto is the outermost Galilean satellite. (Beyond it are eight small moons, which were probably captured from the asteroid belt. These small, outer moons were not investigated by the Voyager missions.) Callisto, like Ganymede, is believed to consist of a rocky core surrounded by a thick mantle of ice. Temperatures at the equator are approximately −150 °C.

The images sent back from the Voyager spacecraft show that Callisto, in contrast to the other Galilean satellites, is densely cratered, somewhat like the highlands of the moon. A glance at *figure 24.45* reveals that the general surface of Callisto is dark, dirty ice but that many of the craters have bright rays and ejecta and bright areas in their centers. This bright material is probably clean, melted ice brought up by the crustal rebound following impact and remelted at the surface. The dense cratering implies that most of Callisto's surface is very old, probably dating back to the period of intense bombardment, which ended about 4 billion years ago.

Aside from the densely cratered terrain, the most striking feature on Callisto is a large multiringed structure. This feature is reminiscent of multiringed basins on the planets of the inner solar system. Important differences are obvious, however, probably because the response of an ice crust to the shock of impact is different from that of rock. The rings on Callisto are more closely spaced, and there is less ejecta radiating from the center of impact. Also, viscous flow of the ice would reduce the topographic relief of the basin and ring structures.

The striking similarities between the surface features of Callisto and those of the moon and Mercury suggest that these three bodies experienced similar events in their early history. These events probably occurred throughout the solar system.

Figure 24.45 The surface of Callisto, shown in this photomosaic assembled from *Voyager 1*'s photographs, is densely cratered and is probably older than the surfaces of the other Galilean satellites. Many of the craters have bright ejecta blankets and ray systems, which formed as clean, melted ice was splashed out by the impact of meteorites. The multiringed basin in the upper left probably resulted from shock waves produced in the ice crust by the impact of a large meteorite. Its rings extend more than 1000 km outward from the basin. The striking similarities between impact features on Callisto and those on the moon, Mercury, and Mars suggest that the planetary bodies of the entire solar system experienced similar events in their early history.

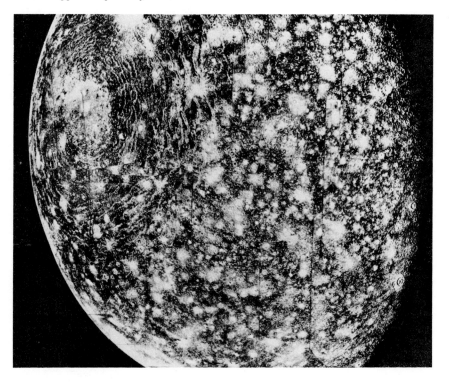

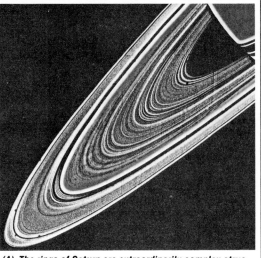

(A) The rings of Saturn are extraordinarily complex structures, with radial ''spokes,'' braided rings, and elliptical features. In this picture, over 95 individual concentric rings can be counted; in images of higher resolution, as many as 1000 separate rings may be present.

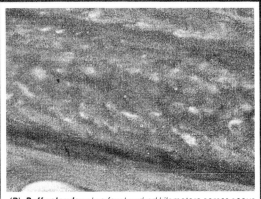

(B) Puffy clouds only a few hundred kilometers across occur in Saturn's northern latitudes and show motion in the atmosphere as they are stretched out into chevron patterns.

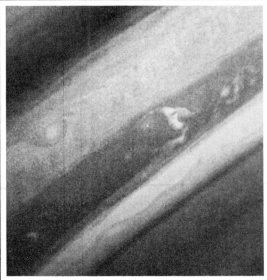

(C) Dark belts and light zones in the Saturnian atmosphere display a variety of cloud features. The small-scale patterns suggest convection currents rising from lower levels.

THE SATURNIAN SYSTEM

Statement

Voyager 1 began its coverage of the Saturnian system in August 1980. In the weeks leading up to its closest approach to Saturn, three months later, the spacecraft witnessed a fascinating display of the planet's atmosphere, rings, and moons. In August 1981, *Voyager 2* greatly supplemented these observations. From a geological point of view, the moons are the most interesting. All but one of them (Titan, the largest satellite in the solar system) are small, icy bodies ranging in diameter from 390 to 1530 km. Various features on several of the moons—fractures, canyons, rilles, and a terrain type called wispy terrain—clearly appear to be the results of significant geologic activity. These small, icy bodies, larger than the asteroid-size moons of Mars but smaller than the moons of Jupiter, constitute a family of planetary objects, which have never before been observed at close range but are now accessible to geologic study.

Discussion

Saturn and Its Rings

Instruments carried by *Voyager 1* and *Voyager 2* recorded a flood of new data concerning Saturn's atmosphere and rings. Some details of these features are shown in *figure 24.46*. In a close-up view, the planet's atmosphere can be seen to be characterized by complex turbulence much like that on Jupiter, although the atmospheric patterns are somewhat less dramatic because the clouds of Saturn are veiled by an overlying haze.

The delicate, shimmering rings of Saturn turn out to be much more fascinating and complex than ever imagined. Previously, astronomers had recognized three principal rings (the *A* ring, the *B* ring, and the *C* ring), but now at least 500, and perhaps as many as 1000, can be distinguished. Even the prominent Cassini division, which had been considered a gap in the ring structure between the *A* ring and the *B* ring, is anything but

Figure 24.46 Saturn, like Jupiter, is a giant sphere of gas (mostly hydrogen and helium) with, possibly, a small rocky core. The rings of Saturn have amazed and intrigued astronomers for over 300 years. Now that we can see them in close-up views, they are even more astonishing.

empty, with numerous ringlets having been revealed within it. Not only are Saturn's rings divided into hundreds of structural components, but there may be differences in composition as well. In addition, numerous radial "spokes", or fingerlike streaks, move across the *B* ring. Time-lapse photography demonstrates that the spokes corotate, perhaps because of the force of Saturn's magnetic field. The newly discovered *F* ring presents a further enigma with what appear to be three separate rings braided around each other along a portion of their orbits.

The Saturnian Satellites

With the exception of the satellite Titan, the moons of Saturn are small, icy bodies, which appear through a telescope only as points of light. Estimates of their diameters, masses, and densities were very uncertain until now. With the photographs from *Voyager 1* and *Voyager 2*, however, the surfaces of these new worlds are available for geologic study. Preliminary reports on the Saturnian satellites present both surprises and mysteries. The diagram in *figure 24.47* displays some comparative date concerning the larger bodies that orbit Saturn, and the photographs in *figure 24.48* illus-

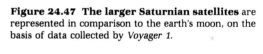

Figure 24.47 The larger Saturnian satellites are represented in comparison to the earth's moon, on the basis of data collected by *Voyager 1.*

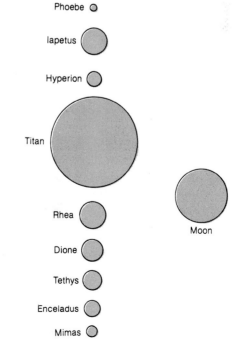

Phoebe

Iapetus

Hyperion

Titan

Rhea

Dione

Tethys

Enceladus

Mimas

Moon

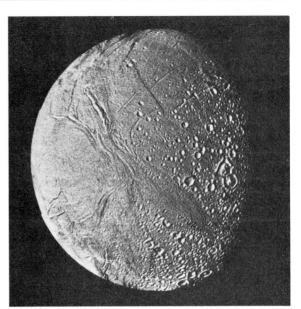

Figure 24.48 Mimas (left) and Enceladus (right) preserve a record of a period of intense bombardment. Mimas bears an enormous impact structure, 100 km across (one-fourth the satellite's diamter), whose prominent rim rests 9000 m above the crater floor. Enceladus resembles the Galilean satellite Ganymede: part of its surface is characterized by impact structures, and other, younger areas are dominated by grooved terrain and long fracture systems. The crust has been split and has separated as a result of internal melting accompanied by extrusions of ice, slush, or water, which form new crust along the rift zone. Enceladus may illustrate how a planet appears in the initial state of plate tectonic activity.

trate some of the major surface features that have been discovered.

The larger satellites (except for Titan) fall into convenient pairs that almost match their order from Saturn outward: Mimas and Enceladus (between 400 and 500 km in diameter), Tethys and Dione (between 1050 and 1120 km in diameter), and Rhea and Iapetus (about 1500 km in diameter).

Mimas. Mimas is heavily cratered everywhere, but its most prominent feature is a huge crater, whose diameter is one-fourth that of the satellite itself (*figure 24.48*). The impact that formed it must have come close to breaking Mimas into fragments. The opposite hemisphere contains large fractures, which may have been produced as shock waves from this impact traveled through the satellite.

Enceladus. In contrast to Mimas, Enceladus has a complex grooved terrain, similar in many ways to that on Ganymede. This terrain may be a result of heat generated by the variable tidal pull of Saturn and Dione (much like the heat produced in Io by its gravitational interaction with Jupiter and Europa). Such heat would soften the interior of Enceladus and could cause a type of volcanic activity that would resurface the satellite.

Dione. With its surface saturated with craters and its large, smooth areas that resemble the lunar

Figure 24.49 Dione is densely cratered. Its largest impact structure, nearly 100 km in diameter, has a prominent central peak. Near the terminator (the line between the illuminated side and the dark side of the satellite) is a series of sinuous valleys, which are probably a result of fracturing in the ice crust. Bright streaks near the northern limb are probably rays from young craters.

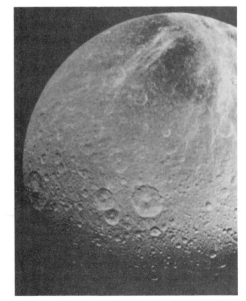

maria, Dione looks much like the earth's moon (*figure 24.49*). The dark areas are crisscrossed with light streaks, which may be extrusions of volatile material, forming a terrain called wispy terrain.

Rhea. The surface of Rhea, photographed in finer detail than any of the other Saturnian satellites, has been found to be saturated with craters, much as Mercury and the earth's moon are (*figure 24.50*). Very large craters, however, which are normally present in the crater populations on other planetary bodies in the solar system, are lacking on Rhea's surface—a fact that may alter our interpretation of the early history of cratering. Two periods of bombardment may have occurred throughout the solar system during its early history, but evidence of only the latter one is preserved on Rhea. The original cratered surface may have been smoothed over by ice flows from within the satellite. Later impact structures can be clearly discerned on this newer surface.

Titan. Titan, the largest satellite in the solar system, appears to be a terrestrial world in deep freeze. It is the only one of Saturn's moons with an atmosphere (*figure 24.51*). Preliminary data indicate that the atmosphere of Titan is five times denser than the earth's, that its atmospheric pressure is 1.5 times that on the earth, and that its temperature is about −175 °C. To the surprise of all, Titan's atmosphere is dominated by nitrogen, with only a smattering of methane (1%), along with such hydrocarbons as propane, ethylene, ethane, and acetylene. These atmospheric conditions are similar to those that are believed to have existed on the earth more than 3 billion years ago.

Figure 24.50 Rhea contains ancient meteorite craters, which closely resemble those of the earth's moon and Mercury. Many have sharp rims and are relatively fresh. Others have subdued rims, indicating greater antiquity.

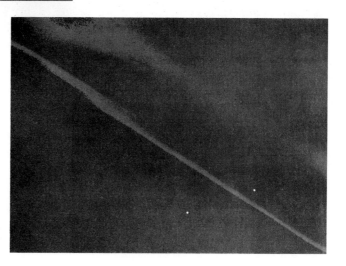

Figure 24.51 Titan is larger than Mercury and possesses a dense atmosphere. The layer of haze merges with a dark "hood" over the North Polar region.

SUMMARY

A vast amount of new knowledge has been obtained about the planets of the solar system as a result of the space program. This new knowledge permits geologists, for the first time, to study details of other planets and compare them with the earth. Some of the more important discoveries and some hypotheses that explain them are summarized below.

Craters. Cratering has been the dominant geologic process in the solar system, and it may have been the dominant process on the earth during the first half-billion years of its history. Crater populations on the moon, Mercury, Mars, Callisto, and perhaps Ganymede are similar and represent similar sequences of events.

Volcanism. The moon, Mercury, Venus, and Mars experienced a thermal event early in their history, after the period of intense bombardment. Mars and Venus have experienced more recent volcanic activity, as is indicated by their great shield volcanoes. The volcanic style and history of the earth and of Io reflect their distinctive tectonic systems.

Tectonic Features. The cirular shape of craters on the moon, Mercury, Venus, Mars, Callisto, and Ganymede indicates that the crusts of these planets have not been deformed by strong compressive forces. Mercury has a global pattern of thrust faults. Mars and Venus have upwarps and graben systems. None of the planets, however, has developed an active tectonic system like the earth's. Ganymede and Europa present some distinctive tectonic styles involving mantles and crusts of ice.

Atmosphere. The moon and Mercury are primitive bodies, and their surfaces have not been changed by atmospheric processes. Mars, in contrast, has a thin atmosphere. Its surface has been modified by stream erosion and deposition, mass movement, wind, ice caps, and ground ice. Venus has a dense atmosphere, but details of its influence on surface features are not yet available.

Internal Structure. To various degrees, all of the planetary bodies appear to have experienced differentiation, but each has a unique, layered internal structure.

Geologic History. A geologic time scale for events on the moon has been established by the principles of superposition and crosscutting relations. These provide a framework of relative time into which the major events of lunar history can be arranged in chronologic order. A geologic time scale has been developed for Mercury and Mars, and a tentative scheme of interplanetary correlation has been established on the basis of crater populations.

The early histories of the moon, Mercury, Venus, Mars, Callisto, and perhaps Ganymede are similar in that, at an early stage in their development, all of these planetary bodies experienced intense bombardment by meteorites, which produced densely cratered terrain, and, somewhat later, the impact of asteroid-size bodies, which formed multiringed basins. The extrusion of flood basalt and a subsequent decline in the rate of cratering were major events on the moon, Mercury, and Mars.

The moon and Mercury remain primitive bodies. Their surfaces have not experienced change from atmospheric erosion or continuing convection in the planets' interiors.

Mars has had a more eventful geologic history, involving (1) the formation of densely cratered terrain, (2) uplift in the Tharsis area and radial fracturing, (3) widespread extrusion of lava, which formed the northern plains, (4) renewed uplift in the Tharsis region and the formation of Valles Marineris, and (5) volcanism and eolian activity.

In planetary development, Venus appears to be more like Mars than any other planetary body, and the history of Venus is probably similar to that of Mars. These two planets appear to have reached an intermediate stage of planetary development, between that of a primitive, impact-dominated body (such as the moon and Mercury) and a tectonically active, water-dominated planet (such as the earth).

Io appears to be the most volcanically active body in the solar system. Its surface is continually being renewed by volcanism.

The small, icy satellites of Saturn appear to have experienced intense meteorite bombardment during an early period of their history, as the other planetary bodies of the solar system did.

The earth, unlike the other planets, has had an active tectonic system throughout its history. Also, the earth's unique atmosphere and its abundant supply of water have played an important role in the development of the planet's surface features. The earth's dynamic systems—its tectonic system and its hydrologic system, both acting to modify its surface—are responsible for the features that distinguish the earth from other planetary bodies.

Key Words

maria	meteorite
terrae	crater
asteroid	ray crater

ejecta blanket
multiringed basin
lineament
rille
wrinkle ridge
lunar basalt
anorthosite
breccia
glass
lunar regolith

mascon
Copernican System
Copernican Period
Eratosthenian System
Eratosthenian Period
Imbrian System
Imbrian Period
Pre-Imbrian System
Pre-Imbrian Period
planetary differentiation

Review Questions

1. What geologic process has been most significant in modifying the surfaces of the moon, Mercury, and Mars?

2. Outline the stages in the production of a crater by the impact of a meteorite. What geologic features are produced by impact?

3. Describe the characteristics of the four major types of craters found on the moon.

4. Explain the origin of a multiringed basin.

5. How are craters modified with time?

6. Describe the major types of lunar rocks.

7. Explain how a geologic time scale was developed for events in the moon's history.

8. Outline the major events in lunar history.

9. Compare and contrast the geology of Mercury with that of the moon.

10. Describe the craters on Mars.

11. Describe the volcanoes on Mars.

12. What tectonic features are found on Mars?

13. Describe the fluvial features on the surface of Mars. How do they compare with fluvial features on the earth?

14. Describe Valles Marineris. How did it originate?

15. What evidence indicates catastrophic flooding on Mars?

16. Describe the surface features generated by wind on Mars.

17. Outline the geologic history of Mars. How is it similar to that of the moon?

18. Compare and contrast the surface features of Venus with those on the earth and Mars.

19. Compare and contrast the surface features of the moons of Jupiter.

20. Explain the origin of volcanism on Io.

21. What is the significance of the major surface features on the Saturnian satellites?

22. How is the earth geologically unique among the planetary bodies of the solar system?

Additional Readings

Allen, J. P. 1972. *Apollo 15:* Scientific journey to Hadley-Apennine. *American Scientist* 60:162–74.

Hartman, W. K. 1977. Cratering in the solar system. *Scientific American* 236(1):84–99.

Journal of Geophysical Research 78(20) (July 10, 1973): 4009–439. The entire issue is devoted to the geology of Mars.

Journal of Geophysical Research 80(17) (June 10, 1975): 2342–514. The entire issue is devoted to the geology of Mercury.

Lowman, P. D. 1972. The geologic evolution of the moon. *Journal of Geology* (80):125–66.

Mason, B. 1971. The lunar rocks. *Scientific American* 255(5):48–58. This paper presents a popular review of the first several groups of lunar samples.

Moore, P., and C. A. Cross: 1977. *The atlas of Mercury.* New York: Crown Publishers.

Mutch, T. A. 1973. *Geology of the moon: A stratigraphic view.* Rev. ed. Princeton, N.J.: Princeton University Press.

Mutch, T. A., et al. 1976. *Geology of Mars.* Princeton, N.J.: Princeton University Press.

National Aeronautics and Space Administration. 1974. The new Mars: The discoveries of *Mariner 9.* NASA SP337. Washington, D.C.: U.S. Government Printing Office.

———. 1976a. Mars as viewed by *Mariner 9.* Rev. ed. NASA SP329/2. Washington, D.C.: U.S. Government Printing Office.

———. 1976b. *Viking 1,* early reports. NASA SP408. Washington, D.C.: U.S. Government Printing Office.

———. 1980. *Voyager 1* encounters Saturn. Washington, D.C.: U.S. Government Printing Ofice.

Nozette, S., and P. Ford. 1981. Venus: A world revealed. *Astronomy* 9(3):6–15.

Science 194(4271) (December 17, 1976): 1274–1353. A large part of this issue is devoted to the Viking missions.

Scientific American 233(3) (September 1975): 1–173. The entire issue is devoted to our new understanding of the solar system.

Shoemaker, E. M. 1964. The geology of the moon. *Scientific American* 221(6):38–47.

Short, N. M. 1975. *Planetary geology.* Englewood Cliffs, N.J.: Prentice-Hall.

Soderblom, L. A. 1980. The Galilean moons of Jupiter. *Scientific American* 242(1):88–100.

Taylor, S. R. 1975. *Lunar science—a post-Apollo view.* Elmsford, N.Y.: Pergamon Press.

GLOSSARY

aa flow A lava flow whose surface is typified by angular, jagged blocks (see diagram). Contrast with *pahoehoe flow.*

ablation Reduction of a glacier by melting, evaporation, iceberg calving, or deflation.

abrasion The mechanical wearing away of a rock by friction, rubbing, scraping, or grinding.

absolute time Geologic time measured in a specific duration of years (in contrast to relative time, which involves only the chronologic order of events).

abyssal Pertaining to the great depths of the oceans, generally 1000 fathoms (2000 m) or more below sea level.

abyssal floor The deep, relatively flat surface of the ocean floor located on both sides of the oceanic ridge. It includes the abyssal plains and the abyssal hills.

abyssal hills The part of the ocean floor consisting of hills rising as much as 1000 m above the surrounding floor. They are found seaward of most abyssal plains and occur in profusion in basins isolated from continents by trenches, ridges, or rises.

abyssal plains Flat areas of the ocean floor, having a slope of less than 1:1000. Most abyssal plains lie at the base of a continental rise and are simply areas where abyssal hills are completely covered with sediment (see diagram).

active plate margin (plate tectonics) The leading edge of a lithospheric plate bordered by a trench.

aftershock An earthquake that follows a larger earthquake. Generally, many aftershocks occur over a period of days or even months after a major earthquake.

agate A variety of cryptocrystalline quartz in which colors occur in bands. It is commonly deposited in cavities in rocks.

aggradation The process of building up a surface by deposition of sediment.

A horizon The topsoil layer in a soil profile.

alcove A large niche or recession formed in a steep cliff.

alluvial fan A fan-shaped deposit of sediment built by a stream where it emerges from an upland or a mountain range into a broad valley or plain (see diagram). Alluvial fans are common in arid and semiarid climates but are not restricted to them.

alluvium A general term for any sedimentary accumulations deposited by comparatively recent action of rivers. It thus includes sediment laid down in river beds, flood plains, and alluvial fans.

alpine glacier A glacier occupying a valley. Synonymous with *mountain glacier, valley glacier.*

amorphous solid A solid in which atoms or ions are not arranged in a definite crystal structure. Examples: glass, amber, obsidian.

amphibole An important rock-forming mineral group of ferromagnesian silicates. Amphibole crystals are constructed from double chains of silicon-oxygen tetrahedra. Example: hornblende.

amphibolite A metamorphic rock consisting mostly of amphibole and plagioclase feldspar.

andesite A fine-grained igneous rock composed mostly of plagioclase feldspar and from 25% to 40% amphibole and biotite but no quartz or K-feldspar. It is abundant in mountains bordering the Pacific Ocean, such as the Andes Mountains of South America, from which the name was derived. Andesitic magma is believed to originate from fractionation of partially melted basalt.

andesite line The boundary in the Pacific Ocean separating volcanoes of the inner Pacific basin, which discharge only basalt, from those near the continental margins, which discharge both andesite and basalt.

angular unconformity An unconformity in which the older strata dip at

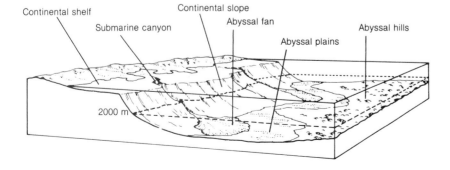

Continental shelf
Submarine canyon
Continental slope
Abyssal fan
Abyssal plains
Abyssal hills
2000 m

a different angle (generally steeper) than the younger strata (see diagram).

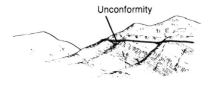

Unconformity

anomaly A deviation from the norm or average.

anorthosite A coarse-grained intrusive igneous rock composed primarily of calcium-rich plagioclase feldspar.

anticline A fold in which the limbs dip away from the axis. After erosion, the oldest rocks are exposed in the central core of the fold (see diagram).

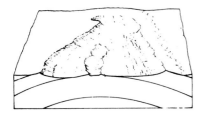

aphanitic texture A rock texture in which individual crystals are too small to be identified without the aid of a microscope. In hand specimens, aphanitic rocks appear to be dense and structureless.

aquifer A permeable stratum or zone below the earth's surface through which ground water moves (see diagram).

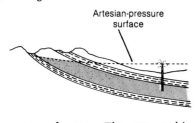

Artesian-pressure surface

arc-trench gap The geographic area in an island arc–deep-sea trench system that separates the arc of volcanoes from the trench. In most cases, the gap is about 100 km wide.

arête A narrow, sharp ridge separating two adjacent glacial valleys.

arkose A sandstone containing at least 25% feldspar.

artesian basin A geologic structural feature in which ground water is confined and is under artesian pressure.

artesian-pressure surface The level to which water in an artesian system would rise in a pipe high enough to stop the flow.

artesian water Ground water confined in an aquifer and under pressure great enough to cause the water to rise above the top of the aquifer when it is tapped by a well.

ash Volcanic fragments the size of dust particles.

ash flow A turbulent blend of unsorted pyroclastic material (mostly fine-grained) mixed with high-temperature gases ejected explosively from a fissure or crater.

ash-flow tuff A rock composed of volcanic ash and dust, formed by deposition and consolidation of ash flows.

asteroid A small, rocky planetary body orbiting the sun. Asteroids are numbered in the tens of thousands; most of them are located between the orbit of Mars and the orbit of Jupiter. Their diameters range downward from 770 km.

asthenosphere The zone in the earth directly below the lithosphere, from 50 to 200 km below the surface. Seismic velocities are distinctly lower in the asthenosphere than in adjacent parts of the earth's interior; therefore, the material in the asthenosphere is believed to be soft and yielding to plastic flow.

astrogeology The study of extraterrestrial bodies by the application of geologic methods and knowledge.

asymmetric fold A fold (anticline or syncline) in which one limb dips more steeply than the other (see diagram).

atmosphere The mixture of gases surrounding a planet. The earth's atmosphere consists chiefly of oxygen and nitrogen, with minor amounts of other gases. Synonymous with *air*.

atoll A ring of low coral islands surrounding a central lagoon (see diagram).

atom The smallest unit of an element. Atoms are composed of protons, neutrons, and electrons.

attitude The three-dimensional orientation of a bed, fault, dike, or other

geologic structure. It is determined by the combined measurements of the dip and the strike of a structure.

axial plane With reference to folds, an imaginary plane that intersects the crest or trough of a fold so as to divide the fold as symmetrically as possible (see diagram).

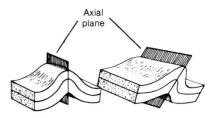

Axial plane

axis **1** (crystallography) An imaginary line passing through a crystal around which the parts of the crystal are symmetrically arranged. **2** (fold) The line where folded beds show maximum curvature. The line formed by the intersection of the axial plane with the bedding surface (see diagram).

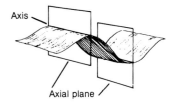

Axis

Axial plane

backswamp The marshy area of a flood plain at some distance beyond and lower than the natural levees that confine the river.

backwash The return sheet flow down a beach after a wave is spent.

badlands An area nearly devoid of vegetation and dissected by stream erosion into an intricate system of closely spaced, narrow ravines.

bajada The surface of a system of coalesced alluvial fans.

bar An offshore, submerged, elongate ridge of sand or gravel built on the sea floor by waves and currents.

barchan dune A crescent-shaped dune, the tips or horns of which point downwind. Barchan dunes form in desert areas where sand is scarce.

barrier island An elongate island of sand or gravel formed parallel to a coast (see diagram).

Barrier island

barrier reef An elongate coral reef that trends parallel to the shore of an island or a continent, separated from it by a lagoon (see diagram).

Partly submerged volcano

basalt A dark-colored, aphanitic (fine-grained) igneous rock composed of plagioclase (over 50%) and pyroxene; olivine may or may not be present. Basalt and andesite represent 98% of all volcanic rocks.

base level The level below which a stream cannot effectively erode. Sea level is the ultimate base level, but lakes form temporary base levels for inland drainage systems.

basement complex A series of igneous and metamorphic rocks lying beneath the oldest stratified rocks of a region (see diagram). In shields, the basement complex is exposed over large areas.

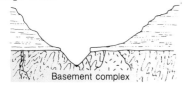

Basement complex

basin **1** (structural geology) A circular or elliptical downwarp. After erosion, the youngest beds are exposed in the central part of the structure. **2** (topography) A depression into which the surrounding area drains.

batholith A large body of intrusive igneous rock exposed over an area of at least 100 km² (see diagram).

bathymetric chart A topographic map of the earth's surface underlying a body of water (such as the ocean floor).

bathymetry The measurement of ocean depths and the charting (mapping) of the topography of the ocean floor.

bauxite A mixture of various amorphous or crystalline hydrous aluminum oxides and aluminum hydroxides, commonly found as a residual clay deposit in tropical and subtropical regions. Bauxite is the principal commercial source of aluminum.

bay (coast) A wide, curving recess or inlet between two capes or headlands.

baymouth bar A narrow, usually submerged ridge of sand or gravel deposited across the mouth of a bay by longshore drift. Baymouth bars commonly are formed by extension of spits along embayed coasts.

beach A deposit of wave-washed sediment along a coast between the landward limit of wave action and the outermost breakers.

beach drift The migration of sediment along a beach caused by the impact of waves striking the shore at an oblique angle.

bed A layer of sediment 1 cm or more in thickness.

bedding plane A surface separating layers of sedimentary rock.

bed load Material transported along the bottom of a stream or river by rolling or sliding, in contrast to material carried in suspension or in solution.

bedrock The continuous solid rock that underlies the regolith everywhere and is exposed locally at the surface. An exposure of bedrock is called an outcrop.

Benioff zone A zone of deep-focus earthquakes that dips away from a deep-sea trench and slopes beneath the adjacent continent or island arc.

berm A nearly horizontal portion of a beach or backshore formed by storm waves. Some beaches have no berms; others have several.

***B* horizon** The solid zone of accumulation underlying the *A* horizon of a soil profile. Some of the material dissolved by leaching in the *A* horizon is deposited in the *B* horizon.

biologic material A general term for material originating from organisms. Examples: fossils (shells, bones, leaves), peat, coal.

biosphere The totality of life on or near the earth's surface.

biotite "Black mica." An important rock-forming ferromagnesian silicate with silicon-oxygen tetrahedra arranged in sheets.

bird-foot delta A delta with distributaries extending seaward and in map view resembling the claws of a bird. Example: the Mississippi Delta.

block faulting A type of normal faulting in which the crust is broken into fault blocks of different elevations and orientations.

blowout A basin excavated by wind erosion.

blueschist A fine-grained schistose rock characterized by high-pressure, low-temperature mineral assemblages and typically blue in color.

boulder A rock fragment whose diameter is more than 256 mm (about the size of a volleyball). A boulder is one size larger than a cobble.

bracketed intrusion An intrusive rock that was once exposed at the surface by erosion and was subsequently covered by younger sediment. The relative age of the intrusion thus falls between, or is bracketed by, the ages of the younger and older sedimentary deposits.

braided stream A stream with a complex of converging and diverging channels separated by bars or islands. Braided streams form where more sediment is available than can be removed by the discharge of the stream.

breaker A collapsing water wave.

breccia A general term for sediment consisting of angular fragments in a matrix of finer particles. Examples: sedimentary breccias, volcanic breccias, fault breccias, impact breccias.

butte A somewhat isolated hill, usually capped with a resistant layer of rock and bordered by talus (see diagram). A butte is an erosion remnant of a formerly more extensive slope.

calcite A mineral composed of calcium carbonate ($CaCO_3$).

caldera A large, more or less circular depression or basin associated with a volcanic vent. Its diameter is many times greater than that of the included vents. Calderas are believed to result from subsidence or collapse and may or may not be related to explosive eruptions.

calving The breaking off of large blocks of ice from a glacier that terminates in a body of water.

capacity The maximum quantity of sediment a given stream, glacier, or wind can carry under a given set of conditions.

capillary A small, tubular opening whose diameter is about that of a human hair.

capillary action The action by which a fluid (such as water) is drawn up into small openings (such as pore spaces in rocks) due to surface tension.

capillary fringe A zone above the water table in which water is lifted by surface tension into openings of capillary size.

carbonaceous Containing carbon.

carbonate mineral A mineral formed by the bonding of carbonate ions (CO_3^{-2}) with positive ions. Examples: calcite ($CaCO_3$), dolomite [$CaMg(CO_3)_2$].

carbonate rock A rock composed mostly of carbonate minerals. Examples: limestone, dolomite.

carbon 14 A radioisotope of carbon. Its half-life is 5730 years.

catastrophism The belief that geologic history consists of major catastrophic events involving processes that were far more intense than any we observe now. Contrast with *uniformitarianism.*

cave A naturally formed subterranean open area, or chamber or series of chambers, commonly produced in limestone by solution activity.

cement Minerals precipitated from ground water in the pore spaces of a sedimentary rock and binding the rock's particles together.

Cenozoic The era of geologic time from the end of the Mesozoic Era (65 million years ago) to the present.

chalcedony A general term for fibrous cryptocrystalline quartz.

chalk A variety of limestone composed of shells of microscopic oceanic organisms.

chemical decomposition Chemical weathering.

chemical weathering Chemical reactions that act on rocks exposed to water and the atmosphere so as to change their unstable mineral components to more stable forms. Oxidation, hydrolysis, carbonation, and direct solution are the most common reactions. Synonymous with *decomposition.*

chert A sedimentary rock composed of granular cryptocrystalline silica.

C horizon The zone of soil consisting of partly decomposed bedrock underlying the *B* horizon. It grades downward into fresh, unweathered bedrock.

cinder A fragment of volcanic ejecta from 0.5 to 2.5 cm in diameter.

cinder cone A cone-shaped hill composed of loose volcanic fragments.

cirque An amphitheater-shaped depression at the head of a glacial valley, excavated mainly by ice plucking and frost wedging (see diagram).

clastic **1** Pertaining to fragments (such as mud, sand, and gravel) produced by the mechanical breakdown of rocks. **2** A sedimentary rock composed chiefly of consolidated clastic material.

clastic texture The texture of sedimentary rocks consisting of fragmental particles of minerals, rocks, and organic skeletal remains (see diagram).

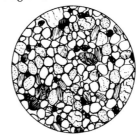

clay Sedimentary material composed of fragments whose diameters are less than $\frac{1}{256}$ mm. Clay particles are smaller than silt particles.

clay minerals A group of fine-grained crystalline hydrous silicates formed by weathering of minerals such as feldspar, pyroxene, or amphibole.

cleavage The tendency of a mineral to break in a preferred direction along smooth planes.

cobble A rock fragment whose diameter is between 64 mm (about the size of a tennis ball) and 256 mm (about the size of a volleyball). Cobbles are larger than pebbles but smaller than boulders.

columnar jointing A system of fractures that splits a rock body into long prisms, or columns (see diagram). It is characteristic of lava flows and shallow intrusive igneous flows.

competence The maximum size of particles that a given stream, glacier, or wind can move at a given velocity.

composite volcano A large volcanic cone built by extrusion of alter-nating layers of ash and lava (see diagram). Synonymous with *strato-volcano.*

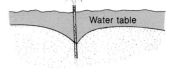

compression A system of stresses that tends to reduce the volume of or shorten a substance.

conchoidal fracture A type of fracture that produces a smooth, curved surface. It is characteristic of quartz and obsidian.

concretion A spherical or ellipsoidal nodule formed by accumulation of mineral matter after deposition of sediment.

cone of depression A conical depression of the water table surrounding a well after heavy pumping (see diagram).

Water table

conglomerate A coarse-grained sedimentary rock composed of rounded fragments of pebbles, cobbles, or boulders.

consequent stream A stream whose course is determined by, or is a direct consequence of, the original slope on which it developed.

contact The surface separating two different rock bodies.

contact metamorphism Metamorphism of a rock near its contact with a magma.

continent A large landmass, from 20 to 60 km thick, composed mostly of granitic rock. Continents rise abruptly above the deep-ocean floor; they include marginal areas submerged beneath sea level. Examples: the African continent, the Australian continent, the Asian continent, the South American continent.

continental accretion The theory that the continents have grown by incorporation of deformed sediments along their margins.

continental crust The type of the earth's crust underlying the continents, including the continental shelves. The continental crust is about 35 km thick; its maximum thickness is 60 km, beneath mountain ranges. Its density is 2.7 g/cm^3, and the velocities of primary seismic waves traveling through it are less than 6.2 km/sec. Synonymous with *sial.* Contrast with *oceanic crust.*

continental drift The theory that the continents have moved in relation to one another.

continental glacier A thick ice sheet covering large parts of a continent. Present-day examples are found in Greenland and Antarctica.

continental margin The zone of transition from a continental mass to the adjacent ocean basin. It generally includes a continental shelf, continental slope, and continental rise.

continental rise The gently sloping surface located at the base of a continental slope (see diagram for *abyssal plains*).

continental shelf The submerged margin of a continental mass extending from the shore to the first prominent break in slope, which usually occurs at a depth of about 120 m.

continental slope The slope that extends from a continental shelf down to the ocean deep. In some areas, such as off eastern North America, the continental slope grades into the more gently sloping continental rise.

convection Movement of portions of a fluid as a result of density differences produced by heating (see diagram).

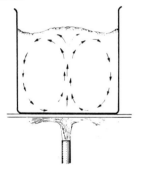

convection cell The space occupied by a single convection current.

convection current A closed system in which material is transported as a result of thermal convection. Convection currents are characteristic of the atmosphere and bodies of water. They are believed also to be generated in the interior of the earth. In the plate tectonic theory, convection within the mantle is thought to be responsible for the movement of tectonic plates.

convergent plate boundary The zone where the leading edges of converging plates meet. Convergent plate boundaries are sites of considerable geologic activity and are characterized by volcanism, earthquakes, and crustal deformation. See also *subduction zone.*

Copernican Period The period of lunar history during which rayed craters, such as Copernicus, and their associated rim deposits were formed (from 2 billion years ago to the present).

Copernican System The youngest system of rocks on the moon, formed during the Copernican Period.

coquina A limestone composed of an aggregate of shells and shell fragments (see diagram).

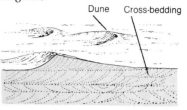

coral A bottom-dwelling marine invertebrate organism of the class of Anthozoa.

core The central part of the earth, ranging from 3380 to 3540 km in diameter, which is surrounded by the mantle.

Coriolis effect The effect produced by a Coriolis force, namely, the tendency of all particles of matter in motion on the earth's surface to be deflected to the right in the Northern Hemisphere and to the left in the Southern Hemisphere.

country rock A general term for rock surrounding an igneous intrusion.

covalent bond A chemical bond in which electrons are shared between different atoms so that none of the atoms has a net charge.

crater An abrupt circular depression formed by extrusion of volcanic material, by collapse, or by the impact of a meteorite (see diagram).

creep The imperceptibly slow downslope movement of material (see diagram).

crevasse 1 (glacial geology) A deep crack in the upper surface of a glacier. 2 (natural levee) A break in a natural levee.

cross-bedding Stratification inclined to the original horizontal surface upon which the sediment accumulated. It is produced by deposition on the slope of a dune or sand wave (see diagram).

Dune   Cross-bedding

crosscutting relations, principle of The principle that a rock is younger than any rock across which it cuts.

crust (planetary structure) The outermost layer, or shell, of the earth (or any other differentiated planet). The earth's crust is generally defined as the part of the earth above the Mohorovičić discontinuity; it represents less than 1% of the earth's total volume (see diagram). See also *continental crust, oceanic crust.*

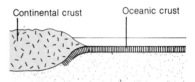

Continental crust   Oceanic crust

crustal warping Gentle bending (upwarping or downwarping) of sedimentary strata.

cryptocrystalline texture The texture of rocks whose crystals are too small to be identified with an ordinary microscope.

crystal A solid, polyhedral form bounded by naturally formed plane surfaces resulting from growth of a crystal lattice.

crystal face A naturally formed smooth plane surface of a crystal.

crystal form The geometric shape of a crystal (see diagram).

crystal lattice A systematic, symmetrical network of atoms within a crystal.

crystalline texture The rock texture resulting from simultaneous growth of crystals.

crystallization The process of crystal growth. It occurs as a result of condensation from a gaseous state, precipitation from a solution, or cooling of a melt.

crystal structure The orderly arrangement of atoms in a crystal.

cuesta An elongate ridge formed on the tilted and eroded edges of gently dipping strata.

daughter isotope An isotope produced by radioactive decay of its parent isotope. The quantity of a daughter isotope continually increases with time.

debris flow The rapid downslope movement of debris (rock, soil, and mud).

debris slide A type of landslide in which comparatively dry rock fragments and soil move downslope at speeds ranging from slow to fast. The mass of debris does not show backward rotation (which occurs in a slump) but slides and rolls forward.

declination, magnetic The horizontal angle between true north and magnetic north at a given point on the earth's surface.

decomposition Weathering by chemical processes. Synonymous with *chemical weathering.*

deep-focus earthquake An earthquake that originates at a depth greater than 300 km.

deep-marine environment The sedimentary environment of the abyssal plains.

deep-sea fan A cone-shaped or fan-shaped deposit of land-derived sediment located seaward of large rivers or submarine canyons. Synonymous with *abyssal cone, abyssal fan, submarine cone.*

deep-sea trench See *trench.*

deflation Erosion of loose rock particles by wind.

deflation basin A shallow depression formed by wind erosion where ground-water solution activity has left unconsolidated sediment exposed at the surface.

degradation The general lowering of the surface of the land by processes of erosion.

delta A large, roughly triangular

body of sediment deposited at the mouth of a river (see diagram).

dendritic drainage pattern A branching stream pattern, resembling the branching of certain trees, such as oaks and maples.

density The measure of concentration of matter in a substance; mass per unit volume, expressed in grams per cubic centimeter (g/cm³).

density current A current that flows as a result of differences in density (see diagram). In oceans, density currents are produced by differences in temperature, salinity, and turbidity (the concentration of material held in suspension).

denudation The combined action of all of the various processes that cause the wearing away and lowering of the land, including weathering, mass wasting, stream action, and ground-water activity.

deranged drainage A distinctively disordered drainage pattern formed in a recently glaciated area. It is characterized by irregular direction of stream flow, few short tributaries, swampy areas, and many lakes (see diagram).

desert pavement A veneer of pebbles left in place where wind has removed the finer material.

desiccation The process of drying out. With reference to sedimentation, the loss of water from pore spaces by evaporation or compaction.

detrital **1** Pertaining to detritus. **2** A rock formed from detritus.

detritus A general term for loose rock fragments produced by mechanical weathering or abrasion.

diastrophism Large-scale deformation involving mountain building and metamorphism.

differential erosion Variation in the rate of erosion on different rock masses. As a result of differential erosion, resistant rocks form steep cliffs, whereas nonresistant rocks form gentle slopes.

differentiated planet A planetary body in which various elements and minerals are separated according to density and concentrated at different

levels. The earth, for example, is differentiated, with heavy metals (iron and nickel) concentrated in the core, lighter minerals in the mantle, and still lighter materials in the crust, hydrosphere, and atmosphere.

differentiation See *magmatic differentiation, planetary differentiation, sedimentary differentiation.*

dike A tabular intrusive rock that occurs across strata or other structural features of the surrounding rock (see diagram).

dike swarm A group of associated dikes.

diorite A phaneritic intrusive igneous rock consisting mostly of intermediate plagioclase feldspar and pyroxene, with some amphibole and biotite.

dip The angle between the horizontal plane and a structural surface (such as a bedding plane, a joint, a fault, foliation, or other planar features).

disappearing stream A stream that disappears into an underground channel and does not reappear in the same, or even in an adjacent, drainage basin. In karst regions, streams commonly disappear into sinkholes and follow channels through caves.

discharge Rate of flow; the volume of water moving through a given cross section of a stream in a given unit of time.

disconformity An unconformity in which beds above and below are parallel (see diagram).

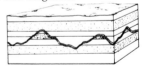

discontinuity A sudden or rapid change in physical properties of rocks within the earth. Discontinuities are recognized by seismic data. See also *Mohorovičić discontinuity.*

disintegration Weathering by mechanical processes. Synonymous with *mechanical weathering.*

dissolution The process by which materials are dissolved.

dissolved load The part of a stream's load that is carried in solution.

distributary Any of the numerous stream branches into which a river divides where it reaches its delta.

divergent plate boundary A plate margin formed where the lithosphere splits into plates that drift apart from one another. Divergent plate boundaries are areas subject to tension, and new crust is generated by igneous activity. Synonymous with *spreading center.* See also *oceanic ridge.*

divide A ridge separating two adjacent drainage basins.

dolomite **1** A mineral composed of $CaMg(CO_3)_2$. **2** A sedimentary rock composed primarily of the mineral dolomite.

dolostone A sedimentary rock composed mostly of the mineral dolomite. Sometimes referred to simply as *dolomite.*

dome **1** (structural geology) An uplift that is circular or elliptical in map view, with beds dipping away in all directions from a central area (see diagram). **2** (topography) A general term for any dome-shaped landform.

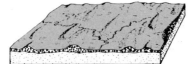

downwarp A downward bend or subsidence of a part of the earth's crust.

drainage basin The total area that contributes water to a single drainage system (see diagram).

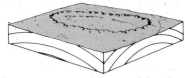

drainage system An integrated system of tributaries and a trunk stream, which collect and funnel surface water to the sea, a lake, or some other body of water.

drift (glacial geology) A general term for sediments deposited directly by glacial ice or in lakes, oceans, or streams as a result of glaciation.

drip curtain A thin sheet of dripstone hanging from the ceiling or wall of a cave.

dripstone A cave deposit formed by precipitation of calcium carbonate from ground water entering an underground cavern.

drumlin A smooth, glacially streamlined hill that is elongate in the direction of ice movement (see diagram). Drumlins are generally composed of till.

dune A low mound of fine-grained material that accumulates as a result of sediment transport in a current system. Dunes have characteristic geometric forms that are maintained as they migrate. Sand dunes are commonly classified according to shape. See also *barchan dune, longitudinal dune, parabolic dune, seif dune, star dune,* and *transverse dune.*

earthquake A series of elastic waves propagated in the earth, initiated where stress along a fault exceeds the elastic limit of the rock so that sudden movement occurs along the fault.

ecology The study of relationships between organisms and their environments.

ejecta Rock fragments, glass, and other material thrown out of an impact crater or a volcano.

ejecta blanket Rock material (crushed rock, large blocks, breccia, and dust) ejected from an impact crater or explosion crater and deposited over the surrounding area.

elastic deformation Temporary deformation of a substance, after which the material returns to its original size and shape. Example: the bending of mica flakes.

elastic limit The maximum stress that a given substance can withstand without undergoing permanent deformation either by solid flow or by rupture.

elastic-rebound theory The theory that earthquakes result from energy released by faulting; the sudden release of stored strain creates earthquake waves.

end moraine A ridge of till that accumulates at the margin of a glacier.

entrenched meander A meander cut into the underlying rock as a result of regional uplift or lowering of the regional base level (see diagram).

environment of sedimentation See *sedimentary environment.*

eolian Pertaining to wind.

eolian environment The sedimentary environment of deserts, where sediment is transported and deposited primarily by wind.

epicenter The area on the earth's surface that lies directly above the focus of an earthquake (see diagram).

epoch A division of geologic time; a subdivision of a period. Example: Pleistocene Epoch.

Eratosthenian Period The period of lunar history during which large craters whose rays are no longer visible, such as Eratosthenes, were formed (from 3.1 billion to 2 billion years ago).

Eratosthenian System The system of rocks formed during the Eratosthenian Period. It is older than the Copernican System but younger than the Imbrian System.

erosion The processes that loosen sediment and move it from one place to another on the earth's surface. Agents of erosion include water, ice, wind, and gravity.

erratic A large boulder carried by glacial ice to an area far removed from its point of origin (see diagram).

escarpment A cliff or very steep slope.

esker A long, narrow, sinuous ridge of stratified glacial drift deposited by a stream flowing beneath a glacier in a tunnel or subglacial stream bed.

estuary A bay at the mouth of a river formed by subsidence of the land or a rise in sea level. Fresh water from the river mixes with and dilutes seawater in an estuary.

eugeocline (plate tectonics) A geocline in which volcanism is associated with clastic sedimentation. Eugeoclines are usually associated with an island arc.

eugeosyncline A geosyncline situated seaward from a continent and characterized by sediments deposited by turbidity currents and derived in part from a volcanic arc.

eustatic change of sea level A worldwide rise or fall in sea level resulting from a change in the volume of water or the capacity of ocean basins.

evaporite A rock composed of minerals derived from evaporation of mineralized water. Examples: rock salt, gypsum.

exfoliation The weathering process by which concentric shells, slabs, sheets, or flakes are successively broken loose and stripped away from a rock mass (see diagram).

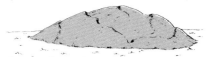

exponential index With reference to mineral or energy reserves, the length of time in which a reserve will be depleted, on the assumption that the rate of consumption will increase exponentially. Contrast with *static index.*

exposure Bedrock not covered with soil or regolith; outcrop.

extrusive rock A rock formed from a mass of magma that flowed out on the surface of the earth. Example: basalt.

faceted spur A spur or ridge that has been beveled or truncated by faulting, erosion, or glaciation.

facies A distinctive group of characteristics within part of a rock body (such as composition, grain size, or fossil assemblages) that differ as a group from those found elsewhere in the same rock unit. Examples: conglomerate facies, shale facies, brachiopod facies.

fan A fan-shaped deposit of sediment. See also *alluvial fan, deep-sea fan.*

fault A surface along which a rock body has broken and been displaced.

fault block A rock mass bounded by faults on at least two sides (see diagram).

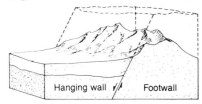

Hanging wall   Footwall

fault scarp A cliff produced by faulting.

faunal succession, principle of The principle that fossils in a stratigraphic sequence succeed one another in a definite, recognizable order.

feldspar A mineral group consisting of silicates of aluminum and one or more of the metals potassium, sodium, or calcium. Examples: K-feldspar, Ca-plagioclase, Na-plagioclase.

felsite A general term for light-colored aphanitic (fine-grained) igneous rocks. Example: rhyolite.

ferromagnesian minerals A variety of silicate minerals containing abundant iron and magnesium. Examples: olivine, pyroxene, amphibole.

fiord A glaciated valley flooded by the sea to form a long, narrow, steep-walled inlet.

firn Granular ice formed by recrystallization of snow. It is intermediate between snow and glacial ice. Sometimes referred to as *névé.*

fissure An open fracture in a rock.

fissure eruption Extrusion of lava along a fissure (see diagram).

flint A popular name for dark-colored chert (cryptocrystalline quartz).

flood basalt An extensive flow of basalt erupted chiefly along fissures. Synonymous with *plateau basalt.*

flood plain The flat, occasionally flooded area bordering a stream.

fluvial Pertaining to a river or rivers.

fluvial environment The sedimentary environment of river systems.

focus The area within the earth where an earthquake originates.

fold A bend, or flexure, in a rock (see diagram).

folded mountain belt A long, linear zone of the earth's crust where rocks have been intensely deformed by horizontal stresses and generally intruded by igneous rocks. The great folded mountains of the world (such as the Appalachians, the Himalayas, the Rockies, and the Alps) are believed to have been formed at convergent plate margins.

foliation A planar feature in metamorphic rocks, produced by the secondary growth of minerals. Three major types are recognized: slaty cleavage, schistosity, and gneissic layering.

footwall The block beneath a dipping fault surface.

foreshore The seaward part of the shore or beach lying between high tide and low tide.

formation A distinctive body of rock that serves as a convenient unit for study and mapping.

fossil Naturally preserved remains or evidence of past life, such as bones, shells, casts, impressions, and trails.

fossil fuel A fuel containing solar energy that was absorbed by plants and animals in the geologic past and thus is preserved in organic compounds in their remains. Fossil fuels include petroleum, natural gas, and coal.

fracture zone **1** (field geology) A zone where the bedrock is cracked and fractured. **2** (oceanography) A zone of long, linear fractures on the ocean floor, expressed topographically by ridges and troughs. Fracture zones are the topographic expression of transform faults.

fringing reef A reef that lies alongside the shore of a landmass (see diagram).

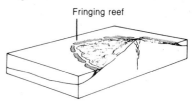

Fringing reef

frost heaving The lifting of unconsolidated material by the freezing of subsurface water.

frost wedging The forcing apart of rocks by the expansion of water as it freezes in fractures and pore spaces. Synonymous with *ice wedging.*

gabbro A dark-colored, coarse-grained rock composed of Ca-plagioclase, pyroxene, and possibly olivine, but no quartz.

gas The state of matter in which a substance has neither independent shape nor independent volume. Gases can readily be compressed and tend to expand indefinitely.

geocline An elongate prism of sedimentary rock deposited in a subsided part of the earth's crust. A modern example is the continental margins of the eastern United States. See also *eugeocline, miogeocline.*

geode A hollow nodule of rock lined with crystals; when separated from the rock body by weathering, it ap-

GLOSSARY **503**

pears as a hollow, rounded shell partly filled with crystals (see diagram).

geologic column A diagram representing divisions of geologic time and the rock units formed during each major period.

geologic cross section A diagram showing the structure and arrangement of rocks as they would appear in a vertical plane below the earth's surface.

geologic map A map showing the distribution of rocks at the earth's surface.

geologic time scale The time scale determined by the geologic column and by radiometric dates of rocks.

geosyncline A subsiding part of the lithosphere in which thousands of meters of sediment accumulate. See also *eugeosyncline, miogeosyncline.*

geothermal Pertaining to the heat of the interior of the earth.

geothermal energy Useful energy that can be extracted from steam and hot water found within the earth's crust.

geothermal gradient The rate at which temperature increases with depth.

geyser A thermal spring that intermittently erupts steam and boiling water.

glacial environment The sedimentary environment of glaciers and their meltwaters.

glacier A mass of ice formed from compacted recrystallized snow that is thick enough to flow plastically.

glass 1 A state of matter in which a substance displays many properties of a solid but lacks crystal structure. 2 An amorphous igneous rock formed from a rapidly cooling magma.

glassy texture The texture of igneous rocks in which the material is in the form of natural glass rather than crystal.

global tectonics The study of the characteristics and origin of structural features of the earth that have regional or global significance.

***Glossopteris* flora** An assemblage of late Paleozoic fossil plants named for the seed fern *Glossopteris*, one of the plants in the assemblage. These

flora are widespread in South America, Africa, Australia, India, and Antarctica and provide important evidence for the theory of continental drift.

gneiss A coarse-grained metamorphic rock with a characteristic type of foliation (gneissic layering), resulting from alternating layers of light-colored and dark-colored minerals. Its composition is generally similar to that of granite.

gneissic layering The type of foliation characterizing gneiss, resulting from alternating layers of the constituent silicic and mafic minerals.

Gondwanaland The ancient continental landmass that is thought to have split apart during Mesozoic time to form the present-day continents of South America, Africa, India, Australia, and Antarctica (see diagram).

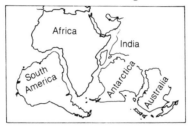

graben An elongate fault block that has been lowered in relation to the blocks on either side (see diagram).

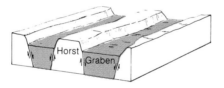

gradation Leveling of the land due to erosion by such agents as river systems, ground water, glaciers, wind, and waves.

graded bedding A type of bedding in which each layer is characterized by a progressive decrease in grain size from the bottom of the bed to the top (see diagram).

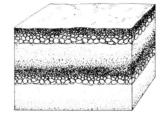

graded stream A stream that has attained a state of equilibrium, or balance, between erosion and deposition, so that the velocity of the water is just great enough to transport the sedi-

ment load supplied from the drainage basin, and neither erosion nor deposition occurs.

gradient (stream) The slope of a stream channel measured along the course of the stream.

grain A particle of a mineral or rock, generally lacking well-developed crystal faces.

granite A coarse-grained igneous rock composed of K-feldspar, plagioclase, and quartz, with small amounts of ferromagnesian minerals.

granitization Formation of granitic rock by metamorphism without complete melting.

gravity anomaly An area where gravitational attraction is greater or less than its normal value.

graywacke An impure sandstone consisting of rock fragments and grains of quartz and feldspar in a matrix of clay-size particles.

groundmass The matrix of relatively fine-grained material between the phenocrysts in a porphyritic rock.

ground moraine Glacial deposits that cover an area formerly occupied by a glacier; they typically produce a landscape of low, gently rolling hills.

ground water Water below the earth's surface. It generally occurs in pore spaces of rocks and soil.

guyot A seamount with a flat top (see diagram).

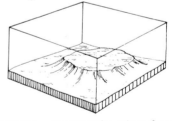

gypsum An evaporite mineral composed of calcium sulfate with water ($CaSO_4 \cdot 2H_2O$).

half-life The time required for half of a given sample of a radioactive isotope to decay to its daughter isotope.

halite An evaporite mineral composed of sodium chloride (NaCl).

hanging valley A tributary valley

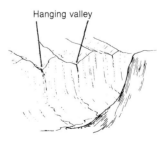

Hanging valley

whose floor lies ("hangs") above the valley of the main stream or shore to which it flows (see diagram). Hanging valleys commonly are created by deepening of the main valley by glaciation, but they can also be produced by faulting or rapid retreat of a sea cliff.

hanging wall The surface or block of rock that lies above an inclined fault plane (see diagram).

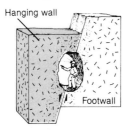

hardness 1 (mineralogy) The measure of the resistance of a mineral to scratching or abrasion. 2 (water) A property of water resulting from the presence of calcium carbonate and magnesium carbonate in solution.

headland An extension of land seaward from the general trend of the coast; a promontory, cape, or peninsula.

headward erosion Extension of a stream headward, up the regional slope of erosion.

heat flow The flow of heat from the interior of the earth.

high-grade metamorphism Metamorphism that occurs under high temperature and high pressure.

hogback A narrow, sharp ridge formed on steeply inclined resistant rock.

horizon 1 (geologic) A plane of stratification assumed to have been originally horizontal. 2 (soil) A layer of soil distinguished by characteristic physical properties. Soil horizons generally are designated by letters (for example, *A* horizon, *B* horizon, *C* horizon).

horn A sharp peak formed at the intersection of the headwalls of three or more cirques (see diagram).

hornblende A variety of the amphibole mineral group.

hornfels A nonfoliated metamorphic rock of uniform grain size,

formed by high-temperature, low-pressure metamorphism. Hornfelses typically are formed by contact metamorphism around igneous intrusions.

horst An elongate fault block that has been uplifted in relation to the adjacent rocks (see diagram).

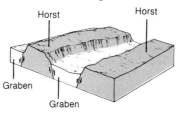

hot spot The expression at the earth's surface of a mantle plume, or column of hot, buoyant rock rising in the mantle beneath a lithospheric plate.

hummock A small, rounded or cone-shaped, low hill or a surface of other small, irregular shapes. A surface that is not equidimensional or ridgelike.

hydraulic Pertaining to a fluid in motion.

hydraulic head The pressure exerted by a fluid at a given depth beneath its surface. It is proportional to the height of the fluid's surface above the area where the pressure is measured.

hydrologic system The system of moving water at the earth's surface (see diagram).

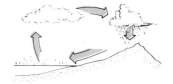

hydrolysis Chemical combination of water with other substances.

hydrosphere The waters of the earth, as distinguished from the rocks (lithosphere), the air (atmosphere), and living things (biosphere).

hydrostatic pressure The pressure within a fluid (such as water) at rest, exerted on a given point within the body of the fluid.

hydrothermal deposit A mineral deposit formed by high-temperature ground water. The high temperature commonly is associated with emplacement of a magma.

ice sheet A thick, extensive body of glacial ice that is not confined to valleys. Localized ice sheets are sometimes called *ice caps.*

ice wedging A type of mechanical weathering in which rocks are broken by the expansion of water as it freezes in joints, pores, or bedding planes. Synonymous with *frost wedging.*

igneous rock Rock formed by cooling and solidification of molten silicate minerals (magma). Igneous rocks include volcanic and plutonic rocks.

Imbrian Period The period of lunar history during which the large multiringed basins, such as Mare Imbrium, were formed and the mare basalts were extruded (from 3.9 billion to 3.1 billion years ago).

Imbrian System The system of rocks formed on the moon during the Imbrian Period.

inclination, magnetic The angle between the horizontal plane and a magnetic line of force.

inclusion A rock fragment incorporated into a younger igneous rock.

intermediate-focus earthquake An earthquake whose focus is located at a depth between 60 and 300 km.

intermittent stream A stream through which water flows only part of the time.

internal drainage A drainage system that does not extend to the ocean.

interstitial Pertaining to material in the pore spaces of a rock. Petroleum and ground water are interstitial fluids. Minerals deposited by ground water in a sandstone are interstitial minerals.

intrusion 1 Injection of a magma into a preexisting rock. 2 A body of rock resulting from the process of intrusion.

intrusive rock Igneous rock that, while it was fluid, penetrated into or between other rocks and solidified. It can later be exposed at the earth's surface after erosion of the overlying rock.

inverted valley A valley that has been filled with lava or other resistant material and has subsequently been eroded into an elongate ridge.

ion An atom or combination of atoms that has gained or lost one or more electrons and thus has a net electrical charge.

ionic bond A chemical bond formed by electrostatic attraction between oppositely charged ions.

ionic substitution The replacement of one kind of ions in a crystalline lattice by another kind that is of similar size and electrical charge.

island A landform smaller than a continent and completely surrounded by water.

island arc A chain of volcanic islands. Island arcs are generally convex toward the open ocean. Example: the Aleutian Islands.

isostasy A state of equilibrium, resembling flotation, in which segments of the earth's crust stand at levels determined by their thickness and density (see diagram). Isostatic equilibrium is attained by flow of material in the mantle.

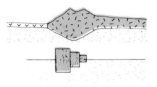

isotope One of the several forms of a chemical element that have the same number of protons in the nucleus but differ in the number of neutrons and thus differ in atomic weight.

joint A fracture in a rock along which no appreciable displacement has occurred.

kame A body of stratified glacial sediment. A mound or an irregular ridge deposited by a subglacial stream as an alluvial fan or a delta.

karst topography A landscape characterized by sinks, solution valleys, and other features produced by ground-water activity (see diagram).

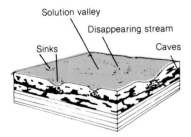

Solution valley
Disappearing stream
Sinks
Caves

kettle A closed depression in a deposit of glacial drift formed where a block of ice was buried or partly buried and then melted.

laccolith A concordant igneous intrusion that has arched up the strata into which it was injected, so that it forms a pod-shaped or lens-shaped body with a generally horizontal floor (see diagram).

lag deposit A residual accumulation of coarse fragments that remain on the surface after finer material has been removed by wind.

lagoon A shallow body of seawater separated from the open ocean by a barrier island or reef.

lamina (*pl.* **laminae**) A layer of sediment less than 1 cm thick.

laminae Plural of *lamina.*

laminar flow A type of flow in which the fluid moves in parallel lines. Contrast with *turbulent flow.*

landform Any feature of the earth's surface having a distinct shape and origin. Landforms include major features (such as continents, ocean basins, plains, plateaus, and mountain ranges) and minor features (such as hills, valleys, slopes, drumlins, and dunes). Collectively, the landforms of the earth constitute the entire surface configuration of the planet.

landslide A general term for relatively rapid types of mass movement, such as debris flows, debris slides, rockslides, and slumps.

lateral moraine An accumulation of till deposited along the side margins of a valley glacier. It accumulates as a result of mass movement of debris on the sides of the glacier.

laterite A soil that is rich in oxides of iron and aluminum formed by deep weathering in tropical and subtropical areas.

Laurasia The ancient continental landmass that is thought to have split apart to form Europe, Asia, North America, and Greenland.

lava Magma that reaches the earth's surface.

leach To dissolve and remove the soluble constituents of a rock or soil.

leachate A solution produced by leaching. Example: water that has seeped through a waste disposal site and thus contains in solution various substances derived from the waste material.

leaching The process by which ground water dissolves and transports soluble components of a rock or soil.

leading edge (plate tectonics) A convergent or active plate margin. Contrast with *trailing edge.*

lee slope The part of a hill, dune, or rock that is sheltered or turned away from the wind. Synonymous with *slip face.*

levee, natural A broad, low embankment built up along the banks of a river channel during floods.

limb The flank, or side, of a fold.

limestone A sedimentary rock composed mostly of calcium carbonate ($CaCO_3$).

lineament A topographic feature or group of features having a linear configuration. Lineaments commonly are expressed as ridges or depressions or as an alignment of features such as stream beds, volcanoes, or vegetation.

liquid The state of matter in which a substance flows freely and lacks crystal structure. Unlike a gas, a liquid retains the same volume independently of the shape of its container.

lithification The processes by which sediment is converted into sedimentary rock. These processes include cementation and compaction.

lithosphere The relatively rigid outer zone of the earth, which includes the continental crust, the oceanic crust, and the part of the mantle lying above the softer asthenosphere.

load The total amount of sediment carried at a given time by a stream, glacier, or wind.

loess Unconsolidated, wind-deposited silt and dust.

longitudinal dune An elongate sand dune oriented in the direction of the prevailing wind.

longitudinal profile The profile of a stream or valley drawn along its length, from source to mouth.

longitudinal wave A seismic body wave in which particles oscillate along lines in the direction in which the wave travels. Synonymous with *P wave.*

longshore current A current in the surf zone moving parallel to the shore. Longshore currents occur where waves strike the shore at an angle: the waves push water and sediment obliquely up the beach, and the backwash returns straight down the beach face, so that the water and sediment follow a zigzag pattern, with net movement parallel to the shore.

longshore drift The process in which sediment is moved in a zigzag pattern along a beach by the swash and backwash of waves that approach the shore obliquely.

low-grade (low-rank) metamorphism Metamorphism that is accomplished under low or moderate temperature and low or moderate pressure.

mafic rock An igneous rock containing more than 50% ferromagnesian minerals.

magma A mobile silicate melt, which can contain suspended crystals and dissolved gases as well as liquid.

magmatic differentiation A general term for the various processes by which early-formed crystals or early-formed liquids are separated and removed from a magma to produce a rock whose composition is different from that of the original magma. Early-crystallized ferromagnesian minerals commonly are separated by gravitational settling, so that the parent magma is left enriched in silica, sodium, and potassium.

magmatic segregation Separation of crystals of certain minerals from a magma as it cools. For example, some minerals (including certain valuable metals) crystallize while other components of the magma are still liquid; these early-formed crystals can settle to the bottom of a magma chamber and thus become concentrated there, forming an ore deposit.

magnetic anomaly A deviation of observed magnetic inclination or intensity (as measured by a magnetometer) from a constant normal value.

magnetic reversal A complete 180° reversal of the polarity of the earth's magnetic field.

mantle The zone of the earth's interior between the Moho discontinuity and the core.

mantle plume A buoyant mass of hot mantle material that rises to the base of the lithosphere. Mantle plumes commonly produce volcanic activity and structural deformation in the central part of lithospheric plates.

marble A metamorphic rock consisting mostly of metamorphosed limestone or dolomite.

mare (*pl.* **maria**) Any of the relatively smooth, low, dark areas of the moon. The lunar maria were formed by extrusion of lava.

maria Plural of *mare.*

mascon A concentration of mass in a local area beneath a lunar mare. The term is derived from *mass concentration.*

mass movement The transfer of rock and soil downslope by direct action of gravity without a flowing medium (such as a river or glacial ice). Synonymous with *mass wasting.*

matrix The relatively fine-grained rock material occupying the space between larger particles in a rock. See also *groundmass.*

meander A broad, looping bend in a river (see diagram).

mechanical weathering The breakdown of rock into smaller fragments by physical processes such as frost wedging. Synonymous with *disintegration.*

medial moraine A ridge of till formed in the middle of a valley glacier by the junction of two lateral moraines where two valley glaciers converge (see diagram).

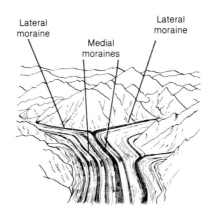

melt A substance altered from the solid state to the liquid state.

mesa A flat-topped, steep-sided highland capped with a resistant rock formation. A mesa is smaller than a plateau but larger than a butte.

Mesozoic The era of geologic time from the end of the Paleozoic Era (225 million years ago) to the beginning of the Cenozoic Era (65 million years ago).

metaconglomerate A metamorphosed conglomerate.

metallic bond A chemical bond in which shared electrons move freely among the atoms.

metamorphic Pertaining to the processes or products of metamorphism.

metamorphic rock Rock formed from preexisting rocks within the earth's crust by changes in temperature and pressure and by chemical action of fluids.

metamorphism Alteration of the minerals and textures of a rock by changes in temperature and pressure and by a gain or loss of chemical components.

meteorite Any particle of solid matter that has fallen to the earth, the moon, or another planet from space.

mica A group of silicate minerals exhibiting perfect cleavage in one direction.

microcontinent A relatively small, isolated fragment of continental crust. Example: Madagascar.

mid-Atlantic ridge The mountain

range extending from north to south down the central part of the Atlantic Ocean floor.

migmatite A mixture of igneous and metamorphic rocks in which thin dikes and stringers of granitic material interfinger with metamorphic rocks.

mineral A naturally occurring inorganic solid having a definite internal structure and a definite chemical composition that varies only within strict limits. Chemical composition and internal structure determine its physical properties, including the tendency to assume a particular geometric form (crystal form).

miogeocline (plate tectonics) A geocline situated near a continental margin, forming a shallow-marine environment of sedimentation for the accumulation of a thick sequence of well-sorted clastic and chemical sediments derived from the continent.

miogeosyncline A geosyncline situated near a continental margin and receiving well-sorted clastic and chemical sediments from the continent, but not associated with volcanism.

mobile belts Long, narrow belts of the continents that have been subjected to mountain-building processes.

Mohorovičić discontinuity The first global seismic discontinuity below the surface of the earth. It lies at a depth varying from about 5 to 10 km beneath the ocean floor to about 35 km beneath the continents. Commonly referred to as the Moho.

monadnock An erosion remnant rising above a peneplain (see diagram).

monocline A bend or fold in gently dipping horizontal strata (see diagram).

moraine A general term for a landform composed of till.

mountain A general term for any landmass that stands above its surroundings. In the stricter geological sense, a mountain belt is a highly deformed part of the earth's crust that has been injected with igneous intrusions and whose deeper parts have been metamorphosed. The topography of young mountains is high, but erosion can reduce old mountains to flat lowlands.

mud crack A crack in a deposit of mud or silt resulting from the contraction that accompanies drying.

mudflow A flowing mixture of mud and water (see diagram).

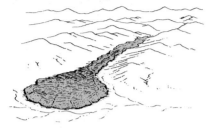

multiringed basin A large crater (more than 300 km in diameter) containing a series of concentric ridges and depressions. Example: the Orientale basin on the moon.

nappe Faulted and overturned folds.

natural arch An arch-shaped landform produced by weathering and differential erosion.

névé Granular ice formed by recrystallization of snow. Synonymous with *firn.*

nodule A small, irregular, knobby, or rounded rock that is generally harder than the surrounding rock.

nonconformity An unconformity in which stratified rocks rest upon eroded granitic or metamorphic rocks (see diagram).

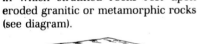

normal fault A steeply inclined fault in which the hanging wall has moved downward in relation to the footwall (see diagram). Synonymous with *gravity fault.*

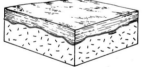

Footwall

Hanging wall

nuée ardente A hot cloud of volcanic fragments and superheated gases that flows as a mass, because it is denser than air (see diagram). Upon cooling, it forms a type of rock called tuff, including ash-flow tuff and welded tuff.

obsidian A glassy igneous rock whose composition is equivalent to that of granite.

ocean basin A low part of the lithosphere lying between continental masses. The rocks of an ocean basin are mostly basalt with a veneer of oceanic sediment.

oceanic crust The type of crust that underlies the ocean basins. It is about 5 km thick, composed predominantly of basalt. Its density is 3.0 g/cm³. The velocities of compressional seismic waves traveling through it exceed 6.2 km/sec. Compare with *continental crust.*

oceanic ridge The continuous ridge, or broad, fractured topographic swell, that extends through the central part of the Arctic, Atlantic, Indian, and South Pacific oceans. It is several hundred kilometers wide, and its elevation above the ocean floor is 600 m or more. It thus constitutes a major structural and topographic feature of the earth.

offshore The area from low tide seaward.

oil shale Shale that is rich in hydrocarbon derivatives. In the United States, the chief oil shale is the Green River Formation, in the Rocky Mountain region.

olivine A silicate mineral with magnesium and iron but no aluminum [(Mg,Fe)₂SiO₄].

oolite A limestone consisting largely of spherical grains of calcium carbonate in concentric spherical layers (see diagram).

ooze (marine geology) Marine sediment consisting of more than 30% shell fragments of microscopic organisms.

ophiolite A sequence of rocks characterized by ultramafic rocks at the base and (in ascending order) gabbro, sheeted dikes, pillow lavas, and

deep-sea sediments. The typical sequence of rocks constituting the oceanic crust.

orogenesis The processes of mountain building.

orogenic Pertaining to deformation of a continental margin to the extent that a mountain range is formed.

orogenic belt A mountain belt.

orogeny A major episode of mountain building.

outcrop An exposure of bedrock.

outlet glacier A tonguelike stream of ice, resembling a valley glacier, that forms where a continental glacier encounters a mountain system and is forced to move through a mountain pass in large streams.

outwash Stratified sediment washed out from a glacier by meltwater streams and deposited in front of the end moraine.

outwash plain The area beyond the margins of a glacier where meltwater deposits sand, gravel, and mud washed out from the glacier (see diagram).

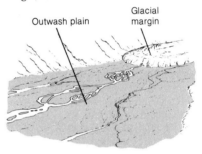

Outwash plain

Glacial margin

overturned fold A fold in which at least one limb has been rotated through an angle greater than 90° (see diagram).

oxbow lake A lake formed in the channel of an abandoned meander.

oxidation Chemical combination of oxygen with another substance.

pahoehoe flow A lava flow with a billowy or ropy surface. Contrast with *aa flow.*

paleocurrent An ancient current, which existed in the geologic past, whose direction of flow can be inferred from cross-bedding, ripple marks, and other sedimentary structures.

paleogeography The study of geography in the geologic past, including the patterns of the earth's surface, the distribution of land and ocean, and ancient mountains and other landforms.

paleomagnetism The study of ancient magnetic fields, as preserved in the magnetic properties of rocks. It includes studies of changes in the position of the magnetic poles and reversals of the magnetic poles in the geologic past.

paleontology The study of ancient life.

paleowind An ancient wind, which existed in the geologic past, whose direction can be inferred from patterns of ancient ash falls, orientation of cross-bedding, and growth rates of colonial corals.

Paleozoic The era of geologic time from the end of the Precambrian (600 million years ago) to the beginning of the Mesozoic Era (225 million years ago).

Pangaea A hypothetical continent from which the present continents originated by plate movement from the Mesozoic Era to the present.

parabolic dune A dune shaped like a parabola with the concave side toward the wind.

partial melting The process by which minerals with low melting points liquefy within a rock body as a result of an increase in temperature or a decrease in pressure (or both) while other minerals in the rock are still solid. If the liquid (magma) is removed before other components of the parent rock have melted, the composition of the magma can be quite different from that of the parent rock. Partial melting is believed to be important in the generation of basaltic magma from peridotite at spreading centers and in the generation of granitic magma from basaltic crust at subduction zones.

passive plate margin (plate tectonics) A lithospheric plate margin at which crust is neither created nor destroyed. Passive plate margins generally are marked by transform faults.

peat An accumulation of partly carbonized plant material containing approximately 60% carbon and 30% oxygen. It is considered an early stage, or rank, in the development of coal.

pebble A rock fragment whose diameter is between 2 mm (about the size of a match head) and 64 mm (about the size of a tennis ball).

pediment A gently sloping erosion surface formed at the base of a receding mountain front or cliff. It cuts across bedrock and can be covered with a veneer of sediment (see dia-

gram). Pediments characteristically form in arid and semiarid climates.

pelagic sediment Deep-sea sediment composed of fine-grained detritus that slowly settles from surface waters. Common constituents are clay, radiolarian ooze, and foraminiferal ooze.

peneplain An extensive erosion surface worn down almost to sea level (see diagram). Subsequent tectonic activity can lift a peneplain to higher elevations.

peninsula An elongate body of land extending into a body of water.

perched water table The upper surface of a local zone of saturation that lies above the regional water table (see diagram).

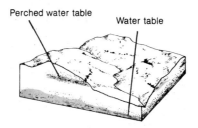

Perched water table   Water table

peridotite A dark-colored coarse-grained igneous rock composed of olivine, pyroxene, and some other ferromagnesian minerals, but with essentially no feldspar and no quartz.

permafrost Permanently frozen ground.

permanent stream A stream or reach of a stream that flows continuously throughout the year. Synonymous with *perennial stream.*

permeability The ability of a material to transmit fluids.

phaneritic texture The texture of igneous rocks in which the interlocking crystals are large enough to be seen without magnification.

phenocryst A crystal that is significantly larger than the crystals surrounding it (see diagram). Phenocrysts form during an early phase in the cool-

ing of a magma when the magma cools relatively slowly.

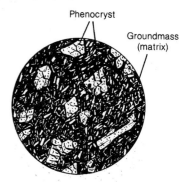

Phenocryst

Groundmass (matrix)

physiographic map A map showing surface features of the earth.

physiography The study of the surface features and landforms of the earth.

pillar A landform shaped like a pillar.

pillow lava An ellipsoidal mass of igneous rock formed by extrusion of lava underwater.

pinnacle A tall, tower-shaped or spire-shaped pillar of rock.

placer A mineral deposit formed by the sorting or washing action of water. Placers are usually deposits of heavy minerals, such as gold.

plagioclase A group of feldspar minerals whose composition ranges from $NaAlSi_3O_8$ to $CaAl_2Si_2O_8$.

planetary differentiation The processes by which the materials in a planetary body are separated according to density, so that the originally homogeneous body is converted into a zoned or layered (shelled) body with a dense core, a mantle, and a crust.

plastic deformation A permanent change in a substance's shape or volume that does not involve failure by rupture.

plate (tectonics) A broad segment of the lithosphere (including the rigid upper mantle, plus oceanic and continental crust) that floats on the underlying asthenosphere and moves independently of other plates.

plateau An extensive upland region.

plateau basalt Basalt extruded in extensive, nearly horizontal layers, which, after uplift, tend to erode into great plateaus (see diagram). Synonymous with *flood basalt.*

plate tectonics The theory of global dynamics in which the lithosphere is believed to be broken into individual plates that move in response to convection in the upper mantle. The margins of the plates are sites of considerable geologic activity.

platform reef An organic reef with a flat upper surface developed on submerged segments of a continental platform.

playa A depression in the center of a desert basin, the site of occasional playa lakes (see diagram).

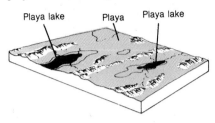

Playa lake   Playa   Playa lake

playa lake A shallow temporary lake formed in a desert basin after a rain.

Pleistocene The epoch of geologic time from the end of the Pliocene Epoch of the Tertiary Period (about 2 million years ago) to the beginning of the Holocene Epoch of the Quaternary Period (about 10,000 years ago). The major event during the Pleistocene was the expansion of continental glaciers in the Northern Hemisphere. Synonymous with *glacial epoch, ice age.*

plucking (glacial geology) The process of glacial erosion by which large rock fragments are loosened by ice wedging, become frozen to the bottom surface of the glacier, and are torn out of the bedrock and transported by the glacier as it moves. The process involves the freezing of subglacial meltwater that seeps into fractures and bedding planes in the rock.

plunge The inclination, with respect to the horizontal plane, of any linear structural element of a rock. The plunge of a fold is the inclination of the axis of the fold.

plunging fold A fold whose axis is inclined from the horizontal.

plutonic rock Igneous rock formed deep beneath the earth's surface.

pluvial lake A lake that was created under former climatic conditions, at a time when rainfall in the region was more abundant than it is now. Pluvial lakes were common in arid regions during the Pleistocene.

point bar A crescent-shaped accumulation of sand and gravel deposited on the inside of a meander bend (see diagram).

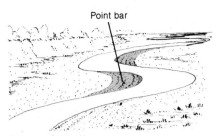

Point bar

polarity epoch A relatively long period of time during which the earth's magnetic field is oriented in either the normal direction or the reverse direction.

polarity event A relatively brief interval of time within a polarity epoch; during a polarity event, the polarity of the earth's magnetic field is reversed with respect to the prevailing polarity of the epoch.

polar wandering The apparent movement of the magnetic poles with respect to the continents.

pore fluid A fluid occupying pore spaces of a rock, such as ground water or liquid rock material resulting from partial melting.

pore space The spaces within a rock body that are unoccupied by solid material. Pore spaces include spaces between grains, fractures, vesicles, and voids formed by dissolution.

porosity The percentage of the total volume of a rock or sediment that consists of pore space.

porphyritic texture The texture of igneous rocks in which some crystals are distinctly larger than others.

porphyry copper Deposits of copper disseminated throughout a porphyritic granitic rock.

pothole A hole formed in a stream bed by sand and gravel swirled around in one spot by eddies (see diagram).

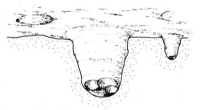

Precambrian The division of geologic time from the formation of the earth (about 4.5 billion years ago) to the beginning of the Cambrian Period of the Paleozoic Era (about 600 million years ago). Also, the rocks formed during that time. Precambrian time constitutes about 90% of the earth's history.

Pre-Imbrian Period The earliest period of lunar history, extending from the formation of the planet

(about 4.5 billion years ago) to the formation of the multiringed basins (about 3.9 billion years ago).

Pre-Imbrian System The system of rocks formed on the moon during the Pre-Imbrian Period. It includes most of the material on the lunar highlands.

pressure ridge An elongate uplift of the congealing crust of a lava flow, resulting from the pressure of underlying and still fluid lava.

primary coast A coast shaped by subaerial erosion, deposition, volcanism, or tectonic activity.

primary sedimentary structure A structure of sedimentary rocks (such as cross-bedding, ripple marks, or mud cracks) that originates contemporaneously with the deposition of the sediment (in contrast to a secondary structure, such as a joint or fault, which originates after the rock has been formed).

primary wave See *P wave.*

pumice A rock consisting of frothy natural glass.

P wave (primary seismic wave) A type of seismic wave, propagated like a sound wave, in which the material involved in the wave motion is alternately compressed and expanded.

pyroclastic Pertaining to fragmental rock material formed by volcanic explosions.

pyroclastic texture The rock texture of igneous rocks consisting of fragments of ash, rock, and glass produced by volcanic explosions.

pyroxene A group of rock-forming silicate minerals composed of single chains of silicon-oxygen tetrahedra. Compare with *amphibole,* which is composed of double chains.

quartz An important rock-forming silicate mineral composed of silicon-oxygen tetrahedra joined in a three-dimensional network. It is distinguished by its hardness, glassy luster, and conchoidal fracture.

quartzite A metamorphosed sandstone.

radioactivity The spontaneous disintegration of an atomic nucleus with the emission of energy.

radiocarbon A radioactive isotope of carbon, C^{14}, which is formed in the atmosphere and is absorbed by living organisms.

radiogenic heat Heat generated by radioactivity.

radiometric dating Determination of the age in years of a rock or mineral by measuring the proportions of an original radioactive material and its decay product. Synonymous with *radioactive dating*.

ray crater A meteorite crater that has a system of rays extending like splash marks from the crater rim (see diagram).

recessional end moraine A ridge of till deposited at the margin of a glacier during a period of temporary stability in its general recession.

recharge Replenishment of the ground-water reservoir by the addition of water.

recrystallization Reorganization of elements of the original minerals in a rock resulting from changes in temperature and pressure and from the activity of pore fluids.

reef A solid structure built of shells and other secretions of marine organisms, particularly coral.

regolith The blanket of soil and loose rock fragments overlying the bedrock.

rejuvenated stream A stream whose erosive power has been renewed by uplift or a lowering of the base level or by climatic changes.

relative age The age of a rock or an event as compared with some other rock or event.

relative dating Determination of the chronologic order of a sequence of events in relation to one another without reference to their ages measured in years. Relative geologic dating is based primarily on superposition, faunal succession, and crosscutting relations.

relative time Geologic time as determined by relative dating, that is, by placing events in chronologic order without reference to their ages measured in years.

relief The difference in altitude between the high and the low parts of an area.

reverse fault A fault in which the hanging wall has moved upward in relation to the footwall; a high-angle thrust fault (see diagram).

rhyolite A fine-grained volcanic rock composed of quartz, K-feldspar, and plagioclase. It is the extrusive equivalent of a granite.

rift system A system of faults resulting from extension.

rift valley 1 A valley of regional extent formed by block faulting in which tensional stresses tend to pull the crust apart. Synonymous with *graben*. 2 The downdropped block along divergent plate margins.

rill A very small stream.

rille An elongate trench of crack-like valley on the moon's surface. Rilles can be sinuous and meandering or relatively linear structural depressions.

rip current A current formed on the surface of a body of water by the convergence of currents flowing in opposite directions. Rip currents are common along coasts where longshore currents move in opposite directions.

ripple marks Small waves produced on a surface of sand or mud by the drag of wind or water moving over it.

river system A river with all of its tributaries.

roche moutonnée An abraded knob of bedrock formed by an overriding glacier. It typically is striated and has a gentle slope facing the upstream direction of ice movement (see diagram).

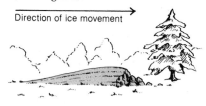

Direction of ice movement

rock An aggregate of minerals that forms an appreciable part of the lithosphere.

rockfall The most rapid type of mass movement, in which rocks ranging from large masses to small fragments are loosened from the face of a cliff.

rock flour Fine-grained rock particles pulverized by glacial erosion and transport.

rock glacier A mass of poorly sorted, angular boulders cemented with interstitial ice. It moves slowly by the action of gravity.

rockslide A landslide in which a newly detached segment of bedrock suddenly slides over an inclined surface of weakness (such as a joint or bedding plane).

runoff Water that flows over the land surface.

sag pond A small lake that forms in a depression, or sag, where active or recent movement along a fault has impounded a stream.

saltation The transportation of particles in a current of wind or water by a series of bouncing movements (see diagram).

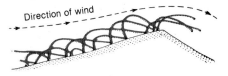

Direction of wind

salt dome A dome produced in sedimentary rock by the upward movement of a body of salt (see diagram).

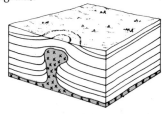

salt-water encroachment Displacement of fresh ground water by salt water in coastal areas, due to the greater density of salt water.

sand Sedimentary material composed of fragments whose diameters range from 0.0625 to 2 mm. Sand particles are larger than silt particles but smaller than pebbles. Much sand is composed of quartz grains, because quartz is abundant and resists chemical and mechanical disintegration, but other materials, such as shell fragments and rock fragments, can also form sand.

sandstone A sedimentary rock composed mostly of sand-size particles, usually cemented by calcite, silica, or iron oxide.

sand wave A wave produced on a surface of sand by the drag of air or water moving over it. Sand waves include dunes and ripple marks.

scarp A cliff produced by faulting or erosion.

schist A medium-grained or coarse-grained metamorphic rock with strong foliation (schistosity) resulting from parallel orientation of platy minerals, such as mica, chlorite, and talc.

schistosity The type of foliation that characterizes schist, resulting from the parallel arrangement of coarse-grained platy minerals, such as mica, chlorite, and talc.

scoria An igneous rock containing abundant vesicles.

sea arch An arch cut by wave erosion through a headland (see diagram).

sea cave A cave formed by wave erosion (see diagram).

sea cliff A cliff produced by wave erosion.

sea-floor spreading The theory that the sea floor spreads laterally away from the oceanic ridge as new lithosphere is created along the crest of the ridge by igneous activity.

seamount An isolated, conical mound rising more than 1000 m above the ocean floor (see diagram). Seamounts are probably submerged shield volcanoes.

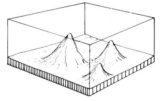

sea stack A small, pillar-shaped, rocky island formed by wave erosion through a headland near a sea cliff (see diagram).

secondary coast A coast formed by marine processes or the growth of marine organisms.

secondary wave See *S wave.*

sediment Material (such as gravel, sand, mud, and lime) that is transported and deposited by wind, water, ice, or gravity; material that is precipitated from solution; and deposits of organic origin (such as coal and coral reefs).

sedimentary differentiation The process in which distinctive sedimentary products (such as sand, shale, and lime) are generated and progressively separated from a rock mass by means of weathering, erosion, transportation, and deposition.

sedimentary environment A place where sediment is deposited and the physical, chemical, and biological conditions that exist there. Examples: rivers, deltas, lakes, shallow-marine shelves.

sedimentary rock Rock formed by the accumulation and consolidation of sediment.

seep A spot where ground water or other fluids (such as oil) are discharged at the earth's surface.

seif dune A longitudinal dune of great height and length.

seismic Pertaining to earthquakes or to waves produced by natural or artificial earthquakes.

seismic discontinuity A surface within the earth at which seismic wave velocities abruptly change.

seismic ray The path along which a seismic wave travels. Seismic rays are perpendicular to the wave crest.

seismic reflection profile A profile of the configuration of the ocean floor and shallow sediments obtained by reflection of artificially produced seismic waves.

seismic wave A wave or vibration produced within the earth by an earthquake or artificial explosion.

seismograph An instrument that records seismic waves.

settling velocity The rate at which suspended solid material subsides and is deposited.

shadow zone (seismology) An area where there is very little or no direct reception of seismic waves from a given earthquake because of refraction of the waves in the earth's core. The shadow zone for *P* waves is between about 103° and 143° from the epicenter.

shale A fine-grained clastic sedimentary rock formed by consolidation of clay and mud.

shallow-focus earthquake An earthquake whose focus is less than 60 km below the earth's surface.

shallow-marine environment The sedimentary environment of the continental shelves, where the water is usually less than 200 m deep.

sheeting A set of joints formed essentially parallel to the surface. It allows layers of rock to spall off as the weight of overlying rock is removed by erosion. It is especially well-developed in granitic rock.

shield An extensive area of a continent where igneous and metamorphic rocks are exposed and have approached equilibrium with respect to erosion and isostasy. Rocks of the shield are usually very old (that is, more than 600 million years old).

shield volcano A large volcano shaped like a flattened dome and built up almost entirely of numerous flows of fluid basaltic lava. The slopes of shield volcanoes seldom exceed 10°, so that in profile they resemble a shield or broad dome.

shore The zone between the waterline at high tide and the waterline at low tide. A narrow strip of land immediately bordering a body of water, especially a lake or an ocean.

sial A general term for the silica-rich rocks that form the continental masses.

silicate A mineral containing silicon-oxygen tetrahedra, in which four oxygen atoms surround each silicon atom.

silicon-oxygen tetrahedron The structure of the ion SiO_4^{-2}, in which four oxygen atoms surround a silicon atom to form a four-sided pyramid, or tetrahedron.

sill A tabular body of intrusive rock injected between layers of the enclosing rock.

silt Sedimentary material composed of fragments whose diameters range from $\frac{1}{265}$ to $\frac{1}{16}$ mm. Silt particles are larger than clay particles but smaller than sand particles.

siltstone A fine-grained clastic sedimentary rock composed mostly of silt-size particles.

sima A general term for the magnesium-rich igneous rocks (basalt, gabbro, and peridotite) of the ocean basins.

sinkhole A depression formed by the collapse of a cavern roof (see diagram).

slate A fine-grained metamorphic rock with a characteristic type of foliation (slaty cleavage), resulting from the parallel arrangement of microscopic platy minerals, such as mica and chlorite.

slaty cleavage The type of foliation that characterizes slate, resulting from the parallel arrangement of microscopic platy minerals, such as mica and chlorite. Slaty cleavage forms distinct zones of weakness within a rock, along which it splits into slabs.

slip face See *lee slope.*

slope retreat Progressive recession of a scarp or the side of a hill or mountain by mass movement and stream erosion.

slump A type of mass movement in which material moves along a curved surface of rupture.

snowline The line on a glacier separating the area where snow remains from year to year from the area where snow from the previous season melts.

soil The surface material of the continents, produced by disintegration of rock. Regolith that has undergone chemical weathering in place.

soil profile A vertical section of soil showing the soil horizons and parent material.

solid The state of matter in which a substance has a definite shape and volume and some fundamental strength.

solifluction A type of mass movement in which material moves slowly downslope in areas where the soil is saturated with water. It commonly occurs in permafrost areas.

solution valley A valley produced by solution activity, either by dissolution of surface materials or by removal or subsurface materials such as limestone, gypsum, or salt.

sorting The separation of particles according to size, shape, or weight. It occurs during transportation by running water or wind.

spatter cone A low, steep-sided volcanic cone built by accumulation of splashes and spatters of lava (usually basaltic) around a fissure or vent.

specific gravity The ratio of the weight of a substance to the weight of an equal volume of water.

spheroidal weathering The process by which corners and edges of a rock body become rounded as a result of exposure to weathering on all sides, so that the rock acquires a spheroidal or ellipsoidal shape (see diagram).

spit A sandy bar projecting from the mainland into open water. Spits are formed by deposition of sediment

moved by longshore drift (see diagram).

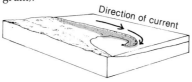

splay A small deltaic deposit formed on a flood plain where water and sediment are diverted from the main stream through a crevasse in a levee.

spreading axis The imaginary axis through the earth about which a set of tectonic plates moves. The motion of a diverging plate can be described as rotation around a spreading axis.

spreading center A plate boundary formed by tensional stress along the oceanic ridge. Synonymous with *divergent plate boundary, spreading edge.*

spreading pole A pole of the imaginary axis about which a set of tectonic plates moves. The spreading poles are the two points at which a spreading axis intersects the earth's surface.

spring A place where ground water flows or seeps naturally to the surface.

stable platform The part of a continent that is covered with flat-lying or gently tilted sedimentary strata and underlain by a basement complex of igneous and metamorphic rocks. An extensive area of a continent where the shield is covered with a veneer of sedimentary rocks.

stack See *sea stack.*

stalactite An icicle-shaped deposit of dripstone hanging from the roof of a cave (see diagram).

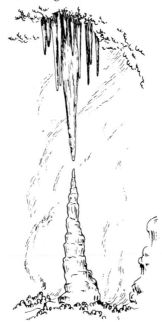

stalagmite A conical deposit of dripstone built up from a cave floor (see diagram).

star dune A mound of sand with a high central point and arms radiating in various directions.

static index With reference to mineral or energy reserves, the length of time in which a known reserve will be depleted, on the assumption that the rate of consumption will remain constant and no new reserves are discovered. Contrast with *exponential index.*

stock A small, roughly circular intrusive body, usually less than 100 km² in surface exposure.

strata Plural of *stratum.*

stratification The layered structure of sedimentary rock.

stratovolcano A volcano built up of alternating layers of ash and lava flows. Synonymous with *composite volcano.*

stratum (*pl.* **strata**) A layer of sedimentary rock.

streak The color of a powdered mineral.

stream load The total amount of sediment carried by a stream at a given time.

stream order The hierarchical number of a stream segment. The smallest tributary has the order number of 1, and successively larger tributaries have progressively higher numbers.

stream piracy Diversion of the headwaters of one stream into another stream. The process occurs by headward erosion of a stream having greater erosive power than the stream it captures (see diagram).

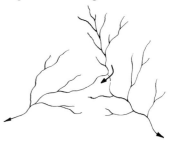

stream terrace One of a series of level surfaces in a stream valley representing the dissected remnants of an abandoned flood plain, stream bed, or valley floor produced in a previous stage of erosion or deposition.

stress Force applied to a material that tends to change its dimensions or volume; force per unit area.

striation A scratch or groove produced on the surface of a rock by a geologic agent, such as a glacier or stream.

strike The bearing (compass direction) of a horizontal line on a bedding plane, a fault plane, or some other planar structural feature (see diagram).

strike-slip fault A fault in which movement has occurred parallel to the strike of the fault (see diagram).

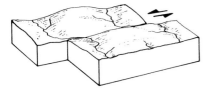

strike valley A valley that is eroded parallel to the strike of the underlying nonresistant strata.

strip mining A method of mining in which soil and rock cover are removed to obtain the sought-after material.

subaerial Occurring beneath the atmosphere or in the open air, with reference to conditions or processes (such as erosion) that occur on the land. Contrast with *submarine* and *subterranean*.

subaqueous sand flow A type of mass movement in which saturated sand or silt flows beneath the surface of a lake or an ocean.

subduction Subsidence of the leading edge of a lithoshperic plate into the mantle.

subduction zone An elongate zone in which one lithoshperic plate descends beneath another. A subduction zone is marked by an oceanic trench, lines of volcanoes, and crustal deformation associated with mountain building. See also *convergent plate boundary*.

submarine canyon A V-shaped trench or valley with steep sides cut into a continental shelf or continental slope.

subsequent stream A tributary

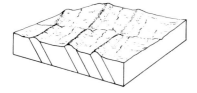

stream that is eroded along an underlying belt of nonresistant rock after the main drainage pattern has been established (see diagram).

subsidence A sinking or settling of a part of the earth's crust with respect to the surrounding parts.

superposed stream A stream whose course was originally established on a cover of rock that is now removed by erosion, so that the stream or drainage system is independent of the newly exposed rocks and structures. The stream pattern is thus superposed, or placed upon, ridges or other structural features that were previously buried (see diagram).

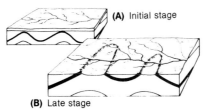

superposition, principle of The principle that, in a series of sedimentary strata that has not been overturned, the oldest rocks are at the base and the youngest are at the top.

surface creep Slow downwind movement of large sand grains by rolling or sliding along the surface due to the impact of smaller, saltating grains.

surface wave (seismology) A seismic wave that travels along the earth's surface. Contrast with *P waves* and *S waves,* which travel through the earth.

suspended load The part of a stream's load that is carried in suspension for a considerable period of time, without contact with the stream bed. It consists mainly of mud, silt, and sand. Contrast with **bed load** and **dissolved load.**

swash The rush of water up onto a beach after a wave breaks.

S wave (secondary seismic wave) A seismic wave in which particles vibrate at right angles to the direction in which the wave travels. Contrast with *P wave.*

symmetrical fold A fold in which the two limbs are essentially mirror images of each other.

syncline A fold in which the limbs dip toward the axis. After erosion, the youngest beds are exposed in the central core of the fold.

talus Rock fragments that accumulate in a pile at the base of a ridge or cliff (see diagram).

tectonic creep Slow, apparently continuous movement along a fault (as opposed to the sudden rupture that occurs during an earthquake).

tectonics The branch of geology that deals with regional or global structures and deformational features of the earth.

tension Stress that tends to pull materials apart.

tephra A general term for pyroclastic material ejected from a volcano. It includes ash, dust, bombs, and other types of fragments.

terminal end moraine A ridge of material deposited by a glacier at the line of maximum advance of the glacier (see diagram).

terra (*pl.* **terrae**) A densely cratered highland on the moon.

terrace A nearly level surface bordering a steeper slope, such as a stream terrace or wave-cut terrace (see diagram).

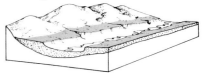

terrae Plural of *terra.*

texture The size, shape, and arrangement of the particles that make up a rock.

thin section A slice of rock mounted on a glass slide and ground to a thickness of about 0.03 mm.

thrust fault A low-angle fault (45°

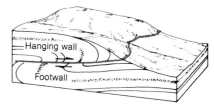

or less) in which the hanging wall has moved upward in relation to the foot-wall (see diagram). Thrust faults are characterized by horizontal compression rather than vertical displacement.

tidal bore A violent rush of tidal water.

tidal flat A large, nearly horizontal area of land covered with water at high tide and exposed to the air at low tide. Tidal flats consist of fine-grained sediment (mostly mud, silt, and sand).

till Unsorted and unstratified glacial deposits.

tillite A rock formed by lithification of glacial till (unsorted, unstratified glacial sediment).

tombolo A beach or bar connecting an island to the mainland (see diagram).

topography The shape and form of the earth's surface.

trailing edge (plate tectonics) A passive or divergent plate margin. Contrast with *leading edge.*

transform fault A special type of strike-slip fault forming the boundary between two moving lithospheric plates, usually along an offset segment of the oceanic ridge (see diagram). See also *passive plate margin.*

transpiration The process by which water vapor is released into the atmosphere by plants.

transverse dune An asymmetrical dune ridge that forms at right angles to the direction of prevailing winds.

travertine terrace A terrace formed from calcium carbonate deposited by water on a cave floor.

trellis drainage pattern A drainage pattern in which tributaries are arranged in a pattern similar to that of a garden trellis (see diagram).

trench (marine geology) A narrow, elongate depression of the deep-ocean floor oriented parallel to the trend of a continent or an island arc.

tributary A stream flowing into or joining a larger stream.

tsunami A seismic sea wave; a long, low wave in the ocean caused by an earthquake, faulting, or a landslide on the sea floor. Its velocity can reach 800 km per hour. Tsunamis are commonly and incorrectly called tidal wides.

tuff A fine-grained rock composed of volcanic ash.

turbidity current A current in air, water, or any other fluid caused by differences in the amount of suspended matter (such as mud, silt, or volcanic dust). Marine turbidity currents, laden with suspended sediment, move rapidly down continental slopes and spread out over the abyssal floor (see diagram).

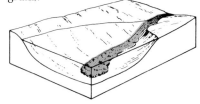

turbulent flow A type of flow in which the path of motion is very irregular, with eddies and swirls.

ultimate base level The lowest possible level to which a stream can erode the earth's surface; sea level.

ultramafic rock An igneous rock composed entirely of ferromagnesian minerals.

unconformity A discontinuity in the succession of rocks, containing a gap in the geologic record. A buried erosion surface. See also *angular unconformity, nonconformity.*

uniformitarianism The theory that geologic events are caused by natural processes, many of which are operating at the present time.

upwarp An arched or uplifted segment of the crust.

valley glacier A glacier that is confined to a stream valley. Synonymous with *alpine glacier, mountain glacier.*

varve A pair of thin sedimentary layers, one relatively coarse-grained and light-colored and the other relatively fine-grained and dark-colored, formed by deposition on a lake bottom during a period of one year (see diagram). The coarse-grained layer is formed during spring runoff, and the fine-grained layer is formed during the winter when the surface of the lake is frozen.

ventifact A pebble or cobble shaped and polished by wind abrasion (see diagram).

vesicle A small hole formed in a volcanic rock by a gas bubble that became trapped as the lava solidified (see diagram).

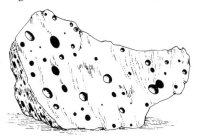

viscosity The tendency within a body to resist flow. An increase in viscosity implies a decrease in fluidity, or ability to flow.

volatile **1** Capable of being readily readily vaporized. **2** A substance that can readily be vaporized, such as water and carbon dioxide.

volcanic ash Dust-size particles ejected from a volcano.

volcanic bomb A hard fragment of lava that was liquid or plastic at the time of ejection and acquired its form and surface markings during flight through the air (see diagram). Volcanic bombs range from a few millimeters to more than a meter in diameter.

volcanic front The line in a volcanic arc system (parallel to a trench) along which volcanism abruptly begins.

volcanic neck The solidified magma that originally filled the vent or neck of an ancient volcano and has subsequently been exposed by erosion (see diagram).

volcanism The processes by which magma and gases are transferred from the earth's interior to its surface.

wash A dry stream bed (see diagram).

water gap A pass in a ridge through which a stream flows (see diagram).

water table The upper surface of the zone of saturation (see diagram).

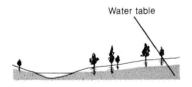

Water table

wave base The lower limit of wave transportation and erosion, equal to half the wavelength.

wave-built terrace A terrace built up from wave-washed sediments. Wave-built terraces usually lie seaward of a wave-cut terrace.

wave crest The highest part of a wave.

wave-cut cliff A cliff formed along a coast by the undercutting action of waves and currents.

wave-cut platform A terrace cut across bedrock by wave erosion. Synonymous with *wave-cut terrace.*

wave-cut terrace See *wave-cut platform.*

wave height The vertical distance between a wave crest and the preceding trough.

wavelength The horizontal distance between similar points on two successive waves, measured perpendicularly to the crest.

wave period The interval of time required for a wave crest to travel a distance equal to one wavelength; the interval of time required for two successive wave crests to pass a fixed point.

wave refraction The process by which a wave is turned from its original direction as it approaches shore at an oblique angle. As a wave approaches shore obliquely, part of it reaches the shallow water near shore while the rest is still advancing in deeper water; the part of the wave in the shallower water moves more slowly than the part in the deeper water.

wave trough The lowest part of a wave, between successive crests.

weathering The processes by which rocks are chemically altered or physically broken into fragments as a result of exposure to atmospheric agents and the pressures and temperatures at or near the earth's surface, with little or no transportation of the loosened or altered materials.

welded tuff A rock formed from particles of volcanic ash that were hot enough that they became fused together.

wind gap A gap in a ridge through which a stream, which is now abandoned as a result of stream piracy, once flowed (see diagram).

wind shadow The area behind an obstacle where air movement is not capable of moving material.

wrinkle ridge A sinuous, irregular, segmented ridge on the surface of a lunar mare, believed to be a result of deformation of the lava.

X-ray diffraction In mineralogy, the process of identifying mineral structures by exposing crystals to a beam of X rays and studying the resulting diffraction patterns.

yardang An elongate ridge carved by wind erosion.

yazoo stream A tributary stream that flows parallel to the main stream for a considerable distance before joining it. Such a tributary is forced to flow along the base of a natural levee formed by the main stream.

zone of aeration The zone below the earth's surface and above the water table, in which pore spaces are usually filled with air (see diagram).

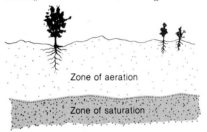

Zone of aeration

Zone of saturation

zone of saturation The zone in the subsurface in which all pore spaces are filled with water (see diagram).

Mineral Identification

Metallic Luster

Gray Streak	**Perfect cubic cleavage;** heavy, Sp. Gr. = 7.6; H = 2.5; silver gray color	GALENA PbS
Black Streak	**Magnetic;** black to dark gray; Sp. Gr. = 5.2; H = 6; commonly occurs in granular masses; single crystals are octohedral	MAGNETITE Fe_3O_4
Black Streak	**Steel gray;** soft, marks paper, greasy feel; H = 1; Sp. Gr. = 2; luster may be dull	GRAPHITE C
Greenish Black Streak	**Golden yellow color;** may tarnish purple; H = 4; Sp. Gr. = 4.3	CHALCOPYRITE $CuFeS_2$
	Brass yellow; cubic crystals; common in granular aggregates; H = 6-6.5; Sp. Gr. = 5; lacks cleavage	PYRITE FeS_2
Reddish Brown Streak	**Steel gray;** black to dark brown; granular, fibrous, or micaceous aggregates; single crystals are thick plates; H = 5-6.5; Sp. Gr. = 5; lacks cleavage	HEMATITE Fe_2O_3
Yellow Brown Streak	**Yellow, brown, or black;** hard structureless or radial fibrous masses; H = 5-5.5; Sp. Gr. = 3.5-4	LIMONITE $Fe_2O_3 \cdot H_2O$

The most diagnostic properties for each mineral are indicated in bold type.

Nonmetallic Luster—Dark Color

HARDER THAN GLASS	Cleavage Prominent	**Cleavage—2 directions nearly at 90°;** dark green to black; short prismatic 8-sided crystals; H = 6; Sp. Gr. = 3.5	PYROXENE GROUP Complex Ca, Mg, Fe, Al silicates Augite most common mineral
		Cleavage—2 directions at approx. 60° and 120°; dark green to black or brown; long prismatic 6-sided crystals; H = 6; Sp. Gr. = 3-3.5	AMPHIBOLE GROUP Complex Na, Ca, Mg, Fe, Al silicates **Hornblende** most common mineral
	Cleavage Absent	**Olive green; commonly in small glassy grains;** conchoidal fracture; transparent to translucent; glassy luster; H = 6.5-7; Sp. Gr. = 3.5-4.5	OLIVINE $(Fe, Mg)_2SiO_4$
		Red, brown, or yellow; glassy luster; conchoidal fracture; commonly occurs in well-formed 12-sided crystals; H = 7-7.5; Sp. Gr. = 3.5-4.5	GARNET GROUP Fe, Mg, Ca, Al silicates
SOFTER THAN GLASS	Cleavage Prominent	**Brown to black; 1 perfect cleavage;** thin, flexible, elastic sheets; H = 2.5-3; Sp. Gr. = 3-3.5	BIOTITE $K(Mg,Fe)_3AlSi_3O_{10}(OH)_2$
		Green to very dark green; 1 cleavage direction; commonly occurs in foliated or scaly masses; nonelastic plates; H = 2-2.5; Sp. Gr. = 2.5-3.5	CHLORITE Hydrous Mg, Fe, Al silicate
		Yellowish brown; resinous luster; cleavage—6 directions; yellowish brown or nearly white streaks; H = 3.5-4; Sp. Gr. = 4	SPHALERITE ZnS
	Cleavage Absent	**Red; earthy appearance;** red streak; H = 1.5	HEMATITE Fe_2O_3 (earth variety)
		Yellowish brown streak; yellowish brown to dark brown; commonly in compact earth masses; H = 1.5	LIMONITE $Fe_2O_3 \cdot H_2O$

The most diagnostic properties for each mineral are indicated by bold type.

Nonmetallic Luster—Light Color

HARDER THAN GLASS	Cleavage Prominent	**Good cleavage in 2 directions at approximately 90°;** pearly to vitreous luster; H = 6-6.5; Sp. Gr. = 2.5	FELDSPAR GROUP **Potassium feldspars:** $KAlSi_3O_8$—Pink, white, or green **Plagioclase feldspars:** $NaAlSi_3O_8$ to $CaAl_2Si_2O_8$—White, blue gray; striations on some cleavage planes
	Cleavage Absent	**Conchoidal fracture; H = 7;** Sp. Gr. = 2.65; transparent to translucent; vitreous luster; 6-sided prismatic crystals terminated by 6-sided triangular faces in well-developed crystals; vitreous to waxy	QUARTZ SiO_2 (silica) **Varieties:** Milky: white and opaque Smoky: gray to black Rose: light pink Amethyst: violet
		Conchoidal fracture; H = 6-6.5; variable color; translucent to opaque; dull or clouded luster	CRYPTOCRYSTALLINE QUARTZ SiO_2 **Varieties:** Agate: banded Flint: dark color Chert: light color Jasper: red Opal: waxy luster
SOFTER THAN GLASS	Cleavage Prominent	**Perfect cubic cleavage;** colorless to white; soluble in water; salty taste; H = 2-2.5; Sp. Gr. = 2	HALITE NaCl
		Perfect cleavage in 1 direction; poor in 2 others; H = 2; white; transparent; nonelastic; Sp. Gr. = 2.3; **Varieties:** Selenite: colorless, transparent; Alabaster: aggregates of small crystals; Satin spar; fibrous, silky luster	GYPSUM $CaSO_4 \cdot 2H_2O$
		Perfect cleavage in 3 directions at approximately 75°; effervesces in HCl; H = 3; colorless, white or pale yellow, rarely gray or blue; transparent to opaque; Sp. Gr. = 2.7	CALCITE $CaCO_3$ (fine-grained crystalline aggregates form limestone and marble)
		Three directions of cleavage as in calcite; effervesces in HCl only if powdered; H = 3.5-4; Sp. Gr. = 2.8; color variable but commonly white or pink rhomb-faced crystals	DOLOMITE $CaMg(CO_3)_2$
		Good cleavage in 4 directions; H = 4; Sp. Gr. = 3; colorless, yellow, blue, green, or violet; transparent to translucent; cubic crystals	FLUORITE CaF_2
		Perfect cleavage in 1 direction, producing thin, elastic sheets; H = 2-3; Sp. Gr. = 2.8; transparent and colorless in thin sheets	MUSCOVITE $KAl_2(AlSi_3O_{10}(OH)_2$
		Green to white; soapy feel; pearly luster; H = 1; Sp. Gr. = 2.8; foliated or compact masses; one direction of cleavage forms thin scales and shreds	TALC $Mg_3Si_4O_{10}(OH)_2$
	Cleavage Absent	**White to red; earthy masses;** crystals so small no cleavage visible; soft; H = 1.2; becomes plastic when moistened; earthy odor	KAOLINITE $Al_4Si_4O_{10}(OH)_8$

The most diagnostic properties for each mineral are indicated by bold type.

Periodic Table of the Elements

KEY

6
C
12.01
Carbon

→ Atomic number (underlined if synthetically prepared)
→ Symbol
→ Atomic weight
→ Name
→ Elements abundant in the earth's crust
→ Elements important geologically

Element	No.	Weight	Name
H	1	1.0080	Hydrogen
He	2	4.003	Helium
Li	3	6.940	Lithium
Be	4	9.013	Berilium
B	5	10.82	Boron
C	6	12.011	Carbon
N	7	14.008	Nitrogen
O	8	16.000	Oxygen
F	9	19.00	Fluorine
Ne	10	20.183	Neon
Na	11	22.991	Sodium
Mg	12	24.32	Magnesium
Al	13	26.98	Aluminum
Si	14	28.09	Silicon
P	15	30.975	Phosphorus
S	16	32.066	Sulfur
Cl	17	35.457	Chlorine
Ar	18	39.944	Argon
K	19	39.100	Potassium
Ca	20	40.08	Calcium
Sc	21	44.96	Scandium
Ti	22	47.90	Titanium
V	23	50.95	Vanadium
Cr	24	52.01	Chromium
Mn	25	54.94	Manganese
Fe	26	55.85	Iron
Co	27	58.94	Cobalt
Ni	28	58.71	Nickel
Cu	29	63.54	Copper
Zn	30	65.38	Zinc
Ga	31	69.72	Gallium
Ge	32	72.60	Germanium
As	33	74.91	Arsenic
Se	34	78.96	Selenium
Br	35	79.916	Bromine
Kr	36	83.80	Krypton
Rb	37	85.48	Rubidium
Sr	38	87.63	Strontium
Y	39	88.92	Yttrium
Zr	40	91.22	Zirconium
Nb	41	92.91	Niobium
Mo	42	95.95	Molybdenum
Tc	43	(99)	Technetium
Ru	44	101.1	Ruthenium
Rh	45	102.91	Rhodium
Pd	46	106.4	Palladium
Ag	47	107.880	Silver
Cd	48	112.41	Cadmium
In	49	114.82	Indium
Sn	50	118.70	Tin
Sb	51	121.76	Antimony
Te	52	127.61	Tellurium
I	53	126.91	Iodine
Xe	54	131.30	Xenon
Cs	55	132.91	Cesium
Ba	56	137.36	Barium
La	57	138.92	Lanthanum
Hf	72	178.50	Hafnium
Ta	73	180.95	Tantalum
W	74	183.86	Wolfram
Re	75	186.22	Rhenium
Os	76	190.2	Osmium
Ir	77	192.2	Iridium
Pt	78	195.09	Platinum
Au	79	197.0	Gold
Hg	80	200.61	Mercury
Tl	81	204.39	Thallium
Pb	82	207.21	Lead
Bi	83	209.00	Bismuth
Po	84	(210)	Polonium
At	85	(210)	Astatine
Rn	86	(222)	Radon
Fr	87	(223)	Francium
Ra	88	(226)	Radium
Ac	89	(227)	Actinium
104	104		(Russian Proposal Unofficial)

Lanthanide series

Element	No.	Weight	Name
Ce	58	140.13	Cerium
Pr	59	140.92	Praseodymium
Nd	60	144.27	Neodymium
Pm	61	(147)	Promethium
Sm	62	150.35	Samarium
Eu	63	152.0	Europium
Gd	64	157.26	Gadolinium
Tb	65	158.93	Terbium
Dy	66	162.51	Dysprosium
Ho	67	164.94	Holmium
Er	68	167.27	Erbium
Tm	69	168.94	Thulium
Yb	70	173.04	Ytterbium
Lu	71	174.99	Lutetium

Actinide series

Element	No.	Weight	Name
Th	90	(232)	Thorium
Pa	91	(231)	Protactinium
U	92	238.07	Uranium
Np	93	(237)	Neptunium
Pu	94	(242)	Plutonium
Am	95	(243)	Americium
Cm	96	(247)	Curium
Bk	97	(249)	Berkelium
Cf	98	(251)	Californium
Es	99	(254)	Einsteinium
Fm	100	(253)	Fermium
Md	101	(256)	Mendelevium
No	102	(253)	Nobelium
Lr	103	257	Lawrencium

CREDITS

Chapter 1, opening page photograph   Courtesy of NASA.

Plates 1–16   National Space Data Center, NASA.

Chapter 2, opening page photograph   U.S. Geological Survey.

Figure 2.2   Modified from P. J. Wyllie, 1976, *The way the earth works* (New York: John Wiley and Sons).

Figure 2.3   After P. J. Wyllie, 1976, *The way the earth works* (New York: John Wiley and Sons).

Figure 2.4   After E. Bullard, 1969, The origin of the oceans, *Scientific American* 221(3):66–75.

Figure 2.5   After B. Isacks, J. Oliver, and L. R. Sykes, 1968, Seismology and the new global tectonics, *Journal of Geophysical Research* 73(18):5855–99.

Figure 2.7   After D. M. Crittenden, 1963, *New data on the isostatic deformation of Lake Bonneville*, U.S. Geological Survey Professional Paper 455-E (Reston, Va: U.S. Geological Survey), p. 9.

Chapter 3, opening page photograph   National Space Data Center, NASA.

Chapter 4, opening page photograph   Courtey of NASA.

Chapter 5, opening page photograph   Courtesy of Myron Best, Brigham Young University.

Figure 5.11   U.S. Geological Survey.

Figure 5.12   Courtesy of Craters of the Moon National Monument, Idaho.

Figure 5.14   Courtesy of Craters of the Moon National Monument, Idaho.

Figure 5.17   Geological Survey of Canada.

Figure 5.20   After H. Williams, F. J. Turner, and C. M. Gilbert, 1954, *Petrography* (San Francisco: W. H. Freeman).

Figure 5.22   U.S. Geological Survey.

Figure 5.24   Modified from Geo-Graphics, Portland, Oregon.

Chapter 6, opening page photograph   W. K. Hamblin.

Figure 6.3   U.S. Geological Survey.

Figure 6.24B   Courtesy of Paul Potter.

Chapter 7, opening page photograph   W. K. Hamblin.

Figure 7.1   National Air Photo Library, Department of Energy, Mines and Resources, Canada.

Chapter 8, opening page photograph   W. K. Hamblin.

Figure 8.3   Modified from D. L. Eicher, 1968, *Geologic time* (Englewood Cliffs, N.J.: Prentice-Hall).

Chapter 9, opening page photograph   Courtesy of Dennis Tasa.

Figure 9.14   After N. M. Strakhov, 1967, *Principles of lithogenesis*, trans. J. P. Fitzemms (Edinburgh: Oliver and Boyd), vol. 1.

Chapter 10, opening page photograph   U.S. Geological Survey.

Figure 10.6   After K. Davis and L. Leopold, 1970, *Water* (Alexandria, Va.: Time-Life Books).

Figure 10.8   After L. Straub, 1942, in *Hydrology*, ed. O. E. Meinzer (New York: McGraw-Hill), p. 625.

Figure 10.10   Courtesy of U.S. Forest Service.

Figure 10.11   After F. Hjulström, 1935, Studies of the morphological activities of rivers as illustrated by the River Fyris, *Uppsala University Geological Institute Bulletin* 25:221–527.

Figure 10.13   Courtesy of Swiss Air.

Figures 10.14–10.15   After L. B. Leopold, 1968, *Hydrology for urban land planning*, U.S. Geological Survey Circular 554, figure 1, p. 3.

Chapter 11, opening page photograph   W. K. Hamblin.

Figure 11.9   U.S. Geological Survey.

Figure 11.11   Modified from D. J. Varnes, 1958, Landslide types and processes, in *Landslides and engineering practices*, ed. E. B. Eckel, Highway Research Board Special Report 29, NAS–NRC 544 (Washington, D.C.: National Academy of Sciences–National Research Council).

Figure 11.12   Courtesy of Swiss Air.

Figure 11.19   After H. N. Pollack, 1969, A numerical model of the Grand Canyon, in *Four Corners Geological Society guidebook*, p. 62.

Figure 11.20   U.S. Geological Survey.

Figure 11.23   National Air Photo Library, Department of Energy, Mines and Resources, Canada.

Figure 11.25   National Air Photo Library, Department of Energy, Mines and Resources, Canada.

Figure 11.26   U.S. Geological Survey.

Figure 11.32   Modified from C. R. Kolb and J. R. Van Lopik, 1966, *Depositional environments of the Mississippi River coastal plain*, U.S. Army Corps of Engineers, Waterways Department, Station Technological Reports 3-483 and 3-484.

Chapter 12, opening page photograph   Courtesy of Levi Hintze, Brigham Young University.

Figure 12.2   U.S. Geological Survey.

Figure 12.14   Modified from J. Shelton, 1966, *Geology illustrated* (San Francisco: W. H. Freeman), figure 358.

Figure 12.15A   After A. N. Strahler, 1951, *Physical geography* (New York: John Wiley and Sons).

Chapter 13, opening page photograph   Winter Park, Florida, sinkhole, May 9, 1981, Barbara Vitaliano/Orlando Sentinel Star.

Figure 13.12   Courtesy of John Shelton.

Figure 13.15   Courtesy of Onadaga Cave, Missouri.

Figure 13.16   Modified from W. J. Schneider, 1970, *Hydrologic implications of solid waste disposal*, U.S. Geological Survey Circular 601-F.

Figures 13.18–13.19   Modified from F. Ward, 1972, The imperiled Everglades, *National Geographic* 141(1):1.

Chapter 14, opening page photograph   National Air Photo Library, Department of Energy, Mines and Resources, Canada.

Figure 14.4   Modified from Lake Gillian and Conn Lake topographic maps, Northwest Territories, Geological Survey of Canada.

Figure 14.5   Courtesy of J. D. Ives.

Figure 14.6   Modified from R. F. Flint, 1957, *Glacial and Pleistocene geology* (New York: John Wiley and Sons).

Figure 14.7   After W. W. Atwood, 1940, *Physiographic provinces of North America* (Boston: Ginn and Company).

Figure 14.12 U.S. Geological Survey.

Figure 14.13 National Air Photo Library, Department of Energy, Mines and Resources, Canada.

Figure 14.15 After A. N. Strahler, 1951, *Physical geography* (New York: John Wiley and Sons).

Figure 14.16 National Air Photo Library, Department of Energy, Mines and Resources, Canada.

Figure 14.20 Compiled from glacial map of the United States east of the Rocky Mountains, Geological Society of America.

Figure 14.21 After R. F. Flint, 1971, *Glacial and Quaternary geology* (New York: John Wiley and Sons).

Figure 14.23 Compiled from R. J. W. Douglas, 1970, *Geology and economic minerals of Canada* (Ottawa: Department of Energy, Mines and Resources, Geological Survey of Canada), p. 691, and R. F. Flint, 1971, *Glacial and Quaternary geology* (New York: John Wiley and Sons), p. 234.

Figure 14.25 National Air Photo Library, Department of Energy, Mines and Resources, Canada.

Figure 14.26 After J. L. Hough, 1958, *Geology of the Great Lakes* (Urbana: University of Illinois Press), and R. J. W. Douglas, 1970, *Geology and economic minerals of Canada* (Ottawa: Department of Energy, Mines and Resources, Geological Survey of Canada), pp. 714–25.

Figure 14.28 After R. F. Flint, *Glacial and Quarternary geology* (New York: John Wiley and Sons), p. 447. Base map courtesy of E. R. Raisz.

Chapter 15, opening page photograph W. K. Hamblin.

Figure 15.6 U.S. Geological Survey.

Figure 15.7 After W. Bascom, 1960, Beaches, *Scientific American* 203(2):81–94.

Figure 15.12 Courtesy of Iceland Tourist Bureau.

Figure 15.15 U.S. Department of Agriculture, ASCS Western Aerial Photo Lab, Salt Lake City, Utah.

Figure 15.17 U.S. Geological Survey.

Figure 15.21 Courtesy of John Shelton.

Chapter 16, opening page photograph Courtesy of Aramco.

Figure 16.3 U.S. Geological Survey.

Chapter 17, opening page photograph U.S. Geological Survey.

Figure 17.1 After A. Wegener, 1915, *The origin of continents and oceans*, English translation of the fourth edition, 1929 (New York: Dover), figure 1.

Figure 17.2 After P. M. Hurley, 1969, The confirmation of continental drift, *Scientific American* 218(4):52–64.

Figure 17.6 After American Association of Petroleum Geologists, 1928, *Theory of continental drift: A symposium* (Tulsa, Okla.: American Association of Petroleum Geologists), figure 2.

Figure 17.7B After R. H. Dott and R. L. Batten, 1971, *Evolution of the earth* (New York: McGraw-Hill), figure 7.4.

Figure 17.9 After A. Cox, G. B. Dalrymple, and R. R. Doell, 1967, Reversals of the earth's magnetic field, *Scientific American* 216(2):44–54.

Figure 17.13 Modified from P. J. Wyllie, 1976, *The way the earth works* (New York: John Wiley and Sons).

Figure 17.15 After B. Isacks, J. Oliver, and L. R. Sykes, 1968, Seismology and the new global tectonics, *Journal of Geophysical Research* 73(18):5855–99.

Chapter 18, opening page photograph U.S. Geological Survey.

Figure 18.3 Modified from J. H. Zumberge and C. A. Nelson, 1972, *Elements of geology*, 3d ed. (New York: John Wiley and Sons).

Figure 18.8 Modified from L. R. Sykes, 1966, The seismicity and deep structure of island arcs, *Journal of Geophysical Research* 71(12):2981–3006.

Chapter 19, opening page photograph U.S. Geological Survey.

Figures 19.2–19.3 Courtesy of Woods Hole Oceanographic Institute.

Figure 19.5 After I. G. Gass, P. J. Smith, and R. C. L. Wilson, eds., 1971, *Understanding the earth* (Cambridge, Mass.: MIT Press).

Figure 19.6 U.S. Geological Survey.

Chapter 20, opening page photograph U.S. Geological Survey.

Figure 20.1 National Geophysical and Solar-Terrestrial Data Center, Boulder, Colorado.

Figure 20.3 Courtesy of John Shelton.

Figure 20.4 LANDSAT photograph, U.S. Department of Agriculture, Salt Lake City, Utah.

Figure 20.20 Modified from J. G. Vedder and R. E. Wallace, 1970, U.S. Geological Survey Map I-574.

Figure 20.21 Photograph courtesy of H. Wailace, U.S. Geological Survey.

Chapter 21, opening page photograph U.S. Geological Survey.

Figure 21.8 Compiled from data in F. P. Shepard, 1963, *Marine geology* (New York: Harper and Row), pp. 321–22.

Chapter 22, opening page photograph National Air Photo Library, Department of Energy, Mines and Resources, Canada.

Figures 22.1–22.2 National Air Photo Library, Department of Energy, Mines and Resources, Canada.

Figure 22.3 After R. S. Deitz, 1972, Geosynclines, mountains, and continent building, *Scientific American* 226(3):30–38.

Figure 22.6 LANDSAT photograph, U.S. Department of Agriculture, Salt Lake City, Utah.

Chapter 23, opening page photograph National Air Photo Library, Department of Energy, Mines and Resources, Canada.

Figure 23.6 Modified from M. K. Hubbert, 1956, Nuclear energy and the fossil fuels, in *Drilling and production practice* (Washington, D.C.: American Petroleum Institute), figure 21.

Figure 23.9 Data from D. H. Meadows, D. L. Meadows, J. Randers, and W. W. Behrens III, 1972, *The limits to growth: A report for the Club of Rome's Project on the Predicament of Mankind* (New York: Universe Books).

Figure 23.10 After D. H. Meadows, D. L. Meadows, J. Randers, and W. W. Behrens III, 1972, *The limits to growth: A report for the Club of Rome's Project on the Predicament of Mankind* (New York: Universe Books).

Chapter 24, opening page photograph Courtesy of NASA.

Figure 24.1 Courtesy of NASA.

Figure 24.2 After E. M. Shoemaker, 1960, Penetration mechanics of high velocity meteorites, illustrated by Meteor Crater, Arizona, *International Geological Congress XXI* 18:418.

Figure 24.3 Courtesy of Jet Propulsion Lab.

Figures 24.4–24.6 Courtesy of National Space Data Center, NASA.

Figures 24.7–24.9 Courtesy of NASA and Garth H. Ladle.

Figure 24.10 Courtesy of C. Klein.

Figure 24.12 After N. Short, 1975, *Planetary geology* (Englewood Cliffs, N.J.: Prentice-Hall).

Figure 24.13 After F. Toksoz et al., 1972, *Lunar Science Abstracts* (January 10–13), p. 670.

Figure 24.14 Courtesy of NASA.

Figure 24.15 Courtesy of Ames Research Center.

Figure 24.16 After E. M. Shoemaker, *Geology of the moon and Project Apollo*, Contribution No. 2037 of the Division of Geological and Planetary Sciences, California Institute of Technology, Pasadena.

Figures 24.18–24.20 Courtesy of Jet Propulsion Lab.

Figure 24.24 After T. A. Mutch et al., 1976, *Geology of Mars* (Princeton, N.J.: Princeton University Press).

Figure 24.25 Courtesy of Jet Propulsion Lab.

Figure 24.26 U.S. Geological Survey.

Figures 24.30–24.35 Courtesy of Jet Propulsion Lab.

Figure 24.36 After T. A. Mutch et al., 1976, *Geology of Mars* (Princeton, N.J.: Princeton University Press).

Figures 24.37–24.46 Courtesy of NASA.

Figures 24.48–24.51 Courtesy of NASA.

INDEX

Page numbers set in italic type refer to illustrations. In entries referring to more than one passage of the text, an asterisk indicates pages on which a definition or an explanation is given.

Student Survey
Hamblin, THE EARTH'S DYNAMIC SYSTEMS, Third Edition

Send us your ideas!

By completing and returning this questionnaire, you can help the author and publisher develop better textbooks. Tell us what you liked about The Earth's Dynamic Systems—*as well as what you think could be improved. We'll consider your comments as we prepare the next edition.*

Your name (optional) _____   School _____

Instructor's name (optional) _____   Course title _____

Your mailing address _____

City _____ State _____ Zip _____

How does *The Earth's Dynamic Systems* compare with other texts you have used?

      Excellent □   Good □   Average □   Poor □   Very Poor □

Circle those chapters you feel were especially well done.

Chapter: 1 2 3 4 5 6 7 8 9 10 11 12 13 14 15 16 17 18 19 20 21 22 23 24

Comments:

Circle those chapters you feel could be improved.

Chapter: 1 2 3 4 5 6 7 8 9 10 11 12 13 14 15 16 17 18 19 20 21 22 23 24

Comments:

Please rate the following:

	Excellent	Good	Average	Poor
1. Understandability and general level of writing style	_____	_____	_____	_____
2. Statement/discussion/summary format	_____	_____	_____	_____
3. Explanations of difficult concepts	_____	_____	_____	_____
4. Correspondence of text organization to your instructor's course organization	_____	_____	_____	_____
5. Use of boldface type to introduce important terms	_____	_____	_____	_____
6. Illustrated glossary	_____	_____	_____	_____
7. Integration of text and illustrations to make the subject easier to understand	_____	_____	_____	_____
8. Usefulness of color satellite photographs in understanding geologic processes	_____	_____	_____	_____
9. Usefulness of color in diagrams in understanding difficult concepts	_____	_____	_____	_____

Did you buy this book new or used?   New ☐   Used ☐

Did you take this course as a geology major?   Yes ☐   No ☐

Do you plan to take additional geology courses?   Yes ☐   No ☐

Did this text/course influence your decision?   Yes ☐   No ☐

Do you plan to keep this text?   Yes ☐   No ☐

Did you purchase the study guide for the text?   Yes ☐   No ☐

Did your instructor test from the text?   Yes ☐   No ☐

What topics did your instructor discuss that were not covered in the text? _____

_____

What chapters in the text were not assigned in the course? _____

Did you take a physical geology lab?   Yes ☐   No ☐

If so, author and title of the lab book used: _____

How well did the lab book coordinate with this text?

Excellent ☐   Good ☐   Average ☐   Poor ☐   Very Poor ☐

GENERAL COMMENTS:

May we quote you in our advertising?   Yes ☐   No ☐

THANK YOU!

To mail, remove this page from the book. Fold so business reply side faces out, and seal open edge. Or, mail in an evelope to the address below. We are sorry, but postal regulations make it difficult for us to use a postal return permit.

WE WILL REFUND DOUBLE YOUR POSTAGE.

Place
Stamp
Here

Burgess Publishing Company

Attn: Geology Editor
7108 Ohms Lane
Minneapolis, Minnesota 55435

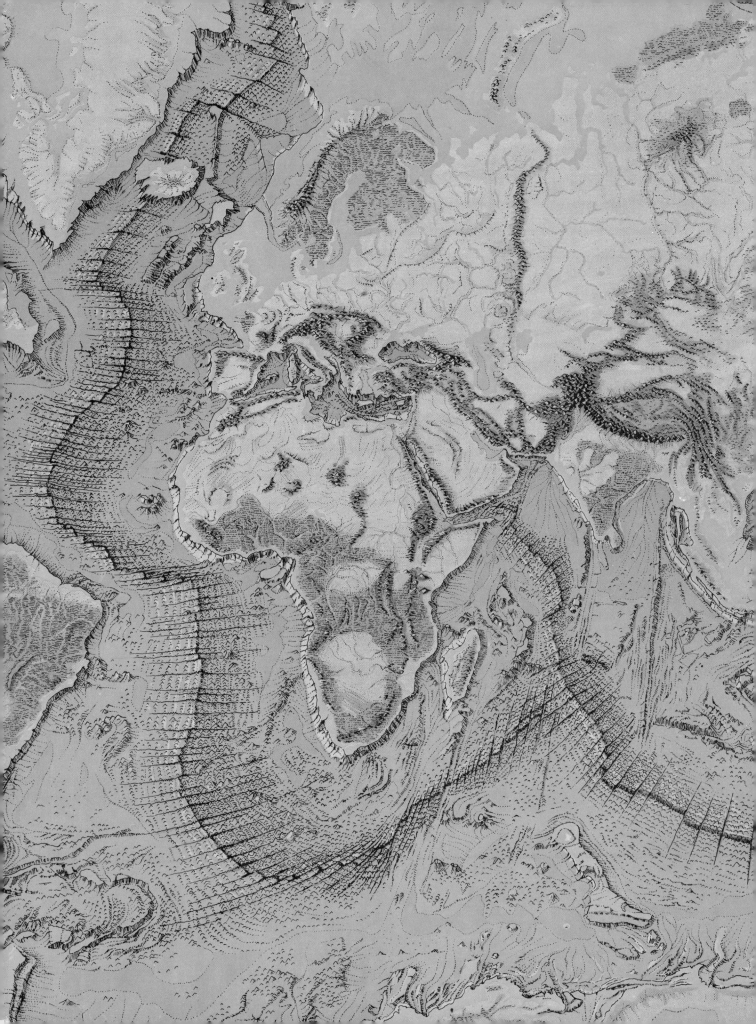